ONTOGENY OF RECEPTORS AND REPRODUCTIVE HORMONE ACTION

Ontogeny of Receptors and Reproductive Hormone Action

Editors:

Dr. Terrell H. Hamilton
Department of Zoology
The University of Texas
Austin, Texas

Dr. James H. Clark
Department of Cell Biology
Baylor College of Medicine
Houston, Texas

Dr. William A. Sadler
Center for Population Research
National Institute of Child Health
and Human Development
National Institutes of Health
Bethesda, Maryland

Dr. W. H. Tam
Dept. of Zoology,
University of Western Ontario,
London, Ontario N6A 5B7.

Raven Press ▪ New York

19 th Sept., 1979.

Raven Press, 1140 Avenue of the Americas, New York, New York 10036

Raven Press, New York 1979

Made in the United States of America

Library of Congress Cataloging in Publication Data

Main entry under title:

Ontogeny of receptors and reproductive hormone action.

   Includes bibliographical references and index.
   1. Estrogen—Receptors. 2. Steroid hormones—
Receptors. 3. Reproduction. 4. Ontogenesis.
I. Hamilton, Terrell H. II. Clark, James H. III. Sad-
ler, William A. [DNLM: 1. Ontogeny. 2. Receptors,
Hormone. 3. Sex hormones. WK102.3 059]
QP572.E85057   599'.01'6   77–92523
ISBN 0–89004–254–3

Preface

Despite our fairly detailed understanding of hormone-receptor interactions in normal adult animals and humans, we know far less about the development of these complex systems. This volume reviews in detail our present understanding of the development of receptor systems and the factors that control this process.

This volume presents material from a variety of hormonally responsive tissues ranging from laboratory animals to humans. Topics discussed include interactions between hormone molecules, receptor molecules, and genetic material, the control by hormones of the growth of hormone-responsive tissues during development, hormonal regulation of the synthesis of specific proteins, and receptor changes during specific pathological states.

This volume will be of interest to endocrinologists, both basic and clinical, and to researchers and clinicians in fields ranging from developmental biology to clinical oncology.

Acknowledgments

A Conference entitled "Ontogeny of Receptors and the Mechanisms of Action of Reproductive Hormones" was held at the University of Texas in Austin, Texas. We acknowledge the encouragement and support provided by the Center for Population Research of the National Institute for Child Health and Human Development in the publication of the proceedings of this Conference. We also acknowledge the support given by Dr. Lorene L. Rogers, President of the University of Texas at Austin, for the use of the Joe C. Thompson Conference Center.

Finally, we express our thanks to the individual participants in this Conference for their cooperation and to Raven Press for ensuring rapid publication of this volume. This is truly a timely treatise on the mechanisms of action of reproductive hormones and how the individual target organs acquire the competence to respond to specific hormones by growth, development, and synthesis of specific proteins necessary for the organs' final and critical reproductive function.

The Editors

Contents

Contributors

Joel Abramowitz
Department of Cell Biology
Northwestern University Medical
  School
Chicago, Illinois 60611

Hollis J. Baker
Department of Biochemisty
University of Illinois
Urbana, Illinois 61801

P. Bagavandoss
Reproductive Endocrinology Program
Department of Pathology
The University of Michigan
Ann Arbor, Michigan 48109

R. H. Benson
Departments of Pharmacology and
  Pediatrics
University of Texas Medical School
  at Houston
Houston, Texas 77025

Lutz Birnbaumer
Department of Cell Biology
Northwestern University Medical
  School
Chicago, Illinois 60611

B. P. Blatti
Department of Biochemistry and
  Molecular Biology
University of Texas Medical School
  at Houston
Houston, Texas 77025

B. R. Brinkley
Department of Cell Biology
Baylor College of Medicine
Houston, Texas 77030

K. L. Campbell
Reproductive Endocrinology Program
Department of Pathology
The University of Michigan
Ann Arbor, Michigan 48109

Kevin J. Catt
Hormonal Regulation
Endocrinology and Reproductive
  Research Branch
National Institute of Child Health and
  Human Development
Bethesda, Maryland 20014

Gary C. Chamness
Department of Medicine
University of Texas Health Science
  Center
San Antonio, Texas 78284

Lawrence Chan
Department of Cell Biology
Baylor College of Medicine
Houston, Texas 77030

James H. Clark
Department of Cell Biology
Baylor College of Medicine
Houston, Texas 77030

L. E. Closs
Ben May Laboratory for Cancer
  Research
University of Chicago
Chicago, Illinois 60637

Sharon L. Day
Department of Cell Biology
Northwestern University Medical
  School
Chicago, Illinois 60611

J. R. Dedman
Department of Cell Biology
Baylor College of Medicine
Houston, Texas 77030

E. R. DeSombre
Ben May Laboratory for Cancer
  Research
University of Chicago
Chicago, Illinois 60637

Roger G. Deeley
Laboratory of Biochemistry
National Cancer Institute
National Institutes of Health
Bethesda, Maryland 20014

Achilles Dugaiczyk
Department of Cell Biology
Baylor College of Medicine
Houston, Texas 77030

Hakan Eriksson
Department of Cell Biology
Baylor College of Medicine
Houston, Texas 77030

R. M. Gardner
Department of Biology
Rochester Institute of Technology
Rochester, New York 14623
formerly
Department of Pharmacology
University of Texas Medical School at
  Houston
Houston, Texas 77025

Roberto E. Garola
Department of Medicine
University of Texas Health Science
  Center
San Antonio, Texas 78284

Frederick W. George
Department of Internal Medicine and
The Eugene McDermott Center for
  Growth and Development
University of Texas Southwestern
  Medical School
Dallas, Texas 75235

Robert F. Goldberger
Laboratory of Biochemistry
National Cancer Institute
National Institutes of Health
Bethesda, Maryland 20014

Roger A. Gorski
Department of Anatomy and Brain
  Research Institute
UCLA School of Medicine
Los Angeles, California 90024

William A. Guyette
Department of Cell Biology
Baylor College of Medicine
Houston, Texas 77030

James W. Hardin
Department of Cell Biology
Baylor College of Medicine
Houston, Texas 77030

Stephen G. Hillier
Reproduction Research Branch
National Institute of Child Health and
  Human Development and Diabetes
  Branch
National Institute of Arthritis,
  Metabolism, and Digestive Diseases
National Institutes of Health
Bethesda, Maryland 20014

Kathryn B. Horwitz
Department of Medicine
University of Texas Health Science
  Center
San Antonio, Texas 78284

Mary Hunzicker-Dunn
Department of Biochemistry
Northwestern University Medical
  School
Chicago, Illinois 60611

J. S. Ireland
Department of Pharmacology
University of Texas Medical School at
  Houston
Houston, Texas 77025

Richard L. Jackson
Department of Cell Biology
Baylor College of Medicine
Houston, Texas 77030

E. V. Jensen
Ben May Laboratory for Cancer
  Research
University of Chicago
Chicago, Illinois 60637

J. A. Jonassen
Reproductive Endocrinology Program
Department of Pathology
The University of Michigan
Ann Arbor, Michigan 48109

Benita S. Katzenellenbogen
*Department of Physiology and
  Biophysics
University of Illinois and
School of Basic Medical Sciences
University of Illinois College of
  Medicine
Urbana, Illinois 61801*

Katrina L. Kelner
*Department of Cell Biology
Baylor College of Medicine
Houston, Texas 77030*

J. L. Kirkland
*Departments of Pharmacology and
  Pediatrics
University of Texas Medical School at
  Houston
Houston, Texas 77025*

T. D. Landefeld
*Reproductive Endocrinology Program
Department of Pathology
The University of Michigan
Ann Arbor, Michigan 48109*

Catherine B. Lazier
*Biochemistry Department
Dalhousie University
Halifax, Nova Scotia,
Canada B3H 4H7*

Ivan Lieberburg
*The Rockefeller University
New York, New York 10021*

Dennis N. Luck
*Department of Biology
Oberlin College
Oberlin, Ohio 44074*

Neil J. MacLusky
*The Rockefeller University
New York, New York 10021*

M. Marcum
*Department of Cell Biology
Baylor College of Medicine
Houston, Texas 77030*

Barry M. Markaverich
*Department of Cell Biology
Baylor College of Medicine
Houston, Texas 77030*

G. Martin-Dani
*Department of Molecular Medicine
Mayo Clinic
Rochester, Minnesota 55901*

A. R. Midgley, Jr.
*Reproductive Endocrinology Program
Department of Pathology
The University of Michigan
Ann Arbor, Michigan 48109*

Robert J. Matusik
*Department of Cell Biology
Baylor College of Medicine
Houston, Texas 77030*

Bruce S. McEwen
*The Rockefeller University
New York, New York 10021*

William L. McGuire
*Department of Medicine
University Health Science Center
San Antonio, Texas 78284*

Anthony R. Means
*Department of Cell Biology
Baylor College of Medicine
Houston, Texas 77030*

Ann L. Miller
*Department of Cell Biology
Baylor College of Medicine
Houston, Texas 77030*

Thomas G. Muldoon
*Medical College of Georgia
Department of Endocrinology
Augusta, Georgia 30902*

Bert W. O'Malley
*Department of Cell Biology
Baylor College of Medicine
Houston, Texas 77030*

Helen A. Padykula
*Laboratory of Electron Microscopy
Wellesley College
Wellesley, Massachusetts 02181*

E. J. Peck, Jr.
Department of Cell Biology
Baylor College of Medicine
Houston, Texas 77030

M. C. Rao
Reproductive Endocrinology Program
Department of Pathology
The University of Michigan
Ann Arbor, Michigan 48109

J. S. Richards
Reproductive Endocrinology Program
Department of Pathology
The University of Michigan
Ann Arbor, Michigan 48109

Jeffrey M. Rosen
Department of Cell Biology
Baylor College of Medicine
Houston, Texas 77030

Griff T. Ross
The Clinical Center
Reproduction Research Branch
National Institute of Child and Human
  Development and Diabetes Branch
National Institute of Arthritis,
  Metabolism, and Digestive Diseases
National Institutes of Health
Bethesda, Maryland 20014

William T. Schrader
Department of Cell Biology
Baylor College of Medicine
Houston, Texas 77030

David J. Shapiro
Department of Biochemistry
University of Illinois
Urbana, Illinois 61801

J. Sidikaro
Department of Ophthalmology
Baylor College of Medicine and
Department of Pharmacology
University of Texas Medical School at
  Houston
Houston, Texas 77025

L. Dale Snow
Department of Cell Biology
Baylor College of Medicine
Houston, Texas 77030

T. C. Spelsberg
Department of Molecular Medicine
Mayo Clinic
Rochester, Minnesota 55901

G. M. Stancel
Department of Pharmacology
University of Texas Medical School at
  Houston
Houston, Texas 77025

Ching Sung Teng
Department of Cell Biology
Baylor College of Medicine
Houston, Texas 77030

Christina T. Teng
Department of Cell Biology
Baylor College of Medicine
Houston, Texas 77030

C. Thrall
Department of Molecular Medicine
Mayo Clinic
Rochester, Minnesota 55901

Susan Upchurch
Department of Cell Biology
Baylor College of Medicine
Houston, Texas 77030

Robert A. Webster
Department of Molecular Medicine
Mayo Clinic
Rochester, Minnesota 55901

P. A. Weil
Department of Biochemistry
Washington University
St. Louis, Missouri 63110
formerly
Department of Biochemistry and
  Molecular Biology
University of Texas Medical School at
  Houston
Houston, Texas 77025

M. J. Welsh
Department of Cell Biology
Baylor College of Medicine
Houston, Texas 77030

Jean D. Wilson
Department of Internal Medicine and
  The Eugene McDermott Center for
  Growth and Development
University of Texas Southwestern
  Medical School
Dallas, Texas 75235

Savio L. C. Woo
Department of Cell Biology
Baylor College of Medicine
Houston, Texas 77030

David T. Zava
Department of Medicine
University of Texas Health Science
  Center
San Antonio, Texas 78284

Anthony J. Zeleznik
Reproduction Research Branch
National Institute of Child Health and
  Human Development and Diabetes
  Branch
National Institute of Arthritis,
  Metabolism, and Digestive Diseases
National Institutes of Health
Bethesda, Maryland 20014

Introduction

Receptors and Reproductive Hormone Action:
A Zeitgeist for the 1980s

How do reproductive hormones regulate the cellular and physiological functions of specific reproductive organs in mammalian and other vertebrate organisms? In mammals and birds, these reproductive organs consist, in general, of the following organs: ovary, testis, uterus, oviduct, and brain (including hypothalamus and the pituitary). Few scientists, let alone the layman, realize that the liver is a reproductive organ in birds, reptiles, amphibians, and other lower vertebrates. The liver synthesizes vitellogenin, the precursor of the yolk proteins. The vitellogenin is secreted into the blood, and transported to the ovaries where it is sequestered into developing oocytes. Within the oocyte, vitellogenin is processed into at least two separate proteins, lipovitellin and phosvitin.

In the early 1960s, research from several laboratories indicated that for at least the sex steroid hormones, a genomic response (DNA-dependent synthesis of specific messenger RNAs and their corresponding proteins) was a fundamental aspect of hormone action in characteristic target organs. From the mid 1960s to the present, a blossoming field of molecular endocrinology has amassed overwhelming evidence for three common features in reproductive hormone action: (i) the primary site of biosynthetic action of the sex steroid hormones is at the level of genetic transcription, (ii) specific recognition and binding of each reproductive hormone (for steroid as well as protein hormones) is necessary for this induced or modulated effect on transcription, and that effects on genetic translation follow thereafter, (iii) specific nonhistone or acidic chromosomal proteins, recognizing and binding in turn the initial "hormone-acceptor" complex, play an important and perhaps direct role in this induction or modulation of genetic transcription by the sex steroid hormones. The role of such chromosomal "acceptor" (or "receptor for the receptor") proteins in organs regulated by the gonadotropic hormones of the anterior pituitary is less clear, but is now a rapidly expanding area of research.

The application of new technologies has permitted two laboratories, in their analysis of the structure of the ovalbumin gene regulated by the sex steroid estrogen, to discover that the gene is divided into several separate pieces. Following the initial transcription of the gene by RNA polymerase, the "parent transcript" is in some manner processed by deletion of the intervening sequences, and the remaining pieces are ligated or connected to form the shorter and smaller functional messenger RNA molecule. These findings have given impetus to the study of the structure of eukaryotic genes in general, as well as to the study of how reproductive hormones can regulate by derepression or modulation the transcription of specific genes. It will be interesting to see in the next several years the extent to which the mode of action of reproductive hormones such as the gonadotropins and other protein hormones parallel in their subcellular actions the actions now known for the sex steroid hormones.

It is in the context of the recent progress in reproductive hormone research summarized above that this book has been organized. We agree that in the next few years, certain specific problems are likely to come under intensive investigation in many laboratories throughout the world. We identify three problems in particular concerning the molecular mode of action of reproductive hormones and how individual reproductive organs (*e.g.,* uterus, testis, ovary, liver, embryonic gonads and sexual ducts, *etc.*) acquire the competence to grow and develop in response to specific hormones.

First: at what stage during prenatal or postnatal development does a specific reproductive organ first acquire the competence to respond to a specific hormone? Does this involve the preceding synthesis of specific receptor proteins and/or specific acceptor chromosomal proteins for the hormone-receptor complex? *Second:* what are the precise molecular events involved in gene activation or modulation by specific receptor or acceptor proteins? In other words, are new receptor or acceptor proteins (cytoplasmic or chromosomal) induced, increased in cellular concentrations, translocated from the cytoplasm or nucleus and/or vice versa, or are these molecules otherwise stabilized or catabolized by hormone action in this process? *Third:* what are the precise molecular interactions involved in the activation or modulation of specific genes by specific steroid or protein hormones? How do these hormones— while bound to receptors or "acceptor-hormone receptor" complexes—interact with DNA molecules, nonhistone or acidic chromosomal proteins, RNA polymerases, and the like, to induce or accelerate the transcription of genes? Are these hormones directly involved in the subsequent processing of "parent" messenger RNA molecules to their functional state and in their transport to the cytoplasm for translation. How do these products of translation affect the regulation of the function of their specific reproductive target organs?

Of course, the chapters in this volume do not answer these problems completely, but they do shed light on recent research developments relevant to their resolutions. Thus, this book not only summarizes recent research progress in two important fields of contemporary research—reproductive biology and regulatory cellular biology—but also presents for the first time in recent years a Zeitgeist, that is, a constellation of research observations, conclusions, and hypotheses that provide a conceptual background for future research in the field of reproductive hormone action. Surveying the chapters of this book, one sees clearly the dominance of the transcriptional theory for steroid hormone action over the translational theory. However, most of the contributors indicate that the two theories of hormone action concerning macromolecular synthesis need not be mutually exclusive, and that future work may well reveal multiple primary biosynthetic effects, with translational effects at the levels of posttranscriptional processing of ribosomal RNA, regulation of the kinetics of genetic translation or peptide synthesis, and membrane function and adenyl cyclase activation.

The research summarized in this book is clearly of significance not only to population, birth control, and reproductive physiology and attendant disorders, but also to the medical field of cancer. Hence, we hope this book will be a valuable treatise to the clinician and the research physician, as well as to the molecular or cellular biologist.

Terrell H. Hamilton
James H. Clark
William A. Sadler

Ontogeny of Receptors and Reproductive Hormone Action,
edited by T. H. Hamilton, J. H. Clark, and W. A. Sadler.
Raven Press, New York 1979.

Genes-in-Pieces

Achilles Dugaiczyk, Savio L. C. Woo, and Bert W. O'Malley

Department of Cell Biology, Baylor College of Medicine, Houston, Texas 77030

In the recent past, codons and start and stop signals for transcription were practically the only tangible sequence elements in a DNA molecule recognized to function in the expression of linearly encoded genetic information for protein. DNA has been considered as a depository of genetic information, undergoing only one round of replication with each cell division, and having only local and limited activities during gene expression. Mapping of DNA has been within the domain of genetics rather than chemistry, mainly because of the longstanding difficulties involved in its sequence determination. But the monotony of a DNA sequence was finally surmounted by the discovery of restriction endonucleases, which are enzymes that cleave DNA at precise sequences and give rise to chemically defined DNA fragments. The availability of such DNA fragments has opened the door for chemical methods of DNA analysis, and new methods of sequence determination quickly followed suit. It was further realized that restriction DNA fragments from various sources could be propagated indefinitely in bacteria after ligation into "recombinant" DNA molecules with DNA vectors such as bacterial plasmid or phage DNAs, a procedure called DNA cloning (7). Thus, cloning of a specific DNA fragment, followed by analysis of its sequence, became the new strategy in attempts to gain further insight into the functioning of eucaryotic genes. This technological triad of restriction enzyme analysis, molecular cloning, and DNA sequence analysis has led to the generation of an entirely new concept of gene architecture.

EXISTENCE OF INTERVENING SEQUENCES OR INTRONS

The earliest studies involved gene coding for ribosomal RNAs, which are repetitive in the genomes of most organisms. The genes that code for 18S, 28S, and 5S ribosomal RNAs turned out to be arranged similarly on a large repeating unit of DNA in *Drosophila* (14,30,43,44), *Xenopus* (2,35), *Zea mays* chloroplast (1), and mouse (8). The transcriptional unit is also composed of "spacer" DNA sequences that separate the individual structural gene sequences coding for these three RNA species, and the entire unit is expressed in a RNA molecule containing all three structural genes plus the spacer sequence as one transcription unit. In some instances, transfer RNA genes were also found to be located between the 18S and 28S rRNA genes as part of the same transcription unit (18,24,25).

Although there is a fair amount of sequence heterogeneity within the individual spacer DNAs separating these repetitive structural genes in the same organism, the arrangement of these structural genes and spacers within the transcription units were not entirely unexpected. What was most surprising is the fact that in some of the repeating units in *Drosophila,* the structural gene for the 28S rRNA was found to be split further at a reproducible point by "intervening" DNA of unknown function (14,30,44). The existence of short but inverted repeat sequences at the termini of these "intervening" sequences, or "introns" (intragenic or intracistronic DNA), had invited speculation that these sequences are translocatable elements similar to procaryotic transposons, a hypothesis yet to be proven.

More recently, it has become possible to construct a physical map of restriction endonuclease cleavage sites in DNA regions flanking a single-copy mammalian structural gene. Briefly, DNA representing the entire genome is cleaved with restriction endonucleases, and the resulting fragments are resolved by electrophoresis in an agarose gel. The restriction DNA fragments are denatured in the gel and transferred onto a nitrocellulose filter by the method of Southern (34). DNA fragments immobilized on the filter are then allowed to hybridize with a radioactively labeled probe specific for the particular gene. After extensive washing, DNA fragments that hybridize with the probe can be detected as labeled bands by autoradiography. This method produces a physical map of a gene as it exists within genomic DNA. When applied to map the rabbit β-globin gene (19) or the chick ovalbumin gene (5,9,21,42), intervening DNA sequences (introns) were found to exist within the structural sequences of both natural genes. These intervening sequences were not represented by corresponding sequences in the mature mRNAs coding for these two proteins.

INTRONS: CLONING ARTIFACTS OR "PSEUDOGENES?"

It happens occassionally that in the process of cloning, *in vivo* recombination events take place that alter the sequence of the DNA used for cloning and at times also that of the vector DNA. There is experimental evidence, however, that introns did not arise in such an event, at least not introns in the cloned ovalbumin gene. Namely, it has been demonstrated that the ovalbumin gene, after having been cloned, had the same restriction cleavage sites within the introns and within structural sequences as the natural gene existing within genomic DNA (10).

The "pseudogene" hypotheses would hold that genes are being inactivated by insertions of extra DNA. This seems to be an unlikely role for intervening sequences, because their presence has been demonstrated in the β-globin gene in the erythropoietic spleen (19) as well as in the ovalbumin gene of chick oviduct (5,9,21,42), tissues actively synthesizing globin and ovalbumin, respectively. Convincing evidence from genetic and sequence analyses have also demonstrated that an intervening sequence of 14 base pairs is present in an active transfer RNA^{Tyr} gene in yeast (15), namely, the mutant SUP4–0 gene, which confers a dominant phenotype on SUP4–0 strains. This sequence is also absent from mature yeast tRNA^{Tyr}. Thus, we can conclude that genes-in-pieces is a fact of biologically active eucaryotic genes and not an artifact of nature.

INTRONS: CONSERVED OR FREE TO DIVERGE?

The spacer sequences separating the 18S, 28S, and 5S rRNA genes in the transcription units of *Drosophila* or *Xenopus* genomes are quite heterogenous in length and sequence (14,30,43,44). The contrary appears to be true for intervening sequences in the yeast tRNA genes.

There are eight unlinked genetic loci in yeast coding for tRNA^{Tyr}. Correspondingly, eight different *Eco*RI fragments of yeast DNA can hybridize to the tRNA^{Tyr}, although *Eco*RI does not cleave the structural gene sequences for tRNA^{Tyr} (29). Because only one tRNA^{Tyr} species is detected in wild-type yeast cells (26), the eight genes are presumed to be identical in the structural region that encodes the mature tRNA sequence. Of the eight genes, three wild-type genes and one tyrosine suppressor mutant gene have been cloned and their sequences determined recently (15). All four genes revealed the unexpected feature of an intervening sequence of 14 base pairs in length, which lies adjacent to the anticodon triplet and does not appear in mature tRNA^{Tyr}. All four genes, although originated from different chromosomal locations, showed a perfect sequence conservation within the structural gene regions encoding for the mature tRNA^{Tyr}. An almost complete sequence divergence was found immediately outside of these regions starting at two base pairs preceding the 5′-end of the tRNA^{Tyr} sequence. The 14 base-pair-long intervening sequences, however, were identical in all four genes, with the exception of only one base pair.

Comparable results are also obtained from the yeast phenylalanine tRNA genes. Three distinct tRNA^{Phe} genes have been cloned and sequenced (41). Two of these have identical flanking DNA sequences and are different from the third tRNA^{Phe} gene. Adjacent to the anticodon triplet, all three tRNA^{Phe} genes have an intervening sequence of 18 (or 19) base pairs, which is longer and is composed of a different sequence from the corresponding intervening sequence in the tRNA^{Tyr} genes (15).

At first glance, it would appear from these examples that intervening sequences in genes are almost as conserved by forces of selection as are the structural gene sequences themselves. Flanking DNA sequences, however, are free to diverge, as is the case in all four tRNA^{Tyr} genes. But on the other hand, two of three tRNA^Phe genes retained sequence homology beyond the structural genes for no obvious reasons, the sequences of which are quite different in the third tRNA^Phe gene. It is not clear at the present time why one stretch of DNA would remain conserved, whereas a similar one at the same position relative to the same gene is free to diverge. Nevertheless, an apparent conservation of an intervening sequence among genes need not necessarily result from a requirement of that sequence for the genes to function.

INTRONS IN HIGHER ORGANISMS

In the BALB/c strain of mice there are two nonallelic β-globin genes that specify two β-globin polypeptides differing from each other by 6 amino acids of a total of 146. These two genes reside in two different *Eco*RI fragments of genomic DNA. Both fragments have been cloned recently, and the genes were compared for sequence homology by endonuclease restriction mapping and electron microscopy of their heteroduplex structures (38,39). It was found that the structural sequences of both genes are interrupted by an intervening DNA sequence of 550 base pairs in length, which occurs immediately after the codon specifying for amino acid 104 of β-globin (36). Of the 7,000 base pairs of DNA sequence in the two genes compared in this manner, only a few hundred base pairs of sequence homology, other than the coding sequence, were found at the flanking regions of both termini of the structural gene, as well as approximately half of the intervening sequence immediately adjacent to the 5′-portion of the structural gene. The significance of conservation of only half of the intervening sequences in these two β-globin genes is not yet understood. The entire sequence, including the intervening sequence, is definitely transcribed into a 15S β-globin mRNA precursor (37). Thus, the processing of this gene transcript must involve elimination of intervening RNA sequences and subsequent rejoining of the structural RNA segments.

If we assume that the two β-globin genes arose from a gene duplication event, then we must conclude that the intervening sequence in the two copies was free to diverge at a faster rate than the structural gene sequence without adversely affecting the functioning of the gene. It follows further that the gene should remain functional if the intervening sequence were eliminated altogether. It could be that the cell is in the process of doing so via evolutionary change. Although the molecular genetics of thalassemias and related disorders are only beginning to emerge (28), it is almost certain that additional genetic disorders must exist that are caused by detrimental changes in the intervening sequences and/or failures in correct splicing of the primary RNA transcript.

Another example of intervening DNA is reported for the gene coding for mouse immunoglobulin light chain (3,40). Two such introns, 93 and 1,250 base pairs in length, were found to be present within the translated portion of this gene. Since no data are presently available on related immunoglobin genes, comparison of sequence homology between intervening sequences in this gene is not yet possible.

HOW MANY TIMES MAY EUCARYOTIC STRUCTURAL GENE BE INTERRUPTED BY INTRONS?

There are no *Eco*RI sites within the structural gene coding for ovalbumin (27), which is a unique DNA sequence in the chick genome (11,45). Nevertheless, a physical map of restriction endonuclease sites within this gene revealed that the major portion of the structural gene resides in three distinct *Eco*RI DNA fragments of 2.4, 1.8, and 9.5 kilobases (kb) in length (5,9–11,21,42,45). This finding has led to the conclusion that there are a minimum of two intervening DNA regions within which the *Eco*RI endonuclease cleavage occurs. All of these *Eco*RI DNA fragments have been cloned using the λgt·WES cloning vector (22) and analyzed for sequence homology with ovalbumin mRNA (10,11,45). Electron microscopic examination of hybrids between the mature mRNA and the cloned natural DNA fragments revealed that the RNA/DNA hybrids exist in three or more noncontiguous segments along the 2.4 kb DNA fragment, leaving two segments of double-stranded DNA interspersed between the RNA/DNA hybrids. Thus, the structural gene sequence within the ovalbumin 2.4 kb DNA is further subdivided into at least three noncontiguous segments, separated by intervening DNA sequences of unknown function. The 2.4 kb DNA fragment contains only about 450 of the 1,859 base pairs of structural gene sequence toward the 5′-end of the gene and is expected to hybridize only to the 5′-terminal region of the mRNA. Indeed, there was always an unhybridized RNA strand originating from one of the two RNA/DNA junctions in the hybrid molecules, representing the 3′-region of the mRNA molecule. Similar R-loop structures were also observed within the 1.8 kb fragment, where mRNA hybridized with two noncontiguous regions of DNA separated by an unhybridized DNA regions, indicating that the structural sequence within the 1.8 kb DNA fragment is subdivided into two portions by an intervening sequence. The electron micrographs further revealed that the 1.8 kb DNA fragment has sequence homology only with the middle section of ovalbumin mRNA, leaving long unhybridized DNA stretches and single-stranded RNAs at both ends of the hybrid structures.

A physical map of the natural ovalbumin gene was constructed subsequently by restriction mapping and Southern hybridization (Fig. 1). A linear correlation was found between the cloned DNA fragments and the corresponding sequences on the mRNA. However, this linearity was interrupted in the natural ovalbumin gene by seven DNA stretches that had no sequence

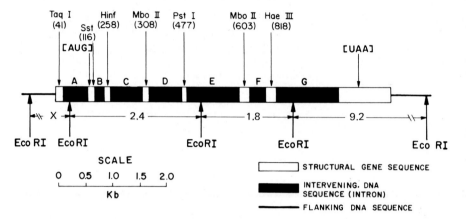

FIG. 1. A map of the organization of structural and intervening DNA sequences comprising the chick ovalbumin gene. Each of the noncontiguous structural sequences is identified by one restriction endonuclease site. The positions of the initiation and termination codons are taken from McReynolds et al. (27).

homology with the mature mRNA (10). This surprising observation was subsequently verified by direct sequence determination of the cloned DNA fragments.

A very complex picture emerged for the sequence organization of the natural ovalbumin gene. The structural sequence is encoded by eight non-contiguous pieces of DNA that are separated by seven intervening sequences of various length. All but one of the introns are located within the peptide-coding portion of the gene. The remaining intron precedes the initiation codon AUG, and there are no introns within the 634 nucleotide pairs of the untranslated region toward the 3′-terminus of the gene. Altogether, more than 7,000 base pairs of DNA are required to code for the mature ovalbumin mRNA of 1,859 bases.

CAN MUTATIONS IN INTERVENING SEQUENCES CREATE GENOTYPIC ALLELES?

It has also been reported that a 1.3 kb *Eco*RI DNA fragment may be allelic with the 1.8 kb DNA (42). This hypothesis was tested directly using the cloned 1.8 kb DNA as a probe to hybridize with total *Eco*RI-digested chick DNA. Since both the 2.4 and 1.8 kb DNA fragments are unique chick DNA sequences (11,45), they are expected to hybridize only with the chick DNA fragments 2.4 and 1.8 kb in length, respectively. The [^{32}P]2.4 kb DNA probe, hybridizing with total *Eco*RI-digested chick DNA, generated only a 2.4 kb hybridizing band. Hybridization with the 1.8 kb DNA probe, however, generated three bands 1.8, 1.3, and 0.5 kb in length, suggesting that the two fragments do represent two alleles of the ovalbumin gene in the chick genome. It appears that this allelic combination results from a divergence in an intervening sequence leading to genetic alleles that can have no phenotypic representation, since a local mutation occurred only in an intron rather than the peptide-coding region of the gene. Such genetic alleles can be demonstrated only by analysis of cloned genes and not by analysis of the mRNAs or proteins themselves.

TRANSCRIPTION OF NATURAL GENES CONTAINING STRUCTURAL AND INTERVENING SEQUENCES

Similar to the β-globin gene, the intervening sequences within the ovalbumin gene are also transcribed in their entirety during gene expression (11,45). Furthermore, the transcription of the intervening sequences is inducible by steroid hormones in a coordinate fashion with the structural gene sequences (31). Using radiolabeled nucleotides, pulse chase experiments in nuclei revealed that the rates of synthesis are almost identical for intron and structural gene sequences (31). Thus, the possibility arises that the entire ovalbumin gene is transcribed into a precursor RNA that is greater than three times the size of the mature mRNA, and the intervening RNA sequences are subsequently enzymatically processed away, yielding mature mRNA by precise excision and proper ligation. Indeed, recent experiments using electrophoresis under denaturing conditions have indicated the presence of discrete oviduct nuclear RNA species up to 40S in size that contain structural sequences and are also capable of hybridizing with the cloned 2.4 and 1.8 kb ovalbumin intron DNA after electrophoresis in denaturing gels (28a).

VIRUSES SHOW EVEN MORE COMPLEX PROCESSING REACTIONS

Splicing of the primary RNA transcript and rejoining of the remaining sequences were first demonstrated in animal viruses. In this respect viral genomes resemble the eucaryotes rather than bacteria, where the primary RNA transcript is the mRNA. In fact, an even more complex picture emerges for the processing mechanisms of viral primary RNA transcripts. In SV40 (4,16) and in adenovirus 2 (6,12,13,20,23), several mRNAs have been identified as messages for specific viral proteins. These mRNAs hybridize to noncontiguous stretches of DNA of their respective genomes and hence arose from those noncontiguous stretches via a contiguous primary RNA transcript. However, several mRNAs that have been found to belong to the same transcription unit and fall under the control of the same promoter share some sequences in their 5'-regions and differ in most of the remaining structural sequences (6,20). Thus, a single DNA transcription unit produces a single primary RNA transcript, which gives rise to one of several possible mRNA molecules by splicing out different regions of the RNA. Furthermore, the splicing occurs in such a way that the resulting mRNAs also share some sequences in their structural regions, yet the resulting proteins do not share amino acid sequences in those regions. Evidently, both integral and non-integral numbers of codons are being spliced out, as a result of which downstream mRNA sequences can be translated in different reading phases. In this way, two (33) or even three (32) proteins can be encoded by the same fragment of viral DNA.

CONCLUDING REMARKS

Previous genetic and sequence analyses of procaryotic genomes have established the existence of structural and regulatory sequences in genes, with the former representing DNA sequences coding for protein and the latter representing DNA sequences interacting with proteins regulating the expression of the structural genes. New physical analyses by restriction mapping, cloning, and sequence determination of eucaryotic genomes has now revealed that genes are composed of additional sequences within this structural region whose function and origin are not presently understood.

The existence of intervening sequences was entirely unpredicted by our prior understanding of eucaryotic genes. From the genes studied to date, it appears that the intervening sequences are quite ubiquitous. It also seems that the entire genes are transcribed as a unit into a primary transcript of RNA, which is subsequently processed (into mature mRNA) by reactions of precise cleavage and ligation.

Eucaryotic proteins are thus being synthesized from these genes-in-pieces. An abundant amount of intracistronic DNA never gets expressed in the phenotype as protein. An eminent possibility for the existence of these introns is that the processing of the primary RNA transcript into mature mRNA may be involved in the posttranscriptional control of the expression of the gene, a time tool for the cell to edit the gene products in order to obtain only those that are required for a particular cell function. A second possibility is that the intervening RNA sequences may be involved in the transport of the RNA from the nucleus to the cytoplasm. A third possibility is that introns may be the result of a functional "expansion" of the eucaryotic genome during evolution by acquiring "extra" DNA. This opens new avenues for the cell to splice together new polypeptides via assembly of various sections of DNA. There are experimental data indicating that two seemingly unrelated proteins can have sequence homology in one part, but yet are unrelated in another part of their polypeptide chains (17), as if they were derived in one part from a common segment of DNA and in another from a different part of the host genome. Once an efficient gene is assembled to code for a successful enzyme, the intervening sequence(s) of its gene may be unnecessary and could be removed over a period of time on grounds of increased efficiency. This hypothesis should be amenable to experimental testing by the removal of an intervening sequence from a gene followed by reintroduction of the structural DNA sequence back into the host cell to see if it will be expressed into the same protein.

In a sense, eucaryotic genomes are inefficient in that the amount of DNA in their intra- and intercistronic regions is far in excess of the genetic information required to code for cellular proteins. It appears as if the process of acquisition of this "extra" DNA was, at least for some time, more rapid than processes that encourage an evolutionary reduction to the necessary minimum for efficient protein synthesis and cellular function. As a result, the eucaryotic genomes may be admittedly inefficient, but on the other hand, they are more plastic and capable of new solutions to evolutionary pressures.

REFERENCES

1. Bedbrook, J. R., Kolonder, R., and Bogorad, L. (1977): Zea mays chloroplast ribosomal RNA genes are part of a 22,000 base pair inverted repeat. *Cell,* 11:739–749.
2. Botchan, P., Reeder, R. H., and Dawid, I. B. (1977): Restriction analysis of the nontranscribed spacers of *Xenopus laevis* ribosomal DNA. *Cell,* 11:599–607.
3. Brack, C., and Tonegawa, S. (1977): Variable and constant parts of the immunoglobulin light chain gene of a mouse myeloma cell are 1250 nontranslated bases apart. *Proc. Natl. Acad. Sci. USA,* 74:5652–5656.
4. Bratosin, S., Horowitz, M., and Laub, O. (1978): Electron microscopic evidence for splicing of SV40 late mRNAs. *Cell,* 13:783–790.
5. Breathnach, B., Mandel, J. L., and Chambon, P. (1977): Ovalbumin gene is split in chicken DNA. *Nature,* 270:314–319.
6. Chow, L. T., Gelinas, R. E., Broker, T. R., and Roberts, R. J. (1977): An amazing sequence arrangement at the 5′ ends of adenovirus 2 messenger RNA. *Cell,* 12:1–8.
7. Cohen, S. N., Chang, A. C. Y., Boyer, H. W., and Helling, R. B. (1973): Construction of biologically functional bacterial plasmids *in vitro Proc. Natl. Acad. Sci. USA,* 70:3240–3244.
8. Cory, S., and Adams, J. M. (1977): A very large repeating unit of mouse DNA containing the 18S, 28S and 5.8S rRNA genes. *Cell,* 11:795–805.
9. Doel, M. T., Houghton, M., Cook, E. A., and Carey, N. H. (1977): The presence of ovalbumin mRNA coding sequences in multiple restriction fragments of chicken DNA. *Nucl. Acids. Res.,* 4:3701–3713.
10. Dugaiczyk, A., Woo, S. L. C., Lai, E., Mace, M. L., McReynolds, L., and O'Malley, B. W. (1978): The natural ovalbumin gene contains seven intervening sequences. *Nature,* 274:328–333.
11. Dugaiczyk, A., Woo, S. L. C., Tsai, M.-J., Lai, E. C., Mace, M. L., and O'Malley, B. W. (1978): Molecular organization and cloning of the ovalbumin gene. In: *Genetic Engineering,* edited by H. W. Boyer and S. Nicosia, pp. 99–107. Elsevier/North Holland, New York.
12. Dunn, A. R., and Hassell, J. A. (1977): A novel method to map transcripts: Evidence for homology between an adenovirus mRNA and discrete multiple regions of the viral genome. *Cell,* 12:23–36.
13. Evans, R. M., Fraser, N., Ziff, E., Weber, J., Wilson, M., and Darnell, J. E. (1978): The initiation sites for RNA transcription in Ad2 DNA. *Cell,* 12:733–739.
14. Glover, D. M., and Hogness, D. S. (1977): A novel arrangement of the 18S and 28S sequences in a repeating unit of *Drosophila melanogaster* rDNA. *Cell,* 10:167–176.
15. Goodman, H. M., Olson, M. V., and Hall, B. D. (1977): Nucleotide sequence of a mutant eukaryotic gene: The yeast tyrosine inserting ochre suppressor *SUP4*-0. *Proc. Natl. Acad. Sci. USA,* 74:5453–5457.

16. Haegeman, G., and Fiers, W. (1978): Evidence for splicing of SV40 16S mRNA. *Nature*, 273:70–73.
17. Hood, J. M., Fowler, A. V., and Zabin, I. (1978): On the evolution of β-galactosidase. *Proc. Natl. Acad. Sci. USA*, 75:113–116.
18. Ikemura, T., and Nomura, M. (1977): Expression of spacer tRNA genes in ribosomal RNA transcription units carried by hybrid Col E1 plasmids in *E. coli. Cell*, 11:779–793.
19. Jeffreys, A. J., and Flavell, R. A. (1977): The rabbit β-globin gene contains a large insert in the coding sequence. *Cell*, 12:1097–1108.
20. Klessing, D. F. (1977): Two adenovirus mRNAs have a common 5′ terminal leader sequence encoded at least 10 kb upstream from their main coding regions. *Cell*, 12:9–21.
21. Lai, E. C., Woo, S. L. C., Dugaiczyk, A., Catterall, J. F., and O'Malley, B. W. (1978): The ovalbumin gene: Structural sequences in native chicken DNA are not contiguous. *Proc. Natl. Acad. Sci. USA*, 75:2205–2209.
22. Leder, P., Tiemeier, D., and Enquist, L. (1977): EK2 derivatives of bacteriophage lambda useful in the cloning of DNA from higher organisms: The λ WES system. *Science*, 196:175:177.
23. Lewis, J. B., Anderson, C. W., and Atkins, J. F. (1977): Further mapping of late adenovirus genes by cell-free translation of RNA selected by hybridization to specific DNA fragments. *Cell*, 12:37–44.
24. Lund, E., and Dahlberg, J. E. (1977): Spacer transfer RNAs in ribosomal RNA transcripts of *E. coli:* Processing of 30S ribosomal RNA *in vitro. Cell*, 11:247–262.
25. Lund, E., Dahlberg, J. E., Lindahl, C. L., Jaskunas, S. R., Dennis, P. P., and Nomura, M. (1976): Transfer RNA genes between 16S and 23S rRNA genes in rRNA transcription units of *E. coli. Cell*, 7:165–177.
26. Madison, J. T., and Kung, H. K. (1967): Oligonucleotides from yeast tyrosine transfer ribonucleic acid. *J. Biol. Chem.*, 242:1324–1330.
27. McReynolds, L., O'Malley, B. W., Nisbet, A. D., Fothergill, J. E., Givol, D., Fields, S., Robertson, M., and Brownlee, G. G. (1978): Sequence of chicken ovalbumin mRNA. *Nature*, 273:723–728.
28. Mears, J. G., Ramirez, F., Leibovitz, D., Nakamura, F., Bloom, A., Konotey-Ahulu, F., and Bank, A. (1978): Changes in restricted human cellular DNA fragments containing globin gene sequences in thalassemias and related disorders. *Proc. Natl. Acad. Sci. USA*, 75:1222–1226.
28a. Nordstrom, J. L. N., Tsai, M.-J., and O'Malley, B. W. (*in preparation*).
29. Olson, M. V., Montgomery, D. L., Hopper, A. K., Page, G. L., Horodyski, F., and Hall, B. D. (1977): Molecular characterization of the tyrosine tRNA genes of yeast. *Nature*, 267:639–641.
30. Pellegrini, M., Manning, J., and Davidson, N. (1977): Sequence arrangement of the rDNA of *Drosophila melanogaster. Cell*, 10:213–224.
31. Roop, D. R., Nordstrom, J. L., Tsai, S. Y., Tsai, M.-J., and O'Malley, B. W. (1978): Transcription of structural and intervening sequences of the ovalbumin gene and identification of potential ovalbumin mRNA precursors. *Cell*, 15:671–685.
32. Shaw, D. C., Walker, J. E., Northrop, F. D., Barrell, B. G., Goodson, G. N., and Fiddes, J. C. (1978): Gene K, a new overlapping gene in bacteriophage G4. *Nature*, 272:510–515.
33. Smith, M., Brown, N. L., Air, G. M., Barrell, B. G., Coulson, A. R., Hutchinson, III, C. A., and Sanger, F. (1977): DNA sequence at the C termini of the overlapping genes A and B in bacteriophage ΦX174. *Nature*, 265:702–705.
34. Southern, E. M. (1975): Detection of specific sequences among DNA fragments separated by gel electrophoresis. *J. Mol. Biol.*, 98:503–517.
35. Spiers, J., and Birnstiel, M. L. (1974): Arrangement of the 5.8S RNA cistrons in the genome of *Xenopus laevis. J. Mol. Biol.*, 87:237–256.
36. Tiemeier, D. C., Tilghman, S. M., Polsky, F. I., Seideman, J. G., Leder, A., Edgell, M. H., and Leder, P. (1978): A comparison of two cloned mouse β-globin genes and their surrounding and intervening sequences. *Cell*, 14:237–245.
37. Tilghman, S. M., Curtis, P. J., Tiemeier, D. C., Leder, P., and Weissman, C. (1978):

The intervening sequence of a mouse β-globin gene is transcribed within the 15S β-globin in mRNA precursor. *Proc. Natl. Acad. Sci. USA,* 75:1309–1313.

38. Tilghman, S. M., Tiemeier, D. C., Polsky, F., Edgell, M. H., Seideman, J. G., Leder, A., Enquist, L. W., Norman B., and Leder, P. (1977): Cloning specific segments of the mammalian genome: Bacteriophage λ containing mouse globin and surrounding gene sequences. *Proc. Natl. Acad. Sci. USA,* 74:4406–4410.

39. Tilghman, S. M., Tiemeier, D. C., Seideman, J. G., Peterlin, B. M., Sullivan, M., Maizel, J. V., and Leder, P. (1978): Intervening sequence of DNA identified in the structural portion of a mouse β-globin gene. *Proc. Natl. Acad. Sci. USA,* 75:725–729.

40. Tonegawa, S., Maxam, A. M., Tizard, R., Bernard, O., and Gilbert, W. (1978): Sequence of a mouse germ-line gene for a variable region of an immunoglobulin light chain. *Proc. Natl. Acad. Sci. USA,* 75:1485–1489.

41. Valenzuela, P., Venegas, A., Weinberg, F., Bishop, R., and Rutter, W. J. (1978): Structure of yeast phenylalanine-tRNA genes: Are intervening DNA segments within the region coding for the tRNA? *Proc. Natl. Acad. Sci. USA,* 75:190–194.

42. Weinstock, R., Sweet, R., Weiss, M., Cedar, H., and Axel, R. (1978): Intragenic DNA spacers interrupt the ovalbumin gene. *Proc. Natl. Acad. Sci. USA,* 75:1299–1303.

43. Wellauer, P. K., and Dawid, I. B. (1977): The structural organization of ribosomal DNA in *Drosophila melanogaster. Cell,* 10:193–212.

44. White, R. L., and Hogness, D. S. (1977): R loop mapping of the 18S and 28S sequences in the long and short repeating units of *Drosophila melanogaster rDNA. Cell,* 10:177–192.

45. Woo, S. L. C., Dugaiczyk, A., Tsai, M.-J., Lai, E. C., Catterall, J. F., and O'Malley, B. W. (1978): The ovalbumin gene: Cloning of the natural gene. *Proc. Natl. Acad. Sci. USA,* 75:3688–3692.

Ontogeny of Receptors and Reproductive Hormone Action,
edited by T. H. Hamilton, J. H. Clark, and W. A. Sadler.
Raven Press, New York 1979.

Immunochemistry of Estrophilin

G. L. Greene, L. E. Closs, E. R. DeSombre, and E. V. Jensen

Ben May Laboratory for Cancer Research, University of Chicago, Chicago, Illinois 60637

The principal features of the interaction of estrogenic hormones with their target cells are now fairly well understood (4,8,16). The hormone binds to an extranuclear receptor protein (estrophilin), inducing its conversion to an active form (Fig. 1). The activated steroid-receptor complex is translocated to the nucleus where it associates with chromatin and alleviates restrictions on RNA synthesis that are characteristic of the hormone-dependent tissue.

Despite our knowledge of the overall pattern of hormone-receptor interaction, detailed understanding of the processes of receptor synthesis, activation, translocation, and nuclear binding is far from complete. In the hope that techniques of immunochemistry might provide new approaches to the study of hormone receptors, we have prepared specific antibodies to estrophilin by immunizing rabbits with highly purified preparations of estrogen-receptor complex from calf uteri and have studied the cross-reactivity of these antibodies with estrophilin from various sources (6).

ANTIBODY PREPARATION

The estradiol-receptor complex used for immunization was prepared by incubating crude calf uterine nuclei with 10 nM tritiated estradiol (5.7 Ci/mmole) in calf uterine cytosol at 30° for 60 min. The nuclear complex was extracted and purified as described elsewhere (1,3,9) by a sequence of salt precipitation, gel filtration through Sephadex G-200, and polyacrylamide gel electrophoresis. During gel filtration the complex loses its tendency to aggregate in media of low ionic strength, and its sedimentation rate changes from 5.2S in sucrose gradients containing 400 mM KCl to 4.8S in either high- or low-salt gradients (Fig. 1). The material used for the immunizations described here contained from 10 to 30% of the radioactivity expected for one estradiol bound to a receptor protein of molecular weight 70,000 daltons, although subsequent preparations of estrogen-receptor complex were obtained that showed 50 to 100% of the theoretical radioactivity.

As described in detail elsewhere (6), 6-month-old male New Zealand white rabbits were immunized with the purified estradiol-receptor complex by the intradermal procedure of Vaitukaitis et al. (20), using 20 μg of re-

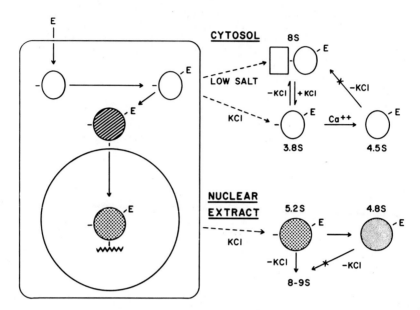

FIG. 1. Schematic representation of interaction pathway of estradiol (E) in uterine cell. Diagram at left indicates extranuclear estradiol-receptor complex undergoing activation and translocation to chromatin in the nucleus. Diagrams at right indicate sedimentation properties of complexes extracted from the cell, before and after losing the ability to aggregate in media of low ionic strength. (From Jensen et al., ref. 9.)

ceptor for the initial injection and 20 to 50 μg for booster injections. Blood was collected at 14-day intervals via the marginal ear vein. In the experiments described here the blood used was drawn from a rabbit who had received six booster injections over a period of 1 year and who had first shown antibody titer approximately 11 months after initial immunization; in subsequent animals receiving antigen of higher purity (i.e., higher specific radioactivity), antibody titer was observed somewhat earlier. A crude immunoglobulin fraction (Ig-i) was prepared from the immune serum by precipitation from 33% saturated ammonium sulfate in 50 mM phosphate buffer and redissolving the washed precipitate in phosphate-buffered saline (18). Immunoglobulin (Ig-n), prepared similarly from the serum of a nonimmunized rabbit, was used as a control.

DEMONSTRATION OF ANTIBODIES TO ESTROPHILIN

The serum of immunized rabbits can be shown to contain specific antibodies to estrophilin by five criteria: double antibody precipitation, adsorption by Sepharose-linked Ig-i, adsorption by protein-A in the presence of Ig-i, sedimentation in sucrose gradients, and elution on gel filtration. In all of these procedures, the radioactivity of bound estradiol serves as a marker for the receptor protein.

TABLE 1. *Interaction of rabbit Ig with tritiated estrogen-receptor complexes of calf uterus*

| Method | Form of ^{3}H-estradiol | % ^{3}H precipitated or bound | | |
		Ig-i	Ig-n	No Ig
Double antibody	Nuclear complex	61	10	
precipitation	Cytosol complex	56	3	
	Free	2	2	
Binding to	Nuclear complex	66	21	17
Sepharose-Ig	Free	6	15	7
Binding to Sepharose-	Nuclear complex	70	17	30
Protein-A	Free	3	5	3

As shown in Table 1, in the presence of Ig-i, but not of Ig-n, a significant amount of estradiol-receptor complex is precipitated by antiserum to rabbit gamma globulin or bound to immobilized *Staphylococcus aureus* protein-A, a substance that reacts specifically with the Fc portion of IgG. Similarly, Ig-i linked to Sepharose binds a much greater proportion of estradiol-receptor complex than does Sepharose-Ig-n or Sepharose alone. In no case is there significant binding or precipitation of estradiol in the absence of the receptor,

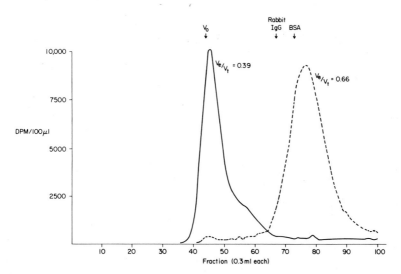

FIG. 2. Elution pattern from Sephadex G-200 of calcium-stabilized (17) estradiol-receptor complex of calf uterine cytosol treated with Ig-i (—) or Ig-n (---). Gel filtration was carried out in 400 mM KCl in 10 mM Tris, pH 7.4. V_e, volume of eluate at which the peak of radioactivity was eluted; V_t, the bed volume of the column (35 ml); V_0, the void volume as determined with blue dextran. Bovine serum albumin (BSA) and rabbit IgG were chromatographed separately in the same column.

indicating that the antibody recognizes estrophilin itself and not the estradiol bound to it.

Because the reaction of Ig-i with estrophilin forms a nonprecipitating product, the interaction can be readily demonstrated by its effect both on the elution properties of the receptor during gel filtration and on its sedimentation properties in sucrose gradients. As shown in Fig. 2, the calcium-stabilized estradiol-receptor complex of calf uterine cytosol is eluted from Sephadex G-200 much earlier in the presence of Ig-i than of Ig-n. A similar shift takes place in the elution of the uncomplexed receptor, indicating that reaction of the antibody with estrophilin does not depend on the presence of the estrogen.

Probably the most informative technique for demonstrating antibody-receptor interaction, and for studying the cross-reactivity with estrophilin from various sources, is the effect of the antibody on the sedimentation rate of the estrogen-receptor complex. As seen in Fig. 3, in the presence of Ig-i, but not Ig-n, the sedimentation of the purified nuclear estradiol-receptor complex, used as the antigen for immunization, is increased from 4.8S to a value greater than 10S. With the crude nuclear complex, which sediments at

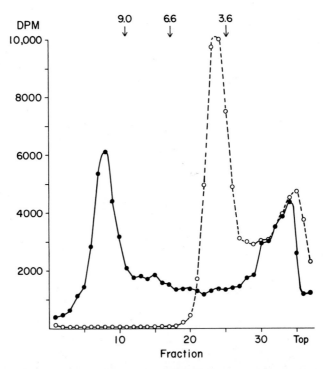

FIG. 3. Sedimentation pattern of highly purified estradiol-receptor complex from calf uterine nuclei on ultracentrifugation at 253,000 × *g* for 15 hr at 2° in 10 to 30% sucrose gradients containing 10 mM KCl in the presence of Ig-i (●) or Ig-n (○).

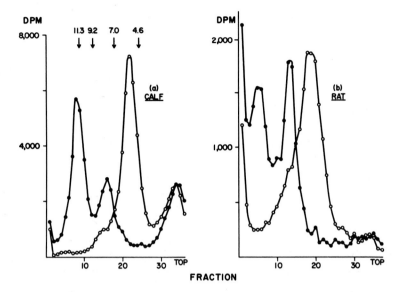

FIG. 4. Sedimentation pattern in 10 to 30% sucrose gradients containing 400 mM KCl of **(a)** extract of calf uterine nuclei after incubation for 60 min at 30° with 20 nM tritiated estradiol in calf uterine cytosol, and **(b)** extract of uterine nuclei from immature rats 4 hr after injection of 100 ng tritiated estradiol *in vivo,* in the presence of Ig-i (●) or Ig-n (○). (From Greene et al., ref. 6.)

5.2S in salt-containing sucrose gradients (Fig. 1), interaction with Ig-i gives rise not only to this rapidly sedimenting product but also to a more slowly moving peak at about 8S (Fig. 4A). These two peaks are also seen with the estradiol-receptor complex extracted from the uterine nuclei of rats injected with tritiated estradiol *in vivo* (Fig. 4B).

CROSS-REACTIVITY OF ESTROPHILIN COMPLEXES

Antibodies to the purified nuclear estradiol-receptor complex of calf uterus react not only with this antigen but also with extranuclear as well as nuclear forms of estrophilin from all reproductive tissues that have been investigated. These include nuclear complexes from calf, rat, rabbit, and sheep uterus and rat endometrial tumor; and extranuclear complexes from calf, rat, mouse, guinea pig, rabbit, and sheep uterus, rat mammary and endometrial tumor, and human breast cancer. In contrast to its reaction with the nuclear estrophilin where two new peaks are produced, Ig-i reacts with the extranuclear estradiol-receptor complex to yield a product that sediments as a single entity in either high-salt or low-salt gradients (Figs. 5 and 6), apparently related to the slower of the two peaks observed with the nuclear complex. However, the calcium-stabilized (4.5 S) form of the extranuclear complex

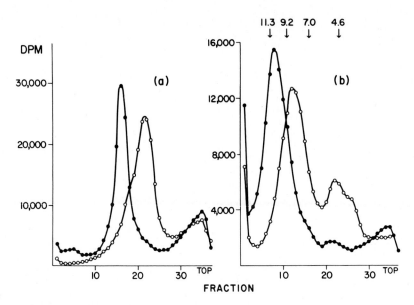

FIG. 5. Sedimentation pattern of rat uterine cytosol plus ³H-estradiol in 10 to 30% sucrose gradients containing **(a)** 400 mM KCl and **(b)** 10 mM KCl, in the presence of Ig-i (●) or Ig-n (○). (From Greene et al., ref. 6.)

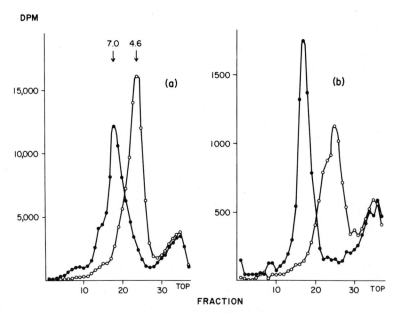

FIG. 6. Sedimentation pattern in 10 to 30% sucrose gradients containing 400 mM KCl of **(a)** calf uterine cytosol and **(b)** human breast cancer cytosol, each containing ³H-estradiol, in the presence of Ig-i (●) or Ig-n (○). (From Greene et al., ref. 6.)

resembles the nuclear complex in showing two rather than one new sedimentation peak in the presence of Ig-i.

Although antibodies to calf uterine estrophilin do not show tissue or species specificity in regard to estrogen receptors, there appears to be great specificity in regard to receptors for other steroid hormones. In the presence of Ig-i, no displacement was observed in the sedimentation properties of the nuclear or extranuclear dihydrotestosterone-receptor complexes of rat prostate or of the extranuclear progesterone-receptor complex of chick oviduct, rabbit uterus, or rat endometrial tumor.

DISCUSSION

Although observations have been reported suggesting the presence of antibodies to estrophilin in the serum of rabbits immunized with partially purified estradiol-receptor complex from calf (2,12) or rat (19) uterine cytosol, the present findings represent the first definitive demonstration of antibodies to a receptor protein for a steroid hormone. The reactivity of the antibody with both the nuclear and extranuclear estradiol-receptor complexes indicates an immunochemical similarity between the two forms of estrophilin and provides supporting evidence for the concept of a two-step interaction mechanism in which the nuclear receptor is derived from the hormone-induced translocation of the extranuclear receptor (5,11). The formation of a larger entity from the reaction of antibody with nuclear as compared to extranuclear estrophilin suggests that the nuclear receptor may contain more antibody binding sites per molecule, consistent with the concept that the nuclear complex may be a dimer of the cytosol complex (13,15).

The fact that the radioactivity of estradiol can be used as a marker for the receptor protein implies that reaction with the antibody does not distort the configuration of the estrophilin to such an extent that it loses its ability to bind steroid. It also appears that interaction with the antibody does not block the hormone-binding site, inasmuch as the same radioactive estradiol-receptor-antibody complex is formed when tritiated estradiol is added to the receptor-antibody complex as when antibody is added to the preformed estrogen-receptor complex.

Antibody to estrophilin provides a reagent that should prove useful in many aspects of receptor research. For the first time one has a potential means for detecting the receptor protein that does not depend on its ability to bind radioactive steroid. The cross-reactivity of Ig-i with estrophilin from human breast cancer, but its lack of reactivity with androgen or progesterone receptors, opens the possibility of a simple radioimmunoassay for the receptor content of breast cancers as a guide to therapy (7,10,14). Ig-i linked to Sepharose or other support may prove to be an efficient tool for the purification of estrophilin, whereas insight into the molecular details of hormone action may come from the application of immunohistological techniques for the precise intracellular localization of nuclear and extranuclear receptor through electron microscopy.

SUMMARY

Immunoglobulin from the serum of rabbits immunized with highly purified estradiol-receptor complex from calf uterine nuclei has been shown to contain specific antibodies to estrophilin by five criteria. Antibodies to calf nuclear estrophilin cross-react with nuclear estradiol-receptor complexes of rat, rabbit, and sheep uterus and rat endometrial tumor, as well as with extranuclear receptor of calf, rat, mouse, rabbit, guinea pig, and sheep uterus, rat mammary and endometrial tumor, and human breast cancer. There is no interaction of the antibody with estradiol itself. The nuclear form of estrophilin appears to bind more immunoglobulin molecules than does the cytosol form. The antibodies do not react with either the nuclear or extranuclear dihydrotestosterone-receptor complexes of rat prostate or with the extranuclear progesterone-receptor complexes of rabbit uterus, chick oviduct, or rat endometrial tumor. These findings indicate an immunochemical similarity among estrophilins from several mammalian species, as well as between nuclear and extranuclear forms of the receptor, but not among receptor proteins for different steroid hormones.

ACKNOWLEDGMENTS

We are grateful to Professors S. Liao and T. C. Spelsberg for supplying radioactive hormone-receptor complexes of rate prostate and chick oviduct, to Dr. G. T. Ross for advice and help in performing the immunizations, and to Susan Margitic and Pamela Wang for technical assistance. This work was supported by grants and contracts from the U.S. Public Health Service (HD 9–2108, HD-07110, CA-02897, CA-09183, and CB-43969), the American Cancer Society (BC-86), the Ford Foundation (690–0109), and the Helena Rubinstein Foundation.

REFERENCES

1. DeSombre, E. R., and Gorell, T. A. (1975): Purification of estrogen receptors. In: *Methods in Enzymology, Vol. 36, part A,* edited by B. W. O'Malley and J. G. Hardman, pp. 349–365. Academic Press, New York.
2. Fox, L. L., Redeuilh, G., Baskevitch, P., Baulieu, E. E., and Richard-Foy, H. (1976): Production and detection of antibodies against the estrogen receptor from calf uterine cytosol. *FEBS Lett.,* 63:71–76.
3. Gorell, T. A., DeSombre, E. R., and Jensen, E. V. (1977): Purification of nuclear estrogen receptors. In: *Proceedings Fifth International Congress of Endocrinology, Vol. 1,* edited by V. H. T. James, pp. 467–472. Excerpta Medica Foundation, Amsterdam.

4. Gorski, J., and Gannon, F. (1976): Current models of steroid hormone action. A critique. *Ann. Rev. Physiol.*, 38:425–450.
5. Gorski, J., Toft, D., Shyamala, G., Smith, D., and Notides, A. (1968): Hormone receptors: studies on the interaction of estrogen with the uterus. *Recent Prog. Horm. Res.*, 24:45–72.
6. Greene, G. L., Closs, L. E., DeSombre, E. R., and Jensen, E. V. (1977): Antibodies to estrogen receptor: immunochemical similarity of estrophilin from various mammalian species. *Proc. Natl. Acad. Sci. U.S.A.*, 74:3681–3685.
7. Jensen, E. V., Block, G. E., Smith, S., Kyser, K., and DeSombre, E. R. (1971): Estrogen receptors and breast cancer response to adrenalectomy. *Natl. Cancer Inst. Monogr.*, 34:55–70.
8. Jensen, E. V., and DeSombre, E. R. (1973): Estrogen-receptor interaction. *Science,* 182:126–134.
9. Jensen, E. V., Mohla, S., Gorell, T. A., and DeSombre, E. R. (1974): The role of estrophilin in estrogen action. *Vitam. Horm.*, 32:89–127.
10. Jensen, E. V., Smith, S., and DeSombre, E. R. (1976): Hormone dependency in breast cancer. *J. Steroid Biochem.*, 7:911–917.
11. Jensen, E. V., Suzuki, T., Kawashima, T., Stumpf, W. E., Jungblut, P. W., and DeSombre, E. R. (1968): A two-step mechanism for the interaction of estradiol with the rat uterus. *Proc. Natl. Acad. Sci. U.S.A.*, 59:632–638.
12. Jungblut, P. W., Hätzel, I., DeSombre, E. R., and Jensen, E. V. (1967): Über Hormon—"Receptoren." Die oestrogenbindenen Prinzipien der Erfolgsorgane. *Coll. Ges. Physiol. Chem.*, 18:58–86.
13. Little, M., Szendro, P., Teran, C., Hughes, A., and Jungblut, P. W. (1975): Biosynthesis and transformation of microsomal and cytosol estradiol receptors. *J. Steroid Biochem.*, 6:493–500.
14. McGuire, W. L., Carbone, P. P., and Vollmer, E. P., editors (1975): *Estrogen Receptors in Human Breast Cancer.* Raven Press, New York.
15. Notides, A., Hamilton, D. E., and Auer, H. E. (1975): A kinetic analysis of estrogen receptor transformation. *J. Biol. Chem.*, 250:3945–3950.
16. O'Malley, B. W., and Means, A. R. (1974): Female steroid hormones and target cell nuclei. *Science,* 183:610–620.
17. Puca, G. A., Nola, E., Sica, V., and Bresciani, F. (1972): Estrogen-binding proteins of calf uterus. Interrelationship between various forms and identification of a receptor-transforming factor. *Biochemistry,* 11:4157–4165.
18. Shiu, R. P. C., and Friesen, H. G. (1976): Interaction of cell-membrane prolactin receptor with its antibody. *Biochem. J.,* 157:619–626.
19. Soloff, M. S., and Szego, C. M. (1969): Purification of estradiol receptor from rat uterus and blockade of its estrogen-binding function by specific antibody. *Biochem. Biophys. Res. Commun.,* 34:141–147.
20. Vaitukaitis, J., Robbins, J. B., Nieschlag, E., and Ross, G. T. (1971): A method for producing specific antisera with small doses of immunogen. *J. Clin. Endocrinol. Metabol.,* 33:988–991.

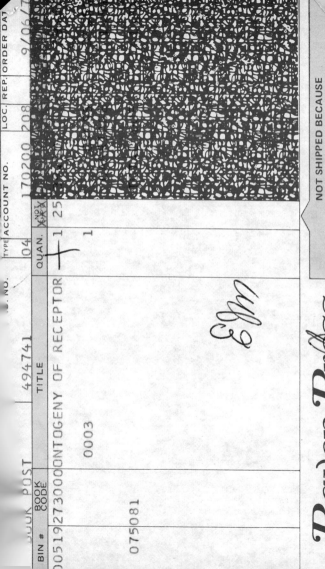

BOOK POST		494741		TYPE	ACCOUNT NO.	LOC.	REP.	ORDER DAT
BIN #	BOOK CODE	TITLE	QUAN.	NO. PACK	04	170200	208	9/06/7
00519	2730000	ONTOGENY OF RECEPTOR	1	25				
075081	0003		1					

Elm 3

Ontogeny of Receptors and Reproductive Hormone Action,
edited by T. H. Hamilton, J. H. Clark, and W. A. Sadler.
Raven Press, New York 1979.

Progesterone Receptor Protein Resolved from ATP-Pyrophosphate Exchange Activity

William T. Schrader and Bert W. O'Malley

Department of Cell Biology, Baylor College of Medicine, Houston, Texas 77030

A recent study (9) has created considerable interest with the report that chick oviduct progesterone receptor B subunit contains an inherent adenosine triphosphate-pyrophosphate (ATP-PPi) exchange activity. Because of the importance such an enzyme activity might have to the understanding of receptor function, we attempted to confirm these observations using the same receptor system. We had available techniques for the resolution of progesterone receptor subunits (14), as well as purification techniques for the intact receptor (6) and the individual A and B subunits (4,13). Finally, an *in vitro* chromatin transcription assay for receptor function (15,16) was available to us for assessing the importance of associated ATP-PPi exchange enzymatic activity to gene regulation.

MATERIALS AND METHODS

Chemicals and Radiochemicals

All chemicals were reagent grade and obtained from J. T. Baker Co. with the following exceptions. Ultrapure Tris buffer was from Schwarz/Mann; ATP (sodium salt) and dithiothreitol were from Sigma; bovine serum albumin (BSA) was from Pentex Laboratories. [1,2-^{3}H$_2$] progesterone (50 Ci/mmoles) and [^{32}P]sodium pyrophosphate (80 mCi/mmoles) were from New England Nuclear. Diethylaminoethyl (DEAE)-cellulose (DE-52) was from Reeve-Angel.

Preparation of Progesterone Receptor

Chick oviduct cytosol was prepared from immature chick oviducts obtained from estrogen-primed immature chicks as described elsewhere (10) in buffer A (10 mM Tris-HCl, pH 8.0/1 mM Na$_2$ ethylenediaminetetraacetic acid/12 mM 1-thioglycerol). Receptors in cytosol were labeled with 10^{-8} M [^{3}H]progesterone for 3 hr at 0°C. Partially purified receptors were prepared by precipitation at 35% saturation ammonium sulfate and then redissolved in 60% of the original volume of buffer A.

Column Chromatography

DEAE-cellulose column chromatography was performed on cytosol and on the ammonium sulfate precipitate as described in earlier publications (10). Cytosol (5 ml) or the precipitated receptors from 35 ml cytosol were applied to a 10-ml DEAE column at 0°C, washed with buffer A and eluted using a 150-ml linear KCl gradient in buffer A from 0 to 0.3 KCl. Fractions (2 ml) were collected, 100 μl aliquots counted for ^{3}H, and the rest saved for the PPi exchange assay.

Sucrose Gradient Ultracentrifugation

Labeled cytosol was analyzed on 5 to 20% sucrose gradients in buffer A containing 0.3 M KCl, exactly as described in earlier publications (10,12). Sedimentation markers run in other experiments showed that 4S, 6S and 8S macromolecules ran in fractions 10, 15, and 20, respectively. Gradients were pierced from the tube bottom and 200-μl fractions collected in a cold room. Aliquots (50 μl) were counted for ^{3}H to detect receptor-hormone complexes. The remaining 150 μl of each sample was saved for PPi exchange assays.

Receptor Subunit Purification

The receptor A subunit purification protocol was as detailed elsewhere (4) and included the following steps in buffer A. Unlabeled oviduct cytosol containing receptor AB dimers (12) was passed through phosphocellulose and DNA-cellulose. It was then labeled with excess [^{3}H]progesterone and precipitated at 35% ammonium sulfate. The precipitate, containing dissociated receptor subunits, was then chromatographed on DEAE-cellulose by KCl gradient elution to isolate receptor A and B subunits. The B protein peak was used in the present experiments without further purification. The A protein peak was chromatographed by KCl gradient elution from a second DNA-cellulose column, followed finally by a second phosphocellulose column. The receptor eluted from this final column was homogeneous as shown by gel electrophoresis in 1% sodium dodecyl sulfate and in acid urea.

ATP-PPi Exchange Activity Assay

The assay was carried out by the method described by Moudgil and Toft (9) with minor modifications. Receptor samples (0 to 100 μl) and reaction mixture were incubated in a final reaction volume of 500 μl at 37°C for 15 min. The reaction mixture contained 10 mM Tris-HCl, pH 8.0/4 mM

MnCl$_2$/2 mM dithiothreitol/20 μg/ml BSA/1 mM ATP/120 nM Na$_4$[^{32}P]PPi (about 15,000 cpm). In addition, duplicate samples contained either 1 mM nonradioactive Na$_4$PPi or an equivalent volume of water. After the incubation, samples were chilled in ice and 400 μl of ice-cold 1% BSA containing 1 mM Na$_4$PPi was added, followed by 200 μl of 10% charcoal-1% dextran T 60 suspension in buffer A. Finally, 2.0 ml of 5% trichloroacetic acid was added. The suspensions were filtered through Reeve-Angel 2.4-cm glass fiber filters to collect the charcoal. The tubes and filters were rinsed with three 1.0-ml washes of 1% trichloroacetic acid. The filters were counted for ^{32}P in a Beckman liquid scintillation counter under conditions that excluded 99% of any ^{3}H counts adsorbed to the filters. Counting efficiency for ^{32}P was nominally 80%. [^{32}P]ATP synthesized in the reaction was determined by subtracting ^{32}P cpm incorporated in the tubes containing excess nonradioactive Na$_4$PPi from the tubes containing H$_2$O.

In the experiments measuring [^{32}P]ATP synthesized by various steps of the receptor purification, aliquots of each sample (5 to 100 μl) were assayed, and the [^{32}P]ATP cpm corrected as described above. The results were plotted graphically to determine the initial slope of enzyme activity. This was necessary because of the presence in some samples of a potent ATPase activity that reduced incorporation when large aliquots of crude receptors were assayed.

RESULTS

The ATP-PPi exchange enzyme activity assay used was essentially that described by Moudgil and Toft (9). Because of the presence of a potent soluble ATPase in oviduct cytosol, the half-life of ATP in these crude extracts is extremely short (W. T. Schrader, *unpublished data*). Thus, the assay was developed using a redissolved protein fraction from oviduct cytosol obtained by precipitation at 50% saturation of ammonium sulfate. This amount of ammonium sulfate quantitatively precipitates progesterone receptors and also brings down a large amount of the exchange enzyme activity. The standard curve of enzyme was linear up to about 35% conversion of ^{32}P into ATP (*unpublished data*). No further characterizations of the activity were carried out.

The initial test of whether the receptor contained ^{32}PPi exchange activity involved the study shown in Table 1. Yields of receptor activity and exchange activity were determined following precipitation of cytosol with ammonium sulfate at either 30 or 50% saturation. The table shows that receptor precipitation occurred in significant amounts at 30% saturation, whereas exchange activity did not precipitate until 50% saturation. This behavior is not consistent with the hypothesis that *all* of the exchange activity was receptor associated. However, since there are many enzymes that can catalyze ATP-PPi exchange, it was possible that the 10% of ex-

TABLE 1. *Differential precipitation of progesterone receptor and ATP-PPi exchange enzyme by ammonium sulfate*

Step	Relative [³H] receptor in fraction (%)	Recovery of exchange enzyme activity (%)
Cytosol	100	100
0–30% Ammonium sulfate pellet	70	10
0–50% Ammonium sulfate pellet	95	90

change activity precipitated at 30% saturation was indeed receptor associated.

To test this possibility, the experiment shown in Fig. 1 was performed. A receptor preparation obtained at 30% saturation of ammonium sulfate was chromatographed on DEAE-cellulose by KCl gradient elution. Equal (100-µl) volumes of each fraction were either counted for [³H]progesterone or assayed for corrected ³²PPi exchange activity. Both receptor A (0.05 to 0.1 M) and B (0.2 M) subunits were clearly distinguished on the elution profile. The ³²PPi exchange activity eluted as a major peak at 0.1 M KCl and a minor shoulder at 0.1 to 0.15 M KCl. Significantly, neither A nor B had substantial ³²PPi exchange activity associated with it. On the basis of this

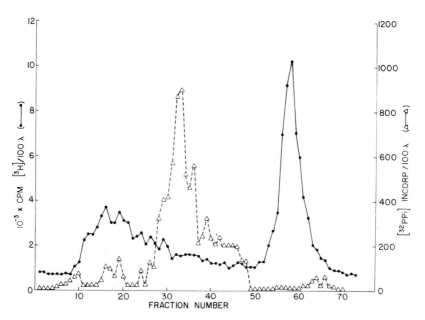

FIG. 1. DEAE-cellulose column chromatography of [³H]progesterone-receptor complexes present in 0 to 30% ammonium sulfate precipitate. Aliquots were counted for [³H] (—●—). Identical aliquots were assayed for ³²PPi exchange activity (--△--).

experiment, the two receptor subunits differed from the exchange activity on the basis of electrostatic charge.

It is possible that the receptor and PPi exchange activities are similar in molecular weight and might therefore be mistakenly identified as parts of the same molecule during gel electrophoresis (9). To test this possibility, [³H]-labeled cytosol was subjected to sucrose gradient ultracentrifugation, and portions of each fraction were assayed for ³²PPi exchange activity. As shown in Fig. 2, the receptor subunits sedimented at about 4S in 0.3 M KCl, as has been reported earlier (10). The ³²PPi exchange activity, however, was clearly separated from the receptors and sedimented at about 6.5S. Less than 1% of the enzyme activity co-sedimented with the receptor subunits. By this criterion as well, the receptors do not have ³²PPi exchange activity.

The foregoing experiments clearly showed that the majority of ³²PPi exchange enzyme activity was independent of progesterone receptor. However, these experiments could not exclude the possibility that a small fraction of the activity would remain in a purified receptor preparation. To test this, the purification protocol used to prepare homogeneous receptor A subunit (4) was followed as shown in Table 2. Aliquots of each step of the purification were assayed for ³²PPi activity and receptor A subunit concentration *via* bound [³H]progesterone. The table shows that during the course of purification of the receptor, ³²PPi exchange activity declined dramatically to less than 0.5% of the activity of crude cytosol. At this level, so little ³²PPi incorporation occurred in the homogeneous A protein preparation that the

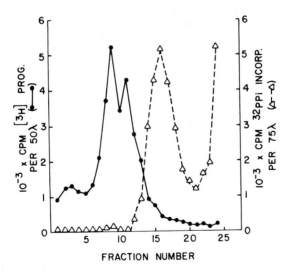

FIG. 2. Sucrose gradient ultracentrifugation of [³H]progesterone receptor in oviduct cytosol. Gradients were run at 45,000 rpm for 16 hr in a Beckman SW-50.1 rotor. Fractions (200 µl) collected at 0°C were assayed for [³H]progesterone (—●—) and for ³²PPi exchange activity (--△--).

TABLE 2. *Receptor A protein separated from ATP-PPi exchange enzyme during purification*

Sample	[³H]prog bound per 100 μl (cpm)	Receptor purifica- tion[c] (−fold, 1×)	Corrected[e] ³²PPi incorporated per 100 μl (cpm)	Activity ratio ³²P cpm/ ³H cpm	Relative activity ratio (−fold, 1×)
Cytosol	34,400[a]		45,600[f]	1.35	
Phosphocellulose drop-through	32,900[a]	1.1	84,000	2.55	0.53
DNA-cellulose drop-through	19,720[a]	1.5	45,000	2.28	0.59
0 to 35% Ammonium sulfate pellet	9,618[b]	20	800	0.083	1/16
DEAE-cellulose receptor A pool	5,147[b]	44	400	0.077	1/17
DEAE-cellulose receptor B pool	9,008[b]	ND	50	0.005	1/270
DNA-cellulose II receptor A pool	11,930[b]	4,900	500	0.042	1/32
Phosphocellulose II receptor A pool	8,030[b]	3,900[d]	50	0.006	1/225

[a] Determined by hormone-binding assay of companion sample.

[b] Hormone associated with receptor after labeling of pooled DNA-cellulose drop-through fraction.

[c] Data from a companion experiment using same protocol but 10 times more tissue (4).

[d] This material in companion experiment was homogenous receptor A subunit as judged by gel electrophoresis in two different denaturing systems.

[e] [³²P]ATP data corrected using subtraction of ³²P bound in presence of 1 mM NaPPi. Aliquots of each sample (5 to 100 μl) assayed; best estimate of activity at 100 μl determined graphically from initial slope.

[f] Determined from ³²P peak area in high-KCl sucrose gradient experiment of Fig. 1.

data must be considered an upper estimate of activity in the preparation. Significantly, the specific activity of ³²PPi exchange (³²P cpm/³H cpm) was only about one-twentieth that reported in other studies on the B protein (9). We conclude that homogeneous receptor A subunit contains no ³²PPi exchange activity.

Since intact progesterone receptor consists of two subunits, it was of interest to determine whether partially purified receptor B subunit contained ³²PPi exchange activity. This was tested by assaying the B subunit pool eluted from the DEAE-cellulose step in the receptor A purification protocol. As shown in Table 2, the receptor B subunit preparation also contained only negligible ³²PPi exchange activity. Although the B protein is not homogeneous at this step, no further purifications were carried out, in view of the low exchange activity remaining. The specific activity of ³²PPi exchange enzyme as defined above was again only about one-twentieth that reported by other studies.

DISCUSSION

The progesterone receptors have been resolved from contaminating ATP-PPi exchange activity by the experiments described above. Thus, the receptors are still envisioned to function by nuclear uptake into chromatin acceptor sites with no known associated activity. The notion that ATP may be involved in the uptake process is a recurrent one, but the role ATP might play remains obscure. It was attractive to think that ATP-Sepharose binding of receptor (8) might indicate a binding site for ATP, but no such site has yet been demonstrated. The mechanism of inhibition of nuclear receptor uptake by orthophenanthroline and rifamycin AF/013 (7) is also unknown. Since progesterone receptors can be separated from Ca^{2+}-dependent proteases (11), cyclic AMP-binding and protein kinase activities (5), and polynucleotide polymerase (2), and do not bind to hen oviduct RNA polymerase II (1) or to ATP-PPi exchange enzymes, we conclude that progesterone receptors probably function by a mechanism independent of these activities. Such a mechanism, involving sequential interaction of the two A and B subunits with the genome, has been proposed by us elsewhere (3,15) and is under continuing study in our laboratories.

SUMMARY

The chick oviduct progesterone receptor was assayed for ATP-PPi exchange enzymatic activity after ammonium sulfate precipitation, chromatography on DEAE-cellulose, sucrose gradient ultracentrifugation, or purification of receptor subunit A protein to homogeneity. In all four experiments the enzymatic activity was completely resolved from the receptor activity. The study concludes that neither the receptor A or B subunit proteins nor the intact A-B complexes contain this enzyme activity as part of their structure.

ACKNOWLEDGMENTS

The authors acknowledge the competent technical assistance of Aron Johnson. This research is supported by National Institutes of Health Grants HD-7857 and HD-8188 to the Baylor Center for Population Research.

REFERENCES

1. Buller, R. E. (1976): Chick Oviduct Progesterone Receptors: Genome Interactions and Functional Significance. Ph.D. Dissertation, Baylor College of Medicine, Houston, Texas.
2. Buller, R. E., Schwartz, R. J., and O'Malley, B. W. (1976): Steroid hormone receptor fraction stimulation of RNA synthesis: A caution. *Biochem. Biophys. Res. Commun.*, 69:106–113.
3. Buller, R. E., Schwartz, R. J., Schrader, W. T., and O'Malley, B. W. (1976):

Progesterone-binding components of chick oviduct. *In vitro* effect of receptor sub-units on gene transcription. *J. Biol. Chem.*, 251:5178–5186.

4. Coty, W. A., Schrader, W. T., and O'Malley, B. W. (1979): Purification and characterization of the chick oviduct progesterone receptor A subunit. *J. Steroid Biochem.* (*In press.*)

5. Keller, R. K., Chandra, T., Schrader, W. T., and O'Malley, B. W. (1976): Protein kinases of the chick oviduct: A study of the cytoplasmic and nuclear enzymes. *Biochemistry*, 15:1958–1967.

6. Kuh, R. W., Schrader, W. T., Smith, R. G., and O'Malley, B. W. (1975): Progesterone-binding components of chick oviduct. Purification by affinity chromatography. *J. Biol. Chem.*, 250:4220–4228.

7. Lohmar, P. H., and Toft, D. O. (1975): Inhibition of the binding of progesterone receptor to nuclei: effects of o-phenanthroline and rifamycin AF/013. *Biochem. Biophys. Res. Commun.*, 67:8–15.

8. Moudgil, V. K., and Toft, D. O. (1975): Binding of ATP to the progesterone receptor. *Proc. Natl. Acad. Sci. USA*, 72:901–905.

9. Moudgil, V. K., and Toft, D. O. (1976): ATP-PPi exchange activity of progesterone receptor. *Proc. Natl. Acad. Sci. USA*, 73:3443–3447.

10. Schrader, W. T. (1975): Methods for extraction and quantification of receptors. *Methods Enzymol.*, 36:187–211.

11. Schrader, W. T., Coty, W. A., Smith, R. G., and O'Malley, B. W. (1977): Purification and properties of progesterone receptors from chick oviduct. *Ann. NY Acad. Sci.*, 286:64–80.

12. Schrader, W. T., Heuer, S. S., and O'Malley, B. W. (1975): Progesterone receptors of chick oviduct: Identification of 6S receptor dimers. *Biol. Reprod.*, 12:134–142.

13. Schrader, W. T., Kuhn, R. W., and O'Malley, B. W. (1977): Progesterone-binding components of chick oviduct. Receptor B subunit protein purified to apparent homogeneity from laying hen oviducts. *J. Biol. Chem.*, 252:299–307.

14. Schrader, W. T., and O'Malley, B. W. (1972): Progesterone-binding components of chick oviduct. Characterization of purified subunits. *J. Biol. Chem.*, 247:51–59.

15. Schwartz, R. J., Kuhn, R. W., Buller, R. E., Schrader, W. T., and O'Malley, B. W. (1976): Progesterone-binding components of chick oviduct. *In vitro* effects of purified hormone receptor complexes on the initiation of RNA synthesis in chromatin. *J. Biol. Chem.*, 251:5166–5177.

16. Tsai, N. J., Schwartz, R. J., Tsai, S. Y., and O'Malley, B. W. (1975): Effect of estrogen on gene expression in the chick oviduct. Initiation of RNA synthesis on DNA and chromatin. *J. Biol. Chem.*, 250:5165–5174.

Ontogeny of Receptors and Reproductive Hormone Action,
edited by T. H. Hamilton, J. H. Clark, and W. A. Sadler.
Raven Press, New York 1979.

Steroid Receptor Interaction with Chromatin

T. C. Spelsberg, C. Thrall, G. Martin-Dani, R. A. Webster, and P. A. Boyd

Department of Molecular Medicine, Mayo Clinic, Rochester, Minnesota 55901

It has become abundantly clear in the past few years that the regulation of gene expression in eucaryotic cells involves, at least in part, the coordinate functioning of a multisubunit RNA polymerase enzyme and components of chromatin (the nucleoprotein complex within eucaryotic cell nuclei). Recent research has been directed toward characterizing the components of chromatin and their biochemical interactions leading to the expression of specific genes in a number of cell systems. With the functional elucidation of the procaryotic repressor/operon regulation system (47), a search was initiated for similar regulators in eucaryotic systems. Due to the vast complexity of the eucaryotic genome and its associated proteins, much difficulty was encountered. Although individual messenger RNAs from unique genes had been isolated and their transcriptional regulation characterized, the monumental job remained to identify and isolate the specific regulatory molecules among the millions of nuclear protein molecules.

To assist in this identification of regulatory molecules, substances that interact with the genetic material to modify specific gene expression were sought. Some of the first physiological substances that were found to regulate cell function are the highly specific and biologically potent steroid hormones. These small hydrophobic compounds enter the cell and, as a major part of their action, interact with the genome to elicit a response via an alteration of gene expression. Thus, steroid hormone systems have become very attractive models for the mechanistic study of eucaryotic gene regulation.

A central dogma has been developed to describe the general mechanism of action of steroid hormones. First, systemic steroid (bound or unbound to serum carrier proteins) penetrates most tissue cells (42,75,92,138). Studies have shown that retention of steroid appears only in certain tissues referred to as target tissues for the particular steroid (125). In the cells of these tissues, specific steroid binding proteins (receptors) have been found for the respective steroid hormones (9,30,75,105,128,129,132). The presence of receptors represents the first level of specificity characteristic of steroid responsive cells. The steroid-free receptors are localized in an unknown compartment of the cell's cytoplasm. Once bound to hormone, the receptor is rendered capable ("activated") for transport to the nucleus. In the nucleus, the steroid-receptor (S-R) complex interacts with the chromosomal material of target cells in an unknown manner, and stimulates the synthesis of specific RNA species (40,62,74,79). The interaction of S-R with the nuclear material of the target cell therefore represents the first nuclear event prior to alteration of gene activity.

Virtually every component in the nucleus has been suggested as the acceptor that binds the S-R. Examples are the nuclear envelope (48,49), the ribonucleoproteins (65), histone proteins (78,107), basic nonhistone proteins (71,94,95), acidic nonhistone proteins (27,59,61,64,108,116,119, 120), and DNA (8,140). There have also been reports of a specific binding of free steroids (not bound to receptor) to nuclear material (49,64,126) suggesting the presence of unbound nuclear receptors. In any case, the chemical identity of the nuclear acceptors has yet to be determined. One basic problem is that the conditions used for the cell-free assays have varied from lab to lab and system to system resulting in conflicting conclusions.

It is the study of steroid hormone interaction with chromatin and chromosomal proteins that has been of interest to this laboratory. This interaction represents the first nuclear event in the steroid-induced alteration in gene activity. We have been attempting to critically evaluate the methodologies involved in analyzing nuclear binding by steroids and to identify, isolate, and characterize the S-R acceptor. The focus of this chapter is on the present knowledge of the nuclear interactions and identification of the chromosomal "acceptor" components for the progesterone receptor in the avian oviduct. For a more thorough discussion of this field in general, the readers are referred to a recent review by the authors (131).

STUDIES IN THIS LABORATORY

In Vivo Studies

The nuclear binding of the steroid–receptor complexes appears to be a prerequisite for steroid effects on target cells. Any interruption in the sequence of formation of S-R complex, translocation of this complex to the nucleus, binding to the chromatin, and alterations in DNA-dependent RNA or protein synthesis, as depicted in Fig. 1, blocks the subsequent morphological and functional responses of the target cells to the hormone. Consequently, correlations between the levels of nuclear binding of steroids and the biochemical response by their target should be possible.

One problem in trying to obtain this correlation is that the nuclear binding of the S-R complexes, like the binding of the steroid to its receptor, is not in a fixed state but rather an oscillating state, constantly changing to maintain an equilibrium between bound and free entities. Figure 2A shows the levels of nuclear binding of [^{3}H]progesterone (109) or [^{3}H]estradiol (T. C. Spelsberg, *in preparation*) in the oviduct of nonwithdrawn and withdrawn chicks, re-

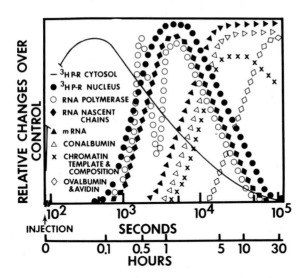

FIG. 1. Composite of the early biochemical responses of a target tissue (oviduct) to a steroid (progesterone or estradiol). The responses to progesterone were usually measured in fully developed oviducts from immature chicks injected for 2–4 weeks with an estrogen. The responses to estrogen and in a few instances progesterone were measured in oviducts from estrogen-treated chicks and withdrawn from estrogen for 2–3 weeks. (—), Levels of labeled progesterone-receptor (P-R) [or estrogen-receptor (E-R)] complex in the cytosol (84,85); (●), levels of labeled P-R or E-R complex bound to nuclear chromatin (57,84,85,89,109,127,134); (0), activity of synthesizing DNA-like RNA (RNA polymerase II) after injection of estradiol; the response to progesterone is chronologically the same but qualitatively different (110); (♦), numbers of newly synthesized nascent RNA chains in the nuclear chromatin after injection of estradiol into estrogen-withdrawn chicks (57,134); (▲), numbers of messenger RNA (for ovalbumin, avidin, and conalbumin) in response to estrogen or progesterone (21, 22,41,89,110); (△), levels (synthesis) of conalbumin in response to estradiol in estrogen-withdrawn chicks (89); (X), changes in chromatin composition and template capacity (for DNA-dependent RNA synthesis) in response to estrogen in estrogen-withdrawn chicks (110); (◇), levels (synthesis) of ovalbumin and avidin in response to estrogen and progesterone (81,83,89).

spectively, following a single subcutaneous injection. There is an immediate increase in the levels of nuclear binding with a short period of equilibrium (between 1 and 2 hr after injection), followed by a rapid decrease. The half lives *in vivo* for the triplex of S-R nuclei for both steroids appears to be about 4 hr (Fig. 2B). Consequently, the exact level of nuclear binding of a steroid hormone is dependent not only on the dose given but on the period of measurement as well. Selecting the equilibrium period of 1 to 2 hr post-injection for measurement, increasing doses of [³H]estrogen (E) or [³H]-progesterone (P) were injected into withdrawn or nonwithdrawn chicks respectively, the nuclei isolated, and the molecules of E or P bound per cell nucleus calculated (109). Plotting the molecules of E or P bound versus the dose injected in Fig. 3A and C, an increasing level of nuclear binding occurs with the increasing dose of hormone shown (109). Inter-

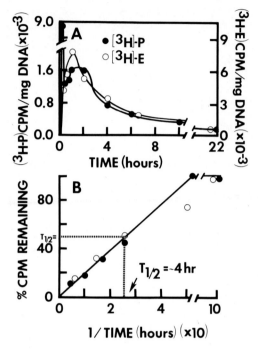

FIG. 2. Binding of [³H]progesterone ([³H]-P) and [³H]estradiol ([³H]-E) to oviduct nuclei *in vivo.* At various intervals after injection, the oviducts were excised from two separate groups of four chicks each, combined, and the nuclei isolated and analyzed for hormone levels as described elsewhere (109). The average values from each of the two groups at each interval are presented. (○), Bound hormone levels in oviduct nuclei from immature chicks, previously treated with diethylstilbestrol (DES) for 2 weeks, and injected (s.c.) with 200 μl [³H]progesterone in sesame oil; (○), bound hormone levels in DES-treated chicks, but withdrawn from the estrogen for 10 days were injected (s.c.) with 200 μl [³H]estradiol in sesame oil. (Data for progesterone adapted with permission from Spelsberg, ref. 109. Data for estradiol from T. C. Spelsberg, *unpublished data.*)

estingly, different classes of binding sites (acceptor sites) with apparent differences in affinities for the estrogen-receptor (E-R) or the progesterone-receptor (P-R) complex are distinguishable. Similar experiments were performed with unlabeled hormones and the effects on synthesizing ribosomal RNA (RNA polymerase I) and synthesizing DNA-like RNA (RNA polymerase II) activities measured (109). As shown in Fig. 3B and D, it is not until the highest affinity classes of sites (~100 sites/cell) are saturated that there is a noticeable change in any RNA polymerase activity. When the two highest affinity classes of nuclear acceptor sites, representing 900 and 4,000 sites/cell for E or 1,000 and 10,000 sites/cell for P are complexed by the steroids, the maximum effect on polymerase activities is achieved. Further increases in doses of E or P followed by further increases in nuclear binding results in no significant alterations in transcription. However, at very high

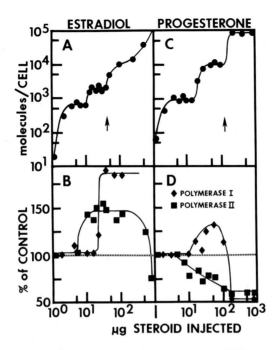

FIG. 3. Effects of dose of progesterone on endogenous RNA polymerase activities and on the binding of the hormone in chick oviduct nuclei. The preparation of the hormones for injections into chicks is described elsewhere (109). The 200 and 1,000 μg doses were administered in the 200 μl sesame oil, and the lower doses in 100 μl. For the [³H]steroid studies, the experiments were carried out as described in the legend to Fig. 2 with the removal of oviducts 2 hr after injection of the [³H]hormones. For the polymerase experiments, groups of six to 10 chicks were injected subcutaneously with a specified amount of the hormone. Two groups were injected with vehicle only and used for determining the noninjected controls. At 2 hr after injection, the oviducts from each group were excised and combined, the nuclei isolated, and the synthesizing ribosomal RNA (RNA polymerase I) (♦) and RNA polymerase II (■) activities assayed as described elsewhere (35). Levels of nuclear bound [³H]estradiol **(A)** and effects of estradiol on RNA polymerase I (♦) and II (■) activities **(B)** in oviducts of immature chicks previously treated with DES for 2 weeks but withdrawn from the estrogen for 10 days. Levels of nuclear-bound [³H]progesterone **(C)** and effects of progesterone on RNA polymerase I and II activities **(D)** in oviducts of immature chicks pretreated for 2 weeks with DES. (Data for progesterone reproduced with permission from Spelsberg, ref. 109. Data for estradiol from T. C. Spelsberg, *unpublished data.*)

levels of E, a marked inhibition of polymerase II activity is observed. In any event, only the two classes of nuclear sites, representing less than 10,000 molecules/cell nucleus, appear to be essential for or directly involved in the alteration of transcription by estrogen or progesterone. It should be mentioned that the responses of the RNA polymerase activities to estradiol in withdrawn chicks are markedly different than the responses to progesterone in nonwithdrawn birds. Interestingly, the doses of each hormone required to fully alter polymerase activities and to bind to the highest affinity classes of sites also result in plasma levels that approximate those measured in adult laying hens (109,119,120).

It takes about 100 μg of each steroid per 100 g bird to obtain a maximal response in the polymerase activity. However, it takes over 5 to 10 times this amount of hormones for maximal protein synthesis and biological effects (79,83,87,88). These discrepancies in optimal doses can be explained by the fact that the responses of the polymerase activities represent one of the earliest events occurring within minutes after nuclear binding of the S-R complexes, whereas responses in protein synthesis occur several steps and hours after this binding. Thus, to maximize the hormone-induced protein synthesis and subsequent biological changes, much more hormone is required to maintain sufficient plasma, cytoplasmic, and nuclear-bound levels that will constantly alter the transcriptional activity over relatively long periods. In short, a sustained nuclear binding of a steroid is required for long range hormone action on target cells. This nuclear binding–physiologic response relationship for steroid hormones has been addressed in detail elsewhere (2,3,25,37,131).

Cell-Free Studies

Rationale

In order to study the translocation and binding of S-R complexes to the cell nucleus, many investigators developed cell free binding assays for assessing the receptor activation and binding to nuclear acceptor sites. We have taken this approach as well because of the following limitations of the *in vivo* or intact cell methods:

1. Only a fraction of the radioactive hormone injected into whole animals localizes in the target organs; consequently, large amounts of expensive [³H]hormones are required to achieve sufficiently labeled nuclei (34,51, 52,108).
2. During standard nuclei isolation procedures, large portions of the nuclear bound S-R complexes appear to be removed from the nuclei (109).
3. Steroid receptors, dissociated from their "native" nuclear binding sites (acceptors) during nuclei isolation may reassociate under nonideal conditions to different sites.
4. The majority of whole cell bound hormones may shift from their initial binding sites (acceptors) to secondary sites.
5. Analysis of the nuclear acceptor sites bound *in vivo* by the S-R complexes are difficult to perform:
   a. Since their interactions are noncovalent, it is not possible to dissociate the acceptors from the chromatin while still bound with the S-R complex.

b. Protease studies to remove the steroid from the nucleus are clouded by the fact that these enzymes may act on the receptor itself thus encouraging the misinterpretation that the acceptor sites are protein in nature.

c. Similarly, DNAse studies may lead to misinterpretation since (i) the DNA sequences may be involved in acceptor activity but may be protected from DNAse actions (thus showing no dissociation of ^{3}H-steroid from the nucleus), or (ii) DNA sequences may represent only one of two or more components of the acceptor site (thus resulting in a dissociation of the steroid from the nucleus).

Cell-free assays eliminate many of the above problems while allowing more controlled conditions and thus greater versatility in experimental design. It is possible, however, that the above reasoning is valid and that only the *in vivo* and not *in vitro* conditions yield a native-like binding of the S-R complex to the nuclear acceptor sites. In the cell-free method, however, the nuclear chromatin can be fractionated, parts removed, parts reconstituted, and the final material assessed for acceptor activity.

Nativeness of the Cell-Free Binding

The usefulness of the cell free binding assay, of course, depends on the "nativeness" of the cell-free binding. At the present time no one can definitely state that the cell-free method is absolutely valid. However, as described in Table 1, there is much evidence that it is. Table 1 lists some of the points supporting and refuting the nativeness of the cell-free binding assays.

The fact that *in vivo* there remains a small percent of total nuclear bound steroid that appears to be tightly bound to nuclear sites and that is not found under *in vitro* conditions, remains mysterious (26,28,73,94). This "covalent-like" nuclear interaction of S-R steroid (extractable with alcohol but not with aqueous buffers containing 0.3 to 1.0 M KCl) is believed by some to be the biologically important interaction. However, others feel that this tightly bound nuclear steroid only represents trapped S-R complexes requiring more thorough extraction (133). It is also possible that this nonextractable radioactivity represents those steroids that have undergone metabolic alterations and covalent linkage to some intracellular metabolite or macromolecules. Absolutely purified nuclei are impossible to obtain. The outer nuclear envelope and closely associated cytoplasmic entities include the smooth endoplasmic reticulum (which contains drug-metabolizing enzymes). Some pharmacologists have reported evidence that isolated nuclei contain drug-metabolizing enzyme activities (99). In intact cells, these activities (nuclear envelope-associated or intracellular) would be active and working, but in cell-free conditions without the proper cofactors, etc., they would be in-

TABLE 1. *Nativeness of the cell-free nuclear binding*

Supporting

1. The receptor binds the steroid under cell-free conditions with the same affinity and specificity as measured under whole cell conditions
2. The S-R complex formed under cell-free conditions has the same physicochemical properties as that complex formed in intact cells
3. Requirements for nuclear uptake and binding in cell-free assays are the same as that required in whole cells
4. Conditions required for activation of the receptor (a prerequisite for nuclear uptake and binding) are the same for both cell-free and whole cell systems
5. Under proper conditions, the cell-free binding results in a similar pattern and level of nuclear binding as does the whole binding.
6. The nuclear triplex of steroid–receptor–chromatin formed under cell-free conditions closely resembles that formed under whole cell conditions with respect to affinity, salt dissociation, and dissociation by certain divalent ions
7. The steroid dissociates from chromatin still complexed to its receptor protein
8. Most importantly, the interaction of an isolated S-R complex with isolated nuclei has been reported to alter RNA polymerase activity and transcription of selected genes in a pattern similar to that which occurs in the contact cell

Opposing

1. There are reports that nuclear binding performed in whole cells, but not in cell-free conditions, results in a small fraction of the steroid receptor that is bound very tightly (nondissociable in 0.3 M KCl). See text, however, for alternate interpretation of these data;
2. Studies on the binding of S-R complex for the nuclear binding sites under whole cell and cell-free binding conditions failed to show competition. However, problems of dissociation or reassociation, or both, of the complex during nuclear isolation and the possibility that the initial binding (i.e., binding to acceptor sites) is only transitory followed quickly by a shifting to "other" sites in whole cell binding could explain the results

---

For complete referencing of the statements in this table, refer to Thrall, et al., ref. 131.

active. Studies to identify the structure of the alcohol-extracted radioactivity are necessary. In order to assess the role of drug metabolism in the tightly bound nuclear steroid phenomenon, a cell-free nuclear binding assay should be performed under conditions that support drug metabolism (e.g., the mixed function oxidases).

Direct Comparison Between Intact and Cell-Free Systems

The strongest support for the nativeness of the cell free binding system are the reports by many laboratories that the steroids bound to nuclei under cell free conditions alter transcription in a manner similar to that of intact cells (4,5,11,16,17,54,55,76,98,103,104). These results not only support the nativeness of the cell free binding but also the fact that isolated nuclei and chromatin contain all the necessary regulatory factors for the hormone induced alterations of gene transcription.

As an example, Fig. 4 shows a pattern of nuclear binding of [³H]proges-

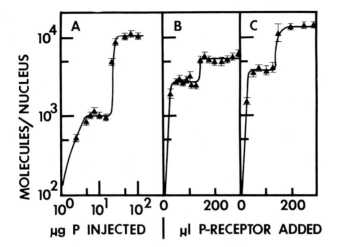

FIG. 4. Comparison of the binding of [³H]progesterone to the high-affinity sites in oviduct nuclei *in vivo* and *in vitro*. The data represents binding to nuclei *in vivo* **(A)** and *in vitro* **(B)**. **(C)**: Binding to isolated chromatin. Molecules per cell nucleus were calculated from the cpm/mg DNA using 2.5 pg DNA/chick oviduct cell. The range and average of three replicates of the analysis for each receptor concentration.

terone using whole-cell and cell-free systems. Although some differences are observed, the patterns are strikingly similar. As discussed earlier, the higher affinity sites, representing between 1,000 and 10,000 sites per cells, appear to be involved in transcriptional alterations (109,119,120). Schwartz and co-workers (103), using a similar cell-free binding system, have shown that P-R complex alters the transcriptional capabilities of oviduct chromatin from withdrawn chicks. We have found that progesterone similarly alters the transcription *in vivo* (110). In short, the cell-free binding of the oviduct P-R to oviduct nuclei and chromatin displays the same requirements and results in the same type of nuclear interaction and transcriptional response, as displayed by the whole cell system.

Conclusions on Cell-Free Binding

In summary, there are many arguments supporting a native-like interaction of steroid hormones with the nuclear acceptor sites *in vitro*. There are also many arguments with supporting evidence refuting such a native interaction. The authors feel, however, that cell-free methods can yield information about the native acceptor sites in target cell nuclei for steroid-receptor complexes. This conclusion is supported by the work of other groups (4,5,11,16,17,54,55,76,98,103,104). The readers are referred to major reviews in the field for further discussion on the subject (36,51,82,90, 97,108,111,131,141).

Evidence for Nucleoprotein as Acceptors for Progesterone in the Avian Oviduct

Early Studies in the Chick Oviduct

This laboratory became interested in the role of nonhistone proteins and DNA in the nuclear acceptor activity for P-R from chick oviduct in 1971 when it was found that these entities displayed the greatest acceptor activity of all the components of chromatin (86,108,111,112,113,115,121,123). The protein fraction containing this acceptor activity (a) is tightly bound to DNA (108,111,113,115); (b) expresses greater activity in target than in nontarget tissues (86,113,121,122); (c) requires reannealing to DNA for activity (86,115); (d) is destroyed by proteases but not nucleases (108, 111); (e) behaves like an acidic chromatin protein and not a histone (108, 111,123); and (f) is found, using immunofluorescence technique, to be localized within the nucleus (nucleoplasm) and not associated with the nuclear envelope or the nucleolus (23). Interestingly, when the histones and some nonhistones are removed from the chromatin, the levels of acceptor activity are enhanced indicating the presence of repressed sites in intact chromatin (86,108,113,115). Subsequent removal of the remaining acidic protein from the DNA results in a loss of much (but not all) of the acceptor activity. The dissociated proteins (called AP_3) which when removed from DNA cause an 80% loss of acceptor activity, also restore acceptor activity to the DNA when reannealed back to the DNA (86,108,113,115). This AP_3 fraction, when selectively removed from or reinstated to whole chromatin, causes a loss or restoration, respectively, of the acceptor activity in that chromatin (86,108,115). Thus, the early work implicated a non-histone acidic protein–DNA complex as the high-affinity nuclear binding component for steroid receptors.

Recent Studies in the Hen Oviduct

Evidence for repressed nuclear acceptor sites.

Figure 5 shows the acceptor activities in the GuHCl extracted chromatins using two methods—centrifugation and chromatin-cellulose (137). Interestingly, as the proteins are removed from the chromatin, the acceptor activity is markedly increased, not lowered. Only extractions with 5 M to 7 M GuHCl result in a lowering of the acceptor activity. Pure DNA displays a binding activity only 20% of that measured at 5 M GuHCl. Thus, a masking or repression of acceptor sites is indicated. We feel that masked sites are the same type as those normally unmasked. This derepression of acceptor sites for progesterone, in target as well as nontarget tissues, by removing certain fractions of chromosomal proteins has been reported previously for the chick

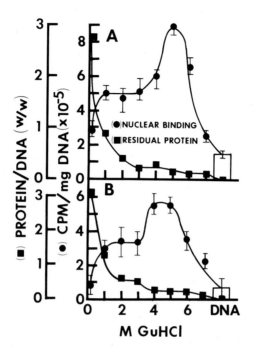

FIG. 5. Binding of [³H]P-R to **(A)** hen oviduct chromatin-cellulose-treated GuHCl or **(B)** hen oviduct chromatin treated with GuHCL and centrifuged. Portions of chromatin-cellulose or chromatin were washed twice in 20 volumes of solution containing 0.1 M HSETOH, 0.005 M sodium bisulfite, 0.01 M Tris-HCl (pH 8.5), and a specified concentration of GuHCl. The cellulose resins were then washed in dilute Tris-EDTA buffer several times and frozen and lyophilized. The free chromatin samples in various concentrations of GuHCl were centrifuged at $10^5 \times g$ for 36 hr. The pellets of residual deoxyribonucleoproteins were resuspended in dilute Tris buffer at 1 mg DNA/ml and dialyzed versus the buffer, and portions of the resin (20 to 50 µg) were then assayed for (■) protein by resuspending the resin in the Tris-ethylenediaminetetra-acetic acid (EDTA) buffer, allowing it to hydrate several hours or overnight with gentle mixing, and assaying for total protein. Residual nucleoproteins were assayed directly. The guanidine-treated resins were tested for their (●) acceptor activity using either 20–25 µg DNA (bound to the cellulose) together with 300 µl of the hormone-receptor complex solution or 100 µg of the free nucleoprotein and 600 µl of the receptor. The binding assays were performed essentially as described previously (137). The average and range of three replicates of the analysis for each assay of the hormone binding are shown. (**A** reproduced with permission from Spelsberg et al., ref. 116; **B**, T. C. Spelsberg, *unpublished data.*)

oviduct (108,111,112) as well as for hen oviduct (116,117, 137). The data obtained on the extent of repression in the latter system as well as in non-target tissues are summarized in Table 2.

The question remains, which fraction or fractions of histone or non-histone proteins are involved in the repression of acceptor sites? As shown in Fig. 6, selectively removing all histone species (fraction I) fails to un-mask acceptor sites (117, 137). Subsequent removal of a fraction of non-histone protein (fraction II) does unmask many sites and results in a five-

TABLE 2. *Extent of masking of high-affinity sites*

Source	Level of binding at saturation (molecules/cell)	% of total sites masked
1. Oviduct nuclei *in vivo*	10,600	58
2. Oviduct nuclei	5,951	76
3. Oviduct chromatin	7,441	71
4. Oviduct chromatin minus histone	9,278	58
5. Oviduct chromatin minus histone, AP, and AP_2	25,290	0
6. DNA	2,055	—
1. Spleen nuclei	60	100
2. Spleen chromatin	48	100
3. Spleen chromatin minus histone, AP, and AP_2	24,106	0
1. Erythrocyte nuclei	357	100
2. Erythrocyte chromatin	416	100
3. Erythrocyte chromatin minus histone, AP_1, and AP_2	15,018	0

These data were taken from values previously reported (118). The number of binding sites measured in each of the chromatin minus histone AP_1, AP_2 preparations was assumed to represent the total sites in the respective chromatin, and thus each of these preparations was assigned 0% masking. The values at saturation for each preparation were taken at 100 µg injected hormone for the *in vivo* binding and 200 µl labeled receptor preparation for the *in vitro* binding. The preparations of the nuclear components as well as the hormone binding assays were performed as described elewhere (108,137). (Reproduced with permission from Spelsberg et al., 118).

fold increase in binding. Interestingly, the reannealing of this fraction (fraction II) back to the nucleoprotein containing the unmasked acceptors results in a repression of the acceptor activity and restoration of the original level of binding (Fig. 6) (G. Martin-Dani and T. C. Spelsberg, *in preparation*). Further studies of this repressor activity are underway.

Evidence for nucleoprotein as the acceptor.

As shown in Fig. 5, removal of the remaining acidic proteins results in a major loss of acceptor sites. These results support those reported previously for chick oviduct chromatin where the acceptor activity (i.e., high-affinity binding) is associated with the acidic chromatin protein fraction, termed AP_3, tightly bound to the DNA (115). When the chromatin-cellulose resin is subjected to a gradient of GuHCl in a solvent containing a high concentration of a reducing agent (HSETOH) at pH 8.0, multiple peaks of protein are eluted from the column (Fig. 7A). As shown in Fig. 5, the majority of the protein is eluted by 4 M GuHCl. The high adsorption observed in Fig. 7A under condition of 4 to 7 M GuHCl extract, which yields only small amounts of protein, is due primarily to the elution of ribonucleoprotein as well as small amounts of DNA. The protein eluted from the column was

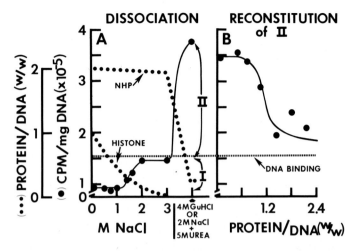

FIG. 6. Acceptor activities in chromatin samples whose proteins have been selectively removed by the designated treatments. **A:** Fraction I represents all histones and ~10% of the nonhistone proteins. Fraction II represents ~60% of the nonhistone proteins. The NaCl and GuHCl solutions contained 0.1 M phosphate buffer, pH 6.0. **B:** Fraction II was reconstituted back to the residual nucleoprotein obtained by the 4 M GuHCl extraction in **A**. The chromatin were attached to cellulose (described in Fig. 6) and extracted and assayed for acceptor activity as described elsewhere (137). (G. Martin-Dani and T. C. Spelsberg, *unpublished data*.) (●), Binding of ^{3}HP-R complex to the residual nucleoprotein after dissociation of proteins from chromatin DNA or after reconstitution of protein Fraction II back to the DNA; (––––), binding to pure DNA; (••••), histone/DNA or nonhistone protein (NHP)/DNA (w/w) in the residual nucleoprotein after salt treatments.

pooled according to the extraction with each unit of molarity of GuHCl (see Fig. 7A) and reannealed to DNA. The resulting nucleoprotein was analyzed for high-affinity binding. As shown in Fig. 7A, only the material that elutes from the column between 5 and 7 M GuHCl contains the high affinity binding activity. Similar studies have been performed using chromatin-hydroxylapatite resins. As shown in Fig. 7B, acceptor activity is also eluted between 5 and 7 M GuHCl with this resin. Table 3 shows the estimated purification (~98,000-fold) at this stage beginning with whole tissue. The acceptor is routinely dissociated from chromatin using stepwise extractions with 0 to 4 M and then 5 to 7 M GuHCl with over 100-fold purification.

To identify the acceptor activity eluted from the resins, the eluting proteins are first reannealed to DNA and the nucleoprotein (NAP) complex then tested for acceptor activity. Analysis of the binding activity without DNA has received little attention in this laboratory since the proteins in this fraction are aggregated in aqueous solutions. When reannealed to DNA, they are soluble. A slowly decreasing gradient of GuHCl (over a 24- to 36-hr period) appears to be the best method for achieving high efficiency reannealing of the protein to the DNA and maximal hormone-receptor binding. Rapid dilution of the GuHCl by the addition of buffer without dialysis (over

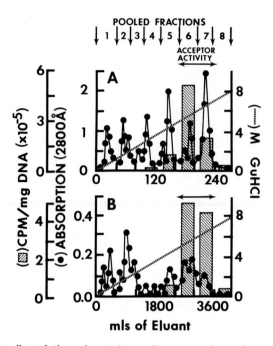

FIG. 7. Selective dissociation of proteins and acceptor from chromatin-cellulose **(A)** and chromatin-hydroxylapatite **(B)** resins. Hen oviduct chromatin-cellulose resin was prepared as described elsewhere (116,137). Twenty grams of this resins containing approximately 60 mg DNA as chromatin was resuspended in 100 ml cold solution containing 0.1 M HSETOH, 0.005 M sodium bisulfite, and 0.01 M Tris-HCl (pH 8.5) and allowed to hydrate for 2–6 hr with gentle stirring at 4°C. The resin was collected on a column and a gradient of 0 to 8 M GuHCl passed through the column within a 4-hr period using constant levels of the buffered solution. Resins of chromatin-hydroxylapatite **(B)** were prepared in the presence of buffered 0.0 M GuHCl. The resin containing 100 mg chromatin DNA and 100 g hydroxylapatite was placed on the column, and 4-ml fractions were collected under a 0 to 7 M GuHCl gradient. Tubes were monitored by adsorption at 280 nm. Fractions were also monitored for conductivity as well as refractive index, and the gradient level of GuHCl was plotted. Fractions were pooled according to their elution with each unit of concentration of GuHCl (1 M, 2 M, 3 M, etc.). Pooled samples were then dialyzed thoroughly against water and lyophilized. The lyophilized materials were resuspended in a small volume of water, homogenized in a Teflon pestle glass homogenizer, assayed for protein (66), and reannealed to pure hen organ DNA. The initial protein:DNA ratio for these reannealings was 0.1. Reannealing consisted of a gradient dialysis from 6.0 M GuHCl in the buffer described above to 0.0 M GuHCl in the same buffer, and finally dialysis against dilute Tris-EDTA buffer. The nucleoproteins were analyzed for acceptor activity by the streptomycin method, subtracting the values obtained with pure DNA. This activity is presented as bar graphs. Total recoverable protein after dialysis and lyophilization was estimated to be 50% of the total protein placed on the column as chromatin-cellulose. (**A** is reproduced with permission from Spelsberg et al., ref. 116; **B** is reproduced with permission from Spelsberg et al., ref. 120).

TABLE 3. *Summary of purification of "acceptor"*

| | Binding at saturation to highest affinity sites | | | |
	DNA cpm/mg	Protein/DNA	Protein pmoles/mg	Purification
Whole tissue	75,000	24.21	0.004	1.0
Nuclei	75,000	3.05	0.972	243.0
Chromatin-cellulose	125,000	2.14	2.3	575.0
Nucleoprotein-cellulose	425,000	0.36	46.7	11,675.0
	Chromatin-cellulose extraction			
GuHCl				
1.0 M	496,546	0.80	24	6,131
2.0 M	452,960	0.31	58	14,434
3.0 M	522,967	0.15	138	34,440
4.0 M	446,927	0.15	115	28,668
5.0 M	829,112	0.13	287	71,823
6.0 M	652,609	0.07	390	97,677
7.0 M	176,602	0.06	116	29,000

(Reproduced with permission from Spelsberg et al., ref. 118.)

a 2-min period) results in little protein bound to DNA and little or no acceptor activity over that of DNA alone.

Figure 8 shows the results of reannealing experiments concerned with varying the ratio of the protein acceptor to DNA. The protein represents the 4 to 7 M GuHCl extract of chromatin-hydroxylapatite resins that involves about 5% of the total nonhistone protein (see Fig. 7B). These reconstituted acceptor-DNA complexes were purified by sedimentation in an ultracentrifuge for 36 hr in 2 mM Tris-HCl + 0.1 M EDTA + 1 mM PMSF buffer (pH 7.5). The pelleted nucleoproteins were then assayed for the protein/DNA ratio (i.e., extent of bound protein) and for the extent of hormone binding. Interestingly, the DNA-bound protein from the acceptor fraction displays a plateau that was reached by adding 0.02 mg protein/mg DNA. As more acceptor protein is added, the plateau is maintained which is then followed by a nonsaturable linear increase in "bound" protein. These results suggest that this small fraction of chromosomal proteins reanneals to specific high-affinity sites or sequences of DNA. Nonspecific binding to DNA or protein aggregation occurs at the higher protein concentrations. The somewhat higher ratios of protein to DNA recovered compared to the starting ratio may represent selective loss of noncomplexed DNA during the reannealing and sedimentation. The binding of the P-R to these nucleoproteins also displays a pattern indicating a specificity for DNA sites (120). When the ratio of starting protein to DNA is low, whereby the plateau of bound protein is achieved, the greatest degree of binding of P-R gesterone-receptor to these nucleoproteins compared to that of pure DNA occurs (Fig. 8). However,

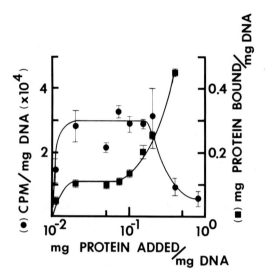

FIG. 8. Effects of varying quantities of the acceptor protein added to DNA on the levels of DNA-bound protein and hormone binding. Acceptor proteins from the 5- to 7-M GuHCl extract from hydroxylapatite-hen oviduct chromatin resin (see Fig. 9B) were added to the hen DNA in varying quantities in the presence of 6 M GuHCl containing 0.01 M phosphate buffer, pH 6.0, 0.005 M HSETOH, 0.005 M PMSF, and 0.001 M EDTA. The DNA concentration was 0.2 mg/ml. The solutions were placed in washed dialysis bags, and the GuHCl was gradually removed over a 24-hr period in a linear gradient in specifically designed containers on a rocker platform. Mixing was performed inside the containers as well as the dialysis tubing by moving air bubbles during the rocking. The dialysate was then replaced for a 12-hr period with 2 mM Tris-HCl, pH 7.5, plus 0.1 mM EDTA. These solutions were centrifuged at a maximum rate of $10^5 \times g$ for 36 hr in an angle rotor. The pelleted DNA-protein was resuspended in the dilute Tris buffer (0.5 mg DNA/ml) and analyzed for protein, DNA, and hormone binding. Pure DNA was also analyzed for hormone binding and the values subtracted from those obtained from the protein-DNA. (Reproduced with permission from Spelsberg et al., ref. 120.)

as the protein/DNA ratio is increased, causing the levels of bound protein to become linear with the added protein, the extent of binding begins to decrease. Thus, the plateau at the lower ratios of protein to DNA may indicate selective, high-affinity sites on the homologous DNA that are preferred by the acceptor proteins that in turn, are preferred by the P-R.

Purification and characterization of the protein co-acceptor.

Further purification in our laboratory of the acceptor activity for progesterone in chick oviduct has recently been achieved with molecular sieve chromatography using 100-cm columns containing agarose, 1.5 M with 6.0 M GuHCl according to the procedure of Mann and Fish (72). Figure 9 shows the pattern of a typical chromatograph of the 5 to 7 M GuHCl extract of chromatin. The eluted material was pooled according to peaks of adsorb-

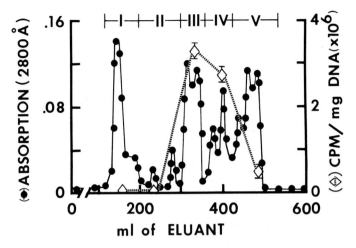

FIG. 9. Agarose 1.5-M molecular sieve chromatography of the acceptor protein fraction in the 5- to 7- M-GuHCl chromatin extract. The acceptor fraction representing the 5- to 7-M-GuHCl extract of chromatin resin was dialyzed against deionized water and lyophilized. The material was resuspended in a 6.0-M GuHCl solution (72) at 2 mg protein/ml. This protein solution was clarified by centrifugation at 2000 *g* for 10 min and then applied to a 2.6 × 94 cm column of agarose 1.5 M. The eluted fractions were pooled according to the (●) absorbing peaks as shown in the figure. The pooled fractions were then reannealed to DNA and assayed for (◇) acceptor activity using the streptomycin assay (137). Values for DNA binding was subtracted from all values obtained with nucleoprotein. The average and range of three replicates of analysis are shown. (Reproduced with permission from Spelsberg et al., ref. 119.)

ing material and reannealed to DNA. The nucleoproteins were then analyzed for acceptor activity. The acceptor activity coelutes with standard molecular weight proteins (histone IV and ribonuclease) in a molecular weight range of 12,000 to 16,000 daltons (see Fig. 10). Polyacrylamide-SDS gel electrophoresis of the fractions from this column are shown in Fig. 11 and gels 4 and 5 reveal only two major bands that migrate with molecular weights between 13,000 and 17,000 daltons (118–120). This molecular weight range is in close agreement with that estimated from molecular sieve chromatography.

Whole chromatin protein or the 4 to 7 M GuHCl extracts from the chromatin hydroxylapatite resin were also subjected to isoelectric focusing on either acrylamide gel plates or Sephadex G-75 resin in 5 M urea. The runs were performed for 16 hr at 2°C after which sections were removed, pH measured, and protein extracted. The extracted protein was quantitated and reannealed to DNA, and the acceptor activity assessed. As shown in Fig. 12A, the acceptor activity of the 4 to 7 M GuHCl extract focuses in a broad region around pH 5 in acrylamide gels. The acceptor activity of a NAP fraction prepared as described elsewhere (108) focuses at pH 6.0 (Fig. 12B). The acceptor activity in whole chromatin or the 4 to 7 M GuHCl extraction was

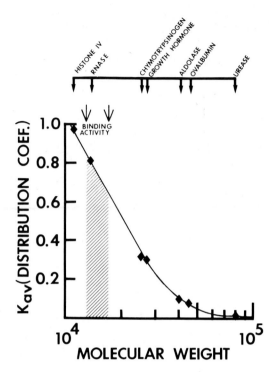

FIG. 10. Selectivity curve for an agarose- 1.5-M column (2.6 × 94 cm) by use of 6.0 M GuHCl buffered at pH 6.0 as a solvent. The standard proteins were purchased from commercial sources, and their subunit molecular weights were obtained from I. M. Klotz and D. W. Darnall, P. L. Biochemicals Chart. The hatched area represents elution of high acceptor activity from the column. (Reproduced with permission from Spelsberg et al., ref. 119.)

more recently analyzed by the more sensitive Sephadex method on a Multiphor apparatus (LKB). In this case, much smaller (0.5 cm) sections could be sampled and thus greater resolution of the acceptor activity achieved. Three peaks of acceptor activity are detected in the regions of pH 5 to 7. The acceptor activity appears to be an acidic protein with indication of different species. The three species probably explain the broad peak of acceptor activity eluting from the agarose 1.5-M column chromatography (Fig. 9) as well as the broad peaks observed in the isoelectric focusing using the acrylamide method, which was less sensitive due to the larger sections taken and lower recovery of protein than the Sephadex method.

Since the higher affinity binding activity of these preparations is destroyed by pronase and since the reannealing to DNA is required for high affinity binding, we conclude that the high affinity binding of the P-R complex *in vitro*, and probably *in vivo*, is due to low molecular weight acidic proteins bound to DNA. These proteins themselves have a high affinity for DNA that yields a rapid and effective approach in their purification.

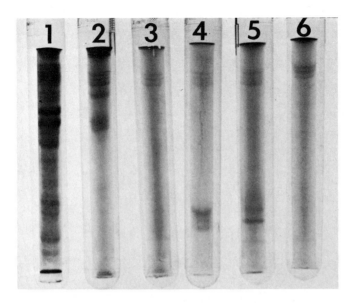

FIG. 11. SDS-polyacrylamide gel electropherogram of purified acceptor activity. Gel 1 represents the proteins eluted from hydroxylapatite-chromatin resins between 5 and 7 M GuHCl. This fraction was then subjected to agarose-1.5 M chromatography (94 cm long) using a solvent containing 6 M GuHCl as described in the legend of Fig. 11. Gels 2, 3, 4, 5, and 6 represent the electropherograms of the fractions I, II, III, IV, and V of the A-1.5 M chromatography depicted in Fig. 11. The acceptor proteins obtained by this approach are seen in gels 4 and 5. Approximately 150 μg of protein were applied to gel 1 and 30 μg to gels 2–6. Protein was applied at the top of the gels. The gels represent 10% acrylamide (9.5 cm long) and were run until the bromphenol blue tracking dye migrated to within 0.5 cm of the bottom. The gels were stained with Coomassie blue.

Problems and Discrepancies

Tissue Specificity of Acceptor Activity

The question has been raised whether or not there is a tissue specificity with respect to the nuclear binding sites (acceptor activity). Since nontarget tissues do not have cytoplasmic receptor for the respective steroid, whether or not their nuclei contain acceptor sites for the steroid is, biologically speaking, probably of no consequence. The question arose from studies of cell-free binding assays wherein nontarget cell nuclei could be exposed to S-R from target cells. Consequently, the problem is really an academic one and not as important as identifying the real acceptor sites or the overall mechanism of action of steroid hormones. In any case, numerous laboratories reported that a variety of target cell nuclei for a variety of steroids contain more acceptor sites than nontarget cell nuclei using cell free binding assays (13,14,15,38,43,44,50, 53,56,58,60,70,80,85,86,93,100,101,102,113,115,119,120,122,137). These reports did not state that nontarget cell nuclei lacked any acceptor sites, but

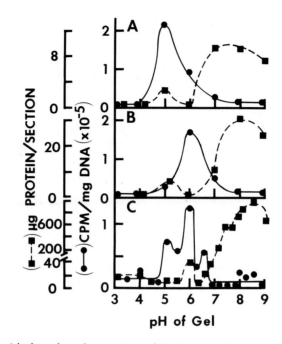

FIG. 12. Isoelectric focusing of acceptor activity in hen oviduct chromatin proteins in the presence of 5 M urea. **A:** Isoelectric focusing in polyacrylamide gels of partially purified acceptor proteins (600 μg) (i.e., the 4- to 7 GuHCl extracts of hen oviduct chromatin described in Fig. 9). **B:** Focusing of 600 μg of nucleoacidic protein in polyacrylamide gels. The proteins were prepared as described elsewhere (113,137). **C:** Focusing of 25 mg of total chromosomal proteins in an LKB Multiphor using Sephadex-G-75. The proteins were prepared by extracting whole chromatin in 7 M GuHCl and centrifuging as described in the legend of Fig. 7. The focusing region in **A–C** involved a pH range of 3–10 and included 5 M urea. In **A** and **B**, sections of the acrylamide gels were pooled based on whole pH units (∼1.5 cm sections). The pooled acrylamide gel sections were homogenized in a Teflon pestle-glass homogenizer in 6-M GuHCl solutions and the acrylamide sedimented by centrifugation. The gels were washed twice more in the GuHCl solution and the latter dialyzed versus water and lyophilized. In **C**, much smaller sections were pooled on 0.3 to 0.5 pH units representing ∼0.5 cm of the Sephadex resins. The Sephadex resins were placed in small columns, the protein eluted with 6 M GuHCl and treated similarly as the acrylamide sample above. The recovery of protein in **A** and **B** was ∼10%. The recovery of protein in **C** was ∼50%. The eluted proteins in each section were quantitated by the method of Lowry (66) or by a Coomassie blue stain technique (12) and then were reannealed to DNA at a ratio of total recovered protein to DNA of 0.1. These nucleoproteins were assayed for acceptor activity by the streptomycin assay as described by Webster et al. (137). (From J. T. Knowler and .T. C Spelsberg, *unpublished data.*)

that they contained fewer numbers of them. Consequently, the term "tissue specific" nuclear binding is erroneously applied; an "enhanced" binding in target tissue nuclei would be more appropriate. A few laboratories have reported, however, that they observed no difference in the level of nuclear binding between target and nontarget cells (20,24,43). The problem in analyzing this discrepancy is that different steroid-target tissue systems and different binding conditions were used. Either of these factors may explain the discrepancy in the literature. It has been reported that nontarget cell chromatin may contain the same number of nuclear acceptors as target cell chromatin but most of the acceptors in the former are masked (117). It is even possible that the acceptors in nontarget cells are different from those in target cells. To answer these questions, the nuclear acceptor sites in target cell nuclei need to be identified and characterized. Antisera prepared against this acceptor (if the acceptor is a protein) could be used to (a) determine whether or not the nontarget cell nuclei contain the same or different type of acceptor for the steroid, and (b) assess how many acceptors are present in the chromatins.

Renaturation of Denatured Acceptor: Reconstitution of NAP

The removal of the components from the DNA in chromatin requires the use of denaturing agents. Solvents containing a high concentration of salt (e.g., 2.0 M NaCl) do not quantitatively remove all chromosomal proteins from the DNA. In past years, the application of chaotropic agents such as urea (10,29,32,33,46,63,68,69,77,114), GuHCl (6,45,63), and formic acid (29), and detergents such as sodium dodecyl sulfate (106) has been carried out in place of, or in addition to, the high salt for complete dissociation of the proteins from the DNA. The possibility of denaturing any acceptor activity, especially if it is a protein, is likely. The only recourse at this point is to renature the activity (a) by methods used to refold denatured enzymes and proteins to their biologically active forms (1,18,67,130,136,142), and (b) by recombining proteins and DNA by methods for the reconstruction of native-like deoxyribonucleoproteins (7,19,31,91,108,111,114,115,119, 120,123,124,139). Fortunately, the methods and conditions used to reconstruct nucleoproteins also renature proteins. In short, if the acceptor activity is identified with DNA, there is little concern over denaturing this activity with the agents used to dissociate proteins. If it is identified with a protein or nucleoprotein, then such macromolecules will probably require refolding to restore nuclear binding (acceptor) activity. If the acceptor is a small protein, like ribonuclease, with no quaternary structure and a relatively simple tertiary structure, it would probably refold spontaneously. Larger molecules or protein–DNA complexes require more involved procedures to renature the activity.

The procedure of reannealing isolated acceptor protein fractions back to DNA for assay has evolved from an experimental basis. In assaying the binding of S-R to chromatin, one deals with NAP complexes. In the sequential removal of protein fractions from chick oviduct chromatin, it was found that the progesterone-acceptor activity was tightly bound to DNA (108,119,137). Reciprocally, the dissociated acceptor species could only be detected by reconstituting it back to the DNA and then assaying for acceptor activity. Since salt–urea or GuHCl was used to extract the acceptor species, the classic reverse gradient reconstitution technique was used to reanneal the acceptor back to DNA in order to assay the acceptor activity. In the most recent work from our laboratory, continuous gradient dialysis from 6 to 0.0 M GuHCl has been used. It is felt that assaying a NAP complex represents a more native situation since it also represents a more native state of the acceptor. Furthermore, the protein extracted under high salt conditions is insoluble in dilute buffer, but is soluble as a NAP complex. This certainly makes it more feasible for assay purposes. Although the procedure was developed in a progesterone/chick oviduct system, the laboratory of Hnilica (61) used similar reconstitution techniques with defined chromatin fractions in an androgen/rat prostate, rat testis, system to successfully delineate the androgen acceptor. It is not known whether these acceptors have specificity for certain DNA sequences or are reannealing in any specific fashion. This is presently being investigated.

Effects of Aging on NAP and DNA Binding

Upon storage of NAP or DNA at 4°C, −20°C or even −80°C, changes in the level of binding occur (T. C. Spelsberg, *unpublished data*). The colder the temperature, the longer the storage period before changes are noticed. The binding to NAP generally is 4- to 5-fold or more than that to DNA. However, with storage, the binding of the P-R to NAP decreases whereas that to pure DNA increases until the levels are equivalent. The action of proteases appears to be the cause of the changes in binding the NAP. However, the cause for such changes on the DNA is unclear. Damage to DNA via ultraviolet light treatment, freezing and thawing, shear, oxidation via peroxides, etc., appears to increase the level of binding to DNA (T. C. Spelsberg, *unpublished data*). Other parameters such as unclean DNA (protein-contaminated) can cause high levels of binding. These problems were solved by lyophilizing NAP and DNA and storing them in a dry state. When needed they are mixed gently with buffer overnight at 4°C which results in their hydration and solubilization. The samples can be stored for many months or even years in this state.

Circannual Variations in Levels and Nuclear Binding by P-R Preparation

Every winter for the past few years, we noticed a loss in the binding of our P-R to acceptor protein–DNA complexes. Figure 13 shows the P-R binding to the NAP and DNA using the winter (February) and summer (August) receptor preparations. It can be seen that the NAP binding by the winter receptor is much less than that by the summer receptor. The level of

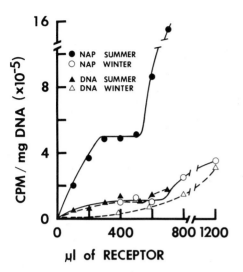

FIG. 13. Binding of two preparations of [³H]P-R to hen oviduct nucleoprotein or DNA. The assays were carried out essentially as described in the legend of Fig. 7B. (●), Binding to nucleoprotein using a [³H]P-R from August; (○), binding to nucleoprotein using a [³H]P-R from February; (▲), binding to pure DNA using a [³H]P-R from August; (△), binding to pure DNA using [³H]P-R from February. (From T. C. Spelsberg, *unpublished data.*)

binding to DNA by these receptors is equivalent. Using the 600 μl of P-R preparations from 1975 to 1977, the levels of binding to acceptor protein reconstituted to DNA was analyzed (Fig. 14). The data represent binding to the reconstituted nucleoprotein over that to pure DNA. As can be seen, a circannual rhythm is observed. Computer analysis of these data by Halberg

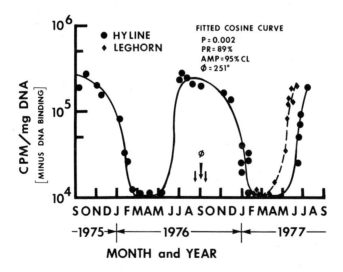

FIG. 14. Binding to reconstituted nucleoprotein using receptor preparations isolated throughout the year. The assays were carried out as described in the legend of Fig. 7B. The points are plotted with respect to the date of the receptor, isolation, and ability of that receptor to bind to the nucleoprotein. (From T. C. Spelsberg and F. Halberg, *in preparation.*)

(Chronobiology Laboratory, University of Minnesota) revealed that ~90% of the points fit a cosine curve within 95% confidence limits. The acrophase (peak period of binding) centers in September with a bathophase (trough) around April. Figure 11A shows similar variations in binding to undissociated NAP. However, there is no such rhythm in the binding to pure DNA (Fig. 15B). Interestingly, the levels of P-R in the cytosol and the amounts of protein in the receptor preparations as well as oviduct weights display the circannual rhythms (Fig. 15C). So once a year, the oviduct weights and P-R levels decrease to about half their usual levels, accompanied by a complete loss of the binding of P-R to the nucleoprotein acceptor sites (over that to DNA). No rhythm in the binding to pure DNA occurs. The binding to NAP decreases to that of pure DNA during the winter. Thus, the levels and nuclear

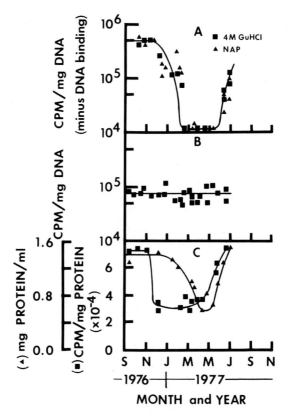

FIG. 15. Binding to DNA and undissociated nucleoprotein using receptor preparations isolated throughout the year. The assays were carried out as described in the legend of Figs. 7B and 15B and (141). **A:** Binding to undissociated nucleoproteins representing residual nucleoprotein after extraction of chromatin with (■) 4.0 M GuHCl, pH 6.0 or (▲) 2.0 M NaCl + 5.0 M urea, pH 6.0. **B:** Binding to pure DNA. **C:** Levels of (■) [³H]P-R (▲) protein in the receptor preparation. (From T. C. Spelsberg and F. Halberg, *in preparation*.)

binding capacities of the progesterone receptor preparations are varying throughout the year. Whether this variation occurs in *Gallus domesticus* of other strains and in other regions of the world or even in other animals remains to be determined. Studies on the possible loss of one of the two molecular species of the progesterone receptor during the winter is being investigated.

RELATED WORK IN OTHER LABORATORIES

Recent studies in the laboratory of Hnilica (61), using techniques very similar to those described above but with target and nontarget tissues of the rat for 5α-dihydrotestosterone, demonstrated results similar to those for the avian oviduct-progesterone system. These investigators observed an enhanced binding to target tissue chromatin compared to nontarget tissue chromatin. They also observed unmasking of acceptor sites by the dissociation of the bulk of the chromosomal proteins from the chromatin. The acceptor activity itself was associated with the nonhistone proteins tightly bound to DNA. These tightly bound proteins of testis or prostate could be dissociated and reconstituted to DNA with recovery of the androgen hormone acceptor activity. The involvement of protein, at least as a component of steroid-nuclear acceptors, is also supported by the early work of Liao (64,135) who found a heat-labile tissue specific chromatin protein to be responsible for high-affinity nuclear binding for 5α-dihydrotestosterone-receptor complex. However, this "protein" is not tightly bound to DNA, being extractable at 0.4 M KCl. The question of whether or not this heat labile protein is an androgen receptor remains to be determined. Gschwendt (38) found two proteins in the chick liver chromatin which bound estrogen; one weakly bound to DNA and the other tightly bound to DNA. The weakly bound one resembled the S-R whereas the tightly bound one had properties of a nonhistone protein-DNA complex. It is possible that the nuclear estrogen-receptor complexes are present in nuclei either in a free state (not bound with estrogen), or in a bound state which exchanges with the labeled estrogen, thus loosely bound and tightly bound E-R complexes are being measured.

There has recently been another approach for isolating and assaying acceptor activity that has resulted in a characterization of very different species of chromosomal protein acceptor. Two groups isolated nuclear fractions by the technique of affinity chromatography and found an active fraction in a 2M NaCl nuclear extract (71,95,96). The fraction was characterized as basic protein relatively large in molecular weight and in bound S-R complexes without the presence of DNA. There was some question of specificity since other proteins displayed similar properties. Our group, on the other hand, chose to assay both the residual nucleoprotein component following extraction and the dissociated components for acceptor activity. An increase in activity after substantial extraction of material from chromatin was observed (indicating some masking of sites) followed by a loss of activity when protein was completely removed from the DNA. Thus, this acceptor activity is very tightly bound to DNA and probably requires DNA for activity. As the acceptor activity is removed from the DNA, it appears in the eluant as acidic protein(s) of small molecular weight.

It is virtually impossible to rationalize the extensive differences in these results, although they were obtained in different systems. Mainwaring et al., (71) and Hnilica's laboratory (61), however, derived different results in the same system, tending to imply that the method of assay largely determines the results. The solution to this diversity of results could be approached by (a) a complete cross-check by the groups involved of the other's methodology in their own system, (b) the preparation of antibody to purified acceptor, and the demonstration that this antibody can specifically block the receptor binding to nuclei or chromatin, or (c) ambitiously, the *in vitro* reconstruction of specific gene activity with steroid receptor, isolated acceptor, and RNA polymerase. These ultimate experiments would indicate which type of isolated acceptor is important in the actual biological response to steroid hormones. For a more thorough analysis of this field, readers are referred to a comprehensive review (131).

CONCLUSION

The acidic nonhistone proteins of avian oviduct chromatin appear to play a significant role in the acceptor activity (nuclear binding sites) for the P-R. Evidence is presented in this chapter implicating these proteins not only in acceptor activity itself but in the repression of the activity. This repressor activity of chromatin nonhistone proteins represents a unique function of negative regulation in contrast to the more common positive regulation (derepression of gene expression) often assigned to these proteins. These data are just beginning to reveal the complexity of the interactions of the nonhistone proteins with each other, with histones, and with the DNA in the chromosomal material. Whether the repressor activity or the acceptor activity involve specific sequences of DNA or alter the DNA helical structure, or both, remain intriguing unanswered questions. Evidence of the masking of acceptor activity and the isolation of a repressing protein fraction suggests a regulation of acceptor activity at the chromosomal level. The masked acceptor sites might represent the masking of hormone-responsive genes, once functional during cytodifferentiation but no longer needed in the mature organ. Nontarget tissue chromatins display almost total masking of progesterone acceptor activity, whereas the developing oviduct displays marked changes in the extent of repression of acceptor sites (118,119). When the acceptor for a particular steroid hormone is isolated, the regulation of its expression and its specificity for DNA sequences will need to be assessed in order to understand its role in the steroid hormone-mediated regulation of gene expression.

ACKNOWLEDGMENTS

This work was supported by Grants HD 9140-B from the National Institutes of Health and the Mayo Foundation.

REFERENCES

1. Ahmad, F., and Salahuddin, A. (1976): Reversible unfolding of major fraction of ovalbumin by guanidine-hydrochloride. *Biochemistry,* 15:5168–5175.
2. Anderson, J. N., Peck, E. J., Jr., and Clark, J. H. (1973): Nuclear receptor-estrogen complex: relationship between concentration and early uterotropic responses. *Endocrinology,* 92:1488–1495.
3. Anderson, J. H., Peck, E. J., Jr., and Clark, J. H. (1975): Estrogen-induced uterine responses and growth-relationship to receptor estrogen binding by uterine nuclei. *Endocrinology,* 98:160–167.
4. Arnaud, M., Beziat, Y., Guilleux, J. C., Hough, A., Hough, D., and Mousseron-Canet, M. (1971): Les récepteurs de l'oestradiol dans l'uterus de génisse stimulation de la biosynthese de RNA in vitro. *Biochim. Biophys. Acta,* 232:117–131.
5. Arnaud, M., Beziat, Y., Borgna, J. L., Guilleux, J. C., and Mousseron-Canet, M. (1971): Le récepteur de l'oestradiol, l'amp cyclique et la RNA polymerase nucleolaire dans l'uterus de génisse. Stimulation de la biosynthesis de RNA in vitro. *Biochim. Biophys. Acta,* 254:241–254.
6. Arnold, E. A., and Young, K. E. (1972): Isolation and partial electrophoretic characterization of total protein from non-sheared rat liver chromatin. *Biochim. Biophys. Acta,* 257:482–496.
7. Axel, R., Melchior, W., Jr., Sollner-Webb, B., and Felsenfeld, G. (1974): Specific sites of interaction between histones and DNA in chromatin (nuclease DNA-electrophoresis). *Proc. Nat. Acad. Sci.,* 71:4101–4105.
8. Baxter, J. D., Rousseau, G. G., Bensen, M. C., Garcea, R. L., Ito, J., and Tomkins, G. M. (1972): Role of DNA and specific cytoplasmic receptors in glucocorticoid action. *Proc. Nat. Acad. Sci.,* 69:1892–1896.
9. Beato, M., Biesewig, D., Braendle, W., and Sekeris, C. E. (1969): On the mechanism of hormone action XV. Subcellular distribution and binding of [I,2-³]cortisol in rat liver. *Biochim. Biophys. Acta,* 192:494–507.
10. Bekhor, I., Kung, G. M., and Bonner, J. (1969): Sequence-specific intraction of DNA and chromosomal protein. *J. Mol. Biol.,* 39:351–364.
11. Beziat, Y., Guilleux, J. C., and Mousseron-Canet, M. (1970): Effect de l'oestradiol et de ses récepteurs sur la biosynthese du RNA dans les nayaux isolés de l'uterus de génisse. *C.R. Acad. Sci. D.,* 270:1620–1623.
12. Bramhall, S., Noack, N., Wu, M., Loewenberg, J. R. (1969): A simple colorimetric method for determination of protein. *Anal. Biochem.,* 31:146–148.
13. Brecher, P. I., Vigerski, R., Wotiz, H. S., and Wotiz, H. H. (1967): An *in vitro* system for the binding of estradiol to rat uterine nuclei. *Steroids,* 10:635–651.
14. Buller, R. E., Toft, D. O., Schrader, W. T., and O'Malley, B. W. (1975): Progesterone-binding components of chick oviduct. VIII. Receptor activation and hormone-dependent binding to purified nuclei. *J. Biol. Chem.,* 250:801–808.
15. Buller, R. E., Schrader, W. T., and O'Malley, B. W. (1975): Progesterone-binding components of chick oviducts 9. Kinetics of nuclear binding. *J. Biol. Chem.,* 250:809–818.
16. Buller, R. E., Schwartz, R. J., Schrader, W. T., and O'Malley, B. W. (1976): Progesterone-binding components of chick oviduct. *In vitro* effect of receptor subunits on gene transcription. *J. Biol. Chem.,* 251:5178–5186.
17. Buller, R. E., Schwartz, R. J., and O'Malley, B. W. (1976): Steroid hormone receptor fraction stimulation of RNA synthesis: a caution. *Biochem. Biophys. Res., Commun.,* 69:106–113.
18. Carlsson, U., Henderson, L. E., and Lindskog, S. (1973): Denaturation and re-

activation of human carbonic anhydrases in guanidine hydrochloride and urea. *Biochim. Biophys. Acta,* 310:376–387.

19. Chae, C. B. (1975): Reconstitution of chromatin: mode of reassociation of chromosomal proteins. *Biochemistry,* 14:900–906.

20. Chamness, G. C., Jennings, A. W., and McGuire, W. L. (1973): Oestrogen receptor binding is not restricted to target nuclei. *Nature [New Biol.],* 241:458–460.

21. Chan, L., Means, A. R., and O'Malley, B. W. (1973): Rates of induction of specific translatable messenger RNAs for ovalbumin and avidin by steroid hormones. *Proc. Nat. Acad. Sci.,* 70:1870–1874.

22. Chan, L., Jackson, R. L., O'Malley, X. W., and Means, A. R. (1976): Synthesis of very low density lipoproteins in the cockeral. Effects of estrogen. *J. Clin. Invest.,* 58:368–379.

23. Chytil, F. (1975): Immunochemical characteristics of chromosomal proteins. *Methods Enzymol.,* 40:191–198.

24. Clark, J. H., and Gorski, J. (1969): Estrogen receptors: an evaluation of cytoplasmic-nuclear interactions in a cell-free system and a method for assay. *Biochim. Biophys. Acta,* 192:508–515.

25. Clark, J. H., Anderson, J. N., and Peck, E. J., Jr. (1973): Nuclear receptor-estrogen complexes of rat uteri: concentration-time-response parameters. *Adv. Exp. Biol. Med.,* 36:15–59.

26. Clark, J. H., and Peck, E. J. (1976): Nuclear retention of receptor-oestrogen complex and nuclear acceptor sites. *Nature,* 260:635–637.

27. Defer, N., Dastugue, B., and Kruh, J. (1974): Direct binding of corticosterone in estradiol to rat liver nuclear non-histone proteins. *Biochimie,* 56:559–566.

28. DeHertogh, R., Ekka, E., Vanderheyden, J., and Hoet, J. J. (1973): Slowly exchangeable pool of estradiol in the rat uterus. *J. Steroid Biochem.,* 4:313–320.

29. Elgin, S. C. R., and Bonner, J. (1970): Limited heterogeneity of the major non-histone chromosomal proteins. *Biochemistry,* 9:4440–4447.

30. Fang, S., Anderson, K. M., and Liao, S. (1969): Receptor proteins for androgens: on the role of specific proteins in selective retention of 17β-hydroxy-5α-androstan3-one by rat ventral prostate *in vivo* and *in vitro. J. Biol. Chem.,* 244:6584–6595.

31. Gadski, R. A., and Chae, C. B. (1976): Mode of reconstitution of chicken erythrocyte and reticulocyte chromatin. *Biochemistry,* 15:3812–3817.

32. Gilmour, R. S., and Paul, J. (1969): RNA transcribed from reconstituted nucleoprotein is similar to natural RNA. *J. Mol. Biol.,* 40:137–139.

33. Gilmour, R. S., and Paul, J. (1970): Role of non-histone components in determining organ specificity of rabbit chromatins. *FEBS Lett.,* 9:242–244.

34. Glascock, R. F., and Hoekstra, W. G. (1959): Selective accumulation of tritium-labelled hexoestrol by the reproductive organs of immature female goats and sheep. *Biochem. J.,* 72:673–682.

35. Glasser, S. R., Chytil, F., and Spelsberg, T. C. (1972): Early effects of oestradiol-17β on the chromatin and activity of the deoxyribonucleic acid-dependent ribonucleic acid polymerases (I and II) of the rat uterus. *Biochem. J.,* 130:947–957.

36. Gorski, J., and Gannon, F. (1976): Current models of steroid hormone action: a critique. *Ann. Rev. Physiol.,* 38:425–450.

37. Gorski, J., and Raker, B. (1974): Estrogen action in the uterus: the requisite for sustained estrogen binding in the nucleus. *Gynecol. Oncol.,* 2:249–258.

38. Gschwendt, M. (1976): Solubilization of the chromatin-bound estrogen receptor from chicken liver and fractionation on hydroxylapatite. *Eur. J. Biochem.,* 67:411–419.

39. Gschwendt, M., and Hamilton, T. H. (1972): The transformation of the cytoplasmic oestradiol-receptor complex into the nuclear complex in a uterine cell-free system. *Biochem. J.,* 128:611–616.

40. Hamilton, T. H., Teng, C. S., and Means, A. R. (1968): Early estrogen action: nuclear synthesis and accumulation of protein correlated with enhancement of two DNA-dependent RNA polymerase activities. *Proc. Nat. Acad. Sci.,* 59:1265–1272.

41. Harris, S. E., Rosen, J. M., Means, A. R., and O'Malley, B. W. (1975): Use of a

specific probe for ovalbumin messenger RNA to quantitate estrogen-induced gene transcript. *Biochemistry,* 14:2072–2081.

42. Harrison, R. W., Fairfield, S., and Orth, D. N. (1974): Evidence for glucocorticoid transport through the target cell membrane. *Biochem. Biophys. Res. Commun.,* 61:1262–1267.

43. Higgins, S. J., Rousseau, G. G., Baxter, J. D., and Tompkins, G. M. (1973): Nuclear binding of steroid receptors: comparison in intact cells and cell-free systems. *Proc. Nat. Acad. Sci.,* 70:3415–3418.

44. Higgins, S. J., Rousseau, G. G., Baxter, J. D., and Tomkins, G. M. (1973): Early events in glucocorticoid action activation of the steroid receptor and its subsequent specific nuclear binding studied in a cell-free system. *J. Biol. Chem.,* 248:5866–5872.

45. Hill, R. J., Poccia, D. L., and Doty, P. (1971): Towards a total macromolecular analysis of sea urchin embryo chromatin. *J. Mol. Biol.,* 61:445–462.

46. Huang, R. C. C., and Huang, P. C. (1969): Effect of protein-bound RNA associated with chick embryo chromatin on template specificity of the chromatin. *J. Mol. Biol.,* 39:365–378.

47. Jacob, F., and Monod, J. (1961): Genetic regulatory mechanisms in the synthesis of protein. *J. Mol. Biol.,* 3:318–356.

48. Jackson, V., and Chalkley, R. (1974): The binding of estradiol-17β and the bovine endometrial nuclear membrane. *J. Biol. Chem.,* 249:1615–1626.

49. Jackson, V., and Chalkley, R. (1974): The cytoplasmic estradiol receptors of bovine uterus. Their occurrence, interconversion, and binding properties. *J. Biol. Chem.,* 249:1627–1636.

50. Jaffe, R. C., Socher, S. H., and O'Malley, B. W. (1975): An analysis of the binding of the chick oviduct progesterone-receptor to chromatin. *Biochim. Biophys. Acta,* 399:403–419.

51. Jensen, E. V., and DeSombre, E. R. (1972): Mechanism of action of the female sex hormones. *Annu. Rev. Biochem.,* 41:203–230.

52. Jensen, E. V., and Jacobson, H. I. (1962): Basic guides to the mechanism of estrogen action. *Rec. Prog. Horm. Res.,* 18:387–414.

53. Jensen, E. V., Numata, M., Smith, S., Suzuki, T., Brecher, P. I., and DeSombre, E. R. Estrogen-receptor interaction in target tissues. *Dev. Biol. Suppl.,* 3:151–171.

54. Jensen, E. V., Brecher, P. I., Numata, M., Mohla, S., and DeSombre, E. R. (1973): Transformed estrogen receptor in the regulation of RNA synthesis in uterine nuclei. *Adv. Enzyme Regul.,* 11:1–16.

55. Jensen, E. V., Mohla, S., Gorell, T. A., and DeSombre, E. R. (1974): The role of estrophilin in estrogen action. *Vit. Horm.,* 32:89–127.

56. Kalimi, M., Beato, M., and Feigelson, P. (1973): Interaction of glucocorticoids with rat liver nuclei. I. Role of the cytosol proteins. *Biochemistry,* 12:3365–3371.

57. Kalimi, M., Tsai, S. Y., Tsai, M. J., Clark, J. H., and O'Malley, B. W. (1976): Correlation between nuclear-bound receptor and chromatin initiation site for transcription. *J. Biol. Chem.,* 251:516–523.

58. King, R. J. B., and Gordon, J. (1972): Involvement of DNA in the acceptor mechanism for uterine oestradiol receptor. *Nature* [*New Biol.*], 240:185–187.

59. King, R. J. B., Gordon, J., and Steggles, A. W. (1969): The properties of a nuclear acidic protein fraction that binds [6,7-^{3}H]oestradiol-17β. *Biochem. J.,* 114:649–657.

60. King, R. J. B., Beard, V., Gordon, J., Pooley, A. S., Smith, J. A., Steggles, A. W., and Vertes, M. (1971): Studies on estradiol-binding in mammalian tissues. *Biosciences,* 7:21–44.

61. Klyzsejko-Stefanowicz, L., Chui, J. F., Tsai ,Y. H., and Hnilica, L. S. (1976): Acceptor proteins in rat androgenic tissue chromatin. *Proc. Nat. Acad. Sci.,* 73:1954–1958.

62. Knowler, J. T., and Smellie, R. M. S. (1971): The synthesis of ribonucleic acid in immature rat uterus responding to oestradiol-17β. *Biochem. J.,* 125:605–614.

63. Levy, S., Simpson, R. T., and Sober, H. A. (1972): Fractionation of chromatin compounds. *Biochemistry,* 11:1547–1554.

64. Liang, T., and Liao, S. (1972): Interaction of estradiol- and progesterone-receptors

with nucleoprotein: heat-labile acceptor factors. *Biochim. Biophys. Acta,* 277:590–594.

65. Liao, S., Liang, T., and Tymoczko, J. L. (1973): Ribonucleoprotein binding of steroid-"receptor" complexes. *Nature [New Biol.],* 241:211–213.

66. Lowry, O. H., Rosebrough, N. J., Farr, A. L., and Randall, R. J. (1951): Protein measurement with the folin phenol reagent. *J. Biol. Chem.,* 193:265–275.

67. Lykins, L. F., Akey, C. W., Christian, E. G., Duval, G. E., and Topham, R. W. (1977): Dissociation and reconstitution of human ferroxidase II. *Biochemistry,* 16:693–698.

68. MacGillivray, A. J., Cameron, A., Krauze, R. J., Rickwood, D., and Paul, J. (1972): The non-histone proteins of chromatin. Their isolation and composition in a number of tissues. *Biochim. Biophys. Acta,* 277:384–402.

69. MacGillivray, A. J., Carroll, D. and Paul, J. (1971): The heterogeneity of the nonhistone chromatin protein from mouse tissues. *FEBS Lett.,* 13:204–208.

70. Mainwaring, W. I. P., and Peterken, B. M. (1971): A reconstituted cell-free system for the specific transfer of steroid-receptor complexes into nuclear chromatin isolated from rat ventral prostate gland. *Biochem. J.,* 125:285–295.

71. Mainwaring, W. I. P., Symes, E. K., and Higgins, S. J. (1976): Nuclear components responsible for the retention of steroid-receptor complexes especially from the standpoint of the specificity of hormone responses. *Biochem. J.,* 156:129–141.

72. Mann, K. G., and Fish, W. W. (1972): Protein polypeptide chain molecular weights by gel chromatography in guanidinium chloride. *Methods Enzymol.,* 26:28–42.

73. Mester, J., and Baulieu, E. E. (1975): Dynamics of oestrogen-receptor distribution between the cytosol and nuclear fractions of immature rat uterus after oestradiol administration. *Biochem. J.,* 146:617–623.

74. Means, A. R., and Hamilton, T. H. (1966): Evidence for depression of nuclear protein synthesis and concomitant stimulation of nuclear RNA synthesis during early estrogen action. *Proc. Nat. Acad. Sci.,* 56:686–693.

75. Milgrom, E., Atger, M., and Baulieu, E.-E. (1970): Progesterone in uterus and plasma. IV. Progesterone receptor(s) in guinea pig uterus cytosol. *Steroids,* 16:714–754.

76. Mohla, S., DeSombre, E. R., and Jensen, E. V. (1972): Tissue-specific stimulation of RNA synthesis by transformed estradiol-receptor complex. *Biochem. Biophys. Res. Commun.,* 46:661–667.

77. Monahan, J. J., and Hall, R. H. (1973): Fractionation of chromatin components. *Can. J. Biochem.,* 51:709–720.

78. Monder, C., and Walker, M. C. (1970): Interactions between corticosteroids and histones. *Biochemistry,* 9:2489–2497.

79. Mueller, G. C., Herranen, A. M., Jervell, K. J. (1958): Studies on the mechanism of action of estrogens. *Rec. Prog. Horm. Res.,* 14:95–139.

80. Musliner, T. A., Chader, G. J., and Villee, C. A. (1970): Studies on estradiol receptors of the rat uterus. Nuclear uptake in vitro. *Biochemistry,* 9:4448–4453.

81. O'Malley, B. W., and McGuire, W. L. (1968): Studies on the mechanism of action of progesterone in regulation of the synthesis of specific proteins. *J. Clin. Invest.,* 47:654–664.

82. O'Malley, B. W., and Means, A. R. (1974): Female steroid hormones and target cell nuclei. The effects of steroid hormones on target cell nuclei are of major importance in the induction of new cell functions. *Science,* 183:610–620.

83. O'Malley, B. W., McGuire, W. L., Kohler, P. O., and Korenman, S. G. (1969): Studies on the mechanism of steroid hormone regulation of synthesis of specific proteins. *Rec. Prog. Horm. Res.,* 25:105–160.

84. O'Malley, B. W., Sherman, M. R., and Toft, D. O. (1970): Progesterone "receptors" in the cytoplasm and nucleus of chick oviduct target tissue. *Proc. Nat. Acad. Sci.,* 67:501–508.

85. O'Malley, B. W., Toft, D. O., and Sherman, M. R. (1971): Progesterone-binding components of chick oviduct. II. Nuclear components. *J. Biol. Chem.,* 246:1117–1122.

86. O'Malley, B. W., Spelsberg, T. C., Schrader, W. T., Chytil, F., and Steggles, A. W. (1972): Mechanisms of interaction of a hormone-receptor complex with the genome of a eukaryotic target cell. *Nature*, 235:141–144.
87. Palmiter, R. D. (1971): Interaction of estrogen, progesterone, and testosterone in the regulation of protein synthesis in chick oviduct. *Biochemistry*, 10:4399–4403.
88. Palmiter, R. D. (1972): Regulation of protein synthesis in chick oviduct. I. Independent regulation of ovalbumin, conalbumin, ovomucoid, and ribonucleic acid synthesis in the oviduct of estrogen-primed chicks. *J. Biol. Chem.*, 248:8260–8270.
89. Palmiter, R. D., Moore, P. B., Mulvihill, E. R., and Emtage, S. (1976): A significant lag in the induction of ovalbumin messenger RNA by steroid hormones: a receptor translocation hypothesis. *Cell*, 8:557–572.
90. Pasqualini, J. R. (1976): *Receptors and Mechanism of Action of Steroid Hormones, Parts I and II*. Marcel Dekker, Inc., New York.
91. Paul, J., Gilmour, R. S., Affara, N., Birnie, G. D., Harrison, B. P., Hell, A., Humphries, S., Windass, J., Young, B. (1973): The globin gene: structure and expression. *Cold Spring Harbor Symp. Quant. Biol.*, 38:885–890.
92. Peck, E. J. Jr., Burgner, J., Clark, J. H. (1973): Estrophile binding sites of the uterus. Relation to uptake and retention of estradiol in vitro. *Biochemistry*, 12:4596–4603.
93. Pikler, G. M., Webster, R. A., and Spelsberg, T. C. (1976): Nuclear binding of progesterone in hen oviduct. Binding to multiple sites in vitro. *Biochem. J.*, 156: 399–408.
94. Puca, G. A., and Bresciani, F. (1968): Receptor molecule for oestrogens from rat uterus. *Nature*, 218:967–969.
95. Puca, G. A., Nola, E., Hibner, U., Circala, G., and Sica, V. (1975): Interaction of the estradiol receptor from calf uterus with its nuclear acceptor sites. *J. Biol. Chem.*, 250:6452–6459.
96. Puca, G. A., Sica, V., and Nola, E. (1974): Identification of a high affinity nuclear acceptor site for estrogen receptor of calf uterus. *Proc. Nat. Acad. Sci.*, 71:979–983.
97. Raspe, G. (1971): *Advances in the Biosciences, Vol. 7*, Pergamon Press, N.Y.
98. Raynaud-Jammet, C., and Baulieu, E.-E. (1969): Action de l'oestradiol in vitro: augmentation de la biosynthese d'acide ribonucléique dans les noyaux utérins. *C.R. Acad. Sci.*, D, 268:3211–3214.
99. Rogan, E. G., Mailander, P., and Calvalieri, E. (1976): Metabolic activation of aromatic hydrocarbons in purified rat liver nuclei: induction of enzyme activities and binding to DNA with and without monooxygenase-catalyzed formation of active oxygen. *Proc. Nat. Acad. Sci.*, 73:457–461.
100. Saffran, J., Loeser, B. K., Bohnett, S. A., and Faber, L. E. (1976): Binding of progesterone receptor by nuclear preparations of rabbit and guinea pig uterus. *J. Biol. Chem.*, 251:5607–5613.
101. Schrader, W. T., and O'Malley, B. W. (1972): Progesterone-binding components of chick oviduct. IV. Characterization of purified subunits. *J. Biol. Chem.*, 247: 51–59.
102. Schrader, W. T., Toft, D. O., and O'Malley, B. W. (1972): Progesterone-binding protein of chick oviduct. VI. Interaction of purified progesterone-receptor components with nuclear constituents. *J. Biol. Chem.*, 247:2401–2407.
103. Schwartz, R. J., Kuhn, R. W., Buller, R. E., Schrader, W. T., and O'Malley, B. W. (1976): Progesterone-binding components of chick oviduct. In vitro effects of purified hormone-receptor complexes on the interaction of RNA synthesis in chromatin. *J. Biol. Chem.*, 251:5166–5177.
104. Schwartz, R. J., Schrader, W. T., and O'Malley, B. W. (1976): Mechanism of steroid hormone action: *in vitro* control of gene expression in chick oviduct chromatin by purified steroid receptor complexes. In: *Juvenile Hormones*, edited by L. I. Gilbert, pp. 530–557. Plenum Press, New York.
105. Sherman, M. R., Corvol, P. L., and O'Malley, B. W. (1970): Progesterone-binding components of chick oviduct. I. Preliminary characterization of cytoplasmic components. *J. Biol. Chem.*, 245:6085–6096.

106. Shirey, T., and Huang, R. C. C. (1969): Use of sodium dodecyl sulfate, alone, to separate chromatin proteins from deoxyribonucleoprotein of *Arbacia punctulata* sperm chromatin. *Biochemistry*, 8:4138–4148.
107. Sluyser, M. (1969): Interaction of steroid hormones with histones *in vitro*. *Biochim. Biophys. Acta*, 182:235–244.
108. Spelsberg, T. C. (1974): The role of nuclear acidic proteins in binding steroid hormones. In: *Acidic Proteins of the Nucleus*, edited by I. L. Cameron and J. R. Jeter, pp. 247–296. Academic Press, New York.
109. Spelsberg, T. C. (1976): Nuclear binding of progesterone in the chick oviduct: multiple binding sites *in vivo* and transcriptional response. *Biochem. J.*, 156:391–398.
110. Spelsberg, T. C., and Cox, R. F. (1976): Effects of estrogen and progesterone in the chick oviduct. *Biochim. Biophys. Acta*, 435:376–390.
111. Spelsberg, T. C., and Toft, D. O. (1976): The mechanism of action of progesterone. In: *Receptors and the Mechanism of Action of Steroid Hormones*, edited by J. R. Pasqualini, Part I, pp. 261–309. Marcel Dekker, Inc., New York.
112. Spelsberg, T. C., Steggles, A. W., and O'Malley, B. W. (1971): Changes in chromatin composition and hormone binding during chick oviduct development. *Biochim. Biophys. Acta*, 254:129–134.
113. Spelsberg, T. C., Steggles, A. W., and O'Malley, B. W. (1971): Progesterone-binding components of chick oviduct. III. Chromatin acceptor sites. *J. Biol. Chem.*, 246:4188–4197.
114. Spelsberg, T. C., Hnilica, L. S., and Ansevin, A. T. (1971): Proteins of chromatin in template restriction. III. The macromolecules in specific restriction of the chromatin DNA. *Biochim. Biophys. Acta*, 228:550–562.
115. Spelsberg, T. C., Steggles, A. W., Chytil, F., and O'Malley, B. W. (1972): Progesterone-binding components of chick oviduct. V. Exchange of progesterone-binding capacity from target to nontarget tissue chromatins. *J. Biol. Chem.*, 247:1368–1374.
116. Spelsberg, T. C., Webster, R. A., Pikler, G. M. (1975): Multiple binding sites for progesterone in the hen oviduct nucleus: evidence that acidic proteins represent the acceptors. In: *Chromosomal Proteins and Their Role in the Regulation of Gene Expression*, edited by G. S. Stein and L. J. Kleinsmith, pp. 153–186. Academic Press, New York.
117. Spelsberg, T. C., Webster, R. A., and Pikler, G. M. (1976): Chromosomal proteins regulate steroid binding to chromatin. *Nature*, 262:65–67.
118. Spelsberg, T. C., Webster, R., Pikler, G., Thrall, C., and Wells, D. (1976): Role of nuclear proteins as high affinity sites ("acceptors") for progesterone in the avian oviduct. *J. Steroid Biochem.*, 7:1091–1101.
119. Spelsberg, T. C., Webster, R. A., Pikler, G. M., Thrall, C. L., and Wells, D. J. (1977): Nuclear binding sites (acceptors) for progesterone in avian oviduct: characterization of the highest affinity sites. *Ann. N.Y. Acad. Sci.*, 286:43–63.
120. Spelsberg, T. C., Thrall, C. L., Webster, R. A., and Pikler, G. M. (1977): Isolation and characterization of the nuclear acceptor that binds the progesterone-receptor complex in hen oviduct. *J. Toxicol. Environ. Health*, 3:309–337.
121. Steggles, A., Spelsberg, T. C., and O'Malley, B. W. (1971): Tissue specific binding *in vitro* of progesterone-receptor to the chromatins of chick tissues. *Biochem. Biophys. Res. Commun.*, 43:20–27.
122. Steggles, A. W., Spelsberg, T. C., Glasser, S. R., and O'Malley, B. W. (1971): Soluble complexes between steroid hormones and target-tissue receptors bind specifically to target-tissue chromatin. *Proc. Nat. Acad. Sci.*, 68:1479–1482.
123. Stein, G. S., Spelsberg, T. C., and Kleinsmith, L. J. (1974): Nonhistone chromosomal proteins and gene regulation. Nonhistone chromosomal proteins may participate in the specific regulation of gene transcription in eukaryotes. *Science*, 183:817–824.
124. Stein, G. S., Mans, R. J., Gabbay, E. J., Stein, J. L., Davis, J., and Adawadkar, P. D. (1975): Evidence for fidelity of chromatin reconstitution. *Biochemistry*, 14:1859–1866.

125. Stumpf, W. E., and Sar, M. (1976): Autoradiographic localization of estrogen, androgen, progestin, and glucocorticosteroid in "target tissues" and "nontarget tissues." In: *Receptors and Mechanisms of Action of Steroid Hormones,* edited by J. R. Pasqualini, part I, pp. 41–84. Marcel Dekker, New York.

126. Sumida, C., Gelly, C., and Pasqualini, J. R. (1974): Étude des complexes ^{3}H-oestradiol-macromolecules dans le noyau du rein de foetus de cobaye. Liaison directe de l'oestradiol a la fraction chromatinsgraph. *C.R. Acad. Sci., (Paris) Ser. D,* 279:1793–1796.

127. Sutherland, R. L., and Baulieu, E.-E. (1976): Quantitative estimates of cytoplasmic and nuclear oestrogen receptors in chick oviduct. Effect of oestrogen on receptor concentration and subcellular distribution. *Eur. J. Biochem.,* 70:531–541.

128. Swaneck, G. E., Chu, L. L. H., and Edelman, I. S. (1970): Stereoscopic binding of aldosterone and renal chromatin. *J. Biol. Chem.,* 245:5382–5389.

129. Talwar, G. P., Segal, J. J., Evans, A., and Davidson, O. W. (1964): The binding of estradiol in the uterus: a mechanism for derepression of RNA synthesis. *Proc. Nat. Acad. Sci.,* 52:1059–1066.

130. Tanford, C. (1968): Protein denaturation. In *Advances in Protein Chemistry, Vol. 23,* edited by C. B. Anfinsen, M. L. Anson, J. T. Edsall, and F. M. Richards, pp. 122–282. Academic Press, New York.

131. Thrall, C., Webster, R. A., and Spelsberg, T. C. (1978): Receptor interaction with chromatin. In: *The Cell Nucleus, Vol. 3,* edited by Harris Busch, pp. 461–529. Academic Press, New York.

132. Toft, D. O., and Gorski, J. (1966): A receptor molecule for estrogens: isolation from the rat uterus and preliminary characterization. *Proc. Nat. Acad. Sci.,* 55:1574–1581.

133. Traish, A. M., Muller, R. E., and Wotiz, H. H. (1977): Binding of estrogen receptor to uterine nuclei. *J. Biol. Chem.,* 252:6823–6830.

134. Tsai, S. Y., Tsai, M. J., Schwartz, R., Kalimi, M., Clark, J. H., and O'Malley, B. W. (1975): Effects of estrogen on gene expression in chick oviduct: nuclear receptor levels and initiation of transcription. *Proc. Nat. Acad. Sci.,* 72:4228–4232.

135. Tymoczko, J. L., and Liao, S. (1971): Retention of an androgen-protein complex by nuclear chromatin aggregates: heat-labile factors. *Biochim. Biophys. Acta,* 252:607–611.

136. Weber, K., and Kuter, D. J. (1971): Reversible denaturation of enzymes by sodium dodecyl sulfate. *J. Biol. Chem.,* 246:4504–4509.

137. Webster, R. A., Pikler, G. M., and Spelsberg, T. C. (1976): Nuclear binding of progesterone in hen oviduct. Role of acidic chromatin proteins in high-affinity binding. *Biochem. J.,* 156:409–418.

138. Williams, D., and Gorski, J. (1974): Equilibrium binding of estradiol by uterine cell suspensions. *Biochemistry,* 13:5537–5542.

139. Woodcock, C. L. F. (1977): Reconstitution of chromatin subunits. *Science,* 195:1350–1352.

140. Yamamoto, K. R., and Alberts, B. M. (1974): On the specificity of the binding of the estradiol receptor protein to deoxyribonucleic acid. *J. Biol. Chem.,* 249:7076–7086.

141. Yamamoto, K. R., and Alberts, B. M. (1976): Steroid receptors: elements of modulation of eukaryotic transcription. *Annu. Rev. Biochem.,* 45:721–746.

142. Yazgan, A., and Henkens, R. W. (1972): Role of zinc (II) in the refolding of guanidine hydrochloride denatured bovine carbonic anhydrase. *Biochemistry,* 11:1314–1318.

Ontogeny of Receptors and Reproductive Hormone Action,
edited by T. H. Hamilton, J. H. Clark, and W. A. Sadler.
Raven Press, New York 1979.

Heterogeneity of Estrogen-Binding Sites in the Rat Uterus

James H. Clark, James W. Hardin, Hakan Eriksson, Susan Upchurch, and E. J. Peck, Jr.

Department of Cell Biology, Baylor College of Medicine, Houston, Texas 70025

For several years investigators have suggested that at least two forms of estrogen-binding sites exist in the uterus (3,8,9,17,20,22,24). One of these sites, the estrogen receptor, has been intensively investigated (2,7,12,13,18). This receptor is a protein macromolecule that binds estrogens in a stereospecific manner and is found in the cytosol of estrogen-sensitive cells. It has a very high affinity for estradiol and is generally considered to exist in uterine cells at a concentration of ~20,000 sites/cell or ~0.5 pmoles/100 μg DNA (1,4,15). The other binding site(s) has received little attention and is often ignored or considered to be due to serum albumin or α-feto protein (17). Early work from the laboratories of Erdos and Korenman indicated that the dissociation of cytosol receptor estrogen complexes took place in two first-order phases (3,9,22). Thus, the dissociation of estradiol could not be visualized as a simple dissociation of the hormone from a single binding site. It is clear that only 50% of the amount of ^{3}H-estradiol in the uterus is bound to the classical estrogen receptor, and we have suggested that the remaining ^{3}H-estradiol is bound to secondary sites, which could include serum albumin, lipids, and other undescribed binding sites (19). In view of this distribution of bound estrogen, we considered it important to evaluate the types of estrogen-binding sites more extensively.

TWO TYPES OF ESTROGEN-BINDING SITES IN UTERINE CYTOSOL

We have observed that saturation analysis of uterine cytosol over a wide range of ^{3}H-estradiol concentration revealed a complex curve for specifically bound ^{3}H-estradiol (10, and Fig. 1A). Although this curve appears to consist of only one binding component, it is actually made up of two binding sites, which can be resolved by Scatchard analysis and by use of the Rosenthal and Feldman method for correction of curved Scatchard plots (11,21,23). One of these sites, Type I (Fig. 1B), conforms to the expected characteristics of the classical estrogen receptor and has a K_d of 0.8 nM and represents 0.6

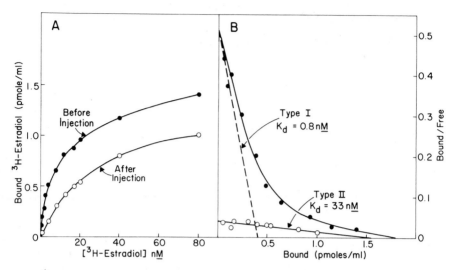

FIG. 1. Saturation analysis of estrogen binding in rat uterine cytosol. **A:** The quantity of specifically bound ^{3}H-estradiol was determined in uterine cytosols from noninjected rats (●) and rats injected with 2.5 μg of estradiol 60 min prior to sacrifice (○). Uteri were homogenized in TE-buffer (0.01 M Tris-HCl, 0.0015 M Na$_2$EDTA, pH 7.9 at 0°C) using a motor-driven glass–glass Duall homogenizer. The homogenate was centrifuged at 800 × g for 10 min in a Beckman JS 7.5 rotor. The supernatant was recentrifuged at 180,000 × g for 30 min and diluted with TE buffer to 25 mg/ml. Cytosol was incubated with ^{3}H-estradiol or ^{3}H-estradiol plus 100-fold excess of DES for 18 hr at 10°C, and the hydroxylapatite assay was performed (24). **B:** Scatchard analysis of the data in **A**. The amount of ^{3}H-estradiol bound to uterine cytosols from estrogen treated animals (○, Type II) was subtracted from the total binding in the system (●) to yield the dashed line labeled Type I.

pmoles/uterus (30 to 35 mg wet weight). In addition, this binding form is depleted from the cytoplasm after an injection of estradiol. The other site, Type II (Fig. 1B), has a lower affinity for estradiol, K_d of 30 nM, but a higher binding capacity of 2.0 pmoles per uterus (30 to 35 mg wet weight). Type II sites are not depleted from the cytosol after an estradiol injection.

In order to examine further the characteristics of these two sites, we have used sucrose density gradient analysis of uterine cytosol. This method has been used extensively for the qualitative and quantitative estimation of estrogen receptors in both normal and abnormal tissue (1–7,21–23). This procedure is usually performed by prelabeling cytosol with ^{3}H-estradiol, followed by ultracentrifugation of the gradients for 12 to 16 hr. It was anticipated that during this 12- to 16-hr period, ^{3}H-estradiol would dissociate rapidly from Type II sites, and hence very little bound hormone would be observed after centrifugation. On the other hand, dissociation from Type I sites is very slow ($T_{1/2} \sim 20$ hr), and hence at least 50% of the bound hormone would still be observed after centrifugation. In order to avoid this problem, we have added ^{3}H-estradiol to the gradient fractions after the centrifugation period and performed the adsorption of receptor to hydroxylapatite (HAP assay) as described in the legend to Fig. 2. As shown in Fig. 2, the quantity of ^{3}H-estradiol that is bound in a specific manner, i.e., the amount inhibited by the presence of a 100-fold molar excess of diethylstilbestrol (DES), on prelabeled gradients is significant in both the 4S and 8S regions. However, neither of these peaks can account for the predicted quantity of Type II binding. Postlabeling of gradient fractions and HAP assay following gradient fractionation reveals the presence of large quantities of an estrogen-binding molecule in the 4S region of the gradient (Fig. 2B). The relative quantities of bound ^{3}H-estradiol in the 8S and 4S regions are compatible with the assumption that the 8S region is made up of Type I sites and Type II sites are found in the 4S region.

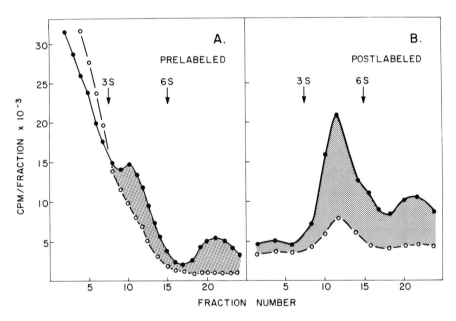

FIG. 2. Sucrose density gradient analysis of Type I and II estradiol binding sites by prelabeled and postlabeled methods. **A:** Prelabeled gradients. Linear 5 to 20% sucrose gradients (4.9 ml) were prepared with a Beckman density gradient former. Sucrose solutions were prepared in TE buffer plus 1 mM dithiothreitol (DTT). Uterine cytosol (250 μl) that had been adjusted to 1 mM DTT was incubated at 4°C fo 60 min with 20 nM ^{3}H-estradiol (●) or the same concentration of ^{3}H-estradiol plus a 100-fold molar excess of DES (○). The cytosol was layered on the gradients and centrifuged at 189,000 × g for 16 hr. **B:** postlabeled gradients. Sucrose density gradients were prepared as described above, and cytosol, either labelled or unlabelled with ^{3}H-estradiol, was centrifuged for 16 hr as described above. After centrifugation either 0.2- or 0.4-ml fractions were collected in tubes that contained ^{3}H-estradiol (20 nM, final concentration). An identical gradient was fractionated into tubes that contained the same concentration of ^{3}H-estradiol plus a 100-fold molar excess of DES. The tubes were incubated for 60 min. at 20°C and the measurement of specific estrogen binding was performed by the HAP assays (10).

Further evidence that the 4S region is made up of Type II sites is shown in Fig. 3. In this experiment various concentrations of ^{3}H-estradiol were used to postlabel gradients and HAP assays were performed. The 8S peak appears to be saturating between 20 and 40 nM; however, attainment of saturation in the 4S region does not occur at these levels. For comparison and to demonstrate that both types of binding sites exist in the mature rat uterus, cytosol from mature ovariectomized rats was examined in an identical fashion. The results are shown in Fig. 3B.

Additional evidence that the 8S and 4S regions of gradients contain Type I and II sites, respectively, was obtained by postlabeling gradients of cytosol from rats that had been injected with estradiol. An injection of 2.5 μg estradiol in immature rats will cause the depletion of the cytosol estrogen receptor (Type I) and should leave Type II in the cytosol. It is clear from the data in Fig. 4 that the 8S region is depleted, whereas the quantity of bound ^{3}H-estradiol in the 4S region does not change. This is in agreement with our previous observation that cytosol from estrogen-injected rats contains only Type II sites when analyzed by the HAP assay (10, and Fig. 1).

The hormone specificity of Type I and II binding sites was examined by

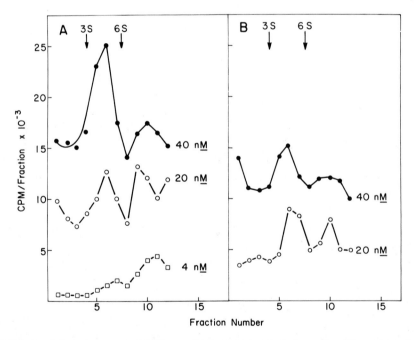

FIG. 3. Postlabeled sucrose density gradient analysis of uterine cytosol from immature and mature castrate rats. **A:** Cytosol from immature rat uteri was prepared as described in Fig. 1. Sucrose density gradients were postlabeled at 4, 20, and 40 nM ^{3}H-estradiol as described in Fig. 2B. The data points represent specifically bound ^{3}H-estradiol. **B:** Uterine cytosol was prepared from mature castrate rats and postlabeled gradient analysis was performed as in **A.**

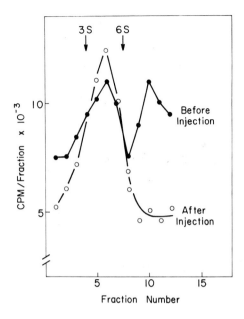

FIG. 4. Effect of estradiol injection on Type I and II binding sites in uterine cytosol. Uterine cytosol was examined by the postlabeled sucrose density gradient analysis, as described in Fig. 2, in noninjected rats (●) and in rats that had been injected 60 min prior to sacrifice (○).

postlabeled gradient-HAP assays and by (direct) HAP assay of high-speed cytosol. The HAP assay of high-speed cytosol showed that estradiol and DES inhibited the binding of ^{3}H-estradiol to both types of sites, whereas progesterone, testosterone, or cortisol did not. Identical specificity of both types of sites is also shown by the gradient postlabeled-HAP technique in Fig. 5.

The presence of estrogen-binding sites in other tissues was examined by postlabeled gradient-HAP assays and by direct HAP assay of high-speed cytosols. The vagina contains large quantities of the Type II binding sites (Fig. 6A), whereas the kidney contains significant, but much lower, quantities (Fig. 6B). This is also true for the spleen (data not shown). In these experiments the animals were injected 60 min before sacrifice with 2.5 μg estradiol; hence very little Type I is present in the vagina. However, the work of several laboratories has demonstrated the presence of a Type I binding site. Saturation analysis of blood serum and cytosols from spleen and kidney by direct HAP assay does not detect any significant specific binding.

These data demonstrate the presence of at least two types of macromolecules that bind estrogen in a stereospecific fashion: Type I has the properties of the cytosol estrogen receptor and Type II has a lower affinity and a higher capacity for the binding of estradiol than Type I. Type II sites also differ from Type I sites because they do not undergo depletion from the cytosol after an estrogen injection. The importance of Type II sites is far-reaching. Their presence interferes with the measurement of Type I sites and can cause an overestimate of Type I or an incorrect identification of

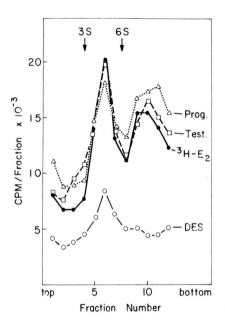

FIG. 5. Hormone specificity of Type I and II estradiol-binding sites. Uterine cytosol from immature rats was examined by the postlabel sucrose density gradient technique as described in Fig. 2. HAP assays were performed with ^{3}H-estradiol (20 nM) or ^{3}H-estradiol plus progesterone (△), testosterone (□), and DES (○). The concentration of each added steroid was 2 μM.

Type II as Type I. An accurate measurement of Type I sites by saturation analysis depends on an adequate assessment of Type II sites. The data shown in Fig. 1 were obtained by incubating cytosol plus hormone for 18 hr at 4°C, which is sufficient time for both types of binding sites to reach equilibrium. Often this is not done, and shorter times are used for the assay. Under these circumstances, Type II sites may not reach equilibrium, resulting

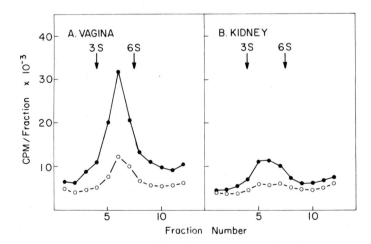

FIG. 6. Tissue specificity of Type I and II estradiol-binding sites. Cytosols were prepared from vagina **(A)** and kidney **(B)** and postlabeled sucrose density gradient analysis was performed as described in Fig. 2.

in an overestimate of Type I sites. The extent of the overestimate will be a function of the equilibrium state of both I and II sites.

The identity of the Type II binding site as a macromolecule with a sedimentation coefficient of 4S on postlabeled gradients also has important implications. Sucrose gradients are usually run by labeling the cytosol before the gradients are centrifuged. During the 12- to 16-hr centrifugation period, hormone dissociation takes place, and hence both types of binding sites will be underestimated depending on the rate of dissociation of the hormone. The HAP assay performed on postlabeled gradients eliminates this problem and creates a highly sensitive assay for both Type I and II binding sites.

A complete analysis of the contribution that is made by Type II binding sites on prelabeled sucrose density gradients is difficult to make because of the various ways in which gradients are performed in different laboratories. In the work reported here, we have used 5 to 20% sucrose with no glycerol added, and we observed significant loss of binding on prelabeled sucrose density gradients. However, other investigators, who have used glycerol-containing gradients, have observed less loss of binding in the 4S region. In addition, the quantity of Type II sites will depend on the concentration of cytosol protein that is layered on the gradient, i.e., more highly concentrated cytosols should yield more Type II on prelabeled gradients. ^{3}H-estradiol binding in the 4S region is usually attributed to α-feto protein (α-FP) in the immature rat uterus, and this serum protein undoubtedly does make some contribution. However, in our experiments DES was used as a competitive inhibitor, and since α-FP has a very low affinity for DES, the Type II binding that we observed cannot be due to α-FP (25). In addition, Type II sites also occur in the adult rat uterus, which does not contain α-FP (25).

The presence of estrogen receptor in human breast tissue is used by several laboratories to determine whether endocrine ablation will be used in the treatment of breast cancer (14,16,26). Since breast tissue is assayed by the prelabeled sucrose density technique, it is obvious from the results shown in Fig. 2 that much of the estrogen-binding capacity of the tissue has been underestimated. The method of postlabeled sucrose density gradients coupled with the HAP assay as introduced here should provide a tool for a more thorough and valid estimate of estrogen-binding sites in breast tissue and thus increase the accuracy and predictability of estrogen receptor assays in breast-cancer therapy.

It is possible that Type II sites are extracellular binding proteins that help to accumulate estrogens. If they are concentrated in the extracellular spaces of the uterus, they could act as effective estrogen-concentrating agents. In addition, they would help maintain the local organ level of free estrogens at high concentration relative to the blood. As discussed above, approximately half of the ^{3}H-estradiol that is found in the uterus after *in vivo* injection or *in vitro* incubation is bound to the estrogen receptor (19). The other half may be distributed between Type II sites and other sites, such as serum albumin. It is also possible that Type II sites represent a precursor form of Type I sites. Thus, one can envision a cytoplasmic reservoir of lower affinity macromolecules that serves to replenish Type I sites after they have been translocated to the nucleus.

TWO CLASSES OF ESTROGEN-BINDING SITES IN UTERINE NUCLEI

As shown in Figs. 1 and 4, Type I cytosol-binding sites are depleted after an injection of estradiol. This depletion is accompanied by the accumulation of these sites in the nucleus and represents the well-known cytoplasmic to nuclear translocation phenomenon (12,13,18). As shown in Fig. 7, saturation analysis of nuclear fractions for estrogen-binding sites by the ^{3}H-estradiol exchange assay reveals a complex picture that involves at least two specific estrogen-binding sites. One site conforms to the Type I site that was depleted from the cytosol and is undoubtedly identical to the classically described estrogen receptor. The other site displays a sigmoidal saturation curve. When the quantity of this second site, which we will call nuclear Type II, is subtracted from the total quantity of nuclear-bound hormone as measured by

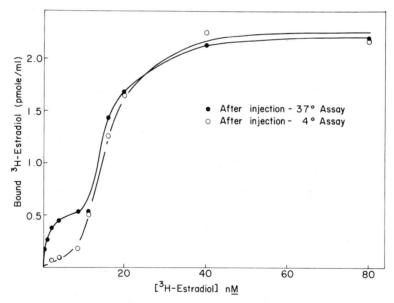

FIG. 7. Saturation analysis of estrogen binding in the nuclear fraction of the rat uterus. Estrogen-binding sites in the nuclear fraction of the rat uterus were determined by the ^{3}H-estradiol exchange assay at 37°C after injection of 2.5 μg of estradiol (●). Specific ^{3}H-estradiol binding was also determined by exchange at 4°C in nuclei from estrogen-treated rats (○).

exchange, one obtains the amount bound to Type I. Scatchard analysis of the nuclear Type I site reveals a K_d of 0.60 nM and a maximal number of sites of 0.36 pmoles/ml (Fig. 8A). These values do not differ significantly from those of the cytosol receptor, Type I, which was depleted by estrogen treatment, and conform to the usually accepted properties of the estrogen receptor.

The nuclear Type II sites do not appear to be identical to the cytosol Type II sites and display cooperative binding behavior with a Hill coefficient of approximately 2 (Fig. 8B). These measurements were made with crude nuclear pellets; therefore, it is possible that the nuclear Type II sites are the result of cytoplasmic contamination. This does not appear to be the case, since purified nuclear preparations also contain these sites (Fig. 9). In this experiment immature rats were injected with 2.5 μg estradiol 1 hr before assay and nuclei were isolated. Specific estrogen binding was measured by ^{3}H-estradiol exchange. Both types of sites are present in nuclei after estradiol treatment; however, neither of these sites appeared in noninjected animals (Fig. 9). Since the cytosol Type II site does not undergo depletion, it appears unlikely that the nuclear Type II sites are derived from them. At this time, the relationship between these two classes of Type II sites is unknown.

Nuclear Type II sites are hormone and tissue specific. Nonestrogen target organs such as spleen and kidney do not have any measurable quantities of this binding site. The data shown in Fig. 10 were obtained with purified

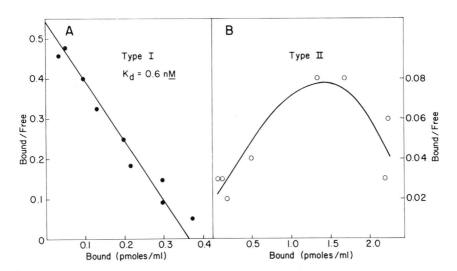

FIG. 8. Scatchard analysis of Type I and II estrogen-receptive sites in the rat uterine nuclear fraction. **A:** Scatchard analysis for Type I nuclear sites. The amount of ^{3}H-estradiol bound to Type I sites was obtained by subtracting the amount of ^{3}H-estradiol bound at 4°C from that observed at 37°C in Fig. 7 (● minus ○). **B:** Scatchard analysis of Type II nuclear sites. The quantity of bound ^{3}H-estradiol that was observed by exchange at 4°C in Fig. 7 was used in this figure.

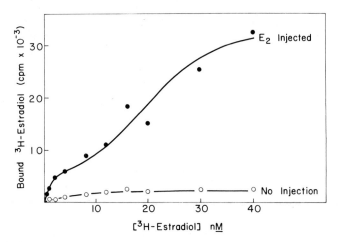

FIG. 9. Saturation analysis of Type I and II binding sites in purified uterine nuclei. Purified uterine nuclei were prepared by the glycerol-triton procedure as described in ref. 10 from noninjected (◯) and estradiol-treated (●) immature rats (2.5 μg estradiol, 60 min prior to sacrifice).

nuclear preparations from the uterus and spleen of estradiol-treated rats. It is clear from these data that spleen nuclei will bind larger quantities of ³H-estradiol; however, all of this binding is of a nonspecific type. Hormone specificity was examined in purified nuclear preparation by exposing them to ³H-estradiol alone or ³H-estradiol plus competitor under exchange conditions. DES inhibits binding to both sites, and nonestrogenic hormones have no inhibitory effect (data not shown). Thus, the nuclear Type II sites also display a binding specificity for estrogens that is similar to Type I estrogen receptors.

These results indicate that at least two types of specific estrogen-binding sites can be found in estrogen target cell nuclei. One of these, Type I, corresponds to the classical estrogen receptor and is probably derived from the cytosol Type I site. A second site, nuclear Type II, does not appear to be derived from the cytosol Type II site, but is found in the nucleus after estrogen injection. The physiological significance of the second sites is unknown; however, the presence of these sites in nuclear fraction has important implications with respect to the validity of receptor measurement. Valid estimates of Type I binding must take into account the contribution that is made by the presence of Type II. Assays that differentiate the two sites are readily accomplished by saturation analyses of nuclear fractions via exchange at both 37 and 4°C. At 37°C, ³H-estradiol exchanges with the Type I sites and binds to Type II sites; hence at 37°C both sites are measured. Since exchange of occupied Type I sites occurs very slowly at 4°C, only Type II sites are measured at this temperature. Subtraction of Type II sites as measured at 4°C from those sites measured via exchange at elevated tem-

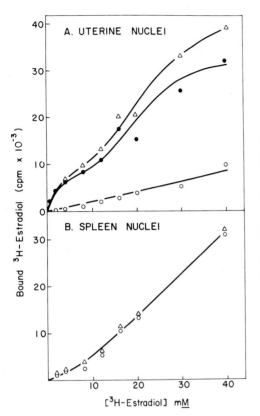

FIG. 10. Estradiol binding in purified nuclei from the uterus and spleen. **A:** Uterine nuclei were prepared as described in Fig. 9 from estradiol-injected rats. Estrogen-binding sites were determined by the ^{3}H-estradiol exchange assay (14). Nonspecific binding (○) was subtracted from total binding (△) to yield the quantity of specifically bound estradiol (●). **B:** Nuclei from spleen were prepared and analyzed as described in **A.**

perature (Types I plus II) yields the contribution made by Type I alone. This evaluation is a necessity if accurate assessments of individual receptor sites are required. Previous work from our laboratory and others has undoubtedly overestimated the quantities of Type I sites in nuclear fractions. As pointed out above, this overestimate is also true for assays of cytosol receptor employing hydroxylapatite or other protein-adsorbant procedures. The charcoal-adsorption assay is less likely to be in error, since long exposure to charcoal strips estradiol from Type II sites (J. H. Clark et al., *unpublished observations*). We have also observed variable quantities of nuclear Type II sites in nontreated crude uterine nuclear fractions. These could be mistakenly identified as Type I sites unless saturation analyses at 4 and 37°C are done.

We do not understand the functional significance of these various forms of estrogen-binding sites; however, it is possible to make some suggestions concerning their function. As indicated earlier, the cytosol Type II site may be involved in the retention of estrogens within the uterus, thus creating an estrogen-rich environment for the Type I sites that translocates estrogen to the nucleus, or cytosol Type II sites may represent a precursor form of

Type I sites. The ability of estrogen to elevate the levels of nuclear Type II sites may represent some phase of the replenishment cycle for cytosol sites. It is also possible that nuclear Type II sites are a component of the machinery for "processing" the Type I nuclear complex. We previously suggested that receptor-estrogen complexes bind to a small number of nuclear sites and undergo "nuclear processing" (5). This processing may be a part of the mechanism by which RNA synthesis is stimulated and the receptor–estrogen complex eventually detached from nuclear sites.

REFERENCES

1. Anderson, J., Clark, J. H., and Peck, E. J., Jr. (1972): Oestrogen and nuclear binding sites: Determination of specific sites by [^{3}H]oestradiol exchange. *Biochem. J.,* 126:561–567.
2. Baulieu, E. E., Atger, M., Best-Belpomme, M., Corvol, P., Courvalin, J. C., Mester, J., Milgrom, E., Robel, P., Rochefort, H., and DeCatalogne, D. (1975): Steroid hormone receptors. *Vitam. Horm.,* 33:649–736.
3. Best-Belsomme, M., Fries, J., and Erdos, T. (1970): Interactions entre l'oestradiol et des sites récepteurs uterins. Donnees cinétiques et d'équilibre. *Eur. J. Biochem.,* 17:425–432.
4. Clark, J. H., and Gorski, J. (1969): Estrogen receptors: An evaluation of cytoplasmic-nuclear interactions in a cell-free system and a method for assay. *Biochim. Biophys. Acta,* 192:508–515.
5. Clark, J. H., and Peck, E. J., Jr. (1976): Nuclear retention of receptor-oestradiol complex and nuclear acceptor sites. *Nature,* 260:635–636.
6. Clark, J. H., and Peck, E. J., Jr. (1977): Steroid hormone receptors: Basic principles and measurement. In: *Hormone Action: Steroid Hormone Receptors,* edited by B. W. O'Malley and L. Birnbaumer. Academic Press, New York (*in press*).
7. Clark, J. H., Peck, E. J., Jr., Hardin, J. W., and Eriksson, H. (1978): Biology and pharmacology of estrogen receptors: Relation to physiological response. In: *Hormone Action: Steroid Hormone Receptors,* edited by B. W. O'Malley and L. Birnbaumer. Academic Press, New York (*in press*).
8. Ellis, D. J., and Ringold, H. J. (1971): The uterine estrogen receptor: A physicochemical study. *The Sex Steroids,* edited by K. W. McKerns, p. 73. Appleton-Century-Crofts, New York.
9. Erdos, T., Bessada, R., and Fries, J. (1969): Binding of oestradiol to receptor-substances present in extracts from calf uterus. *FEBS Lett.,* 5:161–164.
10. Eriksson, H., Upchurch, S., Hardin, J. W., Peck, E. J., Jr., and Clark, J. H. (1978): Heterogeneity of estrogen receptors in the cytosol and nuclear fractions of the rat uterus. *Biochem. Biophys. Res. Commun.,* 81:1–7.
11. Feldman, H. A. (1972): Mathematical theory of complex ligand-binding sites at equilibrium. *Anal. Biochem.,* 48:317–338.
12. Gorski, J., and Gannon, F. (1976): Current models of steroid hormone action: A critique. *Ann. Rev. Physiol.,* 38:425–450.
13. Jensen, E. V., Mohla, S., Gorell, T. A., and DeSombre, E. R. (1974): The role of estrophilin in estrogen action. *Vitam. Horm.,* 32:89–127.
14. Jensen, E. V., Smith, S., and DeSombre, E. R. (1976): Hormone dependency in breast cancer. *J. Steroid Biochem.,* 7:911–917.
15. Katzenellenbogen, J. A., Johnson, H. J. Jr., and Carlson, K. E. (1973): Studies on the uterine, cytoplasmic estrogen binding protein. Thermal stability and ligand dissociation rate. An assay of empty and filled sites by exchange. *Biochemistry,* 12:4091–4099.
16. McGuire, W. L., Carbone, P. P., and Vollmer, E. P., eds. (1975): Estrogen receptors in human breast cancer. Raven Press, New York.
17. Michel, G., Jung, I., Baulieu, E. E., Aussel, C., and Uriel, J. (1974): Two high

affinity estrogen binding proteins of J. Henent specificity in the immature rat uterus cytosol. *Steroids,* 24:437–449.

18. O'Malley, B. W., and Means, A. R. (1974): Female steroid hormones and target cell nuclei. *Science,* 183:610–620.
19. Peck, E. J., Jr., Burgner, J., and Clark, H. J. (1973): Estrophilic binding sites of the uterus. Relation to uptake and retention of estradiol *in vitro. Biochemistry,* 12:4596–4603.
20. Puca, G. A., Nola, E., Sica, V., and Bresciani, F. (1971): Estrogen binding proteins of calf uterus. Partial purification and preliminary characterization of two cytoplasmic proteins. *Biochemistry,* 10:3769–3780.)
21. Rosenthal, H. E. (1967): A graphic method for the determination and presentation of binding parameters in a complex system. *Anal. Biochem.,* 20:525–533.
22. Sanborn, B. M., Rao, B. R., and Korenman, S. G. (1971): Interaction of 17β-estradiol and its specific uterine receptor. Evidence for complex kinetic and equilibrium behavior. *Biochemistry,* 10:4955–4960.
23. Scatchard, G. (1949): The attractions of proteins for small molecules and ions. *Ann. N.Y. Acad. Sci.,* 51:660–672.
24. Steggles, A. W., and King, R. J. (1970): The use of protamine to study oestradiol binding in rat uterus. *Biochem. J.,* 118:695–701.
25. Soloff, M. S., Creange, J. E., and Potts, G. O. (1971): Unique estrogen-binding properties of rat pregnancy plasma. *Endocrinology,* 88:427–432.
26. Wittliff, J. L., Beatty, B. W., Savlow, E. D., Patterson, W. B., and Cooper, R. A. Jr. (1976): Estrogen receptors and hormone dependency in human breast cancer. In: *Recent Results in Cancer Research, Vol. 57,* edited by G. St. Arneault, P. Band, and L. Israel, pp. 59–77. Springer-Verlag, Berlin.

Ontogeny of Receptors and Reproductive Hormone Action,
edited by T. H. Hamilton, J. H. Clark, and W. A. Sadler.
Raven Press, New York 1979.

Regulation of Uterine Responsiveness to Estrogen: Developmental and Multihormonal Factors

Benita S. Katzenellenbogen

*Department of Physiology and Biophysics, University of Illinois, and
School of Basic Medical Sciences, University of Illinois College of Medicine,
Urbana, Illinois 61801*

There are many factors that affect the character and magnitude of tissue responses to estrogens. During early development, the uterus acquires the capacity to respond fully to estradiol. In the fully competent, or mature, uterus other endogenous hormones modulate uterine responsiveness to estrogen. Additionally, synthetic antiestrogens can dramatically alter uterine sensitivity to estrogens.

This chapter focuses on selected aspects of the regulation of uterine responsiveness to estrogen with particular attention to the status and function of the estrogen receptor as a determinant of uterine responsive capacity. The first part of the chapter gives a brief overview of the ontogeny of uterine response to estradiol and addresses the question of the relationship between the acquisition of tissue competence to respond and the presence of receptors in the developing uterus. The second part deals with the influence of other hormones, notably progestins and androgens, on modulation of uterine responsiveness to estrogen. And finally, some studies are presented on the interaction of antiestrogens with the uterus.

ONTOGENY OF UTERINE RESPONSE TO ESTROGEN DURING DEVELOPMENT

Our initial interest in studying the ontogeny of uterine responsiveness to estrogen during early development in the rat stemmed from our concern with the uterine estrogen-induced protein (IP). Since the synthesis of this protein is one of the earliest responses to estrogen (9,18,31) and its induction has been shown proportional to nuclear estrogen receptor at early times (18,37), we felt that this estrogenic response might serve as an early indicator of uterine responsiveness.

Figure 1 shows the relationship between estrogen-stimulated amino acid incorporation into the IP and DNA and soluble protein content of the rat uterus as a function of age (day 0 = day of birth). While there is a progres-

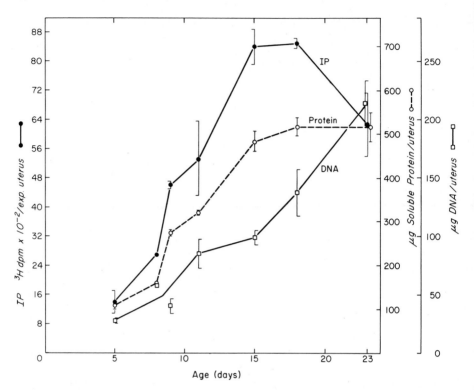

FIG. 1. Magnitude of estradiol-induced IP synthesis as a function of age and its relationship to uterine content of soluble protein and DNA. Rats were injected with 1 µg estradiol/10 g body weight or with control vehicle, and uteri were excised at 1 hr; estradiol-treated uteri were allowed to incorporate [³H]leucine, and controls were allowed to incorporate [¹⁴C]leucine into protein for 2 hr *in vitro* at 37°C. Uterine content of soluble protein and DNA were identical in experimental and control uteri. Determinations employed two groups of experimental and one group of control rats per age (number of uteri per group as stated in text) except at days 8 and 11, which employed one and three experimental groups, respectively, and are expressed as mean ± SEM. (From Katzenellenbogen and Greger, ref. 20.)

sive increase with age in the size of the uterus as measured by DNA and protein content, there is a corresponding and generally parallel increase, at least up to 15 days, in the amount of estrogen-induced IP synthesis per uterus.

When IP synthesis is expressed as a percent of total protein synthesis in the uterus at various ages (Fig. 2, *left top panel*), IP synthesis accounts for 1.6 to 3.1% of total protein synthesis, with near maximal stimulation being obtained before day 10 of development. Figure 2 also shows the uterine response to estradiol in terms of three other estrogen-dependent parameters —24-hr weight gain, increased glucose metabolism, and 3-day uterine weight gain. A single injection of estradiol does not stimulate a 24-hr uterine weight

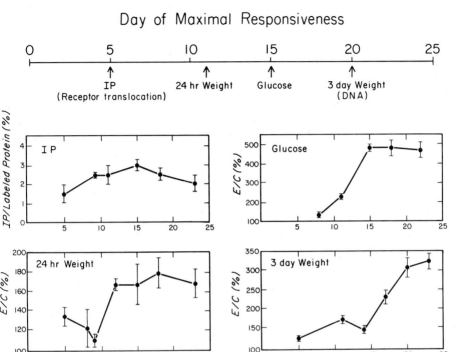

FIG. 2. Acquisition of uterine responsiveness to estradiol during early development. **Top:** Schematic developmental scale depicting the age at which the uterus can respond maximally to estradiol as monitored by different estrogen-dependent parameters (day 0 = the day of birth). **Left top panel:** Estradiol-induced IP synthesis as a percent of newly labeled total uterine-soluble protein at different ages. Rats were injected with estradiol (1 μg/10 g body weight) for 1 hr and labeled with radioactive leucine for 2 hr *in vitro*. Labeled IP in the 27,000 × g uterine supernatant was quantitated by electrophoresis. Newly labeled protein counts in the same supernatant fraction were determined by cold trichloroacetic acid precipitation. IP synthesis is expressed as a percent of the total labeled protein ± SEM ($N = 4$). **Left bottom panel:** One-day uterine wet weight response to estradiol as a function of age. Rats were injected with 1 μg estradiol/10 g body weight or with control vehicle on the day indicated, and uterine wet weight was measured at 24 hr. Determinations employed five to six experimental and five to six control rats per age group. The ratio of estrogen-treated to control wet weight is expressed as percent ± SEM. **Right top panel:** Effect of estradiol on uterine 2-deoxyglucose metabolism as a function of age. Rats were injected with 1 μg estradiol/10 g body weight or with control vehicle; at 2 hr after injection, uteri were excised and incubated *in vitro* with 2-deoxy-D-glucose-1-[^{14}C] for 30 min. 2-Deoxyglucose and 2-deoxyglucose-6-PO$_4$ (DGP) were separated and assayed as described in the text. Determinations employed at least three experimental and three control incubations (number of uteri per incubation as stated in text). The estradiol-stimulated (experimental minus control) DGP production is expressed on a uterine weight basis. Units are pmoles DGP/mg wet weight ± SEM. **Right bottom panel:** Three-day uterine weight response to estradiol as a function of age. Rats were injected with 1 μg estradiol/10 g body weight or with control vehicle once daily on three successive days beginning on the day indicated, and uterine wet weight was determined at 24 hr after the last (third) injection. Determinations employed five to six experimental and five to six control rats per age group. The ratio of estrogen-treated to control wet weights is expressed as percent ± SEM.

gain (Fig. 2, *left bottom panel*) before day 10, but by day 12 and thereafter a maximal response is attained. Increased glucose utilization (Fig. 2, *right top panel*), measured by 2-deoxyglucose phosphorylation, is maximally stimulated only by day 15. However, if one monitors a 3-day weight gain, which is measured after three daily injections of estradiol and is a better measure of true uterine growth, a maximal response is not seen until day 20 (Fig. 2, *right bottom panel*).

Hence, the uterus can respond partially to exogenous estradiol very early, but a full response, including prolonged growth, only becomes apparent during a further period of approximately 2 weeks. This information is summarized at the top of Fig. 2: a maximal IP synthesis response is reached between 5 and 10 days; 24-hr weight change is maximally attained by day 12; maximal stimulation of glucose metabolism is reached by day 15; and a 3-day weight gain involving considerable DNA synthesis and cell division is reached only by day 20. In these rats, puberty is normally attained by day 30 to 35. Extensive studies by Kaye and co-workers (24,41,42) and by Luck et al. (29) have provided complementary and supportive data.

Clark and Gorski (8) have shown that the cytoplasmic estrogen receptor is present at the day of birth; the concentration of cytoplasmic receptor increases to a peak at 10 days of age and then declines slightly. Further studies by Somjen et al. (42) and Luck et al. (29) have shown receptor translocation to the nucleus by day 7.

Interestingly, the developmental pattern of the acquisition of uterine response to estradiol (Fig. 2, *top*) is very similar to the temporal sequence of responses evoked by estradiol in the fully competent (day 22) immature rat uterus (Fig. 3): IP induction, increased glucose utilization, and early weight gain, which occur at early times after estrogen interaction with the fully competent uterus, are also the events that can be stimulated by estradiol early during development. In contrast, long-term uterine growth, involving DNA synthesis and cell division, which occurs at later times after estrogen interaction, is a response that is maximally stimulated by estradiol only in the older (day 20) rat.

From the developmental studies done in several laboratories (20,24,29, 41,42), it appears that the presence of receptor alone is not sufficient to ensure complete uterine responsiveness, because at day 10, when high levels of cytoplasmic receptor are present and are capable of being translocated to the nucleus (29,42), only a few responses, such as IP synthesis, are obtained. Although it appears that during development the uterus acquires a progressive ability to respond fully to estradiol, experimental work so far is insufficient to establish definitively whether the progressive increases in responses observed are actually a true manifestation of changes in the inherent responsiveness of the uterus (such as result from possible changes in nuclear receptor populations or their interaction with chromatin) or progressive changes in steroid pharmacodynamics (including the influence of alpha-

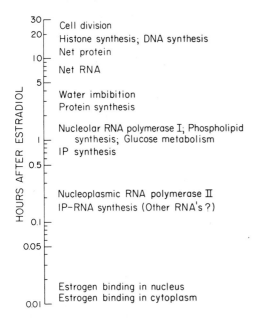

FIG. 3. Temporal sequence of events initiated in the immature or ovariectomized mature rat uterus after *in vivo* injection of estradiol-17β at zero time. Note that time (hours) is on a logarithmic scale (0.01 hr = 0.6 min.). (From Katzenellenbogen and Gorski, ref. 19.)

fetoprotein) that modulate the uterine availability of exogenous estradiol. This is clearly a very interesting area that requires further work.

HORMONAL FACTORS REGULATING RESPONSIVENESS TO ESTROGEN IN THE FULLY COMPETENT RAT UTERUS

There is increasing evidence that uterine sensitivity to estrogen is modulated by several other hormones including progestins, androgens, and possibly thyroid hormones. In the rat uterus, we have found (1) that pretreatment of animals with progesterone decreases their subsequent responsiveness to estrogen, as monitored by the induction of IP synthesis.

In both immature, unprimed (day 20) rats and 3-day estrogen-primed animals, progesterone (administered at zero time) causes a significant inhibition of subsequent estradiol induction of IP synthesis, with maximal inhibition seen by 12 to 24 hr after progesterone treatment (Fig. 4). This reduced rate of IP synthesis appears to be due to the fact that less estrogen receptor is taken up or retained, or both, by uterine nuclei after progesterone (1). Studies (1) have shown this inhibitory effect to be specific to progestins (progesterone and the synthetic progestins norgestrel and chlormadinone acetate); it was not obtained by pretreatment with androgens or glucocorticoids. Further, the antagonistic effect of progesterone appears to be on the rate of synthesis of IP and not on its turnover, as IP turnover either with or without progestin pretreatment of uteri was the same (half-life of over 24 hr).

Studies in several laboratories have also documented the modulation of

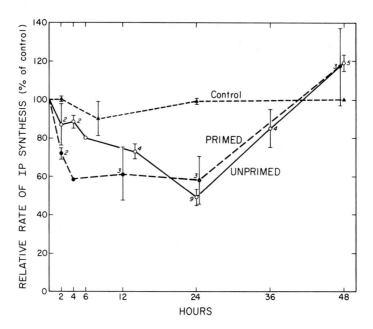

FIG. 4. Time course of the effect of progesterone pretreatment on the relative rate of IP synthesis in immature, unprimed (day 20) rats and in immature, estradiol-primed (5 μg estradiol in saline, once daily for 3 days) rats. Unprimed or estradiol-primed animals were treated with progesterone (2 mg/rat s.c. in sesame oil) for indicated periods of time and then received either estradiol (5 μg/rat s.c. in saline, experimentals) or saline (controls) for 60 min prior to sacrifice. Uteri were excised and labeled with radioactive amino acid ([³H]leucine for estradiol-treated and [¹⁴C]leucine for control uteri) in Eagle's HeLa medium for 2 hr at 37°C. [³H] and [¹⁴C]labeled uteri in each set were homogenized together, and an equal volume of the supernatant fraction of centrifuged homogenates (containing 0.56 experimental plus 0.56 control unprimed uterine equivalents per gel, and 0.38 experimental plus 0.38 control primed uterine equivalents per gel) was electrophoresed on polyacrylamide gels. The radioactivity and ³H:¹⁴C ratio in each gel slice were determined. The counts in IP/experimental uterus were quantitated by the method of Katzenellenbogen and Greger (20). The data is expressed as percent of 1-hr estradiol control, with the cpm in IP/uterus for the control (100% value) being that obtained for unprimed or primed uteri receiving oil pretreatment for designated time periods followed by 1 hr estradiol (and ³H-labeling) or 1 hr saline (and ¹⁴C-labeling). It is seen that pretreatment with oil for up to 48 hr (control line) has no effect on incorporation of counts into the IP band. However, the ³H dpm in IP/experimental uterus is 8,220 dpm/immature uterus and 5,470 dpm/primed uterus. Each point represents the mean ± SEM with the number of groups per point indicated in italics. (From Bhakoo and Katzenellenbogen, ref. 1.)

estrogen receptor level by progesterone, a phenomenon shown in several animals and in uterus as well as oviduct (2,3,13,14,32,44). Progesterone treatment of animals does not appear to affect the preexisting level of cytoplasmic estrogen receptor, but rather affects estrogen receptor replenishment (possibly synthesis, see below).

Figure 5 presents representative data from studies in the immature (day 22) rat uterus. Following an injection of estradiol, there is a rapid rise in

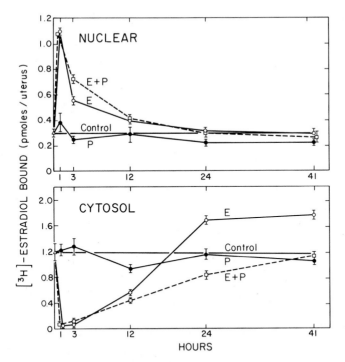

FIG. 5. Content of specific estrogen binding sites present in nuclear (*upper panel*) and cytosol (*lower panel*) fractions of the immature rat uterus as a function of time after a single injection of estradiol (5 µg s.c./rat; E) or progesterone (2 mg s.c./rat; P) or estradiol plus progesterone (E + P) or saline and oil (control utéri) as determined by nuclear and cytosol exchange assays. Each point represents the mean of two closely corresponding determinations in duplicate per point (three uteri/group) and is corrected for nonspecific binding. Cytosol exchange data is after 24 hr of exchange and hence represents "total cytosol sites." Total cytosol sites differ little from unfilled cytosol sites determined by direct binding at 0 to 4°C. The vertical bars represent ± SEM. (From Bhakoo and Katzenellenbogen, ref. 2.)

uterine nuclear receptor, with a parallel fall in cytoplasmic receptor levels. Cytoplasmic receptor is then rapidly restored by processes that are presumed to involve receptor recycling from the nucleus and also *de novo* receptor synthesis (30,39), so that by 24 hr cytoplasmic levels have already exceeded the initial control level. Simultaneous treatment with estradiol and progesterone (Fig. 5) results in a similar movement of receptor into the nucleus in a translocation that closely parallels that seen with estradiol alone. However, the restoration of cytoplasmic sites proceeds more slowly, and the final levels attained are considerably lower.

In the 3-day estrogen-primed rat uterus, a similar inhibitory effect of estradiol plus progesterone on the replenishment of the cytosol estrogen receptor is observed (Fig. 6). Estradiol alone, by 1 hr, translocates most of the estradiol cytoplasmic receptors into the nuclear compartment. Thereafter,

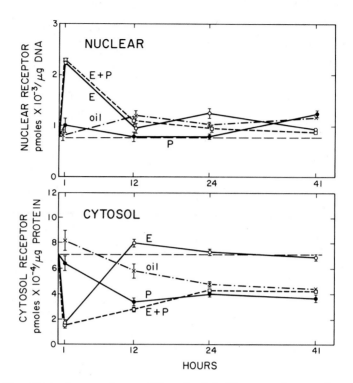

FIG. 6. The effect of progesterone (P), estradiol (E), progesterone plus estradiol (E + P) or oil alone on the distribution and content of estradiol receptor in nuclear (*upper panel*) and cytosol (*lower panel*) fractions of 3-day estradiol-primed rat uteri. Immature female rats (two/group) 20 to 21 days old received 5 μg of estradiol s.c. in saline once daily for 3 days. At 24 hr after the last injection of estradiol, the animals received either P, E, E + P, or oil alone for designated periods of time. Uteri were excised and assayed for nuclear receptor (*upper panel*) and for total cytosol receptor (*lower panel*) by exchange assays. The lightly dashed horizontal line extending across the upper and lower panels represents the nuclear and cytosol receptor levels in primed uteri at zero time. Each point represents the mean ± SEM of three to four determinations. (From Bhakoo and Katzenellenbogen, ref. 2.)

the estradiol binding sites are replenished in the cytoplasm, reach maximal levels by 12 hr after the estradiol injection, and remain at this level until 41 hr, whereas the nuclear receptor levels have decreased to the control level by 12 hr. However, administration of progesterone along with estradiol results in a depressed replenishment of the estradiol cytoplasmic receptor up to 41 hr (compare cytosol estradiol versus estradiol plus progesterone curves), although it does not influence the initial movement of the cytosolic receptor into the nuclear compartment. [It is also clear that in the absence of estradiol treatment (oil, control curve), the levels of the cytosolic receptors in these primed uteri decline markedly by 41 hr. The decline seems to be somewhat faster in the progesterone-treated group than in the oil controls.] In both the immature unprimed rat and in the estrogen-primed rat, these uteri exposed to estradiol plus progesterone, which have lower levels of cytoplasmic receptor, have a reduced responsiveness to subsequent estrogen in terms of uterine weight gain and glucose utilization (2,13,14).

Studies by others (5,30,39) have shown that normal replenishment of estrogen receptor presumably involves both recycling of nuclear sites (cyclo-heximide-insensitive replenishment) and *de novo* receptor synthesis (cyclo-heximide-sensitive replenishment). In an attempt to determine the point at which progesterone acts in the replenishment process, we (2) compared the replenishment after progesterone with either cycloheximide or actinomycin D.

Figure 7 compares the time courses and magnitude of the effects of progesterone, cycloheximide, and actinomycin D on cytosol receptor replenishment by estradiol in 3-day estrogen-primed immature rat uteri. In all cases, animals received a 5-μg estradiol injection at zero time and then were

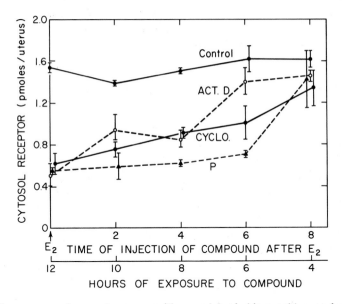

FIG. 7. Time course of progesterone-sensitive, cycloheximide-sensitive, and actinomycin D-sensitive inhibition of estradiol cytosol receptor replenishment in estrogen-primed rat uteri. Immature rats, 20–21 days of age (two/group), were given 5 μg estradiol in saline s.c. once daily for 3 days. At 24 hr, following the third estradiol injection, they received another injection of estradiol (E$_2$; 5 μg s.c.). Animals were then given 200 μg each cycloheximide (CYCLO) or actinomycin D (ACT. D) in saline or saline alone i.p. or 2 mg progesterone (P) in oil s.c. or vehicle (Control) alone at the indicated time at a different site. For longer periods of treatment with the inhibitors (12, 10, and 8 hr), cycloheximide and actinomycin D were reinjected (200 μg/rat) at 6 hr after the first injection to inhibit protein and RNA synthesis up to a minimum at 12 hr. All animals were sacrificed 12 hr after the injection of estradiol; uteri were excised and assayed for cytosol receptor by [³H]estradiol exchange assay. Each point represents the mean plus range of two determinations from separate experiments. (From Bhakoo and Katzenellenbogen, ref. 2.)

exposed to progesterone, cycloheximide, or actinomycin D for periods of 4 to 12 hr with cytosol receptor levels always being determined at 12 hr after the initial estradiol injection.

It is seen that when progesterone or the inhibitors are given either at zero time (along with the estradiol) or up to 4 to 6 hr following estradiol injection (exposure to compound at least 6 to 8 hr), they are effective in decreasing the levels of the estradiol cytoplasmic receptor (to ca. 40 to 50% of the control, estradiol alone, level); however, although exposure for 6 to 8 hr gives maximal inhibition, exposure for only the last 4 hr of the 12-hr period gives little, if any, inhibition of receptor replenishment. The fact that cycloheximide and actinomycin D given within 4 to 6 hr following estradiol are effective in blocking 50 to 60% of the total sites replenished by 12 hr strongly suggests that about 60% of the estradiol cytosolic receptors replenished by 12 hr after a single injection of estradiol are due to *de novo* synthesis. The rest, ca. 40%, may be sites recycled from the nuclear compartment. The progesterone-sensitive time course of inhibition of replenishment of estradiol binding capacity closely parallels that seen for cycloheximide or actinomycin D and suggests that the inhibitory effect of progesterone on the replenishment of the estradiol cytosol receptor is by interference with receptor synthesis per se.

Other studies aimed at elucidating the multihormonal regulation of uterine responsiveness to estrogen have involved androgens and thyroid hormones. Several laboratories have documented the initial observation of Rochefort et al. (35) that pharmacological doses of androgens are capable of interacting with the estrogen receptor system in the uterus and of eliciting responses typically evoked by estrogens, such as nuclear translocation of estrogen receptor and synthesis of the induced protein (11,34,38,40). Cidlowski and Muldoon (6) have also shown that thyroid hormones can influence the level of cytoplasmic estrogen receptor in the uterus, and Gardner et al. (12) have recently made the interesting observation that maintenance of uterine responsiveness to estrogen appears to require continued, possibly direct, thyroid hormone support.

INTERACTIONS OF ANTIESTROGENS WITH THE UTERUS

Antiestrogens are typically nonsteroidal compounds that prevent estrogens from expressing their full effects on estrogen target tissues, and, as such, they antagonize a variety of estrogen-dependent processes, including uterine growth and the growth of estrogen-dependent mammary tumors. They also act to stimulate pituitary gonadotrophin output and subsequent ovulation in certain women by antagonism of estrogen feedback at the level of the hypothalamus and pituitary. It should be noted, however, that these compounds are not pure antagonists and show some estrogenicity themselves. The structures of some of these compounds are seen in Fig. 8. These compounds are related structurally to the better known antiestrogen clomiphene.

Our main interest in working with these compounds has been to try to elucidate some of the molecular aspects of their mode of action. In some early studies (16), we examined the effects of antiestrogens on the sub-cellular distribution of estrogen receptors in the uterus (Fig. 9).

Like estradiol, the antiestrogens U-11,100A (UA) and CI-628 (CI) move cytoplasmic receptor sites into the nucleus. MER-25 appeared as a control compound at the dosage used here. However, although the estradiol receptor complex is lost from the nucleus rather rapidly and cytoplasmic re-ceptor levels are replenished soon thereafter, the antiestrogens retain some receptor in the nucleus and cytoplasmic receptor levels remain depleted for a prolonged period of time. These findings confirmed the earlier reports of Clark et al. (7) and Rochefort et al. (35) demonstrating prolonged nuclear retention and cytoplasmic depletion of receptor following administration of nafoxidine (UA).

Further studies showed that during the period in which antiestrogen has depleted the cytoplasmic receptor level, the uterus is incapable of respond-ing to estradiol as monitored by IP synthesis (Fig. 10) or by uterine weight gain (16). However, after estradiol, the more rapid return of cytoplasmic receptor levels is paralleled by the return of uterine responsiveness to es-trogen.

In other studies we were doing at the time (27), we found that some long-acting estrogens also evoked a retention of nuclear receptor similar to that

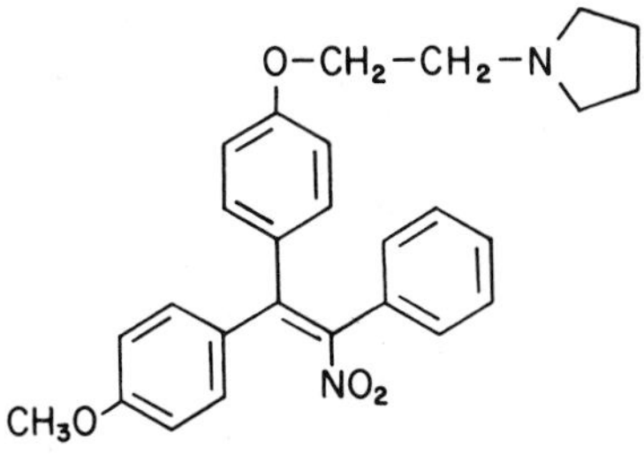

CI-628 (CI)

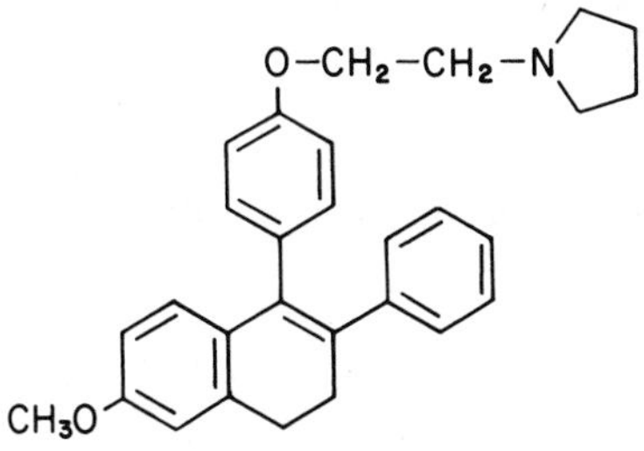

U-11,100A (UA)

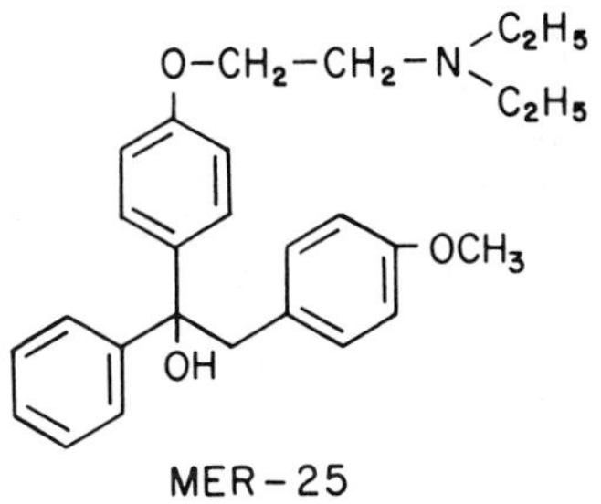

MER-25

FIG. 8. Structures of antiestrogens studied.

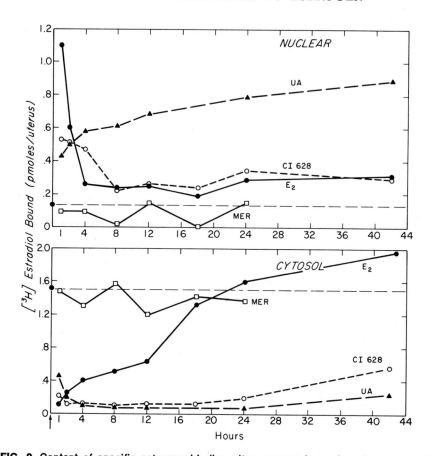

FIG. 9. Content of specific estrogen binding sites present in nuclear (*upper panel*) and cytosol (*lower panel*) fractions of the immature rat uterus as a function of time after a single injection of 17β-estradiol (5 μg s.c./rat) or antiestrogen (50 μg s.c./rat) as determined by the nuclear and cytosol exchange assays. Each point represents the mean of two closely corresponding determinations in duplicate per point (three uteri/group) and is corrected for nonspecific binding. Cytosol exchange data is after 24 hr of exchange and hence represents total cytosol sites. In the upper and lower panels, values on the ordinate connected to lightly dashed lines are those obtained for the control (saline injected) uteri. (From Katzenellenbogen and Ferguson, ref. 16.)

seen with some antiestrogens, causing us to wonder whether some of the effects of antiestrogens might be due to the fact that they are long-acting compounds. Therefore, we decided to compare in detail differences in the actions of long-acting estrogens and antiestrogens on the uterus.

As seen in Fig. 11, we (17) compared the action of the long-acting estrogen 17α-ethinyl-estriol-3-cyclopentyl ether (EE₃CPE) and the long-acting antiestrogen UA and analyzed single- and multiple-injection regimens in studying the effects of these compounds on uterine growth and on estrogen receptor distribution between nuclear and cytoplasmic compartments.

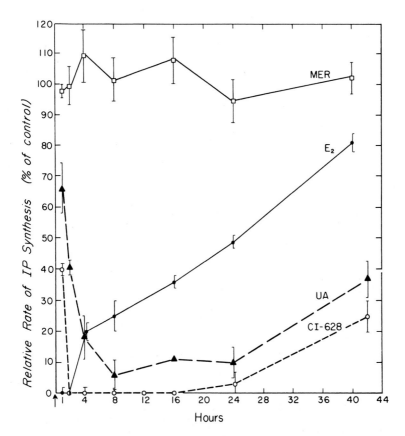

FIG. 10. The effect of an injection of antiestrogen or estradiol on the subsequent ability of estradiol to induce IP synthesis. Note that this figure represents the relative rate of IP synthesis that results from a second injection (5 µg estradiol) given at the indicated time after an initial injection of estradiol or antiestrogen at zero time. (For the response due to estradiol or antiestrogen alone, see text). Immature rats (six/ group) were injected with estradiol (5 µg s.c.) or antiestrogen (50 µg s.c.) at zero time, and at the indicated time thereafter (beginning at 1 hr) rats received an injection of either estradiol (5 µg s.c.) or saline alone. At 1 hr after the second injection, uteri were excised and allowed to incorporate labeled leucine (^{3}H for experimentals and ^{14}C for controls) into protein for 2 hr at 37°C. Control and estradiol-treated uteri in each set were homogenized together. Following centrifugation, the supernatant fraction was separated by polyacrylamide gel electrophoresis, and the relative rate of IP synthesis (experimental/control) was determined by gel analysis. One hundred percent is set as the relative rate of IP synthesis seen in the immature rat uterus at 1 hr after an s.c. injection of 5 µg estradiol. Each point represents the mean ± SEM of two to three determinations employing three experimental and three control rat uteri per determination. Arrow indicates the time of the first injection (estradiol or antiestrogen). (From Katzenellenbogen and Ferguson, ref. 16.)

Following a *single* injection of either compound, both compounds show a gradual movement of receptor to the nucleus (Fig. 11, *middle panels*) and maintenance of elevated levels of nuclear receptor for ca. 48 hr. Both compounds also show a gradual depletion of cytoplasmic receptor sites, and

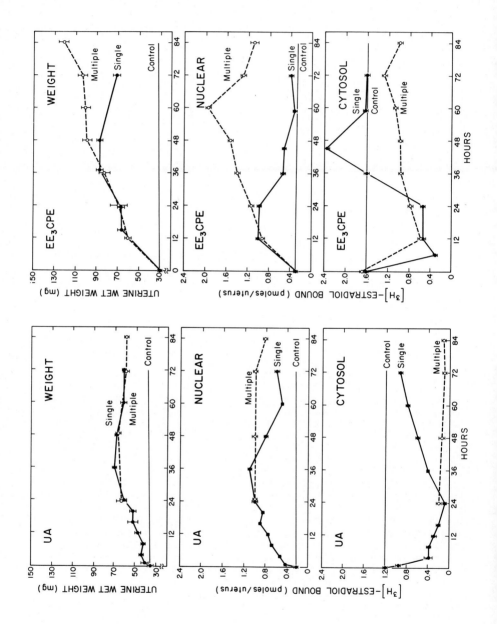

receptor levels remain low for 24 hr, after which time they increase. Both compounds also evoke similar uterine weight increases after a single injection.

Following *multiple* injections of UA, there is a prolonged maintenance of elevated levels of nuclear receptor and a depletion of cytoplasmic receptor, but there is no further increase in uterine wet weight above that elicited by a single injection. In contrast, with multiple injections of EE$_3$CPE, nuclear receptor levels continue to increase beyond the levels seen at 12 hr, and cytosol receptor levels are never fully depleted. Multiple injections of EE$_3$CPE, likewise, result in a continued rise in uterine weight to levels considerably above those evoked by a single injection. It should be noted that the distribution of receptor is very different after multiple injections of the two compounds; after antiestrogen (UA), almost all, over 90%, of receptor is in the nucleus, whereas in the case of the estrogen (EE$_3$CPE), 35 to 50% of the total receptor remains in the cytoplasm.

However, regardless of whether the uterus continues to grow (as with EE$_3$CPE) or stops growing after 24 to 48 hr (as with UA), the receptor content on a cell basis is similar in both cases. Figure 12 shows the uterine content of estrogen receptor expressed on the basis of tissue weight (*upper panels*) and uterine DNA content (*lower panels*) after UA or EE$_3$CPE treatment. Following exposure to UA or EE$_3$CPE, as the tissue increases in size, receptor content on a weight basis decreases to 50% of the control level by 24 hr and remains at about this level, whether the uterus continues to grow (to reach ca. fourfold the control weight, EE$_3$CPE-multiple) or stops growing at ca. twofold the control weight (UA-single or multiple injections). Likewise, when receptor content is expressed on the basis of uterine DNA content, UA and EE$_3$CPE again showed similar time course patterns.

Hence, uterine responsiveness to estrogen and continued uterine growth appear not to be related to the total receptor content of the cell, but rather

---

←

FIG. 11. Temporal effects of single or multiple injections of UA (*left panels*) or 17α-ethinyl-estriol-3 cyclopentyl ether (EE$_3$CPE) (*right panels*) on uterine weight and uterine content of specific estrogen binding sites in the nuclear and cytosol fractions. Immature rats received either (a) a single injection of UA (50 μg) or EE$_3$CPE (5 μg) at zero time (*solid line*) or (b) multiple injections of UA (50 μg every 24 hr) or EE$_3$CPE (5 μg every 12 hr) (dashed line) and were sacrificed at the indicated times. Uteri were weighed, and the content of specific estrogen binding sites present in the nuclear (*middle panel*) and cytosol (*lower panel*) fractions of uteri were determined by the nuclear and cytosol exchange assays. Multiple injections of UA (50 μg) every 12 hr showed receptor and weight patterns similar to that seen after 24 hr injections. Uterine weight values (*upper panel*) are the mean ± SEM with six uteri per point. For nuclear and cytosol receptor content, each value is the mean ± SEM of three determinations per point (three uteri/group) and is corrected for nonspecific binding. (From Katzenellenbogen *et al.*, ref. 17.)

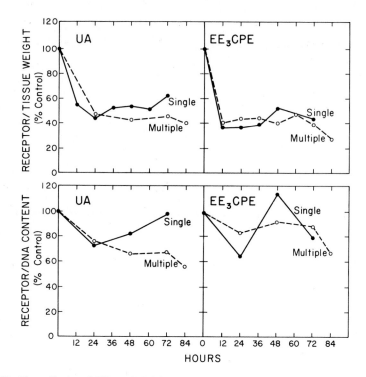

FIG. 12. The effects of UA and EE₃CPE on uterine estrogen receptor levels. Total estrogen receptor (from Fig. 11) is expressed on the basis of uterine tissue weight (from Fig. 11) or uterine DNA content as a function of time after single or multiple injections of UA or EE₃CPE. Treatments are as described in the legend of Fig. 11. Data is expressed as percent of the control (zero time, untreated = 100%) value. (From Katzenellenbogen et al., ref. 17.)

depend upon a proper distribution of receptor within the cell. These studies suggest that the antagonistic action of antiestrogens derives from their ability to effect a marked perturbation in the subcellular distribution of receptor, whereby very little, ca. 10%, of receptor is cytoplasmic and further estrogen receptor accumulation (most likely synthesis) is blocked.

All of the antiestrogen studies that have been discussed to this point have employed unlabeled compounds. With the hope of looking directly at the interaction of antiestrogens with receptor, we prepared one of the anti-estrogens, CI, in tritium-labeled form (22,23). In studies presented below, we have found that the interaction of tritiated CI with the estrogen receptor parallels that of estradiol in many respects.

Figure 13 shows high-affinity estrogen-specific binding of tritiated CI in uterine cytosol. Figure 13A shows CI binding in the presence and absence of unlabeled estradiol, and the difference between these curves, presumed to represent estrogen-specific binding sites, is plotted directly in Fig. 13B, and as a Scatchard plot in Fig. 13C. An equilibrium dissociation constant (K_D)

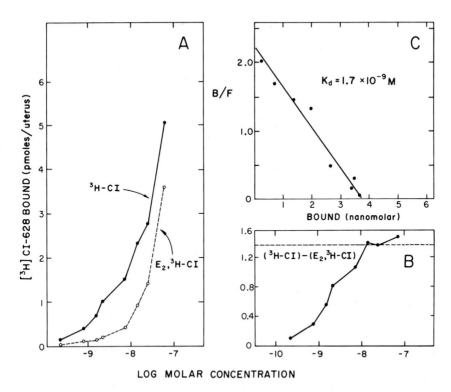

FIG. 13. Interaction of [³H]CI with uterine cytosol estrogen receptor *in vitro.* Cytosol, after exposure to 10^{-6} M unlabeled estradiol or vehicle for 1 hr at 0°C, was then incubated with varying concentrations of [³H]CI for 17 hr at 0°C. Cytosol concentration in the final incubations was 3.0 uterine equivalents/ml. The amount of [³H]CI bound was then determined by charcoal-dextran adsorption (15% v/v, 15 min at 0°C). In all panels, the concentrations of free CI have been corrected for the low-affinity binding. **A:** Binding curves for [³H]CI with and without estradiol pretreatment. **B:** Plot of the difference between the curves for [³H]CI alone and [³H]CI binding after estradiol pretreatment ("estrogen-specific" binding sites). Horizontal dashed line represents the concentration of estrogen-specific binding sites determined in parallel incubations employing [³H]estradiol. **C:** Replot of data of **B** according to the equation of Scatchard. (From Katzenellenbogen et al., ref. 23.)

of 1.7×10^{-9} M is found for the interaction of [³H]CI with the estrogen receptor. In the same sample, the K_D for estradiol was determined to be 1.0×10^{-10} M. Thus, the binding affinity of CI relative to that of estradiol is 6%. This estimate of binding affinity of CI, determined here directly using radiolabeled compound, corresponds closely to that estimated earlier by competitive protein binding methods, 4 to 8% (10,16,21,25). The number of binding sites is also equivalent to that seen with saturating levels of tritiated estradiol (*dashed horizontal line,* Fig. 13B).

Figure 14 shows sucrose density gradient analyses of tritiated CI–cytoplasmic receptor complexes and tritiated estradiol–receptor complexes on low salt gradients. Over 90% of the estradiol–receptor complexes sediment

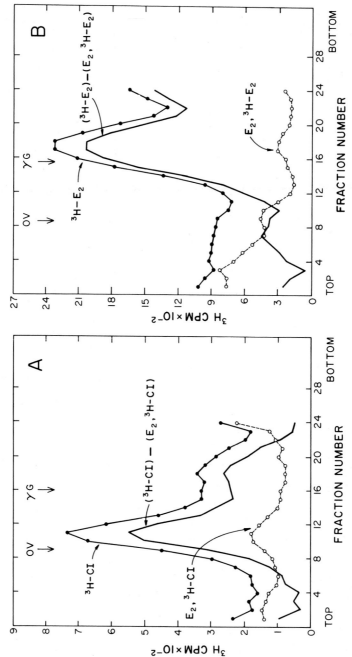

FIG. 14. Low salt sucrose density gradient profiles of cytoplasmic estrogen receptor complexes with [³H]CI or [³H]estradiol. Cytosol (2.14 uteri/ml) was incubated with 3×10^{-8} M [³H]CI (**A**) or 3×10^{-8} M [³H]estradiol (**B**) in the presence or absence of 10^{-6} M unlabeled estradiol. After 8 hr at 0°C, samples were treated with charcoal-dextran (15% v/v, 15 min at 0°C), and 300 μl aliquots were mixed with the ¹⁴C-labeled marker proteins and layered onto gradients. Centrifugation was for 13 hr at 4°C at 246,000 × *g*. (From Katzenellenbogen et al., ref. 23.)

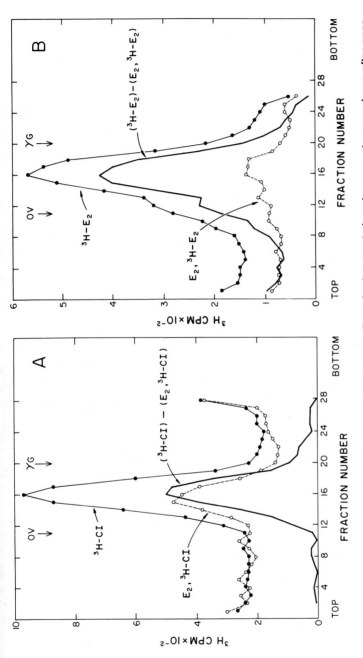

FIG. 15. High salt sucrose density gradient centrifugation profiles of salt-extracted nuclear receptor complexes after exposure to [³H]CI (**A**) or [³H]estradiol (**B**) in vivo. Groups of rats were pretreated for 1 hr in vivo with 5 μg unlabeled estradiol or with vehicle saline alone and then received a subcutaneous injection of 50 μg [³H]CI or 3 μg [³H]estradiol (containing 0.9 μg [³H]estradiol plus 2.1 μg estradiol). At 1 hr after injection, uteri were excised and the three-times washed 800 × g nuclear pellet was extracted with TE buffer containing 0.4 M KCl for 1 hr at 0°C. Extracts were treated with 10% charcoal-dextran prior to addition of ¹⁴C-labeled marker proteins, and 300 μl aliquots (containing 1.6 uterine equivalents, **A**, and 1.0 uterine equivalents, **B**) were layered onto gradients. Centrifugation was for 17 hr at 4°C at 270,000 × g (From Katzenellenbogen et al., ref. 23.)

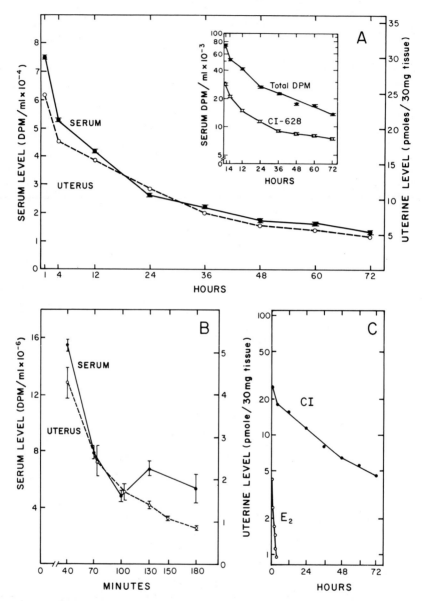

FIG. 16. Rates of clearance of [³H]CI and [³H]estradiol from uterus and serum. **A:** Immature rats (two or three/group) were injected at zero time with 48.6 μg CI/rat containing 0.7 μg (25 μCi) of [³H]CI, and serum and uterine radioactivity were determined at 1 to 72 hr after injection. For serum radioactivity, each point is the mean ± SEM of three determinations from the pooled serum samples (three rats/group). For uterine content, each point is the mean of two determinations. The inset is a semilogarithmic plot of total serum dpm and dpm that co-migrated with CI on thin-layer chromatography. **B:** Immature rats (two/group) were injected at zero time with 3 μg estradiol containing 0.5 μg [³H]estradiol (88 μCi), and serum and uterine radioactivity was determined. Error bars indicate the range of the determinations. **C:** Semilogarithmic plot comparing rates of clearance of CI and estradiol from uterus. Data is taken from **A** and **B** (*dashed lines*). (From Katzenellenbogen et al., ref. 23.)

at 8S, whereas 70% of the tritiated CI–receptor complexes (estrogen-competable sites) sediment at 4.5S, only 30% sedimenting at 8S. These differences in sedimentation behavior may be a manifestation of differences in estrogen and antiestrogen interaction with receptor.

After administration of tritiated antiestrogen *in vivo,* radioactivity can be found associated with specific estrogen receptor sites in uterine nuclei. Salt-extracted nuclear receptor–[³H]CI complexes sediment at 5.4S on high salt containing sucrose gradients, as do [³H]estradiol nuclear receptor complexes (Fig. 15). Thin-layer chromatographic analysis of ethyl acetate extracts of the nuclear receptor peak fractions from sucrose gradients shows that a polar metabolite of CI, and not CI itself, is selectively bound to the 5.4S receptor (23).

Pharmacokinetic studies (Fig. 16) reveal that high levels of CI persist in serum and uterus for long periods of time (half cleared in 18 to 24 hr), whereas estradiol is much more rapidly cleared (half cleared in 30 min).

These results indicate that in many respects [³H]CI and [³H]estradiol interact with the estrogen receptor in a parallel manner; however, certain properties of the antiestrogen receptor complex are different from those of the estradiol receptor complex. Further, it is likely that the prolonged *in vivo* activity of this antiestrogen derives, at least in part, from its slow rate of clearance and that the active agent *in vivo* may be a metabolite of CI.

It is clear that there is still much not yet understood about the basis of antiestrogen action. Hopefully, additional studies, including those presently underway (4,15,26,28,33,36,43,45), will elucidate in detail the nature of the receptor interactions that characterize the agonist/antagonist activities of these compounds in estrogen target tissues.

ACKNOWLEDGMENTS

I am grateful for the support of these studies by United States Public Health Service, National Institutes of Health Grants HD 06726 and CA 18119 and Ford Foundation Grant 700–0333. The contributions of students and collaborators to these studies are greatly appreciated. I also wish to acknowledge the excellent secretarial assistance of the Word Processing Center of the School of Basic Medical Sciences.

REFERENCES

1. Bhakoo, H. S., and Katzenellenbogen, B. S. (1977): Progesterone antagonism of estradiol-stimulated uterine "Induced Protein" synthesis. *Mol. Cell. Endocrinol.,* 8:105–120.
2. Bhakoo, H. S., and Katzenellenbogen, B. S. (1977): Progesterone modulation of estrogen-stimulated uterine biosynthetic events and estrogen receptor levels. *Mol. Cell. Endocrinol.,* 8:121–134.
3. Brenner, R. M., Resko, J. A., and West, N. B. (1974): Cyclic changes in oviductal

morphology and residual cytoplasmic estradiol binding capacity induced by sequential estradiol-progesterone treatment of spayed Rhesus monkeys. *Endocrinology,* 95:1094–1104.

4. Capony, F., and Rochefort, H. (1975): *In vivo* effect of antiestrogens on the localization and replenishment of estrogen receptor. *Mol. Cell. Endocrinol.,* 3:233–251.
5. Cidlowski, J. A., and Muldoon, T. G. (1974): Estrogenic regulation of cytoplasmic receptor populations in estrogen responsive tissues of the rat. *Endocrinology,* 95: 1621–1629.
6. Cidlowski, J. A., and Muldoon, T. G. (1975): Modulation by thyroid hormones of cytoplasmic estrogen receptor concentrations in reproductive tissues of the rat. *Endocrinology,* 97:59–67.
7. Clark, J. H., Anderson, J., and Peck, E. J., Jr. (1973): Estrogen receptor-antiestrogen complex: Atypical binding by uterine nuclei and effects on uterine growth. *Steroids,* 22:707–718.
8. Clark, J. H., and Gorski, J. (1970): Ontogeny of the estrogen receptor during early uterine development. *Science,* 169:76–78.
9. DeAngelo, A. B., and Gorski, J. (1970): Role of RNA synthesis in the estrogen-induction of a specific uterine protein. *Proc. Natl. Acad. Sci. USA,* 66:693–700.
10. Ferguson, E. R., and Katzenellenbogen, B. S. (1977): A comparative study of antiestrogen action: Temporal patterns of antagonism of estrogen stimulated uterine growth and effects on estrogen receptor levels. *Endocrinology,* 100:1242–1251.
11. Garcia, M., and Rochefort, H. (1977): Androgens on the estrogen receptor, II. Correlation between nuclear translocation and protein synthesis. *Steroids,* 29:111–126.
12. Gardner, R. M., Kirkland, J. L., Ireland, J. S., and Stancel, G. M. (1977): Effect of propylthiouracil treatment on the uterine response to estradiol. *Fed. Proc.,* 36:368 (Abstr. 526).
13. Hseuh, A. J. W., Peck, E. J., Jr., and Clark, J. H. (1975): Progesterone antagonism of oestrogen receptor and oestrogen induced uterine growth. *Nature,* 254:337–339.
14. Hseuh, A. J. W., Peck, E. J., Jr., and Clark, J. H. (1976): Control of uterine estrogen receptor levels by progesterone. *Endocrinology,* 98:438–444.
15. Jordan, V. C., Dix, C. J., Rowsby, L., and Prestwich, G. (1977): Studies on the mechanism of action of the nonsteroidal antioestrogen Tamoxifen (I.C.I. 46,474) in the rat. *Mol. Cell. Endocrinol.,* 7:177–192.
16. Katzenellenbogen, B. S., and Ferguson, E. R. (1975): Antiestrogen action in the uterus: Biological ineffectiveness of nuclear bound estradiol after antiestrogen. *Endocrinology,* 97:1–12.
17. Katzenellenbogen, B. S., Ferguson, E. R., and Lan, N. (1977): Fundamental differences in the action of estrogens and antiestrogens on the uterus: Comparison between compounds with similar duration of action. *Endocrinology,* 100:1252–1259.
18. Katzenellenbogen, B. S., and Gorski, J. (1972): Estrogen action *in vitro*. Induction of the synthesis of a specific uterine protein. *J. Biol. Chem.,* 247:1299–1305.
19. Katzenellenbogen, B. S., and Gorski, J. (1975): Estrogen actions on syntheses of macromolecules in target cells. In: *Biochemical Actions of Hormones, Vol. 3,* edited by G. Litwack, pp. 187–243. Academic Press, New York.
20. Katzenellenbogen, B. S., and Greger, N. G. (1974): Ontogeny of uterine responsiveness to estrogen during early development in the rat. *Mol. Cell. Endocrinol.,* 2:31–42.
21. Katzenellenbogen, B. S., and Katzenellenbogen, J. A. (1973): Antiestrogens: Studies using an *in vitro* estrogen-responsive uterine system. *Biochem. Biophys. Res. Commun.,* 50:1152–1159.
22. Katzenellenbogen, B. S., Katzenellenbogen, J. A., and Ferguson, E. R. (1977): Interaction of a radiolabeled antiestrogen (^{3}H-CI-628) with uterine tissue and molecular mechanism of its action. *Proceedings of the 59th Annual Endocrine Society Meeting* (Abstr. 196) p. 154. Lippincott.
23. Katzenellenbogen, B. S., Katzenellenbogen, J. A., Ferguson, E. R., and Krautham-

mer, N. (1978): Antiestrogen interaction with uterine estrogen receptors: Studies with a radiolabeled antiestrogen (CI-628). *J. Biol. Chem.*, 253:697–707.

24. Kaye, A. M., Sheratzky, D., and Lindner, H. R. (1972): Kinetics of DNA synthesis in immature rat uterus: Age dependence and estradiol stimulation. *Biochim. Biophys. Acta*, 261:475–486.

25. Korenman, S. G. (1970): Relation between estrogen inhibitory activity and binding to cytosol of rabbit and human uterus. *Endocrinology*, 87:1119–1123.

26. Koseki, Y., Zava, D. T., Chamness, G. C., and McGuire, W. L. (1977): Estrogen receptor translocation and replenishment by the antiestrogen Tamoxifen. *Endocrinology*, 101:1104–1110.

27. Lan, N. C., and Katzenellenbogen, B. S. (1976): Temporal relationships between hormone receptor binding and biological responses in the uterus: Studies with short- and long-acting derivatives of estriol. *Endocrinology*, 98:220–227.

28. Lippman, M., Bolan, G., and Huff, K. (1976): Interactions of antiestrogens with human breast cancer in long-term tissue culture. *Cancer Treat. Rep.*, 60:1421–1429.

29. Luck, D. N., Gschwendt, M., and Hamilton, T. H. (1973): Oestrogenic stimulation of macromolecular synthesis is correlated with nuclear hormone-receptor complex in neonatal rat uterus. *Nature [New Biol.]*, 245:24–25.

30. Mester, I., and Baulieu, E. E. (1975): Dynamics of oestrogen-receptor distribution between cytosol and nuclear fractions of immature rat uterus after oestradiol administration. *Biochem. J.*, 146:617–623.

31. Notides, A., and Gorski, J. (1966): Estrogen-induced synthesis of a specific uterine protein. *Proc. Natl. Acad. Sci. USA*, 56:230–235.

32. Pavlik, E. J., and Coulson, P. B. (1976): Modulation of estrogen receptors in four different target tissues: Differential effects of estrogen and progesterone. *J. Steroid Biochem.*, 7:369–376.

33. Rochefort, H., and Capony, F. (1977): Estradiol dependent decrease in binding inhibition by antiestrogens (a possible test of receptor activation). *Biochem. Biophys. Res. Commun.*, 75:277–285.

34. Rochefort, H., and Garcia, M. (1976): Androgen on the estrogen receptor, I. Binding and *in vivo* nuclear translocation. *Steroids*, 28:549–560.

35. Rochefort, H., Vignon, F., and Capony, F. (1972): Formation of estrogen nuclear receptor in uterus: Effect of androgens, estrone and nafoxidine. *Biochem. Biophys. Res. Commun.*, 47:662–670.

36. Ruh, T. S., and Baudendistel, L. J. (1977): Different nuclear binding sites for antiestrogen and estrogen receptor complexes. *Endocrinology*, 100:420–426.

37. Ruh, T. S., Katzenellenbogen, B. S., Katzenellenbogen, J. A., and Gorski, J. (1973): Estrone interaction with the rat uterus: *In vitro* response and nuclear uptake. *Endocrinology*, 92:125–134.

38. Ruh, T. S., and Ruh, M. F. (1975): Androgen induction of a specific uterine protein. *Endocrinology*, 97:1144–1150.

39. Sarff, M., and Gorski, J. (1971): Control of estrogen binding protein concentration under basal conditions and after estrogen administration. *Biochemistry*, 10:2557–2563.

40. Schmidt, W. M., Sadler, M. A., and Katzenellenbogen, B. S. (1976): Androgen-uterine interaction: Nuclear translocation of the estrogen receptor and induction of the synthesis of the uterine-induced protein (IP) by high concentrations of androgens *in vitro* but not *in vivo*. *Endocrinology*, 98:702–716.

41. Somjen, D., Kaye, A. M., and Lindner, H. R. (1973): Postnatal development of uterine response to estradiol-17β in the rat. *Dev. Biol.*, 31:409–412.

42. Somjen, D., Somjen, G., King, R. J. B., Kaye, A. M., and Lindner, H. R. (1973): Nuclear binding of oestradiol-17β and induction of protein synthesis in the rat uterus during postnatal development. *Biochem. J.*, 136:25–33.

43. Tsai, T. L., and Katzenellenbogen, B. S. (1977): Antagonism of development and growth of 7,12-dimethylbenz(a)anthracene-induced rat mammary tumors by the antiestrogen U-23,469 and effects on estrogen and progesterone receptors. *Cancer Res.*, 37:1537–1543.

44. West, N. B., Verhage, H. C., and Brenner, R. M. (1976): Suppression of the estradiol receptor system by progesterone in the oviduct and uterus of the cat. *Endocrinology,* 99:1010–1016.
45. Zava, D. T., Chamness, G. C., Horowitz, K. B., and McGuire, W. L. (1977): Human breast cancer: Biologically active estrogen receptor in the absence of estrogen? *Science,* 196:663–664.

Ontogeny of Receptors and Reproductive Hormone Action,
edited by T. H. Hamilton, J. H. Clark, and W. A. Sadler.
Raven Press, New York 1979.

Estrogenic Stimulation of Biochemical and Morphological Differentiation in Rat Uterine Nuclei

James W. Hardin, Barry M. Markaverich, James H. Clark,
and *Helen A. Padykula

*Department of Cell Biology, Baylor College of Medicine, Houston, Texas 77030; and
Laboratory of Electron Microscopy, Wellesley College, Wellesley, Massachusetts 02181

On entering the cytoplasm of target cells, estrogens bind to the cytoplasmic form of the estrogen receptor. The formation of the receptor–hormone complex results in translocation of the complex to the nucleus. The nuclear receptor–hormone complex can bind to either specific high-affinity acceptor sites or low-affinity nonspecific nonacceptor sites on chromatin. Specific binding to the nuclear acceptor sites is thought to initiate the biochemical events that lead to the ultimate expression of estrogen action in target cells (16,19,27).

In previous studies from this laboratory, we have shown that the receptor–hormone complex must be retained in the nucleus of target organs for a critical period of time for ultimate expression of the hormone's effects. To stimulate true uterine growth in the immature rat, the receptor–hormone complex must be retained in the nucleus for a period of at least 6 hr. This long-term retention is presumed to result from receptor–hormone interactions between the receptor–hormone complex and the nuclear acceptor sites. Treatment with short-acting estrogens such as estriol (E_3) do not result in the long-term retention of receptor–hormone complexes when administered as a single injection; consequently, they do not produce true uterine growth. In contrast, treatment with long-acting estrogens such as estradiol (E_2), diethylstilbesterol, or nafoxidine (N) do result in the long-term nuclear retention necessary to stimulate true uterine growth (2–9).

We have investigated several biochemical parameters of estrogen action in isolated nuclei of the immature rat uterus as well as conducted morphological analysis of nuclear and cytoplasmic differentiation of the luminal epithelium.

REGULATION OF TRANSCRIPTIONAL EVENTS

RNA Polymerase Activities and Receptor–Hormone Complex Levels in Isolated Nuclei Following a Single Injection of Hormone

Immature female rats were given a single injection of either 1.0 μg E_2, 1.0 μg E_3, 50 μg N, or a vehicle control. At various times following hormone

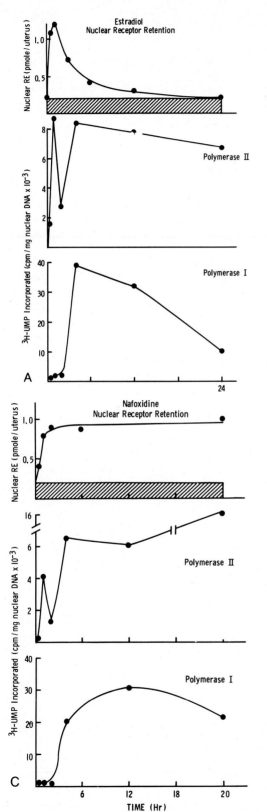
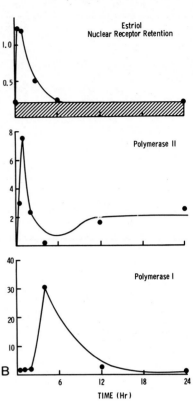

FIG. 1. Relationship of nuclear receptor retention and RNA polymerase activities in isolated uterine nuclei following hormone injections. Immature female rats were injected with either E_2 **(A)**, E_3 **(B)**, or N. **(C)**. At indicated times, animals were sacrificed, uteri were removed, and nuclei isolated by the hexylene glycol procedure (18). The levels of nuclear bound estrogen receptor were measured by the ^3H-estradiol exchange assay (1), and RNA polymerase I and II activities measured as previously described (18).

administration, animals were sacrificed, uteri removed, and nuclei isolated by the hexylene glycol procedure (18). The quantities of receptor–hormone complexes were determined by the ^{3}H-estradiol exchange assay (1) and RNA polymerase activities by differential α-amanitin sensitivity, as previously described (18). The results of these studies are shown in Fig. 1.

All three compounds cause early stimulation (0.5 hr postinjection) of RNA polymerase II activity, followed by an increase in RNA polymerase I activity. However, only E_2 and N cause a secondary rise in RNA polymerase II activity as well as a prolonged stimulation of RNA polymerase I activity. These two compounds also both cause long-term nuclear retention of the estrogen receptor as well as true uterine growth (7). Although E_3 stimulated early increases in polymerase activities, this hormone failed to cause the secondary rise in RNA polymerase II or prolonged stimulation of RNA polymerase I associated with either E_2 or N treatment. This failure of E_3 to stimulate RNA polymerase activities at long times after hormone administration was also reflected by the lack of long-term nuclear retention of the receptor–hormone complex (Fig. 1).

These studies indicate that the prolonged stimulation of RNA polymerase I and a secondary rise in RNA polymerase II activity are necessary for an estrogen to stimulate true uterine growth in the immature rat. The ability of an estrogen to stimulate these activities is correlated with the long-term nuclear retention of the receptor–hormone complex. Although previous studies have suggested that E_2 stimulation of RNA polymerase II activity and subsequent increases in messenger RNA and protein synthesis ultimately lead to the expression of uterine growth (15,22,24), the present studies indicate the biochemical processes involved in estrogen stimulation of uterine growth are more complex. As our studies indicate, it is not a single elevation in polymerase II activity that is associated with the estrogenic properties of the hormone. Rather, estrogen stimulation of true uterine growth results from the ability of the hormone to stimulate a secondary rise in polymerase II and a sustained elevation in polymerase I activity. Hence, in this regard E_2 and N are long-acting estrogens and induce true uterine growth. Conversely, failure of E_3 to stimulate the necessary elevations in polymerase I and II activities prerequisite for true uterine growth correlates with the short-term retention of the E_3–receptor complex in the nucleus. These studies support our earlier findings that to stimulate true uterine growth the receptor–estrogen complex must be retained in the nucleus for at least 6 hr. This retention in the nucleus occurs presumably at acceptor sites. This association appears to result in stimulation of the biochemical processes necessary for true uterine growth.

RNA Polymerase Activities in Hormone-Implanted Animals

The correlation between RNA polymerase activity and nuclear retention of estrogen–receptor complexes was further investigated in studies in which

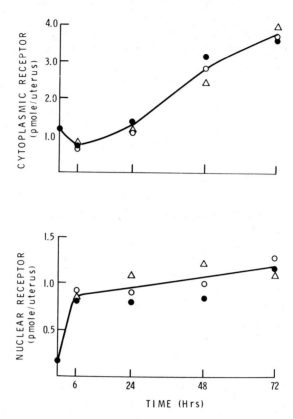

FIG. 2. Nuclear retention of receptor–hormone complexes in implanted animals. Immature female rats were placed under light ether anesthesia and implanted subcutaneously in the nape of the neck with estrogen-containing paraffin pellets prepared by the method of Rudali et al. (29) and as previously described (10). At indicated times, animals were sacrificed and uterine cytoplasmic and nuclear estrogen receptors were measured by the ³H-estradiol exchange assay (1). Animals were implanted with either E_2 (●), E_3 (○), or a combination of the two hormones (△).

E_2 and E_3 were administered to immature female rats as long-acting paraffin hormone implants (29). Administration of E_2 or E_3 in this manner resulted in the long-term nuclear retention of the receptor–hormone complex (10). As receptor is recycled from the nucleus to the cytoplasm, it apparently continually rebinds hormone and recycles to the nucleus. The retention pattern of receptor–hormone complexes following E_2 or E_3 treatment is illustrated in Fig. 2.

We also studied the ability of hormones administered in this manner to stimulate RNA polymerase activities in isolated nuclei. When administered by this route, E_3 was as effective as E_2 and N in stimulating RNA polymerase activities at all times examined (Fig. 3), further supporting the contention that the nuclear retention of the receptor–hormone complex is required to stimulate the biochemical events necessary for true uterine growth. Hence,

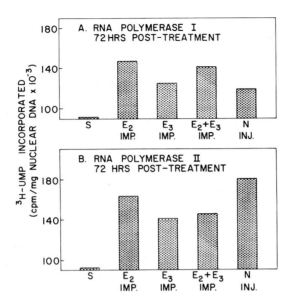

FIG. 3. Long-term stimulation of uterine nuclear RNA polymerase activities in estrogen-implanted rats. Immature female rats were implanted with paraffin pellets (S) or paraffin pellets that contained either E_2 (E_2-IMP), E_3 (E_3-IMP), a combination of the two hormones ($E_2 + E_3$ IMP), or were injected with 50 μg of N (N-inj). Seventy-two hours after treatment, animals were sacrificed, and uteri were removed. Nuclei were isolated as described in Fig. 1. Endogenous nuclear RNA polymerase activities were measured essentially as described in Fig. 1 except for the following modifications. The volume of the reactions were reduced to 50 μl though the final concentrations of all components remained the same. The incorporation of ^{3}H-uridine 5'-monophosphate (^{3}H-UMP) was measured by pipetting a 25-μl aliquot of the reaction mixture on to 2.5 cm DE-81 filter discs and washing and counting as described by Roeder (28).

when nuclear receptor E_3 complexes are continually available for nuclear binding, E_3 is as fully capable of stimulating true uterine growth as E_2.

Changes in Specific Chromatin Initiation Sites Following Administration of Estrogens

The stimulation of RNA synthesis by estrogens in the immature rat uterus appears to be a complicated process. There are numerous components of the transcriptional machinery that might be modified by estrogen administration. It is known that treatment of either the immature or ovariectomized rat with estrogen results in accumulation of RNA (17) reflected by increased endogenous RNA polymerase activities (15,18,33). These increases in RNA polymerase activities do not appear to be due to increase in the number of RNA polymerase molecules during the first 6 hr following hormone administration (13,34). Rather these increases probably reflect alterations in the chromatin template (15,17) or increases in some unknown "factors" that modulate RNA polymerase activity. In the following studies, we have investigated the former possibility.

We used techniques that have recently been adapted from procaryotic systems to allow the quantitation of specific initiation sites for RNA synthesis by bacterial RNA polymerase on eucaryotic chromatin templates (31). These techniques take advantage of the ability of the antibiotic rifampicin to specifically inhibit bacterial RNA polymerase not bound in a highly stable preinitiation complex. These techniques have been previously described in detail (25,31).

Immature female rats were injected with either E_2, E_3, or N as described above. At various times after hormone administration, animals were sacrificed, uteri were removed, and chromatin isolated as described by Spelsberg et al. (30) except the high-salt wash step was eliminated. The final chromatin pellet was suspended in $0.01 \times$ SSC at a final DNA concentration of 0.3 to 0.5 mg/ml. Aliquots of chromatin were preincubated with increasing concentrations of *Escherichia coli* RNA polymerase at 37°C for 30 min. At the end of this period a mixture of nucleotide triphosphates, heparin, and rifampicin was added. The reaction mixtures were incubated for an additional 15 min at 37°C. Reactions were terminated by removing aliquots and spotting them onto DE-81 filters that were washed and counted (28).

Figure 4 illustrates typical data from this type of an assay plotted as a

CHROMATIN

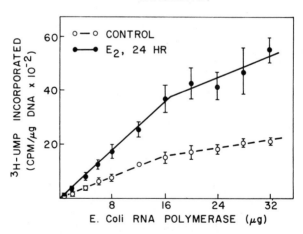

FIG. 4. Titration of rat uterine chromatin with increasing concentrations of *Escherichia coli* RNA polymerase in the presence of rifampicin and heparin. Immature rats were injected with either saline or 1.0 μg of E_2. Twenty-four hours after treatment, animals were sacrificed, uteri removed, and chromatin isolated as described in the text. RNA synthesis in the presence of rifampicin and heparin was carried out essentially as described by Tsai et al. (31). The main modifications were that 1.5 μg uterine chromatin DNA was used as a template and ^{3}H-UMP incorporation was measured by the DE-81 filter assay as described in the legend to Fig. 3. Further details are given in ref. 25. (●), E_2; (○), control.

titration curve. The amount of RNA synthesized increases linearly with increasing enzyme concentration until a transition point is reached. The coordinates of this transition point can be used to calculate the number of high-affinity binding sites for RNA polymerase as well as the number of RNA chains initiated (31). The data in Fig. 4 demonstrate that treatment of immature rats with E_2 24 hr prior to sacrifice significantly increases the number of rifampicin-resistant RNA chains initiated on uterine chromatin as reflected by the large increases in ^{3}H-uridine 5'-monophosphate (^{3}H-UMP) incorporated.

This increase in ^{3}H-UMP incorporated into RNA can be the result of either increased numbers of RNA chains initiated or an increased length of the RNA chains synthesized. To determine whether or not the increased ^{3}H-UMP incorporation in response to estrogen treatment was due to an overall increase in the length of RNA chains, the newly synthesized RNA was sized on sucrose density gradients. Chromatin was isolated from either control or E_2-treated animals and used as a template for RNA synthesis in the presence of rifampicin and heparin. The RNA synthesized under these conditions was isolated by phenol extraction followed by treatment with proteinase K. The size distribution of this isolated RNA, as determined by sucrose density gradient centrifugation, is presented in Fig. 5. There was no significant difference in the size of the RNA product synthesized using either

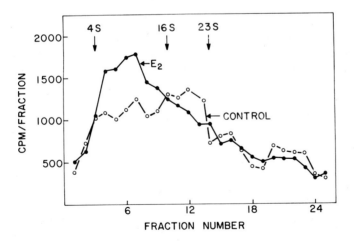

FIG. 5. Sucrose gradient analysis of RNA synthesized on chromatin. Uterine chromatin was isolated from animals injected either with E_2 or with saline 24 hr prior to sacrifice. Five micrograms of chromatin DNA was preincubated with 100 μg of *E. coli* RNA polymerase for 15 min at 37°C. Nucleoside triphosphates, rifampicin, and heparin were added and RNA synthesis allowed to continue for 15 min. RNA was isolated and sized on linear 5–20% sucrose gradients as described by Tsai et al. (31). The chain length of RNA in each fraction was calculated. This value was then used to determine the number average chain length of the RNA. Number average chain lengths for RNA synthesized using E_2 chromatin as template was 390 nucleotides, control chromatin was 420 nucleotides. (●), E_2; (○), control.

control or E_2-treated chromatin as a template. The RNA product synthesized fell into a size range of 390 to 470 nucleotides. Therefore, the increase in RNA synthesis in response to estrogen was the result of increased numbers of specific RNA chain initiation sites rather than an increase in the length of the RNA synthesized.

When DNA was isolated from control and E_2-treated chromatin and the isolated DNA used as a template in the rifampicin challenge assay, the results in Fig. 6 were obtained. There was no apparent difference in the number of initiation sites between treated and untreated samples when DNA was used as a template. It should also be noted that DNA was a much more efficient template than was the chromatin from which it was extracted (compare Fig. 4 with Fig. 6). These data indicate that the structure of chromatin plays an important role in the regulation of RNA synthesis by steroid hormones in target cells (27).

Figure 7 summarizes the effects of a single injection of either E_2, E_3, or N on the numbers of chromatin initiation sites. Chromatin was isolated at each of the indicated time points and the number of initiation sites determined. The most important point from these data is that the numbers of uterine chromatin RNA initiation sites increased and declined temporally with the accumulation-retention and depletion patterns of receptor–hormone complexes. Therefore, it appears that estrogen regulation of chromatin initi-

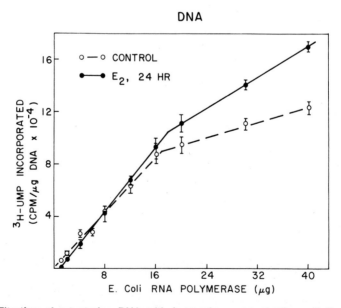

FIG. 6. Titration of rat uterine DNA with increasing concentrations of *E. coli* RNA polymerase in the presence of rifampicin and heparin. DNA was isolated from chromatin as described in the text and in ref. 25. Assays were run as described in Fig. 4 except 0.5 μg of DNA was used as a template. (●), E_2; (○), control.

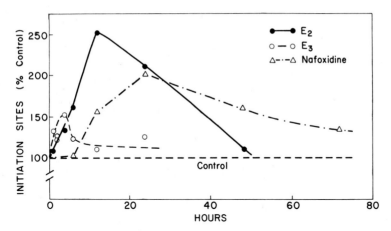

FIG. 7. Time course of chromatin RNA initiation sites following hormone injections. Immature rats were injected with either saline, E_2 (●-●), E_3 (○--○), or N (△-•-△). At the indicated times, animals were sacrificed and uterine chromatin prepared. RNA initiation sites were determined as described in Fig. 4. Control levels represent 42,000 initiation sites/pg DNA.

ation sites following hormone injection was dependent on the retention of receptor–hormone complexes in the nucleus.

Effects of Estrogen Treatment on Uterine Morphology

Recently we began an investigation of uterine morphology following the injection of various estrogenic compounds. This work has been stimulated by observations in our laboratory that injecting neonatal rats with clomiphene (Clomid®) causes multiple reproductive tract abnormalities in the post-pubertal animal (9).

In our morphological studies, immature rats were injected with either vehicle, 1.0 μg E_2, 50 μg N, or implanted with E_2. Animals were sacrificed at 24-hr intervals, uteri removed, and samples prepared for light and electron microscopy by double fixation procedures (12,20,21). The remaining uteri were used for isolation of nuclei by the hexylene glycol procedure. These nuclei were used for the determination of nuclear receptor levels by the ^{3}H-estradiol exchange assay and endogenous RNA polymerase activities as described above. The E_2-implanted and N-treated animals showed sustained elevation in RNA polymerase activities that paralleled the level of receptor–hormone complexes retained in the nucleus (Fig. 3).

The morphology of the uterine luminal epithelial cells was examined by both light and electron microscopy. The height of the luminal epithelial cells was greatly increased over that of control cells by either N or E_2-implant treatments (Fig. 8). This hypertrophy is associated biochemically with long-term nuclear retention of the estrogen receptor and the prolonged stimulation

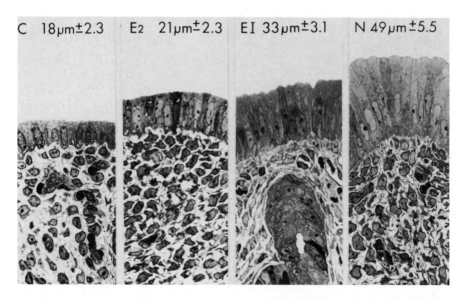

C 18μm±2.3   E2 21μm±2.3   EI 33μm±3.1   N 49μm±5.5

FIG. 8. Immature rat uterine luminal epithelium and stroma 72 hr after treatment with E₂ or N. Immature rats were treated with either saline (C), a single injection of 1.0 μg of E₂, an implant of paraffin–E₂ pellets (E₂I), or a single injection of 50 μg of N. The height of the luminal epithelium was measured in semithin (0.75 μm) Epon sections stained with 0.25% toluidine blue. Epithelial height 72 hr after a single injection of E₂ (21 μm ± 2.3) is comparable to that of the control (18 μm ± 2.3) whereas that after E₂ implant (33 μm ± 3.1) or a single injection of N (49 μm ± 5.5) is distinctly hypertrophied. ×425.

of nuclear RNA polymerase activities. This very active state of uterine cells appears to be one manifestation of the hyperestrogenized uterus. The height of luminal epithelial cells 72 hr after a single injection of E_2 was not significantly different from that of the control tissue, probably representing the regressed state following hyperplasia and hypertrophy that occurs 24 to 48 hr after a single injection of E_2.

At the electron microscopic level, the hypertrophied luminal epithelial cells resulting from a single injection of N or an E_2-implant possess an abundance of free polysomes and rough cisternal endoplasmic reticulum as well as greatly enlarged Golgi systems. These features are all indicative of stimulated protein synthesis (*unpublished data.*)

Hyperestrogenization Induces Nuclear Bodies in Luminal Epithelium

When the ultrastructure of nuclei of luminal epithelial cells following the above treatments was examined, we observed several types of nuclear bodies in uteri from either N-treated or E_2-implanted animals (Fig. 9). Nuclear bodies were quantitated in cells sectioned in the longitudinal plane. Electron microscopic counts were made on ultrathin sections of nuclei referred to as

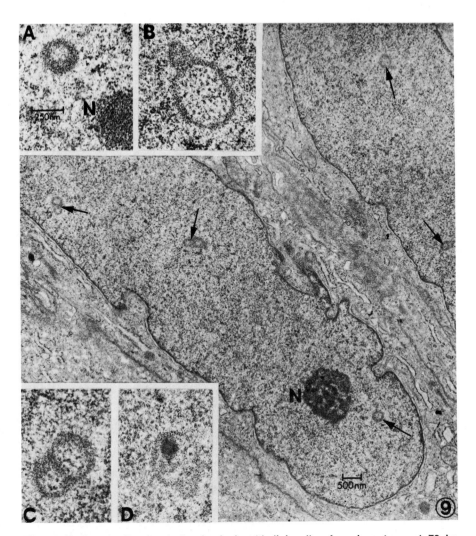

FIG. 9. Nuclear bodies in uterine luminal epithelial cells of an immature rat 72 hr after a single injection of N. In this ultrathin section through two epithelial nuclei (i.e., nuclear "profiles"), there are three nuclear bodies (*arrows*) in the left nucleus and two in the right nucleus. These nuclear bodies are singlets or doublets (as seen at higher magnification in **A** and **C**) that are composed of a filamentous capsule and less dense core. Larger bodies with granular electron opaque inclusions in the core also occur (**B** and **D**) in these cells and have been referred to as granular nuclear bodies (6). Scale bar, 500nm. N, nucleolus. ×11,900. Insets: scale bar, 250 nm; ×42,500.

nuclear "profiles" (Table 1). The number of nuclei that contained nuclear bodies was similar in control and E_2-injected animals (22 and 19%, respectively). The value observed in E_2-injected animals probably reflects the regressed state of the epithelium 72 hr after injection. The percent of nuclei that contained nuclear bodies was significantly elevated in animals which

TABLE 1. *Quantitative morphologic changes in the uterine luminal epithelium*
(immature rat)

	Control	Estradiol single inj.	Estradiol implant	Nafoxidine single inj.
Epithelial height (μm)	18 ± 2.3	21 ± 2.3	33 ± 3.1	49 ± 5.5
% Nuclear profiles				
With nuclear bodies	22 (n = 188)	19 (n = 88)	32 (n = 65)	53 (n = 129)
With multiple nuclear bodies	5 (n = 188)	0 (n = 88)	28 (n = 65)	38 (n = 129)

had been implanted with E_2 (37%) or injected with N (53%). In addition, 38% of the profiles in N-treated animals contained multiple nuclear bodies, as compared to 28% of the positive profiles of E_2-implanted animals and only 5% of the control profiles. No multiple nuclear bodies were seen 72 hr after a single E_2 injection.

Four distinct morphological classes of nuclear bodies were observed during the course of this study (Fig. 9). These included singlets with a filamentous capsule and less dense core (size range 280 to 305 nm), nuclear bodies that contained granules in the core (size range 700 to 1,000 nm), doublets with double cores, and granular bodies with large electron opaque inclusions. The latter three classes were only seen in nuclei of luminal epithelial cells from the hormone-treated animals. Thus, E_2 implants or N injection cause an increase in the number of nuclei that contain nuclear bodies and in the number of nuclear bodies/nucleus and stimulate the appearance of qualitatively different forms of nuclear bodies.

These nuclear bodies are associated with uterine hyperestrogenization. Similar nuclear bodies have been described by Bouteille and co-workers (6,14) and appear to consist of a filamentous protein capsule and a core of either RNA or DNA as determined by cytochemical techniques. Nuclear bodies have been identified in many neoplastic cells (6,14,23). In addition, nuclear inclusions of various types have been seen in the calf adrenal zone fasciculata following adrenocorticotropic hormone treatment (32), and a nucleolar channel system has been described in the endometrial gland cells of the human uterus (26).

The number and complexity of nuclear bodies is increased in hyperestrogenized animals and appear to be present only in luminal epithelial cells. This latter observation appears to relate to recent observations from this laboratory that N preferentially stimulates growth of luminal epithelial cells while having little effect on growth of myometrium and stromal cells (11).

The specific function and composition of nuclear bodies in hyperestrogenized uteri is not known. They appear to be one visible manifestation of hyperestrogenization, a state that appears to be dependent on the long-term nuclear retention of the estrogen receptor as well as prolonged transcriptional activity. Continued study of nuclear bodies is important particularly in light of our recent observations that a single injection of clomiphene or N to neonatal rats results in a number of reproductive tract abnormalities in the adult animal (9). These reproductive tract abnormalities appear to be the result of a hyperestrogenized state during the neonatal period. Since many neoplastic cells contain nuclear bodies that are induced by compounds that induce reproductive tract abnormalities, the continued study of nuclear bodies as a possible precursor state to neoplasia should be of interest.

The interactions of estrogen–receptor hormone complexes with target cell nuclei appear to be very complicated. These interactions appear to be involved in stimulating nuclear transcriptional events. In the hyperestrogenized uterus, these interactions appear to result in the induction of nuclear bodies that in turn may be either involved in the observed increases in transcription or a product of the increased transcriptional activity. The further elucidation of the mechanisms involved when the estrogen receptor–hormone complex interacts with target cell nuclei should be informative in understanding how hormones influence tissue growth.

ACKNOWLEDGMENTS

This work was supported by grants from the American Cancer Society BC-92 and from the National Institutes of Health HD-09209, CA-20605, and HD-08436.

REFERENCES

1. Anderson, J. N., Clark, J. H., and Peck, E. J., Jr. (1972): Oestrogen and nuclear binding sites, determination of specific sites by [³H]-oestradiol exchange. *Biochem. J.,* 126:561–567.
2. Anderson, J. N., Clark, J. H., and Peck, E. J., Jr. (1972): The relationship between nuclear receptor estrogen binding and uterotrophic responses. *Biochem. Biophys. Res. Commun.,* 48:1460–1468.
3. Anderson, J. N., Peck, E. J., Jr., and Clark, J. H. (1973): Nuclear receptor-estrogen complex: Relationship between concentration and early uterotrophic responses. *Endocrinology,* 92:1488–1495.
4. Anderson, J. N., Peck, E. J., Jr., and Clark, J. H. (1974): Nuclear receptor estradiol complex: A requirement for uterotrophic responses. *Endocrinology,* 95:174–178.
5. Anderson, J. N., Peck, E. J., Jr., and Clark, J. H. (1975): Estrogen induced uterine responses and growth: Relationship to receptor estrogen binding by uterine nuclei. *Endocrinology,* 96:160–167.
6. Bouteille, M., Kalifat, S. R., and Delarue, J. (1967): Ultrastructural variations of nuclear bodies in human diseases. *J. Ultrastruct. Res.,* 19:474–486.
7. Clark, J. H., Anderson, J. N., and Peck, E. J., Jr. (1973): Estrogen receptor antiestrogen complex: A typical binding by uterine nuclei and effects on uterine growth. *Steroids,* 22:707–718.
8. Clark, J. H., and Peck, E. J., Jr. (1976): Nuclear retention of receptor-oestrogen complex and nuclear acceptor site. *Nature,* 260:635–637.

9. Clark, J. H., and McCormack, S. A. (1977): Clomid or Nafoxidine administered to neonatal rats causes reproductive tract abnormalities. *Science,* 197:164–165.
10. Clark, J. H., Paszko, Z., and Peck, E. J., Jr. (1977): Nuclear binding and retention of the receptor estrogen complex: Relation to the agonistic and antagonistic properties of estriol. *Endocrinology,* 100:91–96.
11. Clark, J. H., Conn, P. M., and Kalimi, M. (1978): Cell specific agonistic and antagonistic properties of Nafoxidine in the chick oviduct and rat uterus. (*Submitted for publication.*)
12. Clark, J. H., Hardin, J. W., Padykula, H. A., and Cardasis, C. A. (1978): Role of estrogen binding and transcriptional activity in the stimulation of hyperestrogenism and nuclear bodies. (*Submitted for publication.*)
13. Courvalin, J. C., Bouton, M. M., Baulieu, E. E., Nuret, P., and Chambon, P. (1976): Effect of estradiol on rat uterus DNA dependent RNA polymerases: Studies on solubilized enzymes. *J. Biol. Chem.,* 251:4843–4849.
14. Dupuy-Coin, A. M., Kalifat, S. R., and Bouteille, M. (1972): Nuclear bodies as proteinaceous structures containing ribonucleoproteins. *J. Ultrastruct. Res.,* 38:174–187.
15. Glasser, S. R., Chytil, F., and Spelsberg, T. C. (1972): Early effects of oestradiol-17β on the chromatin and activity of the deoxyribonucleic acid-dependent ribonucleic acid polymerase (I or II) of the rat uterus. *Biochem. J.,* 130:947–957.
16. Gorski, J., and Gannon, F. (1976): Current models of steroid hormone action: A critique. *Annu. Rev. Physiol.,* 38:425–450.
17. Hamilton, T. H. (1968): Control by estrogen of genetic transcription and translation. *Science,* 161:649–661.
18. Hardin, J. W., Clark, J. H., Glasser, S. R., and Peck, E. J., Jr. (1976): RNA polymerase activity and uterine growth: Differential stimulation by estradiol, estriol, and Nafoxidine. *Biochemistry,* 15:1370–1374.
19. Jensen, E. V., and DeSombre, E. R. (1972): Mechanism of action of the female sex hormones. *Annu. Rev. Biochem.,* 41:203–230.
20. Kang, Y. H., Anderson, W. A., and DeSombre, E. R. (1975): Modulation of uterine morphology and growth by estradiol-17β and an estrogen antagonist. *J. Cell Biol.,* 64:682–691.
21. Karnovsky, M. J. (1965): A formaldehyde-glutaraldehyde fixative of high osmolarity for use in electron microscopy. *J. Cell Biol.,* 27:137–138A.
22. Knowler, J. T., and Smellie, R. M. S. (1973): The oestrogen stimulated synthesis of heterogenous nuclear ribonucleic acid in the uterus of immature rats. *Biochem. J.,* 131:689–697.
23. Krishan, A., Uzman, B. G., and Hedley-White, E. T. (1967): Nuclear bodies: A component of cell nuclei in hamster tissues and human tumors. *J. Ultrastruct. Res.,* 19:563–572.
24. Luck, D. N., and Hamilton, T. H. (1972): Early estrogen action: Stimulation of the metabolism of high molecular weight and ribosomal RNA's. *Proc. Natl. Acad. Sci. USA,* 69:157–161.
25. Markaverich, B. M., Clark, J. H., and Hardin, J. W. (1978): RNA transcription and uterine growth: Differential effects of estradiol, estriol, and nafoxidine on chromatin RNA initiation sites. (*Submitted for publication.*)
26. More, I. A. R., Armstrong, E. M., McSeveney, D., and Chatfield, W. R. (1974): The morphogenesis and fate of the nucleolar channel system in the human endometrial glandular cell. *J. Ultrastruct. Res.,* 47:74–85.
27. O'Malley, B. W., and Means, A. R. (1974): Female steroid hormones and target cell nuclei. *Science,* 183:610–620.
28. Roeder, R. G. (1974): Multiple forms of DNA-dependent RNA polymerase in *Xenopus laevis:* Isolation and partial characterization. *J. Biol. Chem.,* 249:241–248.
29. Rudali, G., Apiou, F., and Mael, B. (1975): Mammary cancer produced in mice with estriol. *Eur. J. Cancer,* 11:39–41.
30. Spelsberg, T. C., Hnilica, L. S., and Ansevin, A. T. (1971): Proteins of chromatin in template restriction III. The macromolecules in specific restriction of the chromatin DNA. *Biochem. Biophys. Acta,* 228:550–562.

31. Tsai, M. J., Schwartz, R. J., Tsai, S. Y., and O'Malley, B. W. (1975): Effects of estrogen on gene expression in the chick oviduct IV. Initiation of RNA synthesis on DNA and chromatin. *J. Biol. Chem.*, 250:5165–5174.

32. Weber, A., Whipp, S., Usenik, E., and Frommes, S. (1964): Structural changes in the nuclear body in the adrenal zone Fasciculata of the calf following the administration of ACTH. *J. Ultrastruct. Res.*, 11:564–576.

33. Webster, R. A., and Hamilton, T. H. (1976): Comparative effects of estradiol 17β and estriol on uterine RNA polymerases I, II, and III *in vivo. Biochem. Biophys. Res. Commun.*, 69:737–743.

34. Weil, P. A., Sidikaro, J., Stancel, G. M., and Blatti, S. P. (1977): Hormonal control of transcription in the rat uterus. Stimulation of DNA-dependent RNA polymerase III by estradiol. *J. Biol. Chem.*, 252:1092–1098.

Ontogeny of Receptors and Reproductive Hormone Action,
edited by T. H. Hamilton, J. H. Clark, and W. A. Sadler.
Raven Press, New York 1979.

Hormonal Control of Uterine Growth and Regulation of Responsiveness to Estrogen

[a]G. M. Stancel, [c]S. P. Blatti, [a,c]R. H. Benson, [a]*R. M. Gardner,
[a,b]J. L. Kirkland, [a]J. S. Ireland, [a]**J. Sidikaro, and [c]†P. A. Weil

*Departments of [a]Pharmacology, [b]Pediatrics, [c]Biochemistry and Molecular Biology,
The University of Texas Medical School at Houston, Houston, Texas 77025*

Numerous studies in recent years have elucidated a basic outline of the mechanisms by which steroid hormones produce their physiological effects in target tissues (9,13,18,23,28). In all cases the steroids appear to interact with hormone-specific cytoplasmic receptors, and the resultant hormone–receptor complexes subsequently migrate to the nuclear fraction of the target cells where they are thought to interact with specific nuclear chromatin acceptor sites. Alterations in transcription that result from these nuclear interactions appear to be ultimately responsible for producing the specific tissue responses.

As this general mechanism of steroid action has emerged, more recent studies have focused on the relationships between individual biochemical parameters and specific tissue responses and also on the ability of other factors to regulate tissue responsiveness after the initial hormone receptor interaction and nuclear migration. One such system that has been extensively studied is the stimulation of uterine growth by estrogens (13,18).

The overall growth response of the rat uterus to estrogens is a complex process that seems to involve at least two major temporal phases—an early phase and a late phase (13,21,29,32). Of particular interest are recent observations that suggest that different control mechanisms regulate these "early" and "late" uterine responses to estrogens (13,18). In this discussion we focus on a key early event in estrogen action—the stimulation of RNA polymerase activity—and on pharmacological and physiological means of selectively affecting different phases of the overall uterine response.

EFFECT OF ESTROGEN ON RNA POLYMERASE

For studies of hormonal control of transcription, we initially defined the molecular species of RNA polymerases present in the rat uterus and de-

* *Present address:* Department of Biology, Rochester Institute of Technology, Rochester, New York 14623. ** Department of Opthalmology, Baylor College of Medicine, Houston, Texas 77025. † Department of Biochemistry, Washington University, St. Louis, Missouri 63110.

termined their properties. Figure 1 illustrates that four distinct forms of uterine RNA polymerases can be observed by chromatography on diethylaminoethyl (DEAE) Sephadex. In mammalian systems (6,30,39), RNA polymerase I is thought to catalyze the synthesis of ribosomal RNA, polymerase II the synthesis of heterogenous nuclear RNA (6,30,39), and polymerase III the synthesis of transfer RNA and 5S ribosomal RNA (35,37).

As expected on the basis of studies in other systems, these different enzymes show different sensitivities to the inhibitor α-amanitin, as illustrated in Fig. 2. Polymerase II is most sensitive to inhibition; an α-amanitin con-

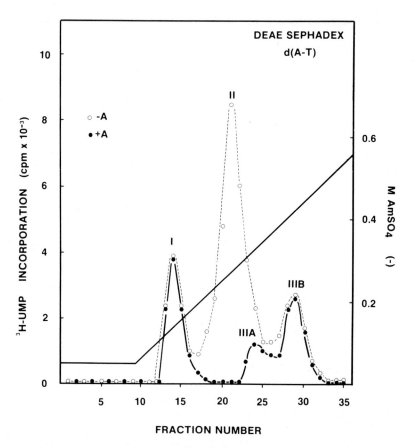

FIG. 1. Diethylaminoethyl (DEAE)-Sephadex chromatography of uterine RNA polymerases. Uterine tissue was homogenized in 0.3 M $(NH_4)_2SO_4$, sonicated, and centrifuged at 100,000 × g. The resulting supernatant was chromatographed and the eluted fractions assayed for activity using poly d(A-T) as template, as previously described (36). The dashed line represents the activity observed in the absence of α-amanitin (○ − A) and the solid line in the presence of 0.5 µg/ml of the inhibitor (● + A). I; RNA polymerase I; II, RNA polymerase II; III A, RNA polymerase III A; III B, RNA polymerase III B.

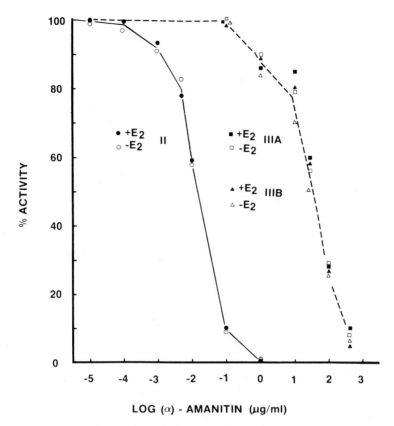

FIG. 2. α-Amanitin sensitivities of solubilized rat uterine RNA polymerases. Peak fractions of RNA polymerases II, IIIA, and IIIB from a DEAE-Sephadex column were assayed in the presence of increasing concentrations of α-amanitin. The open symbols represent enzymes from saline-treated control animals, and the closed symbols represent enzymes from animals treated with 2μg of estradiol 6 hr prior to sacrifice. +E₂, presence of estradiol; −E₂, absence of estradiol; II, RNA polymerase II; III A, RNA polymerase IIIA; IIIB, RNA polymerase IIIB.

centration of 0.006 μg/ml produces 50% inhibition of activity. Polymerases IIIA and IIIB show identical sensitivities to α-amanitin, and a 50% inhibition of activity is produced by a concentration of 18 μg/ml. Polymerase I is essentially unaffected by the highest concentration of the toxin used. Note also that estrogen treatment 6 hr prior to sacrifice of the animals and solubilization of the enzymes do not alter the sensitivities of the enzymes to inhibition by α-amanitin. Other studies also demonstrated the enzyme activities measured in isolated nuclei, with endogenous chromatin as the template, showed identical sensitivities to inhibition by the toxin (36).

We next investigated the effect of estradiol treatment on the activities of uterine RNA polymerases measured in isolated nuclei based on differential sensitivities to α-amanitin, using the endogenous nuclear chromatin as template. The results of these studies (Fig. 3) illustrate that polymerase I and III activities are rapidly stimulated by hormone treatment. Stimulation of polymerase I (7,17,20,31,34) and polymerase III (34) has also been reported by others, but at present there is some controversy concerning the effects of estrogen treatment on polymerase II activity (see 36 for a discussion).

In an attempt to determine the molecular mechanisms responsible for the observed hormonal stimulation of polymerases I and III, we next investigated the effects of estrogen on the total units of the *solubilized,* partially purified enzymes (Fig. 1) present in uterine tissue 6 hr after treatment, a time when the activities measured in *isolated nuclei* were greatly stimulated (Fig. 3). These studies revealed that the total units of polymerases I, II, and III were not altered (36), suggesting that enzyme synthesis is not responsible for the increase in activities observed in isolated nuclei. A similar conclusion has also been reached by Courvalin et al. (14).

Further studies revealed that the properties of the solubilized enzymes were not altered following hormone treatment. The specific properties we measured included divalent metal ion requirements, ionic strength optima, K_M values for adenosine 5′-triphosphate, template preferences [native DNA,

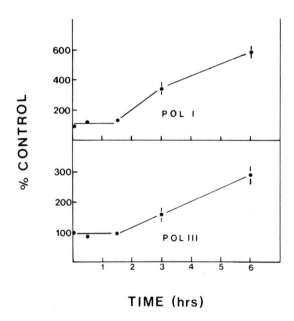

TIME (hrs)

FIG. 3. Time course of estradiol stimulation of uterine RNA polymerases I and III. Following treatment with saline or 2 μg of estradiol, immature female rats were sacrificed at the indicated times and uterine nuclei were prepared and assayed for activity as previously described (36). The data are presented as the percent of saline-treated control values at each time point. POL I, RNA polymerase I; POL III, RNA polymerase III.

denatured DNA, poly d(A-T) and S_I nuclease-treated DNA], elution from DEAE Sephadex, and sedimentation velocity in glycerol-containing sucrose gradients. Taken together these data suggest that the stimulation of RNA polymerase activities observed in isolated nuclei result from hormonal effects on the template activity of uterine chromatin. Such increases have been observed by several workers using *Escherichia coli* RNA polymerase to measure template activity (4,10,17,33).

We next determined the relationship of these relatively early changes in polymerase activities to subsequent physiological responses of the uterus, e.g., long-term uterine growth and cell division. To approach this question we examined the effects of estriol, another compound with estrogenic activity, on the activities of RNA polymerases in uterine nuclei. Estriol was used since its long-term effects on the uterus, such as increases in uterine dry weight (1,3), protein and RNA content (20), and DNA synthesis (8,27), are quite different than those produced by estradiol. Thus estriol does not produce the same long-term (24-hr) responses in the uterus as estradiol. It should be noted, however, that estradiol and estriol produce similar increases in uterine wet weight (1,3) and carbohydrate metabolism (19,27) 3 to 6 hr after administration to immature animals. A comparative study of the two compounds would therefore be expected to help elucidate the relationship between increases in RNA polymerase activities (occurring within 6 hr after treatment) and subsequent tissue responses produced at longer times.

The results of a series of experiments designed to test the relative effects of estradiol and estriol on uterine RNA polymerases (in isolated nuclei) 6 hr after treatment are illustrated in Table 1. As expected estradiol treatment produces large increases in the activities of RNA polymerases I and III. It is also seen that estriol produces essentially the same changes in RNA polymerase activities as estradiol.

This observation illustrates that early changes in RNA polymerase activities and other parameters may be necessary, but are not sufficient to insure that the long-term response of the uterus occurs. Clark and his colleagues

TABLE 1. *Effects of estradiol and estriol on uterine RNA polymerase*

Enzyme	Estradiol	Estriol
RNA polymerase I	499 ± 25 (437–556)	535 ± 29 (487–616)
RNA polymerase III	264 ± 38 (193–367)	294 ± 59 (174–457)

Groups of 10 to 20 animals were treated with estradiol (2 μg), estriol (2 μg), or saline 6 hr prior to sacrifice, and uterine nuclei were prepared. Results for estradiol and estriol are expressed as percent of saline-treated control values ± SEM and represent the means obtained in four separate experiments. Values in parentheses represent the range of values observed in the four experiments.

(1,3,13) originally came to a similar conclusion and suggested that long-term uterine responses require a prolonged retention of estrogen–receptor complexes in uterine nuclei, whereas short-term responses require only a brief nuclear accumulation of hormone–receptor complexes. Since estriol–receptor complexes are retained in uterine nuclei for shorter periods of time than estradiol–receptor complexes (1,13), this would explain the different long-term responses produced by the two estrogens.

These and other studies (5,12) suggested that uterine nuclei contain different acceptor sites for hormone–receptor complexes that regulate the different temporal phases of the overall tissue response. In order to test this hypothesis we decided to use a classic pharmacological approach to determine if we could selectively block either the early or late uterine responses to estradiol. For these studies we employed the antiestrogen nafoxidine (Upjohn).

EFFECT OF ANTIESTROGENS ON THE UTERINE RESPONSE TO ESTRADIOL

Figure 4 illustrates the uterine response to estradiol and two different doses of nafoxidine. Other studies indicated that the 5-μg dose of nafoxidine produced a maximum uterine response and that a 50-μg dose was clearly supramaximal (16). As seen in Fig. 4 both doses of the antiestrogen produce

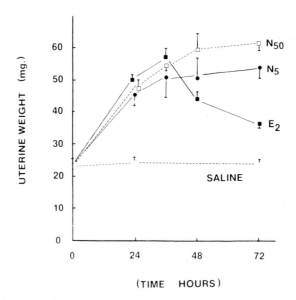

FIG. 4. Time course of the uterine response to estradiol and nafoxidine. Immature female rats were treated with saline, 2 μg of estradiol (E$_2$), 5 μg of nafoxidine (N$_5$), or 50 μg of nafoxidine (N$_{50}$) at the indicated times prior to sacrifice.

a sustained uterine response relative to estradiol. The sustained uterine response to supramaximal doses of nafoxidine has been previously observed and appears to result from a very prolonged nuclear retention of nafoxidine–receptor complexes (11,13,25). This is illustrated in Fig. 5, which depicts the amount of nuclear receptor retained in uterine nuclei 24 hr after treatment with estradiol and both doses of the antiestrogen. It is clear that considerable amounts of receptor are retained in uterine nuclei following treatment with either dose of antiestrogen.

We also examined the amounts of "uncomplexed" *cytoplasmic* receptor available 24 hr after these various treatments. These results (Fig. 6) illustrate that estradiol treatment increases the content of cytoplasmic receptors, but that the decrease in the amounts of cytoplasmic receptor following antiestrogen treatment can be accounted for solely by the nuclear uptake and retention previously observed (Fig. 5), *without* any replenishment of cytoplasmic receptors. This lack of replenishment of cytoplasmic receptors is thought to be the basis for the antiestrogenic activity of nafoxidine and related compounds (11,13,25).

It is particularly interesting to note the distribution of cytoplasmic and nuclear receptors 24 hr after treatment with the 5-μg dose of nafoxidine (Figs. 5 and 6). In this case there is a substantial amount of *cytoplasmic* receptor available (Fig. 6), which could presumably interact with any administered estrogens, but there is also a substantial amount of receptor remaining in the nucleus (Figure 5), which is presumably bound to nuclear acceptor sites since the uterine response is sustained (Fig. 4). We next determined, therefore, whether pretreatment with 5 μg of nafoxidine would selec-

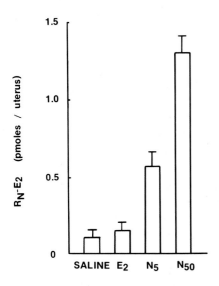

FIG. 5. Content of nuclear estrogen receptors (R_N) following treatment with estradiol and nafoxidine. Immature female rats were treated with saline, 2 μg of estradiol (E_2), 5 μg of nafoxidine (N_5), or 50 μg of nafoxidine (N_{50}) 24 hr prior to sacrifice. Nuclei were isolated, and receptor levels were measured by the nuclear exchange assay (2) as previously described (24).

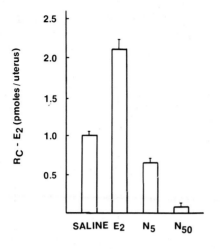

FIG. 6. Content of cytoplasmic estrogen receptors after treatment with estradiol and nafoxidine. Immature female rats were treated as described in the legend to Fig. 5, and uterine cytosol was prepared and assayed for estrogen binding using the hydroxylapatite procedure (38).

tively block either the early or late uterine responses to subsequently administered estradiol.

The effects of nafoxidine pretreatment on uterine responses to estradiol are shown in Figs. 7 and 8. In these studies groups of animals were treated with either saline, estradiol, or 5 or 50 μg of nafoxidine. After 24 hr the animals in each group were subdivided—half the animals received saline (S) and half estradiol (E_2) for either 4 (Fig. 7) or 24 hr (Fig. 8) prior to sacrifice. Figure 7 illustrates that the early uterine response to estradiol is comparable in the animals pretreated with saline, estradiol, or 5 μg of

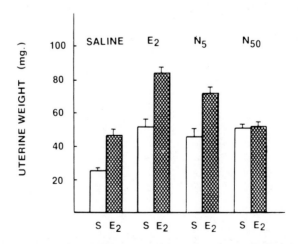

FIG. 7. Effect of nafoxidine pretreatment on the early uterine response to estradiol. Immature female rats were pretreated with saline, 4 μg of estradiol (E_2), 5 μg of nafoxidine (N_5), or 50 μg of nafoxidine (N_{50}) for 24 hr. Half the animals in each group then received saline (S) and half 4 μg of estradiol (E_2) 4 hr prior to sacrifice.

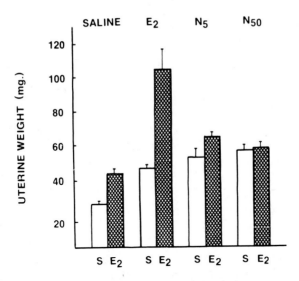

FIG. 8. Effect of nafoxidine pretreatment on the late uterine response to estradiol. Immature female rats were pretreated for 24 hr as described in the legend to Fig. 7. Half the animals in each group then received saline (S) and half 4 μg of estradiol (E₂) 24 hr prior to sacrifice.

nafoxidine. In addition to the increases in wet weight shown in Fig. 7, measurements of uterine glucose metabolism showed a similar pattern of 4-hr responses. As expected animals pretreated with 50 μg of nafoxidine showed no response, since there is no cytoplasmic receptor available (Fig. 6).

Figure 8 illustrates the long-term uterine response to estradiol after a similar pretreatment with either saline, estradiol, or 5, or 50 μg of the antiestrogen. Note that in this case, however, pretreatment with the 5-μg dose of nafoxidine effectively blocks any major long-term response to estradiol. Measurements of uterine glucose metabolism, protein, and RNA content 24 hr after estradiol treatment also demonstrated a similar pattern of responses.

Taken together with other studies (5,12,13,18), these results provide a considerable amount of evidence that uterine nuclei contain different types of acceptor sites that are involved with regulating different temporal phases of the complete uterine response to estrogens. Although this specific hypothesis of multiple acceptor sites certainly requires further experimental substantiation, it nevertheless seems clear that different control mechanisms are involved with regulating the early and late uterine responses to estrogens.

INFLUENCE OF THYROID HORMONE ON THE UTERINE RESPONSE TO ESTRADIOL

In conjunction with the pharmacological studies of antiestrogen action, we also sought to determine whether physiological factors could selectively influence early or late uterine responses to estrogen. Initially we observed that hypophysectomy selectively diminished uterine responses occurring 24 hr after estrogen treatment, but had little if any effect on responses occurring 4 hr after hormone administration (26).

Further studies have revealed that this effect of hypophysectomy is due to hypothyroidism secondary to decreased levels of thyroid-stimulating hormone (15). This is illustrated in Fig. 9, which depicts the uterine response to estradiol in normal animals and in animals made hypothyroid by feeding a diet containing propylthiouracil (PTU). The responses of the two groups are comparable 4 hr after treatment, but the response of the hypothyroid group is clearly diminished at longer times, e.g., 24 hr.

This decreased response 24 hr after hormone treatment is not due to a shift in the dose-response curve for estradiol, but reflects a decrease in the maximum response of the tissue, as illustrated in Fig. 10. Dose-response studies for increases in other uterine parameters, e.g., uterine protein and RNA content, 24 hr after hormone treatment were similar to those observed in Fig. 10 (15).

In order to insure that the decreased long-term uterine responses observed were due to decreased levels of thyroid hormones, it was important to demonstrate that the observed effects could be reversed by treatment with exogenous thyroid hormone at physiologically meaningful doses. The data in Fig. 11 illustrate the effect of T_3 administration to hypothyroid rats on the uterine response to estradiol. These results illustrate several pertinent

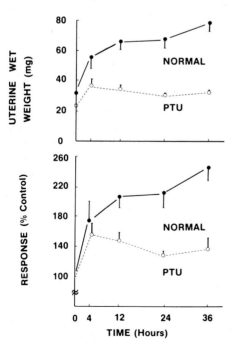

FIG. 9. Time course of action of estradiol (E_2) in normal and hypothyroid rats. Ovariectomized female rats were fed a normal diet (NORMAL) or an iodine-deficient diet containing 0.15% propylthiouracil (PTU) for 30 days. Animals were then treated with 2 μg of estradiol per 100 g body weight at the indicated times prior to sacrifice. The responses are plotted (*bottom panel*) as the percent of the zero time controls in each group.

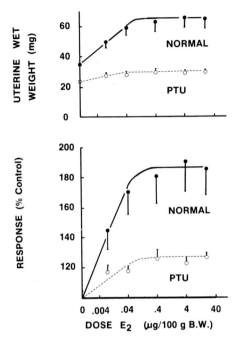

FIG. 10. Dose-response curves for estradiol in normal and hypothyroid (PTU) animals. Normal or hypothyroid rats (see legend to Fig. 9) were treated with the indicated doses of estradiol (E_2) 24 hr prior to sacrifice (*upper panel*). The response (*lower panel*) is expressed as the percent of saline-treated controls for each group.

features. First, T_3 *alone* does *not* elicit any uterine response (see SALINE controls, *upper panel,* Fig. 11). Second, T_3 administration to hypothyroid animals restores the ability of the uterus to respond to estradiol 24 hr after treatment in a dose-dependent manner. The dose response curves for T_3 restoration of estrogen-induced increases in uterine protein and RNA content 24 hr after treatment are also similar.

Other studies revealed that exogenous T_3 can restore the diminished uterine response previously noted in hypophysectomized animals. This observation rules out the possibility that the effect of hypothyroidism merely reflects decreased levels of growth hormone (22) and suggests that T_3 directly affects the uterus to regulate the responsiveness of the tissue to estradiol.

At present, we do not understand the underlying basis for the selective decrease in certain uterine responses to estrogens in hypothyroid animals. It seems unlikely, however, that the effects of hypothyroidism result from alterations of the uterine receptor *per se*. Previous studies indicated that hypophysectomy did not affect the content or properties of uterine estrogen receptors (26), and more recent data indicate that hypothyroidism does not affect the *in vivo* retention patterns of nuclear estrogen–receptor complexes. Furthermore, several early uterine responses to estradiol appear normal (Fig. 9) in hypothyroid animals, and these responses are presumably mediated by the estrogen receptor system. Nevertheless, these studies cannot rule out the possibility that hypothyroidism produces subtle biochemical

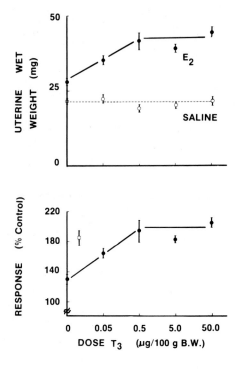

FIG. 11. Thyroid (T_3) replacement of the uterine response to estradiol. Hypothyroid animals (see legend to Fig. 9) were treated with the indicated doses of T_3 for 5 days. Half the animals in each replacement group were then treated with saline (SALINE, *upper panel*) and half with 2 μg of estradiol (E_2-*upper panel*) 24 hr prior to sacrifice. The response (*lower panel*) is expressed as the percent of the saline control at each T_3 replacement level. The open symbol near the ordinate in the lower panel represents the response (as percent of an appropriate saline control group) of euthyroid animals.

alterations in uterine estrogen receptors that account for the effects observed in these experiments.

Based on our results and the extensive studies of numerous other investigators (see 13 and 18 for recent reviews), we have adopted the following model as a working hypothesis to explain the overall uterine response following acute administration of estradiol. Estradiol enters uterine cells, combines with specific cytoplasmic receptors, and migrates to uterine nuclei (9,13,18,23,28). These receptor–hormone complexes initially interact with one type of nuclear acceptor site, and this interaction is sufficient to produce early uterine responses occurring within 3 to 6 hr (1,3,13). An extended nuclear retention of estrogen–receptor complexes for approximately 6 hr (1,13) is then necessary for interactions with a second type of acceptor site to occur. This set of interactions regulates the uterine response occurring 24 to 48 hr after the initial hormone treatment. Thyroid hormones also appear to be required for these long-term uterine responses to occur normally (15), although it is not known at present which molecular event(s) in estrogen action is affected by hypothyroidism.

ACKNOWLEDGMENTS

This research was supported by National Institutes of Health Grants HD 08615 and GM 19494. G. M. Stancel is a recipient of a National Institutes of Health Research Career Development Award. R. M. Gardner is a recipient of a National Institutes of Health Research Fellowship Award (AM 05246). P. A. Weil is a recipient of a Rosalie B. Hite Foundation Predoctoral Fellowship.

REFERENCES

1. Anderson, J. N., Clark, J. H., and Peck, E. J., Jr. (1972): The relationship between nuclear receptor-estrogen binding and uterotrophic response. *Biochem. Biophys. Res. Commun.,* 48:1460–1468.
2. Anderson, J. N., Clark, J. H., and Peck, E. J., Jr. (1972): Oestrogen and nuclear binding sites. *Biochem. J.,* 126:561–567.
3. Anderson, J. N., Peck, E. J., Jr., and Clark, J. H. (1975): Estrogen-induced uterine responses and growth: Relationships to receptor estrogen binding by uterine nuclei. *Endocrinology,* 96:160–167.
4. Barker, K. L., and Warren, J. C. (1966): Template capacity of uterine chromatin: Control by estradiol. *Proc. Natl. Acad. Sci. USA,* 56:1298–1302.
5. Baudendistel, L. J., and Ruh, T. J. (1976): Antiestrogen action: Differentiation of nuclear retention and extractability of estrogen receptor. *Steroids,* 28:223–237.
6. Blatti, S. P., Ingles, C. J., Lindell, T. J., Morris, P. W., Weaver, R. J., Weinberg, J., and Rutter, W. J. (1970): Structure and regulatory properties of eucaryotic RNA polymerase. *Cold Spring Harbor Symp. Quant. Biol.,* 35:649–657.
7. Borthwick, N. M., and Smellie, R. M. S. (1975): The effects of oestradiol-17β on the ribonucleic acid polymerases of immature rabbit uterus. Biochem. J., 147: 91–101.
8. Bujnoch, L. J., Levine, B. D., and Stancel, G. M. (1975): Effects of estradiol, nafoxidine, and CI-628 on uterine weight and DNA synthesis. *Tex. Rep. Biol. Med.,* 33:335.
9. Chan, L., and O'Malley, B. W. (1976): Mechanism of action of sex steroid hormones. *N. Engl. J. Med.,* 294:1372–1381.
10. Church, R. U., and McCarthy, B. J. (1970): Unstable nuclear RNA synthesis following estrogen stimulation. *Biochim. Biophys. Acta,* 199:103–114.
11. Clark, J. H., Anderson, J. N., and Peck, E. J., Jr. (1973): Estrogen receptor antiestrogen complex: Atypical binding by uterine nuclei and effects on uterine growth. *Steroids,* 22:707–718.
12. Clark, J. H., and Peck, E. J., Jr. (1976): Nuclear retention of receptor-estrogen complex and nuclear receptor sites. *Nature,* 260:635–636.
13. Clark, J. H., Peck, E. J., Jr., and Anderson, J. N. (1976): Estrogen receptor binding: Relationship to estrogen-induced responses. *J. Toxicol. Environ. Health,* 1:561–586.
14. Courvalin, J.-C., Bouton, M.-M., Baulieu, E.-E., Nuret, P., and Chambon, P. (1976): Effect of estradiol on rat uterus DNA-dependent RNA polymerase. *J. Biol. Chem.,* 251:4843–4849.
15. Gardner, R. M., Kirkland, J. L., Ireland, J. S., and Stancel, G. M. (1977): The effect of propylthiouracil treatment on the uterine response to estradiol. *Fed. Proc.,* 36:368.
16. Gardner, R. M., and Stancel, G. M. (1977): The effect of antiestrogen pretreatment on the uterine response to estradiol. *The Pharmacologist,* 19:203.
17. Glasser, S. R., Chytil, F., and Spelsberg, T. C. (1972): Early effects of estradiol-17β on the chromatin and activity of the deoxyribonucleic acid-dependent ribonucleic acid polymerase (I + II) of the rat uterus. *Biochem. J.,* 130:947–957.
18. Gorski, J., and Gannon, F. (1976): Current models of steroid hormone action: A critique. *Annu. Rev. Physiol.,* 38:425–450.
19. Gorski, J., and Raker, B. (1974): Estrogen action in the uterus: The requisite for sustained estrogen binding in the nucleus. *Gynecol. Oncol.,* 2:249–258.
20. Hardin, J. W., Clark, J. H., Glasser, J. R., and Peck, E. J., Jr. (1970): RNA poly-

merase activity and uterine growth: Differential stimulation by estradiol, estriol, and nafoxidine. *Biochemistry,* 15:1370–1374.

21. Hechter, O., and Halkerston, I. (1964): On the action of the mammalian hormones. In: *The Hormones,* Vol. 5, edited by G. Pincus, K. Thimann, and E. Astwood, pp. 697–827. Academic Press, New York and London.

22. Hervas, F., Morreale de Escobar, G., and Escobar del Rey, F. (1975): Rapid effects of single small doses of L-thyroxine and triiodo-L-thyronine on growth hormone, as studied in the rat by radioimmunoassay. *Endocrinology,* 97:91–101.

23. Jensen, E. V., and De Sombre, E. R. (1972): Mechanism of action of the female sex hormone. *Annu. Rev. Biochem.,* 41:203–230.

24. Juliano, J. V., and Stancel, G. M. (1976): Estrogen receptors in the rat uterus. *Biochemistry,* 15:916–920.

25. Katzenellenbogen, B. J., and Ferguson, E. R. (1975): Antiestrogen action in the uterus: Biological ineffectiveness of nuclear bound estradiol after antiestrogens. *Endocrinology,* 97:1–12.

26. Kirkland, J. L., Gardner, R. M., Ireland, J. S., and Stancel, G. M. (1977): The effect of hypophysectomy on the uterine response to estradiol. *Endocrinology,* 101: 403–410.

27. Lan, N. C., and Katzenellenbogen, B. J. (1976): Temporal relationship between hormone receptor binding and biological responses in the uterus: Studies with short and long acting derivatives of estriol. *Endocrinology,* 98:220–227.

28. Liao, S. (1975): Cellular receptors and mechanisms of action of steroid hormones. *Int. Rev. Cytol.,* 41:87–172.

29. Mueller, G. C., Herranen, A., and Jervell, K. (1958): Studies of the mechanism of action of estrogens. *Recent Prog. Horm. Res.,* 14:95–139.

30. Reeder, R. H., and Roeder, R. G. (1972): Ribosomal RNA synthesis in isolated nuclei. *J. Mol. Biol.,* 67:433–441.

31. Rutter, W. J., Ingles, C. J., Weaver, R. F., Blatti, S. P., and Morris, P. W. (1972): RNA polymerases and transcriptive specificity in eukaryotes. In: *Molecular Genetics and Development Biology,* edited by M. Sussman, pp. 133–147. Prentice-Hall, Englewood Cliffs, N.J.

32. Szego, C. M., and Roberts, S. (1953): Steroid action and interactions in uterine metabolism. *Recent Prog. Horm. Res.,* 8:419–469.

33. Teng, C. S., and Hamilton, T. H. (1968): The role of chromatin in estrogen action in the uterus, I. The control of template capacity and chemical composition and the binding of H^3-estradiol-17β *Proc. Natl. Acad. Sci. USA,* 60:1410–1417.

34. Webster, R. A., and Hamilton, T. H. (1976): Comparative effects of estradiol-17β and estriol on uterine RNA polymerase I, II, and III *in vivo. Biochem. Biophys. Res. Commun.,* 69:737–750.

35. Weil, P. A., and Blatti, S. P. (1976): HeLa cell deoxyribonucleic acid dependent RNA polymerase: Function and properties of the class III enzymes. *Biochemistry,* 15:1500–1509.

36. Weil, P. A., Sidikaro, J., Stancel, G. M., and Blatti, S. P. (1977): Hormonal control of transcription in the rat uterus. *J. Biol. Chem.,* 252:1092–1098.

37. Weinman, R., and Roeder, R. G. (1974): Role of DNA-dependent RNA polymerase III in the transcription of the t-RNA and 5S RNA genes. *Proc. Natl. Acad. Sci. USA,* 71:1790–1794.

38. Williams, D. W., and Gorski, J. (1972): Kinetics and equilibrium analysis of estradiol in uterus: A model of binding site distribution in uterine cells. *Proc. Natl. Acad. Sci. USA,* 69:3464–3468.

39. Zylber, E. A., and Penman, S. (1971): Products of RNA polymerase in HeLa cell nuclei. *Proc. Natl. Acad. Sci. USA,* 68:2861–2865.

Ontogeny of Receptors and Reproductive Hormone Action,
edited by T. H. Hamilton, J. H. Clark, and W. A. Sadler.
Raven Press, New York 1979.

Synthesis and Processing of Ribosomal RNA by the Rat Uterus During Early Estrogen Action

Dennis N. Luck

Department of Biology, Oberlin College, Oberlin, Ohio 44074

It is well documented that the rate of RNA metabolism is stimulated in the uterus of the immature or ovariectomized adult rat following the administration of estrogen (10,13,27). Moreover, the estrogenic stimulation of uterine RNA synthesis takes place in two phases (9,11). The first response, seen within an hour after hormone administration, involves a stimulation in the synthesis of heterogeneous, mainly high-molecular-weight RNA molecules that appear to contain mRNA species (15–17). To my knowledge, however, no new protein has been identified in the estrogen-stimulated uterus that was not present in the unstimulated organ. The second response is a substantial increase in the rate of synthesis of ribosomal RNA and transfer RNA (2,8,16,20,25). It begins about 1 hr after hormone injection. This estrogen-stimulated production of ribosomal and transfer RNAs depends on the RNA and protein synthesized during the preliminary response (6,13,16). In this chapter I will discuss the effects of estrogen on the synthesis and processing of ribosomal RNA by the uterus of the rat, as well as the regulation of ribosome production.

RIBOSOMAL RNA SYNTHESIS AND PROCESSING

Moore and Hamilton (23) first showed in 1964 that estrogen increases both the rate of synthesis and the cytoplasmic concentration of monomeric ribosomes in the uterus of the ovariectomized rat. Figure 1 shows the effect of estrogen on the appearance of new uterine ribosomes. At 4 hr following administration of the hormone and ³H-uridine, the monomeric ribosomes isolated from the nuclei-free homogenate of the organ and distributed on a sucrose gradient were uniformly labeled. The rate of incorporation of the radioactive precursor into the ribosomal RNA of the ribosomes *in vivo* rose linearly from 4 to 12 hr after hormone treatment. Figure 1 also shows that both the specific activity and the concentration of ribosomes in the cytoplasmic fraction of the uterus were increased by hormone treatment. Figure 2 shows that in the absence of the hormone, but in the presence of the ³H-uridine *in vivo*, the labeling of the relatively few ribosomes extractable

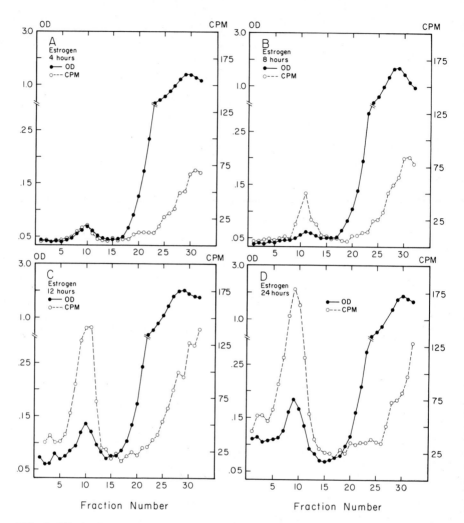

FIG. 1. Effect of estrogen as a function of time on the formation of cytoplasmic ribosomes in the uterus of the ovariectomized rat. All animals received 200μCi ^{3}H-uridine and 10 μg of the hormone at time zero. Each animal was killed at the time indicated, and the monomeric ribosomes of the cytoplasmic fraction were distributed in a linear sucrose gradient that was fractionated for $A_{260\,nm}$ and acid-insoluble radioactivity measurements. (From Moore and Hamilton, ref. 23.)

from the cytoplasm of the estrogen-deficient uterus was most reduced. In other words, in the atrophied organ there is a slow turnover of a very small population of ribosomes.

The first data to suggest that estrogen stimulates not only the synthesis of ribosomal RNA in the uterus of the ovariectomized rat, but also the rate of processing of the precursor ribosomal RNA to the mature 28S and 18S RNA species came from experiments performed in 1972 by Luck and Hamilton

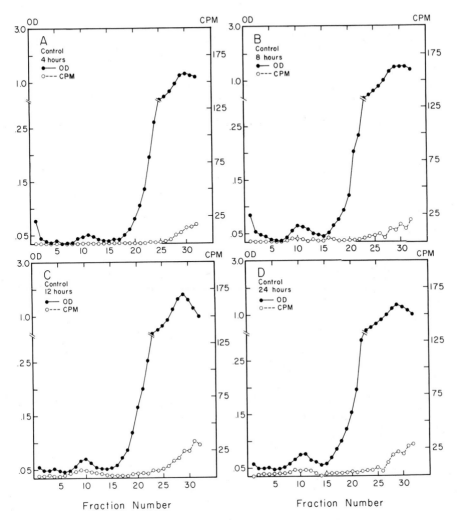

FIG. 2. Control experiments for those described in Fig. 1. The experimental procedure was identical to that described in Fig. 1, except that the animals received only ^{3}H-uridine at zero time. (From Moore and Hamilton, ref. 23.)

(19) in which labeled RNA isolated from uteri of ovariectomized animals was analyzed on sucrose gradients. We developed a procedure (19) for the isolation of undegraded whole-organ RNA that is based on essentially conventional techniques involving the use of sodium dodecyl sulfate, hot phenol and ribonuclease inhibitors. In an early experiment we examined the effect of estrogen, as a function of time, on the radioactivity and absorbancy profiles of sucrose gradient analyses of labeled RNA isolated from uteri of ovariectomized animals that received the homone and 200 μCi of ^{3}H-uridine together at time zero. Table 1 shows the ratios of RNA to DNA determined

TABLE 1. *Time course of the effect of estrogen on the ratio of RNA to DNA and the uptake and incorporation of ^{3}H-uridine into RNA in the uterus of the ovariectomized rat*[a]

Time after treatment (hr)	RNA:DNA ratio	Total-tissue radioactivity (cpm/mg DNA)	Acid-insoluble radioactivity (%)	Specific activity of isolated RNA (cpm/mg)
Animals given ^{3}H-uridine without estrogen				
0.5	0.34	80,900	14	17,500
1	0.36	95,700	18	43,100
2	0.34	128,000	23	52,700
8	0.35	180,500	28	120,000
Animals given estrogen and ^{3}H-uridine				
0.5	0.35	100,400	19	27,800
1	0.38	115,900	24	68,300
2	0.43	166,200	30	81,100
8	0.62	225,700	36	94,800

[a] The details of the experimental procedures employed were as described by Luck and Hamilton, ref. 19.

for the various uterine homogenates, as well as the specific activities of the isolated RNA samples. Estrogen increased the concentration of RNA in the uterus and also the specific activity of the RNA isolated 30 min, 1 hr, or 2 hr after hormone treatment. The higher specific activity of the uterine RNA from animals that received ^{3}H-uridine for 8 hr, compared to that from animals that received both estrogen and the labeled precursor for 8 hr, is a consequence of the much larger amount of unlabeled RNA synthesized later in the course of hormone action. This RNA dilutes the labeled RNA synthesized initially. The uptake of ^{3}H-uridine by the uterus, measured as the total-tissue radioactivity, and the percentage that was acid insoluble were found to increase with hormone treatment time as reported previously.

As mentioned above, the ratios of RNA to DNA in the various uterine homogenates showed that estrogen increased the concentration of RNA in the organ. We have found that this ratio serves as a good criterion for determining whether or not the hormone has been physiologically operative in the uterus. After 2 hr of estrogen stimulation of the uterus of the ovariectomized rat, we routinely find an increased content of RNA compared to the unstimulated control. Other workers (1,13,22) using immature animals have reported no measurable increases in RNA content before 6 hr of hormone treatment. Whether this difference is due to the animals, the solvent for the hormone, the route of injection, or analytical procedures is unclear.

The profiles shown in Fig. 3 are those obtained from sucrose gradient analyses of the isolated RNA samples described in Table 1. The absorbance profiles of all of the RNA samples analyzed were the same. The radioactivity profiles, however, reflected differences in the relative amounts of the

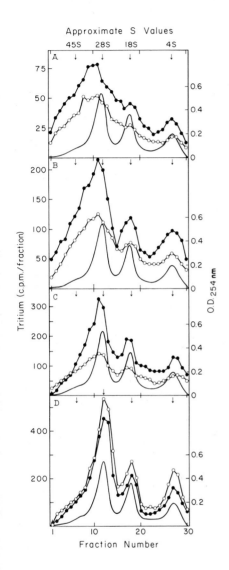

FIG. 3. Radioactivity and absorbancy profiles of RNA isolated from the uterus of the ovariectomized rat as a function of time after administration of estrogen (20 μg) and ³H-uridine (200μCi) together. Animals were killed 30 min **(A)**, 1 hr **(B)**, 2 hr **(C)**, or 8 hr **(D)** later, and the isolated uterine RNA was centrifuged through a linear sucrose gradient as described previously: —○—, radioactivity in the absence of estrogen; —●—, radioactivity in the presence of the hormone; —, $A_{254\,nm}$. (After Luck and Hamilton, ref. 19.)

various classes of labeled RNA present in the uteri at various times after the injection of ³H-uridine. After only 30 min of exposure to ³H-uridine (Fig. 3A), the bulk of the labeled RNA formed had a high molecular weight (>28S). But as the time of exposure to ³H-uridine increased, the amounts of labeled 28S, 18S, and 4S RNA increased, and the amount of labeled RNA of high-molecular-weight decreased (Fig. 3B–D). These observations made in the absence of estrogen are thus in good accord with the well-documented formation of 28S and 18S ribosomal RNAs from a precursor RNA of high molecular weight (21). The radioactivity profiles shown in Fig. 3 also reveal that the rate at which the labeled RNA of high molecular weight decreased, in association with the increased labeling of 28S and 18S RNA, was accelerated by the hormone. This can be quantified in various ways. Seen at its simplest, the ratio of radioactivity in fraction 12 (approximately 28S) to that in fraction 6 (approximately 45S) in Fig. 3A–D increased, respectively 1.3, 1.4, 1.7, and 4.8, in the absence of estrogen, whereas the corresponding ratios are 1.3, 1.6, 2.8, and 4.7 in the presence of estrogen. Apparently, near the end of an 8-hr *in vivo* pulse the pool of labeled precursor is so depleted that hardly any labeled RNA of high molecular weight is synthesized; also, virtually all of these precursor molecules that were synthesized previously appear to have been converted to mature RNA products. These experiments led us to suggest (19) that estrogen stimulates not only the rate of synthesis, but also the rate (or efficiency) of the processing of precursor RNA of high molecular weight (>28S) in the uterus of the ovariectomized adult rat.

In attempts to distinguish between the effects of the hormone on high-molecular-weight methylated RNAs, mainly precursor ribosomal RNA, and high-molecular-weight nonmethylated (or poorly methylated) RNAs, such as messenger RNA and heterogeneous nuclear RNA, further investigations (20) involved the analysis of uterine RNA, doubly labeled *in vivo* with radioactive uridine and (methyl)methionine.

Table 2 shows the effect of estrogen as a function of time on the incorporation of labeled uridine and methionine into uterine RNA of ovariectomized animals that received the radioactive precursors simultaneously 45 min before death. The data obtained demonstrate that estrogen has little or no stimulatory effect on the whole-organ uptake of methionine. This result, which is consistent with other work (2,24), emphasizes the usefulness of methionine as a precursor for studying hormone-stimulated RNA synthesis. Estrogen treatment for 1 hr increased the pool of uridine 21 and 33% in experiments 1 and 2, respectively, and this increase was maintained after 2 and 3 hr of hormone action. The total incorporation of ^{14}C-uridine into whole-organ RNA during the 45-min labeling period was increased 43% by hormone treatment for 3 hr, whereas the corresponding increase in (^{3}H-methyl)methionine was 144%. Furthermore, the results revealed that the ratio of methionine to uridine incorporation increased with time after hormone administration in both experiments, in which the same labeled precursors were used, but with the isotopes reversed. Thus, a greater and greater percentage of the newly synthesized RNA formed in the uterus during the first 3 hr following hormone stimulation is methylated.

The radioactivity and absorbance profiles of sucrose gradient analyses of the isolated RNA samples described in Table 1 are shown in Figs. 4 and 5. The absorbance profiles of all of the RNA samples analyzed were identical to those obtained previously. Figure 4A–D show that incorporation of ^{14}C-uridine into all of the major classes of uterine RNA took place during the 45-min labeling period in both control and hormone-treated animals. Some

TABLE 2. *Effect of estrogen as a function of time on the ratio of RNA to DNA and the uptake and incorporation of labeled uridine and methionine into RNA in the uterus of the ovariectomized adult rat*[a]

Exp.	Hormone treatment	RNA:DNA ratio	Total incorporation into whole-organ RNA (dpm/mg DNA)		Total-tissue radioactivity (dpm/mg DNA)		Specific activity of isolated RNA (dpm/mg)	
			Uridine	Methionine	Uridine	Methionine	^{14}C	^{3}H
1	No hormone	0.34	28,400	27,700 (0.98)[b]	69,000	412,800	5,800	3,600
1	1 hr of hormone	0.36	35,700	41,500 (1.16)	83,800	400,600	14,300	22,200
1	2 hr of hormone	0.42	41,500	54,800 (1.32)	89,500	461,500	19,600	33,100
1	3 hr of hormone	0.47	40,700	67,800 (1.66)	81,500	453,100	20,500	45,000
2	No hormone	0.35	196,400	5,200 (0.02)	835,800	74,900	1,200	49,500
2	1 hr of hormone	0.37	269,900	7,500 (0.02)	1,116,000	82,300	2,500	109,500
2	2 hr of hormone	0.43	395,700	14,800 (0.04)	1,498,000	72,900	3,500	163,400
2	3 hr of hormone	0.47	384,300	19,900 (0.05)	1,486,000	78,400	5,600	181,100

[a] RNA was isolated from the combined uteri of four ovariectomized adult rats that received 20 μg 17β-estradiol in 0.2 ml of 1,2-propanediol (or the solvent alone), intraperitoneally, at the times indicated before sacrifice. All animals in experiment 1 received 50 μCi ^{14}C-uridine (58 Ci/mol) and 250 μCi (^{3}H-methyl)methionine (3.3 Ci/mmol) in 1.0 ml water, intraperitoneally, 45 min before killing. In Experiment 2, the animals received 100 μCi of ^{3}H uridine (8 Ci/mmol) and 50 μCi of (^{14}C-methyl)methionine (55 Ci/mol) in 1 ml water, as in Experiment 1. All other experimental details were as described by Luck and Hamilton, refs. 19,20.

[b] Ratio of incorporation of methionine to uridine.

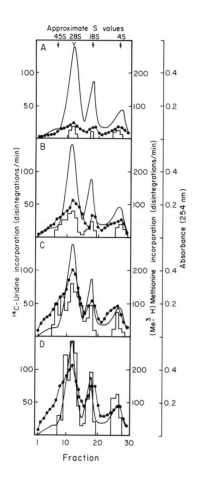

FIG. 4. Sucrose gradient analyses of isolated uterine RNA, labeled *in vivo* with ^{14}C-uridine and (^{3}H-methyl)methionine, as a function of time after administration of estrogen. Experimental details are given in Table 2 (Experiment 1). Rats received either no hormone **(A)**, or estrogen for 1 **(B)**, 2 **(C)**, or 3 hr **(D)**. ●—●, ^{14}C; histogram, ^{3}H; ——, $A_{254\,nm}$. (After Luck and Hamilton, ref. 20).

incorporation of (^{3}H-methyl)methionine occurred in the absence of estrogen (Fig. 4A), the RNA in the 28S, 18S, and 4S fractions being methylated. In the presence of the hormone, however, a substantial amount of methylation of these RNA species, as well as of 45S and 32S RNA, was observed (Fig. 4B–D). Figure 5A also shows that very little methylated RNA was formed by the uteri of animals not treated with the hormone; only the 28S and 4S RNA fractions contained detectable ^{14}C. During the same labeling period, there was a fourfold increase in the amount of methylation in the uteri of animals administered estrogen for 3 hr, and the RNA in the 45S, 32S, 28S, 18S, and 4S fractions was methylated (Fig. 5D).

These results show that little methylated RNA is formed by the atrophied uterus and that the increasing amount of methylated RNA synthesized in the uterus during the first 3 hr following hormone stimulation is mainly 28S and 18S ribosomal RNA and transfer RNA. The smaller quantities of methylated 45S and 32S RNA observed in these experiments are in accord with the

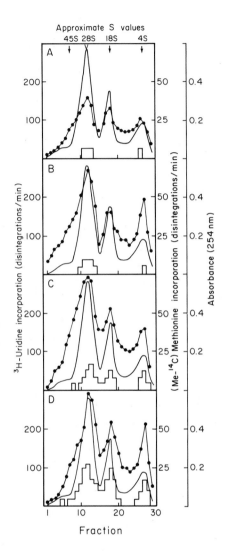

FIG. 5. Sucrose gradient analyses of isolated uterine RNA, labeled *in vivo* with ³H-uridine and (¹⁴C-methyl)methionine, as a function of time after administration of estrogen. Experimental details are given in Table 2 (Experiment 2). Rats received either no hormone **(A)**, or estrogen for 1 **(B)**, 2 **(C)**, or 3 hr **(D)**. —●—, ³H; histogram, ¹⁴C; ——, $A_{254\,nm}$. (After Luck and Hamilton, ref. 20.)

accepted sequence of events in the formation of ribosomal RNA (21). The increased synthesis of ribosomal and transfer RNA commences about an hour after estrogen administration, but is more striking after 2 and 3 hr. These findings are in good agreement with those of Knowler and Smellie (16) and Munns and Katzman (25).

The marked difference in the rate of methylation versus the rate of total RNA synthesis between the estrogen-stimulated uterus and the organ from an ovariectomized animal not treated with the hormone (cf. Fig. 3) suggested that the mechanism by which the formation of ribosomes is accelerated probably involves control of maturation processes, including the rate of methylation and processing of the precursor ribosomal RNA species. Accordingly, in continuing investigations designed to elucidate the effect of estrogen on the metabolism of ribosomal RNA in the uterus of the ovariectomized adult rat, Knecht and Luck (14) analyzed uterine RNA, doubly labeled *in vivo* with (^{3}H-methyl)methionine and ^{14}C-uridine using gel electrophoretic procedures. Hormone treatment times of 0 to 4 hr were employed and RNA samples were isolated from the uteri of ovariectomized animals that received the radioactive precursors 45 min before death. Table 3 shows the effect of estrogen on the uptake and incorporation of (^{3}H-methyl)-methionine and ^{14}C-uridine into uterine RNA, as well as the specific activity of the RNA isolated. Estrogen again had no effect on the whole-organ uptake of methionine during the first 2 hr of stimulation. After 3 and 4 hr of hormone treatment, however, the ^{3}H in the tissue was increased 26 and 45%, respectively. The total incorporation of (^{3}H-methyl)methionine into whole-organ RNA during the 45-min labeling period was increased 134% by hormone treatment for 4 hr. Estrogen also increased the specific activtiy of the ^{3}H-labeled RNA isolated.

The results obtained for the uptake and incorporation of ^{14}C-uridine into the uterus and whole-organ RNA (Table 3) are somewhat at variance with those reported in previous work (20) and shown in Table 2. The difference is most probably explained by the different amounts of the labeled precursor used—50 μCi in the earlier study and 5 μCi in the more recent work. As shown in Table 3 estrogen stimulated the uptake of ^{14}C-uridine by the uterus after 1 hr of treatment, but at later times the size of the pool declined to approximately control values. The specific activities of the isolated ^{14}C-labeled RNA samples followed a similar pattern as a function of time following estrogen treatment. Similar results have been reported previously (24). In good agreement with previous observations (20,25), the incorporation ratio of (^{3}H-methyl)methionine to ^{14}C-uridine increased with time after hormone administration following a lag period of about 1 hr.

The absorbance and radioactivity profiles of the gel electrophoretic analyses of the samples of labeled RNA isolated in the experiment described in Table 3 are shown in Fig. 6. The absorbance profiles of all of the RNA samples analyzed were essentially the same. The ^{14}C profiles (not shown) revealed that uridine had been incorporated into RNA very heterogeneous in size in both the control and hormone-stimulated uterus. No differences in the shapes of the ^{14}C profiles could be attributed to estrogen. The ^{3}H profiles, however, reflected differences in the relative amounts of the various classes of labeled precursor, intermediate, and mature ribosomal RNA species present in the uterus at various times after the injection of the hormone. Incorporation of (^{3}H-methyl)methionine into 45S, 32S, 28S, 21S, and 18S ribosomal RNAs took place during the 45-min labeling period in both control and hormone-treated animals. In the unstimulated uterus the percentage of ^{3}H found in the precursor 45S and 32S ribosomal RNAs totalled 49% of the label incorporated into all ribosomal RNA species. But as the time of the

TABLE 3. Time course of the effect of estrogen on the ratio of RNA to DNA and the uptake and incorporation of (³H-methyl)methionine and ¹⁴C-uridine into RNA in the uterus of the ovariectomized rat[a]

Hormone treatment	RNA:DNA ratio	Total-tissue radioactivity (dpm/mg DNA)		Radioactivity incorporated into whole-organ RNA (%)		Specific activity of isolated RNA (dpm/mg)	
		³H	¹⁴C	³H	¹⁴C	³H	¹⁴C
0	0.34	532,300	10,200	3.9	36	9,200	6,700 (1.4)[b]
1	0.36	535,000	15,100	4.9	32	15,800	11,700 (1.4)
2	0.39	538,900	11,000	5.2	39	18,400	10,200 (1.8)
3	0.42	672,800	11,400	5.5	40	26,300	8,600 (3.1)
4	0.45	770,700	11,900	6.3	40	44,400	9,400 (4.7)

[a] RNA was isolated from the combined uteri of four ovariectomized adult rats that received 20 µg 17β-estradiol, i.p., in 0.2 ml 1,2-propanediol (or the solvent alone) at the times indicated before killing. All animals received 250 µCi L-(³H-methyl)methionine (11 Ci mmoles⁻¹) and 5 µCi ¹⁴C-uridine (55 mCi mmoles⁻¹) in 0.4 ml water, i.p., 45 min before killing. All other experimental details were as described previously.
[b] Ratio of incorporation of methionine to uridine.

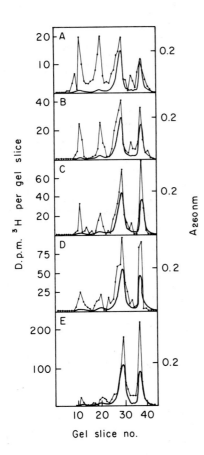

FIG. 6. Effect of estrogen as a function of time on isolated uterine RNA, labeled *in vivo* with (^{3}H-methyl)methionine, and fractionated on polyacrylamide-agarose gels. Experimental details are given in Table 3. Rats received either no hormone **(A)**, or estrogen for 1 **(B)**, 2 **(C)**, 3 **(D)** or 4 hrs **(E)**. —●—, ^{3}H; ——, $A_{260\,nm}$. (After Knecht and Luck, ref. 14.)

TABLE 4. *Effect of estrogen on the incorporation of (^{3}H-methyl)methionine into ribosomal RNAs by the rat uterus*[a]

Hormone treatment time (hr)	Ribosomal RNA fraction (%)			
	45S	32S	28S	18S
0	21	28	32	19
1	19	18	39	24
2	12	11	47	30
3	10	9	51	30
4	5	7	55	33

[a] The percentages in this table are calculated from the radioactivity profiles shown in Fig. 6. The label in the 45S fraction was computed by summing the dpm in gel slices 5–15; the 32S fraction by summing gel slices 16–22; the 28S fraction, gel slices 23–32; and the 18S fraction, gel slices 33–41.

hormone treatment increased, this percentage decreased until, 4 hr after estrogen administration, only 12% of the total amount of methyl-labeled ribosomal RNA formed was present in the 45S and 32S fractions (Table 4). Longer periods of stimulation by the hormone did not significantly alter this distribution of the label among the 45S, 32S, 28S, and 18S fractions. Labeled precursor was available to the uterine cells during the entire 45-min labeling period as evidenced by the presence of (^{3}H-methyl)methionine in the organ at the time of sacrifice of the animals. These data demonstrate that during the first 4 hr of estrogen action in the uterus of the ovariectomized adult rat, the amount of new methylated 28S and 18S ribosomal RNA formed from the methylated 45S precursor per unit time is increased nearly fourfold.

DISCUSSION

The mechanisms responsible for controlling the acceleration of the rate of processing of precursor ribosomal RNA in the hormone-stimulated uterus remain unclear. A number of recent reports, however, have shed light on this problem, and it is interesting to examine the data reviewed here in the context of these studies. Thus, posttranscriptional mechanisms for regulating ribosomal RNA synthesis have been proposed in regenerating liver (3,30), phytohaemagglutinin-stimulated lymphocytes (4,5,29) during compensatory renal hypertrophy (12), and in a mutant of BHK cells (28,34). Cooper (4,5) has concluded that in the resting human lymphocyte considerable amounts of ribosomal RNA are degraded, whereas in the actively growing state most of the 45S precursor ribosomal RNA is efficiently processed to yield functional ribosomes. Hill (12) has also concluded that during compensatory renal hypertrophy there is a decrease in the wastage of precursor ribosomal RNA. Toniolo and Basilico (34) and Ouellette et al. (28) have reported an increased pool of nucleolar and nuclear 45S precursor ribosomal RNA, respectively, in a temperature-sensitive mutant of BHK cells compared with wild-type cells. The mutant cells are unable to synthesize 28S RNA at the nonpermissive temperature, although transcription of 45S precursor ribosomal RNA and the appearance of 18S ribosomal RNA in cytoplasmic 40S subunits remains unimpaired. We have not detected any differences, however, in the relative amounts ($A_{260\ nm}$) of the different ribosomal RNA fractions present in uterine RNA samples isolated from control and hormone-stimulated animals. It is possible that the relatively large amounts of cytoplasmic 28S and 18S ribosomal RNA present in our samples may be masking small differences in the relative contents of 45S and 32S precursor ribosomal RNAs. To date attempts to extract undegraded RNA from endometrial cell nuclei isolated from the uteri of ovariectomized adult rats have been unsuccessful.

Purtell and Anthony (29) have examined the processing of ribosomal

RNA in resting and phytohaemagglutinin-stimulated guinea pig lymphocytes and have concluded, mainly on the basis of electrophoretic patterns, that there are at least two pathways, one of which is stimulated by phytohaemagglutinin. We have routinely observed a methyl-labeled fraction slightly larger than 45S in uterine RNA samples isolated from the control animals. This fraction is not present following estrogen stimulation (see Fig. 6A). Interestingly, Galibert et al. (7,33) have reported that there are three distinct components in the 45S precursor ribosomal RNA fraction of L-cells separable by electrophoresis through 1.7% polyacrylamide gels. Whether this heterogeneity represents a population of slightly different initial precursor molecules or early processing intermediates remains to be elucidated.

Finally, Liau et al. (18) have shown that there are pronounced differences in the activities and characteristics of nucleolar ribosomal RNA methylases of rapidly growing Novikoff ascites tumor cells and resting hepatic cells. These investigators conclude that the synthesis and processing of ribosomal RNA is controlled to a significant extent at the step involving methylation of precursor ribosomal RNA. Ouellette et al. (28) have indicated that the 45S and 32S precursor ribosomal RNA formed by the BHK mutant are substantially undermethylated at the nonpermissive temperature. Luck and Hamilton (20) obtained evidence that has been interpreted to show undermethylation of precursor ribosomal RNA in the atrophied uterus (see ref. 18). Of special interest is the observation of Sharma and Borek (31) that ovariectomy in pigs resulted in a sharp reduction in the total capacity of the uterine transfer RNA methylases, whereas administration of estrogen restored the methylating capacity of extracts of the uterus to normal. More recently, however, Munns et al. (26) demonstrated increased methylation of uterine transfer RNA during the estrogen response in immature rats, but additional experiments indicated that it was largely attributable to an increased rate of synthesis of transfer RNA.

It can only be conjectured at present whether the accelerated rate of processing of precursor ribosomal RNA to 28S and 18S ribosomal RNAs in the estrogen-stimulated uterus is due to more efficient methylation and cleavage, a more efficient processing pathway, or is determined by increased rates of ribosomal RNA synthesis. I think that ribosomal RNA synthesis and processing accelerates in the estrogen-treated uterus only after messenger RNA molecules that code for ribosomal RNA methylases are translated on uterine ribosomes and the enzymes transported to the nucleus. Current work in my laboratory involves attempts to learn some facts about the control of the rate of ribosome formation in the uterus during hormone-stimulated growth. There is a considerable body of evidence that supports the view that the growth rate of mammalian cells is correlated with the rate of ribosome production. An acceleration of the synthesis of ribosomes is an essential early feature not only of the regulation by estrogen of the uterus, but by other growth-promoting and developmental hormones acting in their characteristic target organs as well. It is clear that we need to know what mechanisms trigger and regulate the ribosome-making machinery of mammalian cells.

ACKNOWLEDGMENT

This work was supported in part by research grant PCM 75–07382 from the U.S. National Science Foundation.

REFERENCES

1. Billing, R. J., Barbiroli, B., and Smellie, R. M. S. (1969): The mode of action of estradiol II. The synthesis of RNA. *Biochim. Biophys. Acta,* 190:60–65.
2. Catelli, M., and Baulieu, E.-E. (1976): Estrogen-induced changes of ribonucleic acid in the rat uterus. *Mol. Cell. Endocrinol.,* 6:129–151.
3. Chaudhuri, S., and Lieberman, I. (1968): Control of ribosome synthesis in normal and regenerating liver. *J. Biol. Chem.,* 243:29–33.
4. Cooper, H. L. (1969): Ribosomal ribonucleic acid production and growth regulation in human lymphocytes. *J. Biol. Chem.,* 244:1946–1952.
5. Cooper, H. L. (1973): Degradation of 28S RNA late in ribosomal RNA maturation in nongrowing lymphocytes and its reversal after growth stimulation. *J. Cell Biol.,* 59:250–254.
6. DeAngelo, A. B., and Gorski, J. (1970): Role of RNA synthesis in the estrogen induction of a specific uterine protein. *Proc. Natl. Acad. Sci. USA,* 6:693–700.
7. Galibert, F., Tiollais, P., and Eladari, M. E. (1975): Fingerprinting studies of the maturation of ribosomal RNA in mammalian cells. *Eur. J. Biochem.,* 55:239–245.
8. Hamilton, T. H. (1968): Control by estrogen of genetic transcription and translation. *Science,* 161:649–661.
9. Hamilton, T. H., Teng, C. S., and Means, A. R. (1968): Early estrogen action: nuclear synthesis and accumulation of protein correlated with enhancement of two DNA-dependent RNA polymerase activities. *Proc. Natl. Acad. Sci. USA,* 59:1265–1272.
10. Hamilton, T. H., Teng, C. S., Means, A. R., and Luck, D. N. (1971): Estrogen regulation of genetic transcription and translation in the uterus. In: *The Sex Steroids,* edited by K. W. McKerns, pp. 197–240. Appleton-Century-Crofts, New York.
11. Hamilton, T. H., Widnell, C. C., and Tata, J. R. (1968): Synthesis of ribonucleic acid during early estrogen action. *J. Biol. Chem.,* 243:408–417.
12. Hill, J. M. (1975): Ribosomal RNA metabolism during renal hypertrophy. Evidence of decreased degradation of newly synthesized ribosomal RNA. *J. Cell Biol.,* 64:260–265.
13. Katzenellenbogen, B. S., and Gorski, J. (1975): Estrogen actions on syntheses of macromolecules in target cells. In: *Biochemical Actions of Hormones,* Vol. 3, edited by G. Litwack, pp. 187–243. Academic Press, New York.
14. Knecht, D. A., and Luck, D. N. (1977): Synthesis and processing of ribosomal RNA by the uterus of the ovariectomized adult rat during early estrogen action. *Nature,* 266:563–564.
15. Knowler, J. T. (1976): The incorporation of newly synthesized RNA into nuclear ribonucleoprotein particles after oestrogen administration to immature rats. *Eur. J. Biochem.,* 64:161–165.
16. Knowler, J. T., and Smellie, R. M. S. (1971): The synthesis of ribonucleic acid in immature rat uterus responding to oestradiol-17β. *Biochem. J.,* 125:605–614.
17. Knowler, R. T., and Smellie, R. M. S. (1973): The oestrogen-stimulated synthesis of heterogeneous nuclear ribonucleic acid in the uterus of immature rats. *Biochem. J.,* 131:689–697.

18. Liau, M. C., Hunt, M. E., and Hurlbert, R. B. (1976): Role of ribosomal RNA methylases in the regulation of ribosome production in mammalian cells. *Biochemistry,* 15:3158–3164.

19. Luck, D. N., and Hamilton, T. H. (1972): Early estrogen action: stimulation of the metabolism of high molecular weight and ribosomal RNAs. *Proc. Natl. Acad. Sci. USA,* 69:157–161.

20. Luck, D. N., and Hamilton, T. H. (1975): Early estrogen action: stimulation of the synthesis of methylated ribosomal and transfer RNAs. *Biochim. Biophys. Acta,* 383:23–29.

21. Maden, B. E. H. (1971): The structure and formation of ribosomes in animal cells. *Prog. Biophys. Mol. Biol.,* 22:127–177.

22. Miller, B. G., and Baggett, B. (1972): Effects of 17β-estradiol on the incorporation of pyrimidine nucleotide precursors into nucleotide pools and RNA in the mouse uterus. *Endocrinology,* 90:645–656.

23. Moore, R. J., and Hamilton, T. H. (1964): Estrogen-induced formation of uterine ribosomes. *Proc. Natl. Acad. Sci. USA,* 52:439–446.

24. Munns, T. W., and Katzman, P. A. (1971): Effects of estradiol on uterine ribonucleic acid metabolism. I. In vitro uptake and incorporation of ribonucleic acid precursors. *Biochemistry,* 10:4941–4948.

25. Munns, T. W., and Katzman, P. A. (1971): Effects of estradiol on uterine ribonucleic acid metabolism. II. Methylation of ribosomal ribonucleic acid and transfer ribonucleic acid. *Biochemistry,* 10:4949–4954.

26. Munns, T. W., Simms, H. F., and Katzman, P. A. (1975): Effects of estradiol on uterine ribonucleic acid metabolism. Assessment of transfer ribonucleic acid methylation. *Biochemistry,* 14:4758–4764.

27. O'Malley, B. W., and Means, A. R. (1974): Female steroid hormones and target cell nuclei. *Science,* 183:610–620.

28. Ouellette, A. J., Bandman, E., and Kumar, A. (1976): Regulation of ribosomal RNA methylation in a temperature-sensitive mutant of BHK cells. *Nature,* 262:619–621.

29. Purtell, M. J., and Anthony, D. D. (1975): Changes in ribosomal RNA processing paths in resting and phytohemagglutinin-stimulated guinea pig lymphocytes. *Proc. Natl. Acad. Sci. USA,* 72:3315–3319.

30. Riggs, A. J., and Webb, T. E. (1972): Regulation of ribosome formation in regenerating rat liver. *Eur. J. Biochem.,* 27:136–144.

31. Sharma, O. K., and Borek, E. (1970): Hormonal effect on transfer ribonucleic acid methylases and on serine transfer ribonucleic acid. *Biochemistry,* 12:2507–2513.

32. Tata, J. R. (1968): Hormonal regulation of growth and protein synthesis. *Nature,* 219:331–337.

33. Tiollais, P., Galibert, F., and Boiron, M. (1971): Evidence for the existence of several molecular species in the "45S fraction" of mammalian ribosomal precursor RNA. *Proc. Natl. Acad. Sci. USA,* 68:1117–1120.

34. Toniolo, D., and Basilico, C. (1976): Processing of ribosomal RNA in a temperature sensitive mutant of BHK cells. *Biochim. Biophys. Acta,* 425:409–418.

Ontogeny of Receptors and Reproductive Hormone Action,
edited by T. H. Hamilton, J. H. Clark, and W. A. Sadler.
Raven Press, New York 1979.

Gonadotropin Induction of Granulosa Cell Luteinization

K. L. Campbell, P. Bagavandoss, J. A. Jonassen, T. D. Landefeld,
*M. C. Rao, J. S. Richards, and A. R. Midgley, Jr.

*Reproductive Endocrinology Program, Department of Pathology,
The University of Michigan, Ann Arbor, Michigan 48109*

Luteinization of granulosa cells is a complex process closely associated with the physiologic, morphologic, and biochemical changes that result in ovulation and transformation of the graafian follicle into the corpus luteum. The early ovulatory events involve changes in follicular fluid, dissociation of adhering granulosa cells, and rupture of the follicular basement membrane. Each of these events facilitates exit of the oocyte and cumulus from the follicle. Subsequent events leading to corpus luteum formation require 2 to 3 days and involve complete breakdown of the basement membrane, granulosa cell hypertrophy, thecal and stromal invasion of the interior aspect of the former follicle, and markedly enhanced progesterone production.

The trigger for these processes is the surge of luteinizing hormone (LH) and the binding of this LH by the preovulatory follicle. Although it is known that the bound LH leads to increased cyclic AMP production and subsequent, enhanced steroidogenesis, the actual biochemical mechanisms which lead to ovulation and luteinization of granulosa and thecal cells are essentially unknown. Here we will describe some of our recent efforts at understanding the biochemical mechanisms by which LH [or its functionally equivalent, stable counterpart, human chorionic gonadotropin (hCG)] induces granulosa cell luteinization.

Mechanisms describing LH action must ultimately explain how a granulosa cell is transformed into a luteal cell. These mechanisms must indicate how the morphologic change is set in motion. Prior to LH induction of luteinization, granulosa cells have a sparse cytoplasm heavily populated with ribosomes, prominent, dilated, rough endoplasmic reticulum, mitochondria with lamelliform cristae, small Golgi, and a nucleus containing heavy peripheral heterochromatin with no distinguishable nucleoli (9,17,21,28). The cells interdigitate and communicate extensively with each other by cellular processes and an extensive array of gap junctions (2,3,16,17,19–21,27,28). Luteal cells, however, possess abundant cytoplasm containing extensive smooth endoplasmic reticulum, many lipid droplets, mitochondria with villiform cristae, and prominent Golgi. Their nuclei have little heterochromatin

---

* *Present address:* Department of Medicine, Gastroenterology Section, University of Chicago, Chicago, Illinois 60637.

and very prominent, multiple nucleoli (17,21). Although the cells are still connected by gap junctions, these complexes are much less prominent (1,8,16,31).

The cytologic changes occurring during luteinization indicate a shift from primarily protein synthetic functions to highly active steroidogenic functions. Although these changes are consequences of LH action, and as such are not themselves the biochemical mechanisms set in motion by the binding of LH, they stand as signposts pointing to at least some of the morphogenetic processes which must be initiated by LH. Thus, LH must initiate changes in (a) membrane characteristics, (b) nuclear function, and (c) cytoplasmic activity.

CHANGES IN MEMBRANE CHARACTERISTICS

Formation of Blebs

At the morphologic level we have examined follicular cells by scanning electron microscopy during the course of follicular growth and subsequent to induction of luteinization by hCG (6,7). For this purpose hypophysectomized, immature, female rats were treated with estradiol for 3 days (1.5 mg/day), and with highly purified human follicle-stimulating hormone

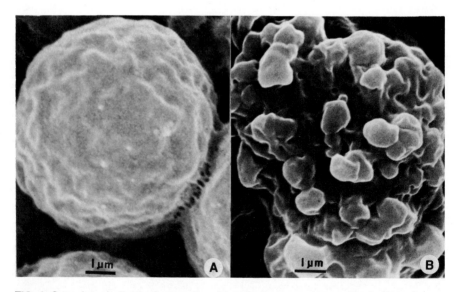

FIG. 1. Scanning electron micrographs of two typical granulosa cells. **A.** Hypophysectomized rat treated with estradiol for 3 days followed by purified FSH for 2 days. The surface is relatively smooth. **B.** Same as **A** except that the rat from which this cell was obtained received 5 IU hCG 12 hr previously. Note the presence of prominent blebs protruding from the surface.

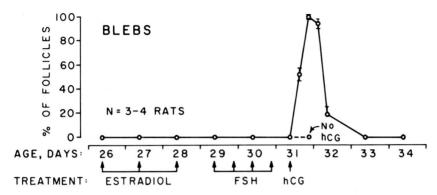

FIG. 2. The incidence of blebs on granulosa cells in follicles from hypophysectomized rats at various times during hormone treatment. The vertical bars represent one standard error of the mean.

(hFSH) for 2 days (1 μg, b.i.d.). The hFSH had been treated with chymotrypsin to remove residual LH activity, and was kindly supplied by Dr. Leo Reichert (LER-8/117).

Granulosa cells which stayed relatively smooth during treatment with estradiol and FSH (Fig. 1A) underwent dramatic changes in membrane topography after administration of hCG. In particular, the adluminal and cumulus cells of the graafian follicles exhibited pleomorphic blebs on their free surfaces (Fig. 1B). Transmission electron micrography showed the blebs to be rich in free ribosomes and vesicles. Over the next 48 hours the blebs disappeared while the cells continued to undergo luteinization. A semi-quantitative time course for the appearance and disappearance of the blebs is shown in Fig. 2. Dekel et al. (18) have also reported the presence of numerous blebs on the surface of postovulatory cumulus cells.

The observed formation of blebs is reminiscent of the response of thyroid cells to thyroid-stimulating hormone (TSH) (41). In that study formation and shedding of blebs appeared to be associated with a loss of RNA. Such a mechanism could be useful for the elimination of protein synthetic machinery that is no longer needed. During the differentiation of the granulosa cells in the avascular graafian follicle, this process could involve a smaller expenditure of energy than would a complete turnover of ribosomes, mRNA, and translational factors.

Changes in Gap Junctions

During the early period of luteinization, the number and size of gap junctions between granulosa cells (abutment junctions) appear to undergo a pronounced decline (8,16). This decline coincides with an initial rise and subsequent fall in the incidence of intracellular, or annular, gap junctions that are formed by interdigitation of the cells (8,16). These annular junctions are believed to be associated with granulosa cell cytoplasmic exchange or mutual autophagy of cellular processes (8,16,19,28). Consequently, the increase in the number of annular junctions may imply an active degradation of the existing abutment gap junctions, whereas the ensuing decrease may imply a decrease in the number of cell processes being developed. This overall loss of gap junctions is presumed to be partially responsible for the decrease in adhesiveness of granulosa cells observed following LH or hCG administration. Thus, the number of dispersed, viable granulosa cells obtained by applying pressure to whole ovaries increases dramatically several hours following hCG injection. Although this does not necessarily imply dissociation of gap junctions, application to ovaries of treatments known to uncouple and disrupt gap junctions results in a similar increase in viable, dispersed cells (11). It is uncertain whether these junctional changes serve to mediate or are an end result of LH action. However, they do indicate, along with the bleb formation described above, that the granulosa cell membrane is activated at times very close to initial binding of LH (or hCG).

Receptor Changes

To obtain information as to the content of, and changes in, the receptor for LH as well as for the hormone itself, we have quantified the number of LH receptor sites per granulosa cell during sequential administration of estradiol, FSH, and LH (hCG) to hypophysectomized, immature, female rats

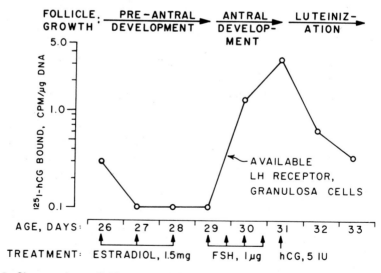

FIG. 3. Changes in available receptor for LH associated with granulosa cells expressed from follicles at various times during the course of the indicated hormonal treatment. Available receptor was quantified by the specific binding of ^{125}I-hCG and is expressed on a log scale as binding per µg DNA.

(34,35,37,38). As shown in Fig. 3, the number of available receptor sites per cell rises rapidly during the administration of FSH and falls precipitously within 12 to 18 hr after administration of hCG. This fall is not due to an inability to measure the sites because of occupancy of the receptors by the injected hCG. Inclusion of a small trace of labeled hCG with the luteinizing dose indicated that the injected hormone can only be occupying a very small number of sites relative to the total at any time during this decline in available receptors. The sites which are lost begin to be replenished within 2 to 3 days, but only if prolactin is provided (38,39), as would occur if pseudo-pregnancy or pregnancy were to ensue. Coordinated with this loss of LH receptor is a similar hCG-induced loss of FSH receptor and an increase in prolactin receptor (34,35,37–39). Thus, in addition to structurally evident changes in the granulosa cell surface, hCG induces changes in at least three receptors within the plasma membrane.

Metabolism of Receptor-Bound hCG

A number of mechanisms could account for the relatively rapid loss of receptor sites. Among them are shedding, decreased synthesis, masking, degradation within the membrane, and endocytotic internalization. Although direct evidence for or against any of these mechanisms being causative is not available, indirect evidence favors degradative and internalization mechanisms. Thus, studies we and others have done using electron microscopic autoradiography to follow the distribution of ^{125}I-hCG have revealed the presence of some grains within the target cell (4,12,13,22,23). These grains are associated with subplasmalemmal vesicles as early as 60 min following intravenous injection (23). Further, subsequent analyses have indicated that these vesicle-associated grains tend to move further into the cell with the passage of time, and the labeled vesicles then appear to fuse with lysosomes (12,13,23). Interpretation of such studies is complicated, however. The radioiodine is known to be incorporated almost exclusively in the α-subunit of hCG. Thus, it is the distribution of the label on the α-subunit which is being followed, and whether this distribution corresponds with that of the intact subunit or hormone cannot be directly ascertained. Furthermore, the number of grains inside the cell is always substantially less than that on the plasma membrane during the critical early hours following hCG binding. Therefore, the functional significance, if any, of the radioactivity observed to enter the cell during this period is unclear. These data do not allow us to decide whether binding and subsequent generation of cyclic AMP are the only important initial events in LH action, or if LH acts to induce much of the luteinization process by entering the cell. They do, however, indicate that at least a fraction of the hormone molecules bound do enter the cell either intact or modified. Once inside they could conceivably act to alter genotypic expression.

As a different approach to this problem we have used an approach first described by Morgan et al. (29). Instead of labeling the intact hormone, we labeled the α- and β-subunits individually with different radioisotopes of iodine. As expected, the separately labeled subunits were incapable of binding to receptor. However, they were capable of recombining, and the recombined hormone was able to bind specifically to receptor.

By injecting the dual-labeled hCG into immature, female rats primed with estrogen and FSH, and following the time course of uptake and loss of each label in several tissues and cellular subfractions, we were able to establish that the radioactivity associated with the β-subunit of hCG is preferentially retained by the post–30,000 $\times$ g particulate fraction of homogenized granulosa cells, whereas the radioactivity associated with the α-subunit is preferentially lost (10). Changes in the ratio of the β-subunit label to that of the α-subunit label over time for supernatant and particulate fractions of various tissues in one representative experiment are shown in Fig. 4. Radioactivity ratios in the adrenal compartments, follicular fluid, and ovarian interstitial/thecal (residual) tissue supernatants closely followed the ratios in sera. Liver (not shown) and kidney, tissues that actively degrade the hormone, initially concentrated the β-subunit–associated label, but then rapidly cleared it. The free subunits were also concentrated by these tissues, suggesting that this is a scavenging mechanism, i.e., nonspecific. On the other hand, the selective uptake and retention shown by the granulosa cells, and to a much lesser extent by the residual tissue, appears to be specific; it did not occur with unrecombined or free subunits, and persisted for a long period. Furthermore, this concentration of activity appears to be associated with vesicles or membrane components that are labile to freeze-thawing, since storage and homogenization of granulosa cells in 20% glycerol shifted the accumulation of the β-subunit label from the particulate to the cytosolic portion of the cell. Similar results were obtained in other experiments when the two subunits were labeled with the opposite radioisotopes and then recombined.

The time course of this label retention corresponds to that seen with high resolution autoradiography by Conn et al. (13) and by Amsterdam et al. (4) for the movement of radioactivity from the surface of target cells to a clearly cytoplasmic position. These results suggest that, after hCG or LH binds to receptor located at the membrane surface, at least a part of the α-subunit is lost while part or all of the β-subunit remains bound to, or enclosed in, a subcellular particle. It is possible that some fragments of the hormone are generated which are the actual triggers for some of the actions of the hormone. The temporal association of the internalization of hormone with morphologic events such as blebbing and biochemical events such as loss of receptor and uncoupling remains to be clarified, but the results suggest that hormone and/or receptor internalization may play an important role in the

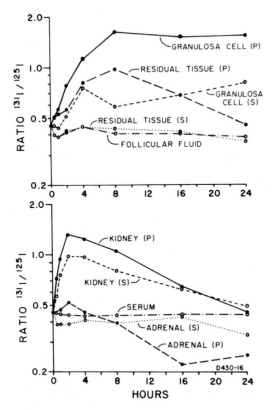

FIG. 4. Changes over time in the ratio of $^{131}I/^{125}I$ in various tissues. Immature rats pretreated with estradiol and FSH received an intravenous injection of dual-labeled hCG prepared by recombining ^{131}I-labeled β-subunit with ^{125}I-α-subunit. *P* represents a 30,000 × *g* pellet and *S* the respective supernatant. All data have been corrected for channel overlap, changes due to half-life, and differences in geometry. See text for interpretation.

mechanism of LH action. Further, the observation that the changes were most pronounced in granulosa cells suggests that selective metabolism of hCG may have special pertinence to granulosa cell luteinization.

Uncoupling

Target cells exposed transiently to a subsaturating concentration of hormone often lose their ability to respond to a subsequent administration of the hormone, a process termed desensitization. For example, Hunzicker-Dunn and Birnbaumer have demonstrated that, after administration of hCG to immature rats primed with pregnant mare serum gonadotropin (PMSG),

LH-stimulatable adenylate cyclase activity is not reduced at 2 hr, is partially reduced at 6 hr, and is completely lost at 24 hr (25). To gain further insight into mechanisms that might account for such changes, we have studied the ability of hCG and of FSH to induce desensitization in granulosa cells of hypophysectomized rats treated with estrogen and FSH (40). An acute injection of hCG can induce refractoriness to further stimulation of cyclic AMP accumulation by a second injection of hCG or FSH given within 2 hr (Fig. 5). This occurs at a time when available receptors have not declined. Furthermore, FSH can also desensitize the periovulatory follicle to either

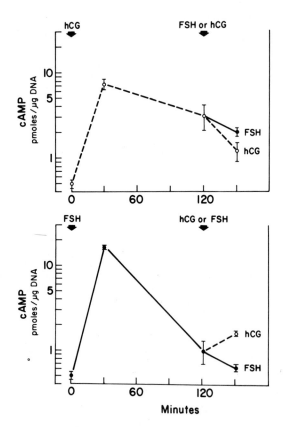

FIG. 5. Immature, hypophysectomized rats, primed with estradiol and FSH, were injected intravenously at time zero with either 50 IU hCG (top panel) or 5 μg FSH (bottom panel). After 2 hr they were challenged with a second injection of either gonadotropin, and killed 30 min later. At each indicated time point groups of rats were killed, ovaries removed, granulosa cells expressed and rapidly boiled, and cyclic AMP quantified in the boiled extracts by radioimmunoassay. The results, expressed as pmoles of cyclic AMP/μg DNA, are presented on a log scale. Note that the first injection of either hormone stimulated a major increase in concentration of cyclic AMP when measured 30 min later. However, at 2 hr, when the initial response had declined, a subsequent injection of the same or the other gonadotropin had little effect.

FSH or hCG. These findings imply that both gonadotropin receptors may share the cyclase or at least exert similar control over its activity and/or that of phosphodiesterase. In view of the absence of losses in receptor availability at the time of this desensitization, it appears that the gonadotropins do not induce desensitization by acutely changing receptor availability. A number of reports (5,14,15,24,32) have shown that adenylate cyclase activity and steroidogenesis each remain depressed for several days following a desensitizing dose of LH or hCG. Accordingly, these reports attribute desensitization to the loss of receptors seen following LH or hCG injection, probably by internalization or degradation as discussed above. This can explain the loss of hormone-sensitive adenylate cyclase activity at the later times, but it is probably not the explanation for changes such as those we have observed occurring at earlier time intervals. Thus, part of the mechanism of LH action probably involves acute stimulation of adenylate cyclase followed closely by desensitization and later by a loss of hormone receptors.

CHANGES IN NUCLEAR FUNCTION

To understand LH action on granulosa cells in greater detail, LH-dependent events need to be sharply defined. In particular, changes in turnover of macromolecules must occur since the granulosa cells differentiate both structurally and functionally into luteal cells in response to LH (or hCG). In an attempt to define some of these changes, we have initiated experiments designed to reveal changes in macromolecular synthesis subsequent to administration of hCG to immature rats primed with estrogen and FSH.

To study changes in DNA synthesis, the incorporation *in vivo* of tritiated thymidine into granulosa cells has been studied autoradiographically (33). A marked decline in the number of cells incorporating thymidine was observed after LH administration as compared to the number seen beforehand. This is shown graphically in Fig. 6 where the granulosa cell labeling index has been plotted throughout the course of hormonal stimulation. In contrast to the initial stimulating effects of estrogen and FSH, hCG initially acted to halt DNA synthesis. Although substantial incorporation of thymidine occurs subsequently, autoradiographic analysis reveals that this later incorporation is into invading stromal cells and not into luteinizing granulosa cells. It would appear that luteinization is accompanied by irreversible cessation of granulosa cell proliferation. Luteal cells do not incorporate thymidine into nuclear DNA even at much later times when the corpus luteum is well established. The LH (hCG)-induced inhibition of DNA synthesis may not solely be a response to elevated concentrations of cyclic AMP; the state of cellular differentiation and possibly other mechanisms may be more important. Thus, administration of FSH to rats with predominantly preantral follicles caused a marked elevation in granulosa cell cyclic AMP concentra-

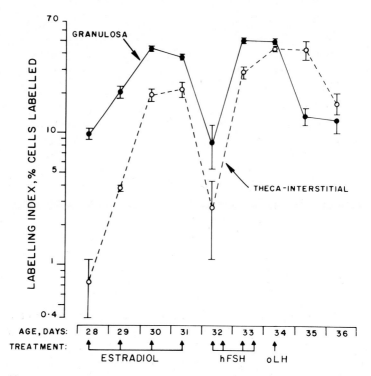

FIG. 6. Changes in the percentage of granulosa and theca-interstitial cells labeled with tritiated thymidine at various times during treatment with estradiol and FSH as determined by autoradiography. The labeled thymidine (15 μCi/g body weight) was injected 1 hr prior to death. (Adapted from Rao et al., ref. 33.)

tions and, if anything, led to enhanced cell division (33,36). This effectively irreversible inhibition of DNA synthesis supports the interpretation that the luteal cell is the product of a stable differentiation involving a change in genomic expression.

CHANGES IN CYTOPLASMIC ACTIVITY

If transcriptional changes occur, it should be possible to demonstrate such changes by studying the synthesis of specific proteins (26). For this purpose, we chose the two-dimensional electrophoretic analysis of O'Farrell (30), a procedure that can reveal most labeled proteins as discrete spots in an autoradiogram. Granulosa cells from immature rats primed for 2 days with estradiol and FSH were removed at various times after an intravenous injection of a luteinizing dose of hCG (5 IU), and incubated *in vitro* for 1 hr in Earle's balanced salts containing 300 μCi of ^{35}S-methionine. After incubation the cells were washed and sonicated. The extract was submitted to isoelectric focusing in the first dimension, followed by SDS electrophoresis in the second dimension. After drying and autoradiography, examination of

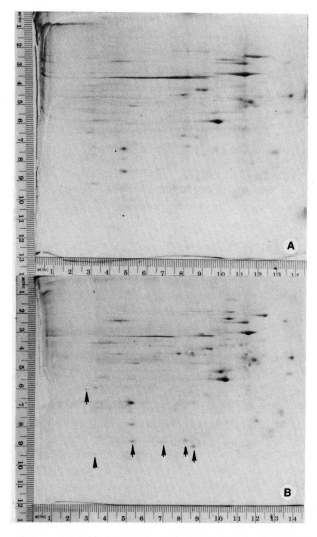

FIG. 7. Autoradiograms obtained following two-dimensional electrophoresis of total granulosa cell lysate. Cells were isolated from rats pretreated with estradiol and FSH, and then incubated for 1 hr in 1 ml Earle's balanced salts with 300 μCi ^{35}S-methionine. Following incubation the cells were washed, sonicated, and subjected to two-dimensional electrophoresis. The top panel **(A)** shows a representative pattern obtained with cells from rats that had not received hCG. **B** illustrates a typical pattern obtained 3 hr following the injection of 5 IU of hCG. The arrows point to 6 proteins not present in **A.**

the gels revealed a time-dependent shift in incorporation of label into several spots. Representative gels from time zero and 3 hr after hCG administration are shown in Fig. 7. By 30 min after hCG injection, at least two new proteins appeared on the gels, and continued to be synthesized at 24 hr after the hCG injection. By 3 hr after treatment with hCG, four additional proteins could be identified, three of which continued to be synthesized at 24 hr after hCG administration.

The demonstration that protein synthetic patterns begin to change within 30 min after hCG injection suggests that these changes may constitute some of the initial events which lead to luteinization. However, until the hCG-induced proteins and their respective roles are identified, no definitive mechanism for LH action can be described.

SUMMARY

Much remains to be learned concerning the process of luteinization. Nevertheless, current evidence is compatible with the possibility that LH acts to induce luteinization according to the following sequence:

(a) Binding to granulosa cells of antral follicles via a membrane-associated receptor that is coupled to adenylate cyclase and that is lost from the membrane subsequent to a single hormone binding event—either by internalization or degradation.

(b) Causing a transient increase in intracellular cyclic AMP that may bind to the regulatory subunit of one or several protein kinases, thereby activating or inactivating several other proteins by phosphorylation.

(c) Causing a change in the synthesis or processing of several proteins whose actions may be responsible for subsequent events.

(d) Causing an uncoupling of adenylate cyclase from gonadotropin receptors either directly or functionally; e.g., by inhibiting adenylate cyclase or enhancing phosphodiesterase activity or changing substrate compartmentalization, thereby isolating the cell from further gonadotropin-induced surges of cyclic AMP.

(e) Causing endocytosis of LH-receptor complexes with subsequent differential loss of part or all of the α-subunit and possible entrance of a portion of the gonadotropin into the cytoplasm, perhaps leading to genomic interactions.

(f) Initiating changes in the cell surface including the formation of adluminal blebs.

(g) Diminishing cell-to-cell linkage and communication by decreasing the number and size of gap junctions, changes that could serve to free the cells from inhibitory influences or the biochemical demands of other cells.

(h) Causing DNA synthesis and cell division to halt, thereby freeing energy stores for other types of synthesis.

(i) Stimulating continued functional and structural metamorphosis of the granulosa cells into luteal cells.

Clearly, this proposed sequence is incomplete, but it provides a framework for the design of future experiments on this fascinating, hormonally controlled differentiation.

REFERENCES

1. Albertini, D. F., and Anderson, E. (1975): Structural modifications of lutein cell gap junctions during pregnancy in the mouse. *Anat. Rec.,* 181:171–194.
2. Albertini, D. F., and Anderson, E. (1974): The appearance and structure of intercellular connections during the ontogeny of the rabbit ovarian follicle with particular reference to gap junctions. *J. Cell Biol.,* 63:234–250.
3. Albertini, D. F., Fawcett, D. W., and Olds, P. J. (1975): Morphological variations in gap junctions of ovarian granulosa cells. *Tissue Cell,* 7:389–405.
4. Amsterdam, A., Lamprecht, S. A., and Nimrod, A. (1978): Temporal relationship between desensitization of granulosa cells to human chorionic gonadotropin and internalization of receptor-bound hormone. *Endocrinology* (Suppl.), 102:1, 244a.
5. Ascoli, M., and Puett, D. (1978): Degradation of receptor-bound human choriogonadotropin by murine Leydig tumor cells. *J. Biol. Chem.,* 253:4892–4899.
6. Bagavandoss, P., and Midgley, A. R., Jr. (1977): Topographical changes in granulosa cell plasma membrane in response to FSH and hCG. *Biol. Reprod. (Suppl.),* 16:1, 50a.
7. Bagavandoss, P., Richards, J. S., and Midgley, A. R., Jr. (1977): Morphological changes in the rat granulosa cell surface during differentiation induced by estradiol and gonadotropins. In: *35th Ann. Proc. Electron Microscopy Society of America,* edited by G. W. Bailey, p. 484. Claitor's Publishing Division, Boston.
8. Bjersing, L., and Cajander, S. (1974): Ovulation and the mechanism of follicle rupture. IV. Ultrastructure of membrana granulosa of rabbit Graafian follicles prior to induced ovulation. *Cell Tissue Res.,* 153:1–14.
9. Bjorkman, N. (1962): A study of the ultrastructure of the granulosa cells of the rat ovary. *Acta Anat. (Basel),* 51:125–147.
10. Campbell, K. L., Landefeld, T. D., Bagavandoss, P., Quasney, M. W., Sanders, M. M., Byrne, M. D., Jonassen, J. A., and Midgley, A. R., Jr. (1978): Selective processing of the subunits of human chorionic gonadotropin (hCG) by target cells. *J. Cell Biol.,* 79:197a.
11. Campbell, K. L., and Midgley, A. R., Jr. (1978): Functional characteristics of granulosa cells obtained by sequential treatment with EGTA and hypertonic sucrose. *Biol. Reprod. (Suppl.),* 18:1, 63a.
12. Chen, T. T., Abel, J. H., Jr., McLellan, M. C., Sawyer, H. R., Diekman, M. A., and Niswender, G. D. (1977): Localization of gonadotropic hormones in lysosomes of ovine luteal cells. *Cytobiologie,* 14:412–420.
13. Conn, P. M., Conti, M., Harwood, J. P., Dufau, M. L., and Catt, K. J. (1978): Internalization of gonadotrophin-receptor complex in ovarian luteal cells. *Nature,* 274:598–600.
14. Conti, M., Harwood, J. P., Dufau, M. L., and Catt, K. J. (1977): Effect of gonadotropin-induced receptor regulation on biological responses of isolated rat luteal cells. *J. Biol. Chem.,* 252:8869–8874.
15. Conti, M., Harwood, J. P., Hsueh, A. J. W., Dufau, M. L., and Catt, K. J. (1976): Gonadotropin-induced loss of hormone receptors and desensitization of adenylate cyclase in the ovary. *J. Biol. Chem.,* 251:7729–7731.
16. Coons, L. W., and Espey, L. L. (1977): Quantitation of nexus junctions in the granulosa cell layer of rabbit ovarian follicles during ovulation. *J. Cell Biol.,* 74: 321–325.
17. Crisp, T. M., and Denys, F. R. (1975): The fine structure of rat granulosa cell cultures correlated with progestin secretion. In: *Electron Microscopic Concepts of Secretion,* edited by M. Hess, pp. 3–33. Wiley, New York.
18. Dekel, N., Kraicer, P. F., Phillips, D. M., Sanchez, R. S., and Segal, S. J. (1978): Cellular associations in the rat oocyte–cumulus cell complex. Morphology and ovulatory changes. *Gamete Res.,* 1:47–57.
19. Espey, L. L., and Stutts, R. H. (1972): Exchange of cytoplasm between cells of the membrana granulosa in rabbit ovarian follicles. *Biol. Reprod.,* 6:168–175.
20. Fletcher, W. H. (1978): Intercellular junctions in ovarian follicles. A possible functional role in follicle development. In: *Ovarian Follicular Development and Func-*

tion, edited by A. R. Midgley, Jr. and W. A. Sadler, pp. 113–120. Raven Press, New York.

21. Fletcher, W. H., and Everett, J. W. (1973): Ultrastructural reorganization of rat granulosa cells on the day of proestrus. *Anat. Rec.,* 175:320a.

22. Han, S. S. (1978): Preliminary observations on time-dependent intrafollicular distribution of ^{125}I-hCG. In: *Ovarian Follicular Development and Function,* edited by A. R. Midgley, Jr. and W. A. Sadler, pp. 107–111. Raven Press, New York.

23. Han, S. S., Rajaniemi, H. J., Cho, M. I., Hirshfield, A. N., and Midgley, A. R., Jr. (1974): Gonadotropin receptors in rat ovarian tissue. II. Subcellular localization of LH binding sites by electron microscopic radioautography. *Endocrinology,* 96:589–598.

24. Hsueh, A. J. W., Dufau, M. L., and Catt, K. J. (1977): Gonadotropin-induced regulation of luteinizing hormone receptors and desensitization of testicular 3'5'-cyclic AMP and testosterone responses. *Proc. Natl. Acad. Sci.,* 74:592–595.

25. Hunzicker-Dunn, M., and Birnbaumer, L. (1976): Adenylyl cyclase activities in ovarian tissues. IV. Gonadotropin-induced desensitization of the luteal adenyl cyclase throughout pregnancy and pseudopregnancy in the rabbit and the rat. *Endocrinology,* 99:211–222.

26. Landefeld, T. D., Campbell, K. L., and Midgley, A. R., Jr. (1978): Rapid hCG-induced changes in the synthesis of specific granulosa cell proteins prior to the appearance of luteinization. *J. Cell Biol.,* submitted.

27. Merk, F. B., Albright, J. T., and Botticelli, C. R. (1973): The fine structure of granulosa cell nexuses in rat ovarian follicles. *Anat., Rec.,* 175:107–126.

28. Merk, F. B., Botticelli, C. R., and Albright, J. T. (1972): An intercellular response to estrogen by granulosa cells in the rat ovary. An electron microscope study. *Endocrinology,* 90:992–1007.

29. Morgan, F. J., Kaye, G. I., and Canfield, R. E. (1974): Characterization of preparations of radioiodinated human chorionic gonadotropin. *Isr. J. Med. Sci.,* 10:1263–1271.

30. O'Farrell, P. H. (1975): High resolution two-dimensional electrophoresis of proteins. *J. Biol. Chem.,* 250:4007–4021.

31. Quatacker, J. (1975): Cell contacts in rabbit corpora lutea. *J. Ultrastruct. Res.,* 50:299–305.

32. Rajaniemi, H., Jaaskelainen, K., and Hyvonen, T. (1978): Regulation of LH(hCG)-receptor and adenylate cyclase in luteinizing granulosa cells. In: *Ovarian Follicular and Corpus Luteum Function,* edited by C. P. Channing, J. M. Marsh, and W. A. Sadler. Plenum Press, New York (*in press*).

33. Rao, M. C., Midgley, A. R., Jr., and Richards, J. S. (1978): Hormonal regulation of ovarian cellular proliferation. *Cell,* 14:71–78.

34. Rao, M. C., Richards, J. S., Midgley, A. R., Jr., and Reichert, L. E., Jr. (1977): Regulation of gonadotropin receptors by luteinizing hormone in granulosa cells. *Endocrinology,* 101:512–523.

35. Richards, J. S., Ireland, J. J., Rao, M. C., Bernath, G. A., Midgley, A. R., Jr., and Reichert, L. E., Jr. (1976): Ovarian follicular development in the rat: Hormone receptor regulation by estradiol, follicle stimulating hormone, and luteinizing hormone. *Endocrinology,* 99:1562–1570.

36. Richards, J. S., Jonassen, J. A., Rolfes, A. I., Kersey, K., and Reichert, L. E., Jr. (1978): Cyclic AMP, luteinizing hormone receptor, and progesterone during granulosa cell differentiation: Effects of estradiol and follicle stimulating hormone. *Endocrinology* (*in press*).

37. Richards, J. S., and Midgley, A. R., Jr. (1976): Protein hormone action: A key to understanding ovarian follicular and luteal cell development. *Biol. Reprod.,* 14:82–94.

38. Richards, J. S., Rao, M. C., and Ireland, J. J. (1978): Actions of pituitary gonadotrophins on the ovary. In: *Control of Ovulation,* edited by D. B. Crighton, pp. 197–216. Butterworths, Sevenoaks, Great Britain.

39. Richards, J. S., and Williams, J. J. (1976): Luteal cell receptor content for prolactin

(PRL) and luteinizing hormone (LH): Regulation by LH and PRL. *Endocrinology,* 99:1571–1581.

40. Williams, J. J., and Richards, J. S. (1978): Granulosa cell desensitization: Reciprocal actions of FSH and hCG. *Biol. Reprod. (Suppl.),* 18:1, 62a.

41. Zeligs, J. D., and Wollman, S. H. (1977): Ultrastructure of blebbing and phagocytosis of blebs by hyperplastic thyroid epithelial cells *in vivo. J. Cell Biol.,* 72:584–594.

Ontogeny of Receptors and Reproductive Hormone Action,
edited by T. H. Hamilton, J. H. Clark, and W. A. Sadler.
Raven Press, New York 1979.

Gonadotropic and Steroid Hormone Interactions During Follicle Growth in Mammalian Ovaries

*Griff T. Ross, Stephen G. Hillier, and Anthony J. Zeleznik

Reproduction Research Branch, National Institute of Child Health and Human Development and Diabetes Branch, National Institute of Arthritis, Metabolism and Digestive Diseases, National Institutes of Health, Bethesda, Maryland 20014

At first glance it would appear that the title of this chapter is not consistent with the subject matter of this section on Ontogeny of Receptors in Ovary and Testes. However, I think it is apparent that the processes of growth and differentiation are similar whether one considers growth and development of follicles a function of the pre- or postnatal age of the organism or alternatively of the age of a follicle once it begins to mature. In this context the subject matter of this presentation will be found to be consistent.

In Tables 1 and 2 the matrices summarize the information discussed here. The columns in Table 1 represent events as they transpire during the growth and development of an individual follicle prior to ovulation. The rows represent hormones currently regarded as playing some role in these processes. Superficial examination of the matrices indicates several features worthy of consideration at the outset. First, it is obvious that not every hormone has been discussed. Second, in Table 1 the column labeled "Oocyte growth" and the rows labeled "Prol." and "Prog." are all wasted since every cell contains a question mark, indicating that information is not available. Similarly, in Table 2 cells in the columns labeled "Oocytes" and "other IC cells" and in the row labeled "Prog." contain only question marks.

Beginning with the processes that characterize follicular maturation, it is clear that the oocyte grows as a part of the maturation of every follicle. Growth is completed well in advance of the achievement of maximal diameter of the follicle complex as a whole. Thus, in most mammalian ovaries, the oocyte grows only minimally after the midsagittal diameter of the follicle reaches 150 to 200 μm (10). After the oocyte ceases to grow, the follicle continues to enlarge until the time of ovulation. Initiation of growth of the oocyte appears not to be dependent entirely on pituitary hormones since

---

* Currently, Deputy Director, The Clinical Center, National Institutes of Health, Bethesda, Maryland 20014.

TABLE 1. Hormone actions on preovulatory follicles

Hormone	Oocyte growth	Granulosa cell prolif.	Thecal hypertrophy	AMPS secretion	Antrum formation	LH receps. in gran. cell	Atresia	Steroid hormone secretion		
								Estr.	Andr.	Prog.
FSH	?	↑→	?	↑	↑	↑→	↓→	↑	?	↑
LH	?	↑	↑	↑†	?	↑	↑↓	↑	↑	↑
Prol.	?	?	?	?	?	?	?	?	?→	?
Estr.	?	↑→	?	?	?	↑	↑↓	?	?	?
Andr.	?	↑	?	?	0	0	↑↓	↑	?	↑
Prog.	?	?	?	?	?	?	?	?	?	?

↑, stimulation; ↓, inhibition; ?, role of hormone not known; 0, hormone not essential; †, LH may modulate the qualitative nature of AMPS secretion; AMPS, acid-mucopolysaccharides; FSH, follicle-stimulating hormone; LH, luteinizing hormone; Prol, prolactin; Estr., estrogen; Andr., androgen; Prog., progesterone.

TABLE 2. *Receptor profiles in ovarian cells*

Hormone	Oocytes	Granulosa cells	Theca cells	Thecal/ IC cells	Other IC cells
FSH	?	? & +	?	?	?
LH	?	? & +	+	+	?
Prol.	?	? & +	?	?	?
Estr.	?	+	?	?	?
Andr.	?	+	?	?	?
Prog.	?	?	?	?	?

+, receptor demonstrated; ?, receptor never demonstrated; IC, interstitial; Prol., prolactin; Estr., estrogen; Andr., androgen; Prog., progesterone.

progression of growth of the oocyte continues in the hypophysectomized mammal (26). However, in hypophysectomized mammals the progression of growth would seem to be attenuated compared to that in the intact mammal. Hertz has shown that this growth persists in ovaries of newborn rats that are transplanted under the renal capsule of hypophysectomized castrated male rats (7). These data are consistent with the concept that although the rates may be different, initiation of follicular growth and maturation can occur very well in the total absence of any pituitary hormonal stimulation.

The first change observed in the progression from the primordial to the primary follicle consists in cuboidal transformation in the shape of the granulosa cell (19). This change is accompanied by initiation of secretion of the zona pellucida and initiation of mitotic activity among the granulosa cells, which replicate many times during the preantral stage of follicle growth (3). Both follicle-stimulating hormone (FSH) and estrogens appear to act synergistically in stimulating granulosa cell proliferation (5). Luteinizing hormone (LH) stimulates follicular growth and maturation by stimulating the interstitial compartment of the ovary to secrete androgens that serve as substrate for aromatization, giving rise to estradiol 17β, which stimulates granulosa cell proliferation. If one provides an exogenous source of estrogen adequate to stimulate granulosa cell proliferation and superimposes on that stimulus the effects of an interstitial cell-stimulating hormone such as LH or human chorionic gonadotropin (hCG), one observes not only atresia but also an inhibition of the growth of nonatretic follicles (12). In the estrogen-primed, interstitial cell-stimulating hormone-treated, hypophysectomized, immature female rat, the mean diameter of nonatretic follicles is smaller than that of follicles in animals receiving the estrogen alone. Again, the role of prolactin in granulosa cell proliferation is unknown. Estrogen, as previously observed, synergizes with FSH in stimulating granulosa cell proliferation. Irrespective of whether one stimulates endogenous ovarian secretion of androgens or, alternatively, provides an exogenous source of androgen in the estrogen-primed hypophysectomized immature female rat, the consequences of these two maneuvers are identical.

The role of progesterone in these processes remains unknown.

The next event among changes in the morphology of the growing follicle is hypertrophy of the theca (19). This alteration is referred to in the literature as theca luteinization or epithelioid transformation of the theca interna cells from a spindle-shaped cell to an epithelioid type of cell with foamy cytoplasm. The role of FSH in stimulating hypertrophy of thecal cells remains unknown. LH or any variety of interstitial cell-stimulating hormone stimulates that hypertrophy dramatically. The role of prolactin, estrogen, androgen, and progesterone in this process is also unknown.

Following hypertrophy of the theca, the next event in follicular maturation is massive granulosa cell proliferation, secretion of acid-mucopolysaccharides, chondroitin sulfates, and antrum formation. In pulse chase experiments *in vivo,* radioactively labeled sulfur ($^{35}SO_4$) has been shown to traverse the cytoplasm of the granulosa cells and to be deposited in the form of acid-mucopolysaccharides in antral fluid (28). FSH has been shown to stimulate incorporation of $^{35}SO_4$ into acid-mucopolysaccharides *in vivo* and *in vitro* (20). The role of LH seems to be to modulate the qualitative nature of acid-mucopolysaccharide secretion. Whether that be a direct effect of LH acting on the granulosa cell or an action mediated by steroid hormones is not known. No other hormone has been shown to participate in these processes, so that it becomes a reasonably specific biochemical marker of the stimulatory effect of FSH on the granulosa cell. This secretory process is a concomitant of antrum formation and rapid preovulatory follicular enlargement.

Zeleznik and Midgley and their colleagues have shown that, coincident with antrum formation, interstitial cell-stimulating hormone or LH bind specifically to membrane receptors on granulosa cells of such follicles (29). Appearance of these receptors results from exposure to FSH in the presence of estradiol (21). Thus, estrogen and FSH act together to stimulate the development of LH receptors in granulosa cells of antral follicles. Again, LH can be said to act in two ways: (a) to stimulate interstitial cell secretion of androgens as a substrate for estrogen production, in which case the effect is stimulatory, and (b) in concert with prolactin, to modulate LH receptors (22).

Zeleznik and Hillier have shown that androgen is not required for the process of generating LH receptors (30). In their experiments, treatment of estrogen-primed hypophysectomized rats with flutamide, a competitive inhibitor of cytosolic receptor-binding of testosterone and dihydrotestosterone, failed to effect the generation of LH receptors in granulosa cells. A second model indicates that a receptor-dependent activity of androgens may not be required for any of the processes of follicular growth and maturation. The homozygous T_{fem} mouse, which is deficient in cytosolic androgen receptors, ovulates, becomes pregnant, and bears litters (16). Although the length of the reproductive life of these mice seems to be shortened when compared to that of either their heterozygous or wild homozygous counterparts, nonetheless, at least transiently they have the capacity to ovulate and reproduce.

Concomitant with all of the processes occurring up to the point of antrum formation and the development of ovarian follicles in rats, a large proportion of the maturing follicles undergo processes collectively referred to as atresia (9). In these processes, although the fate of other components of the follicular complex varies, the oocyte inevitably dies. FSH seems to inhibit atresia, and LH, by stimulating androgen synthesis which in turn stimulates the production of estradiol 17β, inhibits atresia in preantral follicles (6). When estrogens are provided, then androgens synthesized in response to interstitial cell-stimulating hormone seem to stimulate atresia (13). The role of prolactin in the atretigenic processes remains unknown as does the role of progesterone.

In accompaniment with all the morphological changes occurring prior to the time of ovulation, sex steroid hormone secretion is stimulated. Quantities of sex steroid hormones have been measured in the ovarian venous effluent of monotocus cyclic ovulators, such as the human and other primates (2,18). Estrogens, androgens, and progestogens are higher in the venous effluent from the ovary containing the dominant follicle destined to ovulate in that cycle. Thus, that follicle either secretes these steroids or directs their secretion by other follicles. Estradiol secretion by the ovary, measured in terms of peripheral plasma levels or urinary levels of the hormone, has been used as a marker of the extent of maturation of follicles during ovulation induction in human subjects (27). FSH seems to induce aromatase activity, and LH stimulates the secretion of androgen, which serves as a substrate for the enzyme (4). The role of androgens and FSH in stimulating secretion of androgens is not clear. However, granulosa cells from preantral follicles secrete progesterone proportionate to levels of FSH added to the medium (1). Simultaneous addition of androgens to the medium enhances the stimulatory effect of FSH. Although LH acts synergistically with FSH to stimulate the production of estradiol 17β *in vivo* (11) by stimulating the synthesis and secretion of androgens, these effects cannot be reproduced *in vitro* in granulosa cells isolated from preantral follicles since these cells appear to lack LH or hCG receptors. The physiological role, if any, of androgens stimulating progesterone secretion by granulosa cells in preantral follicles remains to be determined. Again the role of prolactin in this process is controversial. At a recent meeting of the Society for the Study of Reproduction, reasonable data were presented to indicate that ovarian androgen secretion *in vivo* is inhibited by treating the animal with estradiol. The role of estrogen in regulating progesterone secretion is unknown. Androgens serve as a substrate for estrogen production by isolated granulosa cells incubated *in vitro,* as indicated above, and enhance both basal secretion and the effect of any other substance that stimulates secretion of progesterone by such cells (1,8,15). Finally, progesterone effect has not been discovered.

Events transpiring during the pre- and postovulatory periods of follicular growth and maturation have been excluded deliberately. Midgley in a later chapter in this volume discusses how hormones interact during these periods in regulating receptor content of the various follicular components.

Interactions of steroid and peptide hormones in regulating follicle growth and atresia presume the existence of receptors for these substances. Moreover, receptor profiles of cells comprising the follicular complex and other hormonally responsive ovarian cells should provide some evidence for the sites of hormone actions and interactions. This information is summarized in Table 2.

Little, if anything, is known concerning oocyte receptors for hormones. Granulosa cells have membrane receptors for FSH, but when these receptors are acquired in the development of a given follicle is unknown (14). Moreover, the postnatal age of the host at which membrane receptors for FSH are acquired is yet to be learned. The examination of hormone binding by homogenates of whole ovaries provides no information on the localization of these among the several cell types in the ovary. Within the limits of sensitivity of methods for detection, no FSH receptors exist in thecal cells (17). Interstitial cells derived from thecal cells of follicles having undergone atresia may or may not have FSH receptors. The receptor profiles of other varieties of interstitial cells remain unknown. Before antrum formation, granulosa cells have no demonstrable LH receptors (17). In contrast, thecal cells usually have demonstrable membrane receptors for LH or hCG (17). Receptors for prolactin appear late in the course of the development of the follicle, and these are discussed by Midgley.

The cytosolic fraction of granulosa cells contains estrogen receptors that appear to participate in nuclear translocation of the hormone (23). The presence or absence of such receptors in thecal and other varieties of interstitial cells has not been determined. Granulosa cells from preantral follicles have cytosolic receptors for androgens that seem to participate again in nuclear translocation of the steroid (24,25). The presence or absence of these receptors in any compartment of the ovary has escaped detection so far. Thus, steroid hormone effects mediated by cytosolic receptors would appear to be confined to granulosa cells.

Considering the material contained in the two matrices discussed, it is apparent that much remains to be learned concerning hormonal control of follicular growth and atresia. Moreover, if one assumes these effects to be mediated by receptors for the hormones, then there are many gaps in our knowledge concerning distribution of these molecules among the cells that constitute the follicular complex. Furthermore, it remains to be shown whether cyclic variation in receptor activity of follicular components occurs.

It is hoped that drawing attention to these "blank spots" will stimulate a systematic study of these substances.

REFERENCES

1. Armstrong, D. T., and Dorrington, J. H. (1976): Androgens augment FSH induced progesterone secretion by cultured rat granulosa cells. *Endocrinology,* 99: 1411–1414.
2. Baird, D. T., and Frazier, I. J. (1975): Concentration of estrone and estradiol 17β in follicular fluid and ovarian venous blood of women. *Clin. Endocrinol (Oxf.),* 4:259–266.
3. Chiquone, A. D. (1966): The development of the zona pellucida of the mammalian ovum. *Am. J. Anat.,* 106:149–169.
4. Dorrington, J. H., Moon, Y. S., and Armstrong, D. T. (1975): Estradiol-17β synthesis in cultured granulosa cells from hypophysectomized immature rats; stimulation by follicle stimulating hormone. *Endocrinology,* 97:244–247.
5. Goldenberg, R. L., Vaitukaitis, J. L., and Ross, G. T. (1972): Estrogen and follicle stimulating hormone interactions on follicle growth in rats. *Endocrinology,* 90:1492–1498.
6. Harman, S. M., Louvet, J-P., and Ross, G. T. (1975): Interaction of estrogen and gonadotropins on follicular atresia. *Endocrinology,* 96:1145–1152.
7. Hertz, R. (1963): Pituitary independence of the prepubertal development of the ovary of the rat and the rabbit and its pertinence to hypoovarianism in women. In: *The Ovary,* edited by H. G. Grady and D. E. Smith. Williams & Wilkins, Baltimore.
8. Hillier, S. G., Knazek, R. A., and Ross, G. T. (1977): Androgenic stimulation of progesterone production by granulosa cells from preantral ovarian follicles: Further *in vitro* studies using replicate cell cultures. *Endocrinology,* 100:1539–1549.
9. Ingram, D. L. (1962): Atresia. In: *The Ovary,* edited by S. Zuckerman, A. M. Mandl, and P. Eckstein, pp. 247–273. Academic Press, London.
10. Lintern-Moore, S., Peters, H., Moore, G. P. M., and Faber, M. (1974): Follicular development in the infant human ovary. *J. Reprod. Fertil.,* 39:53–64.
11. Lostroh, A. J., and Johnson, R. E. (1966): Amount of interstitial cell-stimulating hormone and follicle stimulating hormone required for follicular development, uterine growth and ovulation in the hypophysectomized rat. *Endocrinology,* 79: 991–996.
12. Louvet, J-P., Harman, S. M., and Ross, G. T. (1975): Effects of human chorionic gonadotropin, human interstitial stimulating hormone, and human follicle stimulating hormone on ovarian weights in estrogen-primed hypophysectomized immature female rats. *Endocrinology,* 96:1179–1186.
13. Louvet, J-P., Harman, S. M., Schreiber, J. R., and Ross, G. T. (1975): Evidence for a role of androgens in follicular maturation. *Endocrinology,* 97:366–372.
14. Louvet, J-P., and Vaitukaitis, J. L. (1976): Induction of FSH receptors in rat ovaries by estrogen-priming. *Endocrinology,* 99:758–764.
15. Lucky, A. W., Schreiber, J. R., Hillier, S. G., Schulman, J. D., and Ross, G. T. (1977): Progesterone production by cultured preantral rat granulosa cells: Stimulation by androgens. *Endocrinology,* 100:128–133.
16. Lyon, M. F., and Glenister, P. H. (1974): Evidence from Tfm-o that androgen is inessential for reproduction in female mice. *Nature,* 247:366–367.
17. Midgley, A. R., Jr. (1972): Gonadotropin binding to frozen sections of ovarian tissue. In: *Gonadotropins,* edited by B. B. Saxena, C. G. Beling, and H. M. Gandy, pp. 248–260. Wiley (Interscience), New York.
18. Mikhail, G. (1970): Hormone secretion by the human ovary. *Gynecol. Invest.,* 1:5–20.
19. Mossman, H. W., and Duke, K. L. (1973): *Comparative Morphology of the Mammalian Ovary.* Univ. of Wisconsin Press, Madison.
20. Mueller, P., Schreiber, J. R., Lucky, A. W., Schulman, J. B., Rodbard, D., and Ross, G. T. (1978): Follicle stimulating hormone stimulates ovarian synthesis of

acid mucopolysaccharide in the estrogen-stimulated hypophysectomized immature female rat. *Endocrinology. (In press.)*

21. Richards, J. S., Ireland, J. J., Rao, M. C., Bernath, G. A., Midgley, A. R., Jr., and Reichert, L. E., Jr. (1976): Ovarian follicular development in the rat: Hormone receptor regulation by estradiol, follicle stimulating hormone, and luteinizing hormone. *Endocrinology,* 99:1562–1570.

22. Richards, J. S., and Midgley, A. R., Jr. (1976): Protein hormone action: A key to understanding ovarian follicular and luteal cell development. *Biol. Reprod.,* 14:82–94.

23. Richards, J. (1975): Estradiol receptor content in rat granulosa cells during follicle development: Modification by estradiol and gonadotropins. *Endocrinology,* 97:1174–1184.

24. Schreiber, J. R., Reid, R., and Ross, G. T. (1976): A receptor-like testosterone-binding protein in ovaries from estrogen-stimulated hypophysectomized immature female rats. *Endocrinology,* 98:1206–1213.

25. Schreiber, J. R., and Ross, G. T. (1976): Further characterization of a rat ovarian testosterone receptor with evidence for nuclear translocation. *Endocrinology,* 99:590–596.

26. Smith, P. E. (1939): The effect on gonads of the ablation and implantation of the hypophysis and the potency of the hypophysis under various conditions. In: *Sex and Internal Secretions,* edited by E. Allen, C. H. Danforth, and E. A. Doisy, pp. 931–965. Williams & Wilkins, Baltimore.

27. Tredway, D. R., Goebelsmann, U., Thorneycroft, I. H., Sribyatta, B., and Mishell, D. R., Jr. (1973): Comparison of serum and urinary estrogen as aids in monitoring HMG therapy. In: *Gonadotropin Therapy in Female Infertility,* edited by E. Rosenberg, pp. 209–215. Excerpta Medica, Amsterdam.

28. Zachariae, F. (1957): Studies on the mechanism of ovulation. *Acta Endocrinol. (Kbh.),* 26:215–253.

29. Zeleznik, A. J., Midgley, A. R., and Reichert, L. E. (1974): Granulosa cell maturation in the rat; increased binding of human chorionic gonadotropin following treatment with follicle stimulating hormone *in vivo. Endocrinology,* 95:818–825.

30. Zeleznik, A. J., Hillier, S. G., Knazek, R. A., and Ross, G. T. (1977): FSH-stimulated granulosa cell development in the rat ovary: An examination of the role of androgens. *Tenth Annual Meeting of the Society of the Study of Reproduction, Austin, Texas.* (Abstr. 76.)

Ontogeny of Receptors and Reproductive Hormone Action,
edited by T. H. Hamilton, J. H. Clark, and W. A. Sadler.
Raven Press, New York 1979.

Ontogeny of the Corpus Luteum: Regulatory Aspects in Rats and Rabbits

Lutz Birnbaumer, Sharon L. Day, *Mary Hunzicker-Dunn, and Joel Abramowitz

*The Department of Cell Biology, Baylor College of Medicine, Houston, Texas 77030, and *The Department of Biochemistry, Northwestern University Medical School, Chicago, Illinois 60611*

OVERVIEW OF FOLLICULAR MATURATION, LUTEINIZATION, AND CORPUS-LUTEUM LIFE-SPAN REGULATION IN RATS AND RABBITS

A simplified scheme of the ovarian development of follicles and of the effects of androgens (T), estrogen (E$_2$), follicle-stimulating hormone (FSH) both in tonic and "surge" concentrations, as well as the involvement of prolactin (PRL) and luteinizing hormone (LH) is presented in Fig. 1.

Follicles develop to the preantral stage under the influence of estrogens, and they respond to FSH in the presence of continued estrogen supply by undergoing a series of mitotic cycles, developing an antrum, forming intra-follicular fluid, and becoming "ovulable." Preantral follicles contain FSH receptors (52,61,68) and a FSH-responsive adenylyl cyclase (44,68), and it is assumed that their response to FSH is mediated by cAMP. Proof for this mediation is still missing, however. Two important characteristics of follicle maturation are the induction of receptors for LH (14,74) and their coupling to adenylyl cyclase (9,38,46,48,58). Mature follicles respond to tonic (low) levels of LH by secreting a mixture of steroids consisting of estradiol and variable amounts of androgens and progesterone. The cellular source of these steroids, as well as their relative proportion, varies with the species studied (13,26,28,29,30,54). The estrogen secreted by follicles acts as a positive feedback regulator on the hypothalamus and the pituitary (4), such that hypothalamic luteinizing hormone-releasing hormone (LHRH) initiates an ovulatory LH surge (1,4,22). Mature follicles then respond to ovulatory (high) levels of LH by initiating ovulation, initiating luteinization of both granulosa and theca cells, stimulating prostaglandin (PG) formation, and secreting, in a transient manner, large quantities of steroids, primarily estrogen and androgens. All of these effects have been shown to be mediated and mimicked by cAMP (15,56,57,86).

In rats and rabbits ovulation occurs about 10 to 14 hr after the LH surge.

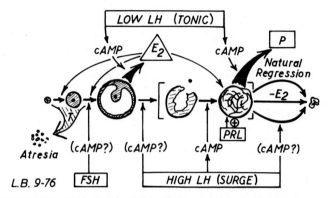

FIG. 1. A simplified scheme representing the development of ovarian follicles and CL. The interactions of androgens (T), estrogens (E₂), FSH, LH, and Prl, as well as the possible involvement of cAMP, are indicated.

Although all of the initial effects of LH, as well as most of the "late" effects of LH, such as stimulation of PG synthesis and luteinization, have been mimicked by cAMP or its dibutyryl derivative (51,62), ovulation per se has not yet been obtained except *in situ* and with LH.

Occupancy of LH receptors, stimulation of adenylyl cyclase, formation of cAMP and accumulation of intracellular steroids also lead rather rapidly to a refractory state with respect to further stimulation of cAMP formation by LH (58) and to profound alterations in the steroidogenic pattern (34,35, 65,79), such that either by the time of ovulation (28) or shortly afterwards (40,88), the newly forming luteal cells are steroidogenically quiescent, i.e., no longer synthesize steroids nor yet produce progesterone.

Formation of corpora lutea (CL) does not depend on ovulation, as shown by the fact that inhibition of PG synthesis does not interfere with luteinization yet blocks ovulation and leads to "CL" (luteinized follicles) with enclosed ova (2,3). Depending on the species involved, new CL start producing progesterone (P_4) either rapidly (rat) or rather slowly (rabbit). In the rat, P_4 secretion is evident (although by no means maximal) about 48 hr after ovulation; in the rabbit, P_4 secretion does not become significant until 3 to 4 days after ovulation (12,79).

Activation and maintenance of a corpus luteum involves the interplay of various hormones. The details of this interplay are by no means clear and appear to vary from species to species. In the rabbit, which ovulates in response to a combined surge of LH and PRL that does not include FSH (21, 25), newly formed CL do not make P_4 and are initially devoid of estradiol (E₂) receptors; within 3 to 4 days, however, E₂ receptors develop and become dependent on this steroid (63,64). This initial development may be dependent on PRL, which among other things may also be responsible for LH-receptor induction in the luteal tissue. However, definitive experiments in this direction are lacking. After day 5, rabbit CL are totally and constantly dependent on exogenous (extra-CL) E_2 supply for their survival. The required E_2 originates in the follicles, as shown by Keyes and Nalbandov (43). A dose of 1 to 2 μg E_2 twice daily will maintain viable CL in a hypophysectomized rabbit and is all that is needed in fully developed CL (73). Thus, such a dose will maintain CL that are located either in ovaries of normal rabbits whose follicles have been removed by X-ray irradiation (43) or underneath the kidney capsule of ovariectomized rabbits (72). No other steroid precursor of estradiol will substitute for it, for rabbit CL lack aromatase activity (27).

In the absence of the establishment of a pregnancy, rabbit CL persist for about 16 days and then regress. The reasons for their regression are not understood. The life-span of such a CL of pseudopregnancy (PSP) can be extended either by constant administration of high levels (20 to 40 μg/day) of E_2 (82) or by hysterectomizing the animal (78). Furthermore, uterine "injury" (as represented by the introduction of a thread into the uterine wall) improved the effectiveness of E_2 treatment in prolonging CL life-span. These maneuvers suggest that a uterine signal may be involved in the regulation of luteolysis in the rabbit. Although such a signal has been identified in the ewe as $PGF_{2\alpha}$ (59), the same type of identification has not yet been made in the rabbit. It is tempting, however, to suggest that either a uterine PG or a luteal PG may be responsible for initiation of luteolysis in the rabbit.

Implantation in the rabbit occurs on the seventh day after ovulation and leads to pregnancy. The life-span of the CL of pregnancy is 32 days. Also, the signal for luteal regression at the end of pregnancy, which precedes parturition by about a day to a day-and-a-half (12), has not yet been identified. Neither has the luteotropic substance responsible for maintenance of the CL during pregnancy been identified. The E_2 supply to the CL of PSP does not seem to differ from that to the CL of pregnancy, and it is therefore likely that maintenance of the rabbit CL of pregnancy beyond 16 days of age must be dependent on a product from the maternal-fetal unit.

Much more is known about hormones involved in CL maintenance in the rat. Thus, the CL of the cycle (which, as mentioned earlier, synthesize and secrete some P_4 throughout diestrus 1 and then regress) are rescued from demise by the pituitary hormone PRL (80). Secretion of PRL is initiated by reflex after cervical stimulation on prostrus and is episodic, i.e., occurs in two daily surges: one at about 0500 hr and the other at about 1700 hr. In intact animals these PRL surges continue for 10 to 11 days and then stop within a 1- to 2-day period. Coincident with the termination of the PRL surges, the CL, which had been actively secreting P_4 during the time of PSP, regress (81). These surges can be blocked by ergocryptine analogs, such as CB-154, which act at the pituitary level and mimic the action of dopamine, a neuroinhibitor of PRL release. Timing experiments showed that a CB-154 block after mating results in CL regression (23).

Implantation occurs on day 4 in the rat, and CL are maintained beyond the time of PSP by a placental luteotropin present in serum of day 12, pregnant rats. Thus, although during the first half of pregnancy the CL is under the control of pituitary PRL, during the second half it appears to be under the control of a placental luteotropin (32,67). During the transition time, days 8 to 11, the CL is strictly dependent on E_2 for its survival, and E_2 supply is in turn dependent on pituitary LH. Thus, administration of anti-LH during days 8 to 11 of pregnancy results in CL regression and abortion (67), and CL of pregnant rats that have been hypophysectomized and hysterectomized on day 10 will be maintained by an initial regime of E_2 followed by day 12 pregnant rat serum (32).

Radioreceptor assays of pregnant rat serum fractionated over Sephadex G-100 columns for PRL-like activities (heterologous rabbit mammary gland assay), revealed the presence of two types of cross-reacting substances, both dubbed "rat choriomammotropins" (rCMs) by Friesen's laboratory (41). One of these, having a higher apparent molecular weight, is secreted mainly between days 11 and 13, and may be the placental luteotropin capable of supporting the rat CL on day 12. The other, having a lower apparent molecular weight, may also be a luteotropic substance, in addition to being a mammotropin, but definitive identification of biological activities with the cross-reacting material is still missing. In any event, it appears clear that rat CL of pregnancy are first maintained by pituitary PRL, then by ovarian estradiol, and finally by "placental" luteotropin(s).

Of interest is a recent report by Takahashi et al. (85) indicating that the LH dependency, and hence E_2 dependency, of rat CL in midpregnancy (days 8 through 11) appears to be conditioned by LH itself. Thus, in a model of rats with 16-day-long cycles due to the presence of an additional pituitary underneath the kidney capsule, CL life-span was extended to more than 18 days after hypophysectomy on day 0 or after anti-LH treatment from day 1 through day 4, followed by hypophysectomy on day 5, but was shortened to less than 9 days after hypophysectomy on day 4.

As in the rabbit, the signal responsible for triggering CL regression shortly before parturition has not been identified. It may be that an intraluteal rise of $PGF_{2\alpha}$ occurs towards the end of pregnancy, but this still needs to be established. Interestingly, high (luteolytic) LH is capable of raising intraluteal $PGF_{2\alpha}$ levels, as shown by Demers et al. (20). The decrease of P_4 levels in the serum of pregnant rats starts on days 17 to 18, i.e., 3 to 4 days before parturition (66), indicating that luteolytic signals may have to be looked for at relatively early times.

In summary, follicles mature under the influence of E_2 and FSH and acquire sensitivity to LH. LH, via cAMP, triggers an intrafollicular steroid surge, an intrafollicular PG surge, ovulation, and luteinization. Newly formed CL persist for variable lengths of time depending on the species studied and whether the animal cycles establish a PSP or a pregnancy. In both rats and rabbits, the initial development of CL (cycle or PSP) appears to be independent of E_2. Complex interactions of CL with LH, PRL, and steroid hormones lead to maintenance of CL. In the rabbit, E_2 appears to be essential from day 5 throughout the CL life-span, but additional factors that are likely to be of importance need identification. In rats, PRL, E_2, and placental luteotropin(s) are required sequentially for CL maintenance throughout pregnancy. Additional feedback loops, such as suggested by the LH requirement at the beginning of CL life-span for the development of E_2 dependency at midpregnancy, also need to be identified. Finally, although PGs may be involved in the triggering of the natural regression of CL, especially at the end of pregnancy, much more work is needed in rats and rabbits to substantiate this concept, and surely other endocrine factors also play a crucial role.

An effect of LH on rat and rabbit CL not yet mentioned is its luteolytic action if presented to the CL in large concentrations. Since LH therefore seems to have both positive and negative actions on CL and since most of the actions of LH can be mimicked by cAMP, our laboratory explored the properties of LH-sensitive adenylyl cyclase activity in both follicles and CL. The purposes were (a) to describe the physiologic regulation of adenylyl cyclase and it responsiveness to LH; (b) to describe the correlations between changing levels of enzyme activity and responsiveness and activity states of follicles and CL; and (c) to gain new and perhaps novel insights into the regulation of corpus luteum life-span that would not be detected if LH-specific binding and progesterone output were the only parameters measured, since LH stimulability of adenylyl cyclase is clearly a measure of the capacity of LH receptor to couple and elicit a response.

METHODOLOGY

In all experiments that follow in which adenylyl cyclase activities were determined in crude homogenates, the assay medium contained high (3.0 mM) ATP, 5.0 mM $MgCl_2$, 1.0 mM EDTA, 1.0 mM cAMP, an ATP regenerating system consisting of 20 mM creatine phosphate, 2 mg/ml creatine kinase and 0.02 mg/ml myokinase, 25 mM Tris-HCl, pH 7.0, the indicated additives, and 20 µl of homogenate in a final volume of 50 µl. For details on the development of the conditions used and further processing of incubated samples, see Birnbaumer and Yang (5) and Birnbaumer et al. (6). Homogenates were prepared as described in detail elsewhere (37,38,66). In brief, the structures of interest (follicles or CL) were dissected, placed in cold Krebs-Ringer bicarbonate buffer until homogenization (20 min to 2 hr), blotted, weighed, and subjected to gentle homogenization in 20 volumes of 27% w/w sucrose, 1 mM EDTA, and 10 mM Tris-HCl, pH 7.5. Assays were performed within 30 min of homogenization.

The need to dissect individual structures (follicles or CL) is illustrated in

Fig. 2. Immature (26-day-old) rats were treated with 3 IU of pregnant mare serum gonadotropin (PMSG), and both their whole ovaries or their maturing follicles and recently ovulated CL were assayed for adenylyl cyclase activities. No significant changes in activities were observed when whole ovaries were homogenized, but large and significant changes were observed in those structures that underwent developmental changes: follicular maturation and luteinization. Therefore, all studies presented below, except when stated otherwise, were performed on carefully dissected follicles and CL.

In experiments where serum progesterone levels were determined, blood was collected either from the marginal ear vein (rabbits) or upon sacrifice from the jugular (rat and rabbits) and allowed to clot at room temperature. Serum obtained by centrifugation of the clotted blood was then stored at $-20°C$ until it was analyzed. For analysis, aliquots were extracted with petroleum ether and the organic phase containing progesterone was saved and evaporated. Progesterone was then taken up into 500 μl (50 mM) of phosphate-buffered saline (0.15 M NaCl) containing 0.10% w/v gelatin and assayed by radioimmunoassay using antiserum #337 kindly supplied by

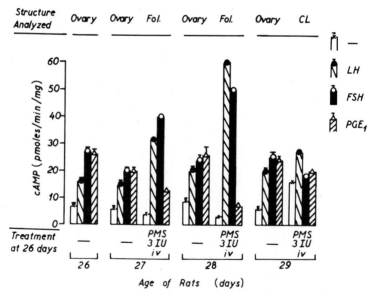

FIG. 2. Effect of PMSG on adenylyl cyclase activities in ovarian follicles and resulting CL from prepubertal rats. Charles River, outbred, received at 20 days of age, were injected at 26 days with 3 IU PMSG, i.v. (0800 hr). Animals were sacrificed at 10 A.M. on days 27, 28, and 29. Follicles and/or CL were dissected and handled as previously described. The adenylyl cyclase activities in homogenates of follicles and CL are compared to those of total ovaries from rats receiving no injections (—). Single points represent one assay in which follicles and/or CL from 10 rats were used. Means ± SEM are shown where triplicate assays in which ovaries from 3 rats were used in each assay. (From Hunzicker-Dunn and Birnbaumer, ref. 38.)

Dr. G. Niswender. Recovery of progesterone was calculated from the recovery of tracer [³H]progesterone added to the serum before extraction. Samples were not chromatographed. The validation of this radioimmunoassay using antiserum #337 has been published by Gibori et al. (32).

RESPONSIVENESS TO LH AND FSH OF FOLLICULAR AND CL ADENYLYL CYCLASE IN RATS, RABBITS, AND PIGS

As shown in Fig. 3, small immature antral follicles from pig ovaries contain an adenylyl cyclase system that responds better to FSH than to LH. Maturation of these follicles, assessed by their increase in size, leads to loss of FSH-responsive adenylyl cyclase relative to LH responsiveness and to development of a highly responsive LH-sensitive adenylyl cyclase system.

Thus, development of the ovulability of pig follicles correlates with the appearance of an LH-responsive cyclase. In view of Channing and Kammerman's (14) studies showing the appearance of LH-specific receptor sites with increasing size of pig follicles, it would seem that ovulability is associated not only with a general "maturing" of the follicle, with replication of granulosa cells, with protein secretion into antrum, and with steroid secretion into follicular fluid, but also with synthesis *and* coupling of LH receptors to adenylyl cyclase. The fall in FSH responsiveness from small to medium follicles may reflect either loss of FSH-responsive adenylyl cyclase or a relative increase of basal and LH-responsive adenylyl cyclase without absolute loss of the FSH-sensitive system. Upon transition to large follicles, FSH-

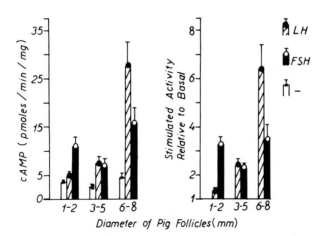

FIG. 3. Adenylyl cyclase activity and responsiveness to LH in homogenates of antral pig follicles of various sizes. Ovaries were obtained on the assembly line from nonpregnant, slaughterhouse pigs and were immediately placed in iced KRB. Follicles were dissected from the ovaries less than 3 hr after animal slaughter and were popped, homogenized, and assayed. (From Bockaert, et al., ref. 9, and Hunzicker-Dunn, *unpublished*).

responsive cyclase clearly increased. The significance of this is not clear, however. For one, since *all* the follicles were homogenized, it is not clear that all of the FSH-responsive system stems from the same cells that also respond to LH; for another, some of the FSH response can be accounted for by contaminating LH in the FSH-NIH-S9 used in these assays.

Mature ovulable follicles of rabbits also contain a highly responsive LH-sensitive adenylyl cyclase (Fig. 4, activities at $t = 0$). This finding agrees with previous indirect findings by Marsh et al. (58), who showed that large (1.5- to 2.0-mm) follicles from estrous rabbits respond to an *in vitro* incubation with LH by accumulating large quantities of cAMP (assessed by an ATP-prelabeling technique). Interestingly, Marsh et al. (58) also showed that this capacity of cAMP accumulation was lost shortly after treatment of estrous rabbits with ovulatory doses of human chorionic gonadotropin (hCG). As shown in Fig. 4, the appearance of such a refractory state to LH stimulation after hCG is selective for gonadotropin stimulation. Responsiveness to either PGE_1 or fluoride ion, known to nonspecifically stimulate all animal membrane-bound mammalian adenylyl cyclases (7,84), is unaltered. Two characteristics of the development of this re-

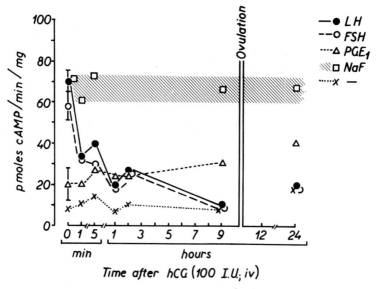

FIG. 4. Effect of the administration of an ovulatory dose of hCG on the adenylyl cyclase activities in rabbit ovarian follicles and resulting CL formed after ovulation. Estrous rabbits were given 100 IU hCG, i.v., and sacrificed by cervical dislocation at the times indicated on the figure. The ovaries were removed, and the follicles and/or CL were dissected, homogenized, and assayed. The "0 min" point represents Graafian follicles dissected from ovaries of estrous rabbits that received no injection. Single points represent one assay (duplicate incubations) in which a minimum of two rabbits were used. Means ± SEM are shown where two to six such assays were performed. (From Hunzicker-Dunn and Birnbaumer, ref. 37.)

fractory or desensitized state of the rabbit follicular adenylyl cyclase should be noted: (a) it appears very rapidly, and (b) it is complete 9 hr after the hCG injection. Figure 5 shows that not only exogenous hCG but also the natural LH surge induced by coitus result in an adenylyl cyclase system that is desensitized to gonadotropin stimulation. In addition, it can be seen that although newly formed (1-day-old) CL contain an adenylyl cyclase system that is unresponsive to stimulation, 2- and 3-day-old CL already contain a responsive system.

Based on the properties of the follicular versus luteal LH-responsive adenylyl cyclase activities—such as a 10-fold difference in sensitivity to LH, different nucleotide requirements for optimal coupling, and the appearance of highly responsive epinephrine-sensitive adenylyl cyclase in CL— it would appear that the cyclase system in CL is a different one from the one in follicles rather than the same one having regained responsiveness after total desensitization. LH responsiveness in rabbit CL of pregnancy (Fig. 5), as well as in rabbit CL of PSP (37), persists for as long as pregnancy (32 days) or PSP (16 days) lasts in this species. It was of interest to investigate how the levels of LH-sensitive adenylyl cyclase compared to a measure of functional activity of the CL, such as is reflected by serum P_4 levels. Figure 6 shows such a comparison and strongly suggests that after implantation the LH-sensitive adenylyl cyclase system should play a role in the moment-to-moment regulation of CL progesterone secretion.

Final maturation of rat follicles, occurring between diestrus 2 and proestrus, is also accompanied by the appearance of a highly responsive LH-sensitive adenylyl cyclase that appears between noon of diestrus 2 and noon of proestrus (Fig. 7). This LH-responsive cyclase does not persist in the cycling rat. Associated with the LH surge on the afternoon of proestrus, the cyclase

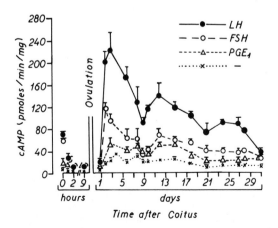

FIG. 5. Effect of mating on adenylyl cyclase activities in rabbit ovarian follicles and resulting CL. Estrous rabbits were mated to two fertile and experienced bucks and were sacrificed at times indicated. (From Hunzicker-Dunn and Birnbaumer, ref. 37.)

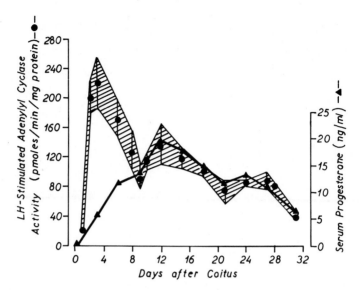

FIG. 6. Relationship between LH-stimulated adenylyl cyclase activity in homogenates of dissected ovarian CL and serum progesterone levels (Challis et al, ref. 12) obtained 1 to 31 days after mating. (From Hunzicker-Dunn and Birnbaumer, ref. 37.)

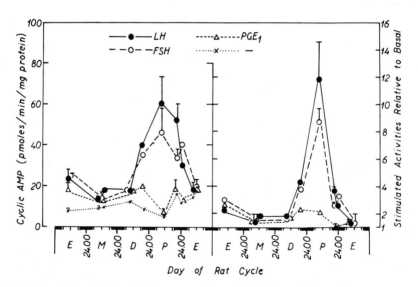

FIG. 7. Adenylyl cyclase activities measured in homogenates of ovarian follicles obtained from rats with a 4-day estrous cycle. Rats (Charles River, outbred) exhibiting at least two consecutive 4-day estrous cycles, as determined by vaginal smears, were used (lights on from 0500 to 1900 hr). Rats were sacrificed at the indicated times on each day of the estrous cycle. Ovaries were placed in iced KRB, and follicles and CL were dissected with the aid of a dissecting microscope and homogenized. Single points represent one assay in which follicles from five rats were used. Means ± SEM are shown where two or three such assays were performed. (From Hunzicker-Dunn and Birnbaumer, ref. 38.)

system in the thus resulting preovulatory follicles becomes desensitized. Blockade of the LH surge with pentothal anesthesia of the hypothalamus results in preservation of the LH sensitivity (38), indicating that the desensitization seen in the experiment of Fig. 7 is due to the endogenous LH surge.

Since this desensitization to LH in follicles is a consequence of exposure to ovulatory concentrations of gonadotropins, the question arises as to whether it is a necessary requisite for successful ovulation to occur. Data obtained in the rat (Fig. 7) indicate this not to be so, since (a) desensitization of follicle LH-sensitive adenylyl cyclase in this species is slow in developing, and (b) at a time at which ovulations have already occurred, the LH-stimulated activities are still quite elevated. Thus, the rat differs in its desensitizing response from the rabbit in that it is much slower. Newly formed rat CL, dissected on the day of estrus, are totally desensitized and show no remaining LH response (Figs. 7 and 8). Figure 8 shows the variations of LH-stimulated adenylyl cyclase expressed both in absolute activities and in activity relative to basal values in the rat CL of the cycle. It should be noted that the significantly increased LH-stimulable activity as found in estrus is observed only on diestrus 1 (metestrus on the figure) and that the relative stimulation never exceeds twofold. Compared to activities obtained in the CL of PSP or pregnancy (which are two- to sixfold higher), the CL of the cycle achieve (develop) only a minimally active adenylyl cyclase system. This correlates positively with the fact that serum P_4 levels due to the CL of the cycle are also very low when compared to those obtained in PSP or pregnancy. Figure 9 shows LH-stimulable adenylyl cyclase activity in CL of both

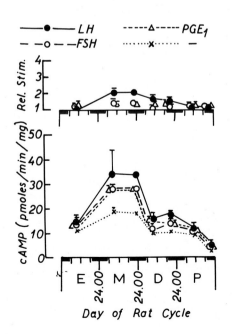

FIG. 8. Adenylyl cyclase activities measured in homogenates of ovarian CL obtained from rats with a 4-day estrous cycle. Rats (Charles River, outbred) exhibiting at least two consecutive 4-day estrous cycles, as determined by vaginal smears, were used (lights on from 0500 to 1900 hr). Rats were sacrificed at the indicated times on each day of the estrous cycle. Ovaries were placed in iced KRB and follicles and CL were dissected with the aid of a dissecting microscope, homogenized, and assayed for adenylyl cyclase activities. Single points represent one assay in which CL from five rats were used. Means ± SEM are shown where two or three such assays were performed. (From Hunziker-Dunn and Birnbaumer. ref. 38.)

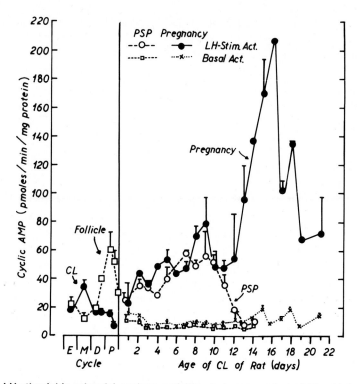

FIG. 9. LH-stimulable adenylyl cyclase activities in homogenates of CL obtained from pseudopregnant and pregnant rats and in follicles and CL obtained from rats exhibiting 4-day estrous cycles. For data on cycling rats, see legend to Figs. 7 and 8. PSP was induced by stimulating the cervix with a glass rod at 1700 to 1800 hr on the evening of proestrus and was confirmed by the presence of a leukocytic vaginal smear. Pregnancy was induced by placing females on the afternoon of proestrus with males known to be fertile and was confirmed by the presence of sperm in the vaginal smear on the morning of estrus. The day following mating or PSP induction was counted as day 1. Single points represent one assay (duplicate incubation) in which a minimum of two rats were used. Means ± SEM are shown where three or four such assays were performed. (From Hunzicker-Dunn and Birnbaumer, ref. 38.)

PSP and pregnancy. The activity in CL of PSP persists for as long as this type of CL is known to be functional, decaying between days 11 and 13, coincident with the lack of appearance of placental support due to rat placental luteotropin. Presence of an implant (pregnancy) has a "rescuing" action not only on the functional (P_4-secreting) aspects of CL (32,67), but also on the LH-stimulable adenylyl cyclase of these CL. A comparison of LH-stimulable activity and serum P_4 levels and the ratio between serum P_4 levels and serum LH levels throughout pregnancy are shown on Fig. 10. The P_4/LH ratio was calculated to see whether it would yield a pattern resembling that of LH-stimulable adenylyl cyclase, thereby testing for the possibility that LH and the responsible effector system might play a constant

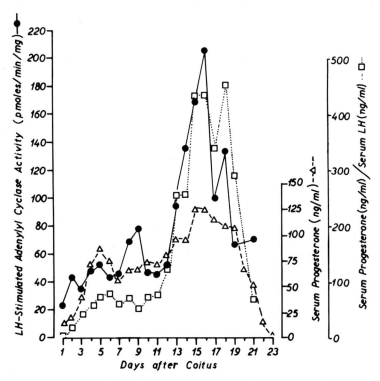

FIG. 10. Comparison of LH-stimulated adenylyl cyclase activity measured in homogenates of CL from pregnant rat ovaries to serum progesterone levels measured during pregnancy and to progesterone levels corrected for the daily fluctuations in serum LH. (From Morishige et al., ref. 66.)

rate-limiting, and hence regulatory, role in P_4 secretion by the CL. A quite close correlation of the patterns was found in this way, suggesting that in both the rabbit and the rat, the moment-to-moment regulation of CL is under LH control. Since the LH-stimulable activities and the serum P_4 and LH values stem from two separate laboratories, further correlations and possible cause-and-effect relationships between circulating LH and CL activity will have to come either from a reexamination in which all measurements will be made on the same rats or from different types of experiments. As part of this last approach, the question arose of whether LH stimulability would be extended in several of the situations known to extend the functional life-span of the CL. This was explored in rats and rabbits. Among the situations tested were hysterectomy of PSP rabbits (6) and PRL injection into cycling rats (Fig. 11). Both manipulations lead to the presence of LH-stimulable adenylyl cyclase on days where the CL would otherwise not have had such elevated activity and also would have lost its P_4 secreting capacity.

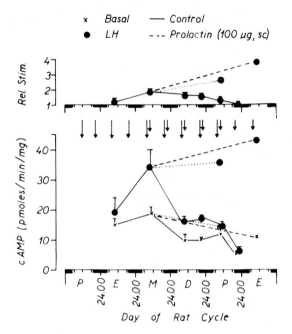

FIG. 11. Effect of the administration of ovine PRL on adenylyl cyclase activity in rat CL of the cycle. Rats exhibiting at least two consecutive 4-day estrous (*E*) cycles received PRL (100 μg, s.c.) every 12 hr starting on proestrus (*P* and *long arrows*) or diestrus 1 (*M* and *short arrows*) and sacrificed on proestrus or estrus, respectively. The adenylyl cyclase activity in homogenates of CL are compared to those of CL from rats that received no injections. Means ± SEM are shown where triplicate assays in which CL from three rats were used in each assay. (From Hunzicker-Dunn and Birnbaumer, ref. 38.)

It appears, therefore, that the presence of LH-sensitive adenylyl cyclase may be a marker for two ovarian functions: (a) maturation and appearance of ovulable follicles and (b) functional steroidogenic capacity of mature CL.

DESENSITIZING ACTION OF LH IN CL OF RATS AND RABBITS: LUTEOLYTIC ACTION OF LH

In view of the fact that ovulatory doses of LH and hCG caused the appearance of a refractory state in preovulatory follicles, and in view of the fact that large doses of LH had been described by Greep (33), as well as by Rothchild (75) and Stormshak and Casida (83), to be luteolytic in the rabbit, it was of interest to explore (a) whether luteolytic doses of gonadotropin would be deleterious to the CL adenylyl cyclase system, i.e., cause desensitization to LH action, and (b) what the characteristics of this desensitizing effect would be in terms of the time required for its appearance and reversibility. As shown in Fig. 12, the injection of a large dose of hCG (100 IU) into pregnant rabbits results in loss of LH-responsive adenylyl cyclase

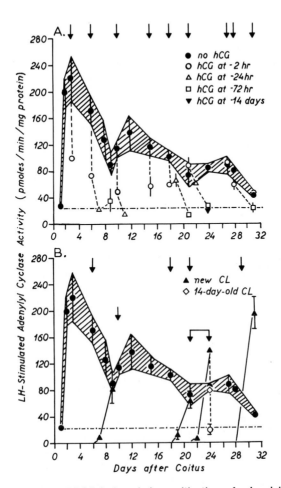

FIG. 12. A: Time courses of hCG-induced desensitization of adenylyl cyclase to LH stimulation in pregnant rabbits. hCG (100 IU/3.5 to 4.5-kg rabbit, i.v.) was injected on the days indicated by the *arrows* and LH-stimulated adenylyl cyclase activity was then determined in excised CL at the times indicated after hCG injection (In one instance the activity was determined 14 days after hCG-induced desensitization.) Dotted lines represent the time course for desensitization of CL adenylyl cyclase. **B:** Same experiment. LH-stimulated adenylyl cyclase activity in newly induced CL in pregnant rabbits and desensitization of the new CL by hCG in late pregnancy. hCG (100 IU/3.5 to 4.5-kg rabbit, i.v.) was injected at the days indicated by the *arrows,* and LH-stimulated adenylyl cyclase activity was determined in newly formed CL 1 and 3 days after the induction of ovulation (▲). In one set of rabbits, a second hCG injection was given (*double arrow*) and the activity of the newly formed, 3-day-old CL was determined 2 hr after the second (desensitizing) hCG injection (○). In another set of pregnant rabbits, hCG was injected on day 10 (*lowered arrow*), and adenylyl cyclase activity was determined in the new CL 14 days later (day 24 of pregnancy) (◇). The main curve (*hatched area*) in both **A** and **B** represents LH-stimulated adenylyl cyclase activity ± SEM in CL of pregnant control rabbits that received no treatment. (From Hunzicker-Dunn and Birnbaumer, ref. 39.)

activity shortly afterwards. Several interesting findings emerged from this study: (a) the rate at which desensitization occurs in pregnant rabbits varies with the progress of pregnancy and is much slower between days 20 to 30 than between days 11 to 20; (b) desensitization, once developed, does not appear to reverse, a finding that correlates with the fact that desensitization in CL is followed by functional luteolysis (interruption of P_4 secretion) and morphologic luteolysis (conversion of CL into corpora albicantia); and (c) our rabbits did not abort. This last finding indicated that (a) the desensitizing and luteolytic dose of hCG caused ovulation and formation of new CL, and (b) newly formed CL apparently acquired their capacity to synthesize and secrete progesterone more rapidly than the original CL. As will be shown below, the rapid activation of new CL is "pregnancy dependent," since the mere presence of high progesterone levels at the time of ovulation and initiation of luteinization, such as seen in 12-day-old hCG-treated pseudopregnant rabbits, is not sufficient to accelerate the rate of appearance of P_4 secretion. Interestingly, the LH-sensitive adenylyl cyclase activity in CL newly formed in pregnant rabbits, measured 72 hr after hCG, was found to be similar to the activity that would have been found in the original CL at that time of pregnancy, rather than being similar to that seen in original 72-hour-old CL (Fig. 13). On the other hand, 72-hr-old CL in 24-day-pregnant rabbits desensitized in response to hCG at a rate that corresponds to their age rather than to the time of pregnancy (Fig. 13). Thus, it is clear that the LH-sensitive cyclase system in rabbit CL is not a fixed preprogrammed function;

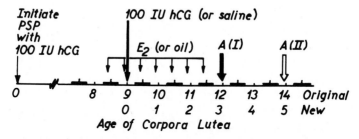

FIG. 13. Treatment schedule for the investigation of the effects of estradiol treatment and withdrawal on normal and desensitized CL function and adenylyl cyclase activity. PSP was initiated in estrous New Zealand white rabbits (3.5 to 4.5 kg) that had littered at least once. Three groups of 15 animals received twice daily either 0.1 ml sesame oil (control), 1.5 μg E_2 in 0.1 ml sesame oil (low E_2), or 15.0 μg E_2 in 0.1 ml sesame oil (high E_2); injections were started at 2200 hr on day 8 and continued at 12 hr intervals. The last injection was at 1000 hr of day 11. As indicated by the *arrows*, 6 rabbits of each group received 0.5 ml saline with 100 IU hCG on day 9 between 1100 and 1300 hr, and the remainder received 0.5 ml saline alone. On day 12 from each group 6 control animals and 3 animals treated with hCG on day 9 were sacrificed, and their CL dissected, weighed, homogenized, and assayed for adenylyl cyclase activities and progesterone content. The rest of the rabbits representing animals from which E_2 had been withdrawn for 2 days were sacrificed between 1000 and 1100 hr on day 14, and the same measurements were made. (From S. L. Day and L. Birnbaumer, *unpublished*.)

rather it is a function under complex control by factors present in pregnant animals that determine the absolute activities in CL and the rate at which hCG causes desensitization, as evidenced by the slower rate seen in 20- to 30-day-old CL. The data indicate that the establishment of a more slowly responding system requires more than 72-hr, and investigations are currently underway to identify factor(s) that is (are) responsible for this phenomenon.

LH and hCG cause desensitization of the CL LH-sensitive adenylyl cyclase in rats as well (39). The rate at which desensitization is established in this species is similar in PSP and in early and late pregnancy, and is comparable to that seen in the rabbit during the last 10 days of pregnancy.

Thus, in both species (rats and rabbits) LH is luteolytic and its luteolytic effect is preceded by desensitization of the adenylyl cyclase system to LH stimulation.

ASPECTS OF ESTROGEN ACTION IN THE RABBIT CL: UNCOUPLING EFFECTS ON LH-STIMULABLE ADENYLYL CYCLASE

As mentioned earlier, there is ample evidence that because of their lack of estrogen-synthesizing capacity (27), rabbit CL are absolutely dependent for their survival on an exogenous supply of estrogen. The estrogen is normally supplied by surrounding follicles, which secrete it in response to tonic stimulation by low concentrations of LH. In the absence of LH and/or follicles, 1 to 3 μg estradiol per day are sufficient to maintain viable CL in the rabbit (43,73). In exploring the possible mechanism by which hCG and LH exert their desensitizing action on the adenylyl cyclase system in the rabbit CL, the possibility was considered that one of the natural roles of estrogen might be to maintain an LH-sensitive cyclase system and that both luteolysis and loss of LH-stimulable adenylyl cyclase in the CL observed after hCG treatment was due to interruption of the follicular supply of E_2.

Table 1 shows that the desensitizing action of hCG in PSP rabbit CL could not be prevented by continued treatment of the animals with a physiologic dose of E_2 twice daily starting 15 hr prior to the hCG injection. In view of a report by Stormshak and Cassida (83) that higher doses of E_2 were effective in preventing hCG-induced morphological luteolysis in rabbits, it became of interest to test whether 15 μg E_2 twice daily protected the rabbit corpus luteum adenylyl cyclase against the desensitizing action of hCG. In addition to adenylyl cyclase activities, serum progesterone (functional activity of CL) and CL weight (structural integrity of CL) were also tested. The protocol is shown in Fig. 13. Rabbits were treated for three-and-a-half days with oil (controls), 1.5 μg E_2 in oil twice daily, or 15μg E_2 in oil twice daily starting at 2200 hr of day 8 of PSP (day after hCG = day 1). One-half of each group received 100 IU hCG/3.5- to 4.5-kg rabbit, i.v., between 1300 and

TABLE 1. *Lack of effect of E_2 treatment on desensitizing action of hCG on adenylyl cyclase of rabbit CL of PSP*[a]

Treatment	Time after hCG (days)	Adenylyl cyclase activity	
		Basal	LH
		(pmoles cAMP/min/mg)	
Control	—	25	124
hCG[b]	3	22	67
hCG + E_2[c]	3	35	58

[a] Age of CL at autopsy: 9 days.
[b] hCG: 75 IU/3.5 to 4.5 kg rabbit, i.v.; 9 A.M. on day 6.
[c] E_2: 1.5 μg/rabbit every 12 hr starting 24 hr prior to hCG.
From Hunzicker-Dunn and Birnbaumer, ref. 39.

1500 hr of day 9. Ear vein bleedings were taken daily, and animals were sacrificed by cervical dislocation between 1000 and 1100 hr of day 12, at which time old and new CL were dissected, weighed, homogenized, and assayed for adenylyl cyclase activities. As shown in Fig. 14, E_2 treatments per se had no apparent deleterious effect on serum P_4 levels. Neither low nor high E_2 was capable of fully preventing functional luteolysis due to hCG desensitization, even though the hCG plus high E_2 treatment resulted in

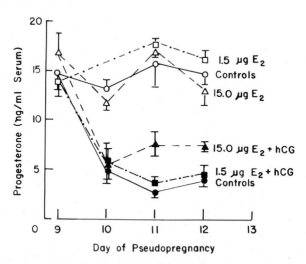

FIG. 14. The effect of estradiol treatment in the presence and absence of hCG treatment on serum progesterone levels in the pseudopregnant rabbit. Animals treated as indicated in Fig. 13 were bled from the marginal ear vein between 1300 and 1500 hr on days 9, 10, and 11, and between 0800 and 1000 hr on day 12. Serum progesterone was measured by radioimmunoassay as described in the text for days 9 and 12, $n = 6$, for days 10 and 11, $n = 3$. Means ± SEM are represented. (From S. L. Day and L. Birnbaumer, *unpublished*).

somewhat higher serum P_4 levels 2 and 3 days after hCG than did hCG alone or hCG plus low E_2. Thus, high levels of E_2 appeared to afford partial protection against functional luteolysis, but further experiments are required to confirm this on a statistical basis. CL were found to be structurally maintained after hCG only when simultaneously treated with high E_2 (Fig. 15). Three-day-old new CL were all smaller than the 12-day-old CL and showed no difference in weight with E_2 treatment (Fig. 15). On the basis of these data and of previous experiments establishing a close positive correlation between LH-sensitive adenylyl cyclase levels and luteal function, one might predict that LH-stimulable cyclase levels would be high in all CL of non-hCG treated animals, partially maintained in CL of animals treated with hCG plus high E_2, and low in CL of animals treated with hCG only or hCG plus low E_2. However, as shown in Fig. 16, LH-stimulable activity was high only in control animals, and animals that had received physiologic E_2 doses only. Surprisingly, administration of 15 μg E_2 twice daily was deleterious to the LH-sensitive adenylyl cyclase system.

Thus, under conditions where P_4 secretion by the CL was unaltered, and no macroscopic changes occurred (e.g., CL weight was unaffected), high E_2 treatment resulted in a selective loss of the LH-stimulable adenylyl cyclase activity. Experiments in the rat suggest that an E_2 treatment that extends CL life-span of "superovulated" rats by several days, as assessed by determination of serum P_4, has no deleterious effect on the LH/hCG receptor content of the E_2 supported CL (49). It appears, therefore, that a treatment that shows no effect on either receptor or final output may result in an uncoupling and alteration of the intermediary regulatory effector system. In agreement with findings of Holt et al. (36), withdrawal of exogenous E_2 supply from animals having received pharmacologic (but not physiologic)

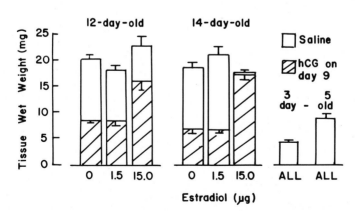

FIG. 15. Effect of estradiol treatment on CL weight after a lytic dose of hCG. Pseudopregnant rabbits were treated as indicated in Fig. 13. After sacrifice, 12- or 14-day-old CL, and where appropriate 3- or 5-day-old CL, were dissected and weighed. Means ± SEM are represented. (From Day and Birnbaumer, ref. 19.)

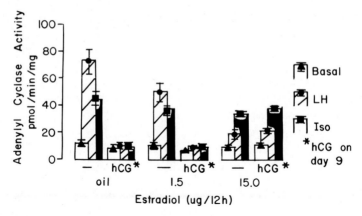

FIG. 16. The effect of estradiol treatment on adenylyl cyclase activities in 12-day-old CL from pseudopregnant rabbits that received either saline or a lytic dose of hCG on day 9 as described in Fig. 13. After sacrifice on day 12, CL homogenates were made and assayed as described in the text. When present in the assay, LH (NIH-LH-S19) was 10 µg/ml and isoproterenol was 10^{-4} M. Means ± SEM are represented. (From S. L. Day and L. Birnbaumer, *unpublished*).

E_2 resulted in decreased serum P_4 levels as assessed 2 days afterwards (Fig. 17). This indicated that although P_4 secretion in animals treated with high E_2 appeared normal, the uncoupled system that resulted had become dependent on E_2 in a manner that does not occur normally, i.e., such that physiologic LH levels could no longer act on the CL and possibly also not on the follicles. Although untested, it seems likely that not only CL adenylyl cyclase but also follicular adenylyl cyclase became uncoupled during the treatment with high E_2, thus leaving the CL in "double jeopardy": no E_2 from follicles and no "regular" LH-modulated cAMP within themselves.

Withdrawal of high E_2 did not result in recovery of the CL LH-stimulable

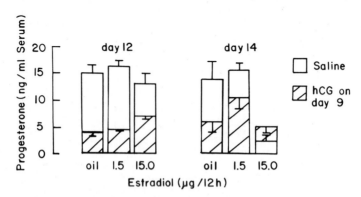

FIG. 17. Serum progesterone levels from pseudopregnant rabbits treated with estradiol and hCG as described in Fig. 13. Serum was collected from the marginal ear vein and assayed for progesterone by radioimmunoassay as described in the text. Means ± SEM are represented. (From S. L. Day and L. Birnbaumer, *unpublished* and ref. 19.)

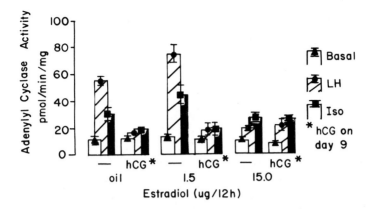

FIG. 18. The effect of a 2-day withdrawal of estradiol treatment on adenylyl cyclase activities in 14-day-old CL from pseudopregnant rabbits that received either saline or a lytic dose of hCG on day 9 as described in Fig. 13. After sacrifice on day 14, CL homogenates were made and assayed as described in the text. When present in the assay, LH (NIH-LH-S19) was 10 μg/ml and isoproterenol (Iso) was 10^{-4} M. Means $\pm$ SEM are represented. (From S. L. Day and L. Birnbaumer, *unpublished*).

adenylyl cyclase (Fig. 18). The experiments shown in Figs. 14 and 16–18 also included controls and tests for effects of low doses of E_2 (1.5 μg twice daily), which indicated that both treatment and withdrawal of low E_2 had no detectable effect on either serum P_4 or the CL adenylyl cyclase system. Furthermore, the effect of high E_2 appears to be specific for the LH-stimulability or the cyclase, for no significant changes were induced either in basal activity or in isoproterenol stimulability (Fig. 16, open and black bars).

As predicted from serum P_4 levels (Fig. 14), and in accordance with previous findings, hCG-induced desensitization is not prevented by low E_2 treatment. Since the animals were sacrificed 3 days after hCG, luteolysis is well advanced and desensitization (originally specific for LH stimulation) has extended into general loss of stimulable adenylyl cyclase (Fig. 16), as indicated by loss of isoproterenol-stimulable activity. The luteolytic effect of hCG was not prevented by low E_2, but appeared to be blocked by high E_2. Thus, although high E_2 had resulted in loss of LH-stimulable activity with retention of isoproterenol stimulability, treatment with hCG resulted in no further changes (Fig. 16) when tested 3 days after the administration of the otherwise luteolytic dose of hCG (Fig. 16). The mechanism(s) of this "protecting" effect is of obvious interest and deserves further investigation.

PERMANENCE OF DESENSITIZATION: LACK OF REVERSAL IN RAT AND RABBIT CL

Under no circumstance has reversal of desensitization induced by a high dose of hCG been observed in either rat or rabbit CL. This of course is not surprising, since desensitization is associated first with functional and then structural luteolysis. One example of the lack of reversal of desensitization in rabbit CL was presented earlier (Fig. 12), where CL in a 24-day-pregnant rabbit that had been desensitized on day 10 of pregnancy showed LH-stimulable activities that were low and comparable to those found in surrounding interstitial tissue. The experiments shown in Figs. 16 and 18 also indicate that once desensitized, the rabbit CL adenylyl cyclase in pseudopregnant rabbits does not reacquire LH-stimulable activity. Figure 19 presents an experiment with superovulated "pseudopregnant" rats, showing that in this species also, desensitization of the CL cyclase to LH stimulation is permanent. It is interesting to note that in this animal model, luteal cells are likely to be viable up to 7 days after the desensitizing dose of hCG, since both isoproterenol-stimulable and fluoride-stimulable activities were preserved until that time. In the rat it appears that morphologic luteolysis is more delayed with respect to functional luteolysis than in the rabbit. That luteolysis had indeed occurred is indicated by the serum P_4 levels in these rats (Fig. 20).

The findings shown in Fig. 19 are in apparent contradiction to those of

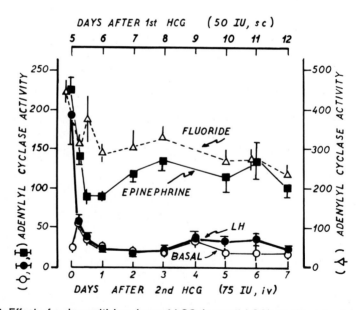

FIG. 19. Effect of a desensitizing dose of hCG (second hCG) on the adenylyl cyclase activities of CL induced in immature rats subjected to PMSG treatment (1600 hr on day 26 of age) and a first hCG treatment 65 hr afterwards. Rats were sacrificed at the indicated times after the second hCG treatment, their original CL were dissected, homogenized, and assayed for adenylyl cyclase activities in the absence and presence of either LH (10 μg/ml), epinephrine (10^{-4} M), or NaF (10 mM). Each point represents the mean ± SEM. (From M. Hunzicker-Dunn, S. L. Day, and L. Birnbaumer, *unpublished*).

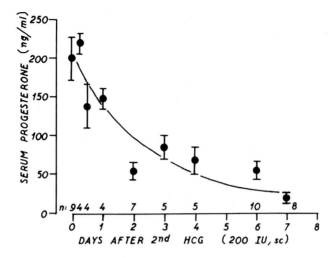

FIG. 20. Effect of a desensitizing dose of hCG (second hCG) on serum progesterone levels in immature rats subjected to PMSG treatment (1600 hr on day 26 of age) and a first hCG treatment 65 hr afterwards. Rats were sacrificed at the indicated times after the second hCG treatment. The rats were bled and serum progesterone levels were determined by radioimmunoassay. Each point represents the mean ± SEM. (From M. Hunzicker-Dunn, S. L. Day, and L. Birnbaumer, *unpublished*).

Catt's laboratory (17), which reported that LH-stimulable adenylyl cyclase in rat luteal tissue desensitized under identical conditions with even higher (200 IU/rat) doses of hCG returns within 4 to 6 days and is restored to control values by day 7. As shown in Fig. 21, this apparent discrepancy can be accounted for by the fact that return of activity can only be seen if total ovarian activities are determined, as opposed to activities in the dissected original CL. Data from Catt's laboratory on "luteal" functions are routinely obtained on total ovary and, in view of the findings reported here, are difficult to interpret.

To some extent it may seem surprising to find such a drastic difference between values obtained on dissected CL (which correlate positively with serum P_4 levels, cf., Fig. 19 versus Fig. 20) and values obtained on total ovary of a superovulated ovary, which at the moment of injection of the desensitizing dose of hCG is likely to be 90+% luteal tissue. However, histologic examination of the model used by Catt and collaborators (17), which was followed in the experiments presented here, indicated that whereas the ovary subjected to desensitization was indeed mostly luteal tissue, a large number of small antral follicles were also present, and hCG administration led to (a) further development of these follicles into mature large antral follicles, (b) subsequent partial and total luteinization of these follicles with or without concomitant ovulation, and (c) development by day 3 or 4 of a fair number of large haemorrhagic follicles as well as new antral follicles. Thus, by day 7, the day on which total ovarian data would indicate return

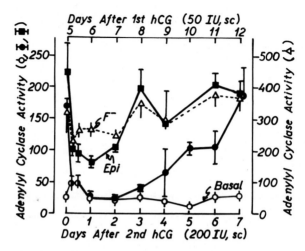

FIG. 21. Effect of a desensitizing dose of hCG (second hCG) on the adenylyl cyclase activities of whole ovaries in immature rats subjected to PMSG treatment (1600 hr on day 26 of age) and a first hCG treatment 65 hr afterwards. Rats were sacrificed at the indicated times after the second hCG treatment, and their ovaries homogenized and assayed for adenylyl cyclase activities in the absence and presence of either LH (10 μg/ml), epinephrine (10^{-4} M) (*Epi*), or NaF (10 mM) (*F*$^-$). Each point represents the mean $\pm$ SEM. (From M. Hunzicker-Dunn, S. L. Day, and L. Birnbaumer, *unpublished*).

to "normality" in terms of LH receptors and LH-stimulable adenylyl cyclase, the actual morphologic makeup of the hCG-treated ovaries is one of a heterogeneous tissue containing old (desensitized) CL; new CL (some of which contain enclosed ova), haemorrhagic follicles at various stages of regression (atresia); new, large, and presumably ovulable antral follicles; and a fair amount of preantral and antral follicles at various stages of development. Since immature rats treated with PMSG to develop large quantities of their follicles do so within 55 to 57 hr, the time at which they also have an endogenous gonadotropin surge (77), it may be that one reason for the asynchronous response of the hCG-ovulated ovary to the second (desensitizing) dose of hCG is the fact that the ovulatory hCG was administered 65 hr after PMSG and hence some 7 to 10 hr after the endogenous release of a mixture of LH, FSH, and PRL. This mixture can be assumed to initiate ovulation in some, but not all, of the mature follicles, to initiate desensitization and lutenization in some, but again not all, of the follicles, to promote further maturation of follicles, and to leave the ovary ill-prepared for the massive hCG dose that is supposed to synchronize and ovulate *all* follicles. This raises the question as to which cells of the total ovarian mass treated in this manner contain which features, i.e., secrete progesterone, respond by making cAMP, contain LH/hCG receptors, contain LH-stimulable adenylyl cyclase, contain FSH-stimulable cyclase, and so forth. Clearly, further studies should be carried out (a) with a better synchronized model [such as used by

Lee and Ryan (50), who administered the ovulatory dose of hCG 56 hr after PMSG, coincident with the endogenous gonadotropin surge] and (b) on properly dissected CL. The dissection step is especially important if isolated cells are to be prepared, regardless as to whether the effects of low or high doses of hCG are to be studied.

ON THE "ACTIVATION" OF CL

As mentioned earlier, newly formed CL in pregnant rabbits must become activated rapidly—in fact, more rapidly than the original CL—to synthesize and secrete P_4. This conclusion stems from the indirect observation that our hCG-desensitized pregnant rabbits did not abort. Interestingly, neither in PSP rabbits nor in superovulated PSP rats did newly formed CL become activated. Thus, in the rabbit, P_4 levels in serum dropped after hCG on day 9 and remained low and below control until day 14 (Fig. 17). Normal CL of PSP became activated by day 5, as seen from serum P_4 levels (26). As can also be seen from the data in Fig. 17, 1.5 μg E_2 twice daily did not aid in activating new CL during the first 3 days after their formation, since serum P_4 levels as seen on day 14 (5 days after formation of the new CL) were not significantly higher in rabbits treated with only low E_2 than those in rabbits treated with hCG. This of course agrees with the observations of Miller and Keyes (63) that development of rabbit CL is independent of E_2 during its first 4 to 5 days of life, and with the observation of Mills (64) that newly formed CL lack E_2 receptors. It is easy to speculate that in rat CL, activation may not have occurred because of a lack of PRL and that a pregnancy-related luteotropin must exist in the rabbit to account for rapid activation of its CL. However, a clear definition and identification of CL-activating factors are still missing.

FINAL COMMENTS AND SOME WORKING HYPOTHESES

Questions arise as to what may be the cause-and-effect relationships between stimulation of adenylyl cyclase by LH, desensitization of adenylyl cyclase to LH stimulation, and LH-induced luteolysis. Although a clear answer based on experimental evidence is not yet available, it can be speculated that a possible sequence of events involves the following: (a) LH causes an intracellular cAMP surge and, through prolonged occupancy of the receptors (20 min to 1 hr), leads to an uncoupling of the receptor cyclase system; (b) the uncoupling of the receptor leads to a deficiency in tonic effects of circulating LH, which in turn leads to a decrease in P_4 secretion following the initial rise; (c) the extraordinarily high levels of cAMP formed due to massive receptor occupancy and cyclase stimulation initiate in luteal cells, as they do in lymphocytes (10,18), melanocytes (16), and skin cells (87), "terminal differentiation" such that (i) the uncoupling of the receptor system, which may initially have been reversible, becomes irreversible, (ii) a steroidogenic block appears that can no longer be overcome by cAMP and (iii) the cell looses its function and either dies or becomes incorporated into the ovarian stroma.

Using this sequence of events as a working hypothesis, several phases or stages of LH-induced effects can be described that vary with time. The most significant aspect of these stages that needs to be considered in terms of accounting for the apparently wide discrepancies and differences among species is the time span during which they are reversible. Thus, while uncoupling may be reversible in the beginning (stage 1), it soon becomes irreversible (stage 2). Although the decreased P_4 production that follows the appearance of uncoupling should be "cured" by cAMP, analogs of cAMP, treatment with phosphodiesterase inhibitors, or choleratoxin (stage 1), the cAMP-induced steroidogenic block (stage 3) should not be "curable." While initial uncoupling need not be associated with receptor loss (stage 1), later stages may well include clearance of receptors from the plasma membrane either by shedding or internalization (stages 2 to 3). Although initial desensitization (uncoupling) can be expected to be independent of protein and RNA synthetic events (stage 1), later changes may well depend on such events, if as assumed, cAMP initiates "terminal differentiation" (stages 3 and 4).

If we accept this sequence of events as plausible, then the apparent discrepancies and differences observed in various systems can be accounted for. Thus, one might predict that reversible desensitization of the adenylyl cyclase system should precede irreversible loss of LH-stimulable activity; but if the initial desensitization reaction is very slow, proceeding at a rate that is of the same order of magnitude with which the cell initiates clearing of occupied LH receptors from its plasma membrane, then the initial reversible change might not be observed, despite it being an obligatory first step. Furthermore, since initial desensitization is assumed to be due to uncoupling and mechanistically not dependent on immediate loss of receptors, situations may be found in which desensitization has occurred without loss of LH receptors, such as described by Lamprecht et al. (47) in rat follicles subjected to 24-hr organ culture with LH, or in which receptor loss (clearing from membrane) proceeds concomitant with desensitization, such as the model described by Conti et al. (17) for the "pseudopregnancy" of the superovulated rat after hCG desensitization. One might also predict that desensitization and hence the decrease in cAMP synthesizing capacity of the cell should go hand-in-hand with the appearance of functional luteolysis, i.e., loss of progesterone secretion. In fact, functional luteolysis may well lag behind desensitization, since it appears that cAMP may be compartmentalized in cells (8,24,45,55, 70) and that cells are capable of eliciting increases in intracellular cAMP levels that are between 10 and 100 times higher than those needed to elicit maximal steroidogenic responses (11,60). A decreased capacity to produce cAMP would therefore not be expected to become rate limiting until it is well advanced.

Roles of E_2 and LH

The essential supportive action of E_2 on CL has been documented overwhelmingly, and there is no doubt that removal of E_2 results in regression of CL both in rabbits and rats. In rabbits this can be seen throughout the whole of the life-span of the CL, whereas in rats this action of E_2 can be detected only towards the end of PSP and during days 7 to 11 of pregnancy. As mentioned earlier, rat CL differ from rabbit CL in that they contain the machinery for synthesizing their own E_2 from androgen precursors. They resemble each other, though, in that neither can synthesize E_2 from P_4. In view of what is known about the mechanism of action of steroid hormones (53), and of the fact that the appearance of E_2 dependency correlates with the appearance of E_2 receptors in CL (63,64), it would appear that the role of E_2 is one of regulating gene expression and thereby assuring synthesis of the necessary enzymes involved in both CL life and function. It was surprising to us that replacement therapy, such as carried out by Gibori et al. in maintaining CL in hypophysectomized and hysterectomized 10-day-pregnant rats, was found to require up to 100 μg E_2/day (32). It has been proposed that the exceedingly large concentrations of E_2 needed to maintain the CL, as well as other ovarian tissues (69), are physiologic, since the ovary is the site of origin of E_2, and its components are normally exposed to a higher E_2 gradient than, say, the uterus, whose supply depends on systemic (serum) E_2.

But it has been demonstrated in rats that the CL is very sensitive to E_2 and actually does not require high levels of the steroid for survival. This was elegantly shown in 1971 by Keyes (42), who maintained pregnancy from days 7 through 14 in ovariectomized rat with ectopic CL using as little as 0.1 μg E_2/day. Requirement for high E_2 for maintenance of CL in hypophysectomized and hysterectomized rats, such as shown by Gibori et al. (32), must therefore signal the absence or deficiency of one or more factors, in addition to E_2. Since in experiments with rats a common denominator to high E_2 requirement is LH "deficiency" due to hypophysectomy and since LH is capable of directly acting on the rat CL (38), it is reasonable to assume that the normal regulatory complex, responsible for both CL maintenance and CL function, is a combination of (low) E_2 and (low) LH, the former acting through high-affinity cytosolic receptors by regulating gene expression and the later acting through high-affinity membrane receptors by regulating adenylyl cyclase activity and formation of intracellular cAMP.

As a plausible working hypothesis on the LH-E_2 interplay in developed CL, we should like to propose that whereas E_2 has the "single" task of providing the enzymes necessary for P_4 synthesis and the machinery for P_4 secretion, LH has the "double" task of (a) directly regulating (via cAMP) the activity of P_4 synthesis and P_4 secretion and (b) indirectly providing (also via cAMP) for enough E_2 supply from surrounding follicles either in the form of aromatizable androgen (rats) or as finished E_2 (rats and rabbits). We feel that high (pharmacologic) E_2 treatments are needed in the absence of LH because E_2, though capable of "inducing" more P_2 synthetic enzymes, is incapable of activating the secretion of P_4. In support of this view, we observed in preliminary experiments that desensitized, high E_2 treated CL have "normal" P_4 content (indicative of sufficient synthesis), but do not maintain "normal" serum P_4 levels (indicative of deficient secretory activity) (see Table 2). It should be noted that studies by Rubin and collaborators (31, 76) on the effect of ACTH on cat adrenal fasciculata cells upon hormonal stimulation have shown both the appearance of high-density granules that may be of the secretory type and secretion of protein concomitant with steroid release. Thus, the assumption that steroid hormones exit cells by simple diffusion (71) rather than via exocytosis of secretory vesicle has by no means been adequately documented.

We have tried to present some of our views on the ontogeny of corpus luteum and some aspects of its regulation. Many different lines of experimentation are open for research, since our knowledge is still scant. We need a better definition of "luteotropic" complexes and have few, if any, experimentally supported ideas about the nature of signals responsible for natural luteal regression. From our own studies and reasonings, it follows that elucidation of these questions requires a battery of well-applied determinations that not only include various types of steroidal and nonsteroidal hormones (P_4, E_2, PRL, PRL-like, LH, and FSH) and hormone receptors of various kinds (LH, PRL, FSH, E_2), but also some of the intervening steps between receptor occupancy and final output, such as accumulation of secretory products and, of course, hormone-stimulable adenylyl cyclase.

TABLE 2. *Effect of Estradiol and hCG treatment on tissue and serum progesterone in the day 12 PSP rabbit*

Treatment[a]	Tissue P_4		Serum P_4	
	n	pg/μg protein	n	ng/ml
Control	6	135 ± 15	6	14.9 ± 1.4
hCG	3	12 ± 1	6	4.0 ± 0.7
E-30 + hCG	3	123 ± 10	6	6.9 ± 0.5

[a] When treated, rabbits received either hCG alone on day 9 or hCG on day 9 and estradiol (15 μg E-30 in oil twice daily) from the evening of day 8 through the evening of day 11 plus hCG on day 9. Animals were sacrificed on day 12 and serum and CL progesterone (P_4) determined by radioimmunoassay. For further details, see Fig. 13 and text. Values represent means ± SEM of the number of determinations (n) shown.
From Day and Birnbaumer, ref. 19.

ACKNOWLEDGMENT

This research was supported in part by NIH Grants HD-06513, HD-09581, HD-07495, and HD-111356.

REFERENCES

1. Arimura, A., and Schally, A. V. (1971): Augmentation of pituitary responsiveness to LH-releasing hormone (LH-RH) by estrogen. *Proc. Soc. Exp. Biol. Med.,* 136: 290–293.
2. Armstrong, D. T., and Grinwich, D. L. (1972): Blockade of spontaneous and LH induced ovulation in rats by indomethacin, an inhibitor of prostaglandin biosynthesis. *Prostaglandins,* 1:21–28.
3. Armstrong, D. T., Grinwich, D. L., Moon, Y. S., and Zamecnik, J. (1974): Inhibition of ovulation in rabbits by intrafollicular injection of indomethacin and prostaglandin F antiserum. *Life Sci.,* 14:129–140.
4. Barraclough, C. A. (1973): Sex steroid regulation of reproductive neuroendocrine processes. In: *Handbook of Physiology, Sect. 7: Endocrinology,* Vol. 2, Part 1, edited by R. O. Greep, pp. 29–56. American Physiological Society, Washington, D.C.
5. Birnbaumer, L., and Yang, P. Ch. (1974): Studies on receptor-mediated activation of adenylyl cyclases. III. Regulation by purine nucleotides of the activation of adenylyl cyclases from target organs for prostaglandins, luteinizing hormone neurohypophyseal hormones and catecholamines. Tissue and hormone-dependent variations. *J. Biol. Chem.,* 249:7867–7873.
6. Birnbaumer, L., Yang, P.-Ch., Hunzicker-Dunn, M., Bockaert, J., and Duran, J. M. (1976): Adenylyl cyclase activities in ovarian tissues. I. Homogenization and conditions of assay in Graafian follicles and corpora lutea of rabbits, rats, and pigs; regulation by ATP and some comparative properties. *Endocrinology,* 99:163–184.
7. Birnbaumer, L. (1973): Hormone-sensitive adenylyl cyclases; useful models for studying hormone receptor functions in cell-free systems. *Biochim. Biophys. Acta* 300:129–158.
8. Birnbaumer, L., Bockaert, J., Hunzicker-Dunn, M., Plisca, V., and Glattfelder, A. (1976): On the modes of regulation of intracellular cyclic AMP: desensitization of adenylyl cyclase to hormonal stimulation and compartmentalization of cyclic AMP. In: *Eukaryotic Cell Function and Growth,* edited by J. E. Dumont, B. L. Brown, and N. J. Marshall, pp. 43–66. Plenum Press, New York.
9. Bockaert, J., Hunzicker-Dunn, M., and Birnbaumer, L. (1976): Hormone-stimulated desensitization of hormone-dependent adenylyl cyclase: Dual action of luteinizing hormone on pig Graafian follicle membranes. *J. Biol. Chem.,* 251:2653–2663.
10. Bourne, H. R., Coffino, P., Melmon, K. L., Tomkins, G. M., and Weinstein, Y. (1975): Genetic analysis of cyclic AMP in a mammalian cell. *Adv. Cyclic Nucleotide Res.,* 5:771–786.
11. Catt, K. J., and Dufau, M. L. (1973): Spare gonadotrophin receptors in rat testis. *Nature [New Biol.],* 244:219–221.
12. Challis, J. R. G., Davies, I. J., and Ryan, K. J. (1973): The concentration of progesterone, estrone and estradiol-17β in the plasma of pregnant rabbits. *Endocrinology,* 93:971–976.
13. Channing, C. P., and Coudert, S. P. (1976): Contribution of granulosa cells and follicular fluid to ovarian estrogen secretion in the rhesus monkey *in vivo. Endocrinology,* 98:590–597.
14. Channing, C. P., and Kammerman, S. (1973): Characteristics of gonadotrophin receptors of porcine granulosa cells during follicle maturation. *Endocrinology,* 92: 531–540.
15. Channing, C. P., and Seymour, J. F. (1970): Effects of dibutyryl cyclic-3'5'-AMP and other agents upon luteinization of porcine granulosa cells in culture. *Endocrinology,* 87:165–169.

16. Chen, S-T., Wahn, H., Turner, W. A., Taylor, J. D., and Tchen, T. T. (1974): MSH, cyclic AMP, and melanocyte differentiation. *Recent Prog. Horm. Res.,* 30:319–345.

17. Conti, M., Harwood, J. P., Hsueh, A. J. W., Dufau, M. L., and Catt, K. J. (1976): Gonadotropin-induced loss of hormone receptors and desensitization of adenylate cyclase in the ovary. *J. Biol. Chem.,* 251:7729–7731.

18. Daniel, V., Litwack, G., and Tomkins, G. M. (1973): Induction of cytolysis of cultured lymphoma cells by adenosine 3',5' cyclic monophosphate and the isolation of resistant variants. *Proc. Natl. Acad. Sci. USA,* 70:76–79.

19. Day, S. L., and Birnbaumer, L. (1978): Effect of estradiol (E) on corpus luteum (CL) function and adenylyl cyclase (AC) in control and hCG-treated rabbits, p. 22. In: *Abstracts of the 11th Annual Meeting of the Society for the Study of Reproduction,* Champaign/Urbana, Ill.

20. Demers, L. M., Behrman, H. R., and Greep, R. O. (1973): Effects of prostaglandins and gonadotropins on luteal prostaglandin and steroid biosynthesis. *Adv. Biosci.,* 9:701–707.

21. Desjardins, C., Kirton, K. T., and Hafs, H. D. (1967): Anterior pituitary levels of FSH, LH, ACTH and prolactin after mating in female rabbits. *Proc. Soc. Exp. Biol. Med.,* 126:23–26.

22. Docke, F., and Dorner, G. (1965): The mechanism of induction of ovulation by oestrogens. *J. Endocrinol.,* 33:491–499.

23. Dohler, K. D., and Wuttke, W. (1974): Total blockade of phasic prolactin release in rats: Effect on serum LH and progesterone during the estrous cycle and pregnancy. *Endocrinology,* 94:1595–1600.

24. Dufau, M. L., Horner, K. A., Hayashi, K., Tsuruhara, T., Conn, P. M., and Catt, K. J. (1978): Actions of choleragen and gonadotropin in isolated leydig cells. Functional compartmentalization of the hormone-activated cyclic AMP response. *J. Biol. Chem.,* 253:3721–3729.

25. Dufy-Barbe, L., Franchimont, P., and Faure, J. M. A. (1973): Time courses of LH and FSH release after mating in the female rabbit. *Endocrinology,* 92:1318–1321.

26. Eaton, L. W., Jr., and Hilliard, J. (1971): Estradiol-17β, progesterone and 20α-hydroxygpregn-4-en-3-one in rabbit ovarian venous plasma. I. Steroid secretion from paired ovaries with and without corpora lutea; effect of LH. *Endocrinology,* 89:105–111.

27. Elabum, D. J., and Keyes, P. L. (1976): Synthesis of 17β-estradiol by isolated ovarian tissues of the pregnant rat: aromatization in the corpus luteum. *Endocrinology,* 99:573–579.

28. Erickson, G. F., and Ryan, K. J. (1976): Stimulation of testosterone production in isolated rabbit thecal tissue by LH/FSH, dibutyryl cyclic AMP, PGF$_{2\alpha}$, and PGE$_2$. *Endocrinology,* 99:452–458.

29. Fortune, J. E., and Armstrong, D. T. (1978): Hormonal control of 17β-estradiol biosynthesis in proestrous rat follicles: Estradiol production by isolated theca *versus* granulosa. *Endocrinology,* 102:227–235.

30. Fortune, J. E., and Hansel, W. (1977): Androgen production by isolated components of bovine preovulatory follicles, p. 39. In: *Abstracts of the 10th Annual Meeting of the Society for the Study of Reproduction,* Champaign/Urbana, Ill.

31. Gemmell, R. T., Laychock, S. G., and Rubin, R. P. (1977): Ultrastructural and biochemical evidence for a steroid-containing secretory organelle in the perfused cat adrenal gland. *J. Cell. Biol.,* 72:209–215.

32. Gibori, G., Antczak, E., and Rothchild, I. (1977): The role of estrogen in the regulation of luteal progesterone secretion in the rat after day 12 of pregnancy. *Endocrinology,* 100:1483–1495.

33. Greep, R. O. (1938): The effect of gonadotropic hormones on the persisting corpora lutea in hypophysectomized rats. *Endocrinology,* 23:154–163.

34. Hillensjo, T., Hamberger, L., and Ahren, K. (1977): Effect of androgens on the biosynthesis of estradiol-17β by isolated periovulatory rat follicles. *Mol. Cell. Endocrinol.,* 9:183–193.

35. Hilliard, J., and Eaton, L. W. (1971): Estradiol-17β, progesterone and 20-hy-

droxypregn-4-en-3-one in rabbit ovarian venous plasma. II. From mating through implantation. *Endocrinology*, 89:522–527.

36. Holt, J. A., Keyes, P. L., Brown, J. M., and Miller, J. B. (1975): Premature regression of corpora lutea in pseudopregnant rabbits following the removal of polydimethylsiloxane capsules containing 17β-estradiol. *Endocrinology*, 97:76–82.

37. Hunzicker-Dunn, M., and Birnbaumer, L. (1976): Adenylyl cyclase activities in ovarian tissues. II. Regulation of responsiveness to LH, FSH, and PGE₁ in the rabbit. *Endocrinology*, 99:185–197.

38. Hunzicker-Dunn, M., and Birnbaumer, L. (1976): Adenylyl cyclase activities in ovarian tissues. III. Regulation of responsiveness to LH, FSH, and PGE₁ in the prepubertal, cycling, pregnant and pseudopregnant rat. *Endocrinology*, 99:198–210.

39. Hunzicker-Dunn, M., and Birnbaumer, L. (1976): Adenylyl cyclase activities in ovarian tissues. IV. Gonadotrophin-induced desensitization of the luteal adenylyl cyclase throughout pregnancy and pseudopregnancy in the rabbit and the rat. *Endocrinology*, 99:211–222.

40. Kalra, S. P., and Kalray, P. S. (1974): Temporal interrelationships among circulating levels of estradiol, progesterone and LH during the rat estrous cycle: Effects of exogenous progesterone. *Endocrinology*, 95:1711–1718.

41. Kelly, P. A., Shiu, R. P. C., Robertson, M. C., and Friesen, H. G. (1975): Characterization of rat chorionic mammotropin. *Endocrinology*, 96:1187–1195.

42. Keyes, L. P. (1973): Maintenance of postimplantation-pregnancy in the rat in the presence of ectopic corpora lutea: requirement for ovarian follicles and estrogen. *Biol. Reprod.*, 8:618–624.

43. Keyes, P. L., and Nalbandov, A. V. (1967): Maintenance and function of corpora lutea in rabbits depend on estrogen. *Endocrinology*, 80:938–946.

44. Kolena, J., and Channing, C. P. (1972): Stimulatory effects of LH, FSH and prostaglandins upon cyclic 3′–5′-AMP levels in porcine granulosa cells. *Endocrinology*, 90:1543–1550.

45. Kuo, J. F., and DeRenzo, E. C. (1969): A comparison of the effects of lipolytic and antilipolytic agents on adenosine 3′:5′-monophosphate levels in adipose cells as determined by prior labeling with adenine-8-¹⁴C. *J. Biol. Chem.*, 244:2252–2260.

46. Lamprecht, S. A., Zor, U., Tsafriri, A., and Lindner, H. R. (1973): Action of prostaglandin E₂ and of luteinizing hormone of ovarian adenylate cyclase, protein kinase, and ornithine decarboxylase activity during postnatal development and maturity in the rat. *J. Endocrinol.*, 57:217–233.

47. Lamprecht, S. A., Zor, U., Salomon, Y., Koch, Y., Ahren, K., and Lindner, H. R. (1977): Mechanism of hormonally induced refractoriness of ovarian adenylate cyclase to luteinizing hormone and prostaglandin E₂. *J. Cyclic Nucleotide Res.*, 3:69–83.

48. Lee, C. Y. (1976): The porcine ovarian follicle: III. Development of chorionic gonadotropin receptors associated with increases in adenyl cyclase activity during follicle maturation. *Endocrinology*, 99:42–48.

49. Lee, C. Y., and Ryan, R. J. (1974): Estrogen stimulation of human chorionic gonadotropin binding by luteinized rat ovarian slices. *Endocrinology*, 95:1691–1693.

50. Lee, C. Y., and Ryan, R. J. (1971): The uptake of human luteinizing hormone (hLH) by slices of luteinized rat ovaries. *Endocrinology*, 89:1515–1523.

51. Lindner, H. R., Amersterdam, A., Salomon, Y., Tsafriri, A., Nimrod, A., Lamprecht, S. A., Zor, U., and Koch, Y. (1977): Intraovarian factors in ovulation: determinants of follicular response to gonadotropins. *J. Reprod. Fertil.* 51:215–235.

52. Louvet, J. P., and Vaitukaitis, J. L. (1975): Induction of follicle stimulating hormone (FSH) receptors in rat ovaries by estrogen priming, p. 135. In: *Abstracts of the 57th Annual Meeting of the Endocrine Society*, Champaign/Urbana, Ill.

53. Louvet, J.-P., Harman, S. M., and Ross, G. T. (1975): Effects of human chorionic gonadotropin, human interstitial cell stimulating hormone and human follicle-stimulating hormone on ovarian weights in estrogen-primed hypophysectomized immature female rats. *Endocrinology*, 96:1179–1186.

54. Makris, A., and Ryan, K. J. (1975): Progesterone, androstenedione, testosterone, estrone, and estradiol synthesis in hamster ovarian follicle cells. *Endocrinology,* 96:694–701.

55. Manganiello, V. C., Murad, F., and Vaughan, M. (1971): Effects of lipolytic and antilipolytic agents on cyclic 3′,5′-adenosine monophosphate in fat cells. *J. Biol. Chem.,* 246:2195–2202.

56. Marsh, J. (1971): The effect of prostaglandins on the adenyl cyclase of the bovine corpus luteum. *Ann. N.Y. Acad. Sci.,* 180:416–425.

57. Marsh, J. M. (1975): The role of cyclic AMP in gonadal function. In: *Advances in Cyclic Nucleotide Research,* Vol. 6, edited by P. Greengard and G. A. Robison, pp. 137–199. Raven Press, New York.

58. Marsh, J. M., Mills, T. M., and LeMaire, W. J. (1973): Preovulatory changes in the synthesis of cyclic AMP by rabbit Graafian follicles. *Biochim. Biophys. Acta,* 304:197–202.

59. McCracken, J. A., Barickowski, B., Carlson, J. C., Green, K., and Samuelsson, B. (1973): The physiological role of prostaglandin $F_{2\alpha}$ in corpus luteum regression. *Adv. Biosci.,* 9:599–624.

60. Mendelson, C., Dufau, M. L., and Catt, K. J. (1975): Gonadotropin binding and stimulation of cyclic adenosine 3′,5′-monophosphate and testosterone production in isolated Leydig cells. *J. Biol. Chem.,* 250:8818–8823.

61. Midgley, A. R., Jr. (1973): Autoradiographic analysis of gonadotropin binding to ovarian tissue sections. In: *Receptors for Reproductive Hormones,* edited by B. W. O'Malley and A. R. Means, pp. 365–378. Raven Press, New York.

62. Miller, J. B., and Keyes, P. L. (1974): Initiation of luteinization in rabbit Graafian follicles by dibutyryl cyclic AMP *in vitro. Endocrinology,* 95:253–259.

63. Miller, J. B., and Keyes, P. L. (1978): Transition of the rabbit corpus luteum to estrogen dependence during early luteal development. *Endocrinology,* 102:31–38.

64. Mills, T. M., and Osteen, K. G. (1977): 17β-Estradiol receptor activity and progesterone and 20α-hydroxy-pregn-4-en-3-one content of the developing corpus luteum of the rabbit. *Endocrinology,* 101:1744–1750.

65. Mills, T. M., and Savard, K. (1973): Steroidogenesis in ovarian follicles isolated from rabbits before and after mating. *Endocrinology,* 92:788–791.

66. Morishige, W. K., Pepe, G. J., and Rothchild, I. (1973): Serum luteinizing hormone, prolactin and progesterone levels during pregnancy in the rat. *Endocrinology,* 92: 1527–1530.

67. Morishige, W. K., and Rothchild, I. (1974): Temporal aspects of the regulation of corpus luteum function by luteinizing hormone, prolactin and placental luteotrophin during the first half of pregnancy in the rat. *Endocrinology,* 95:260–274.

68. Moyle, W. R., Erickson, G., Bahl, P. O., and Gatowski, J. (1978): Role of the carbohydrates in the actions of pregnant mares serum gonadotropin on rat leydig and granulosa cells. *J. Biol. Chem. (in press).*

69. O'Malley, B. W., and Means, A. R. (1974): Female steroid hormones and target cell nuclei. *Science,* 183:610–620.

70. Pliska, V., Glattfelder, A., and Birnbaumer, L. (1972): Regulation of cellular cAMP levels: a compartment model. *Experientia,* 28:750.

71. Porter, K. R., and Bonneville, M. A. (1967): *Fine Structure of Cells and Tissues,* pp. 75–76. Lea & Febiger, Philadelphia.

72. Rennie, P. (1968): Follicular activity in the maintenance of homotransplanted rabbit corpora lutea. *Endocrinology,* 83:314–322.

73. Rennie, P. (1968): Luteal-hypophyseal interrelationship in the rabbit. *Endocrinology,* 83:323–328.

74. Richards, J. S., and Midgley, A. R. (1976): Protein hormone action: a key to understanding ovarian follicular and luteal development. *Biol. Reprod.,* 14:82–94.

75. Rothchild, I. (1965): Interrelations between progesterone and the ovary, pituitary, and central nervous system in the control of ovulation and the regulation of progesterone secretion. *Vitam. Horm.,* 23:209–327.

76. Rubin, R. P., Sheid, B., McCauley, R., and Laychock, S. G. (1974): ACTH-induced

protein release from the perfused cat adrenal gland: evidence for exocytosis? *Endocrinology,* 95:370–378.

77. Sasamoto, S., and Kennan, A. L. (1973): Endogenous gonadotropin requirements for follicular growth and maintenance in immature rats pretreated with PMS. *Endocrinology,* 93:292–296.

78. Scott, R. S., and Rennie, P. I. C. (1970): Factors controlling the life-span of the corpora lutea in the pseudopregnant rabbit. *J. Reprod. Fertil.,* 23:415–422.

79. Smith, M. S., Freeman, M. E., and Neill, J. D. (1975): The control of progesterone secretion during the estrous cycle and early pseudopregnancy in the rat: prolactin, gonadotropin and steroid levels associated with rescue of the corpus luteum of pseudopregnancy. *Endocrinology,* 96:219–226.

80. Smith, M. S., McLean, B. K., and Neill, J. D. (1976): Prolactin: the initial luteotropic stimulus of pseudopregnancy in the rat. *Endocrinology,* 98:1370–1377.

81. Smith, M. S., and Neill, J. D. (1976): Termination at midpregnancy of two daily surges of plasma prolactin initiated by mating in the rat. *Endocrinology,* 98:696–701.

82. Spies, H. G., Hilliard, J., and Sawyer, C. H. (1968): Pituitary and uterine factors controlling regression of the corpora lutea in intact and hypophysectomized rabbits. *Endocrinology,* 83:291–299.

83. Stormshak, F., and Casida, L. E. (1965): Effects of LH and ovarian hormones in corpora lutea of pseudopregnant and pregnant rabbits. *Endocrinology,* 77:337–342.

84. Sutherland, E. W., and Rall, T. W. (1962): Adenyl cyclase. I. Distribution, preparation, and properties. *J. Biol. Chem.,* 237:1220–1227.

85. Takahashi, M., Shiota, K., and Suzuki, Y. (1978): Preprogramming mechanism of luteinizing hormone in the determination of the lifespan of the rat corpus luteum. *Endocrinology,* 102:494–498.

86. Tsafriri, A., Lindner, M. R., Zor, U., and Lamprecht, S. A. (1972): *In vitro* induction of meitoic division in follicle-enclosed rat oocytes by LH, cyclic AMP, and prostaglandin E_2. *J. Reprod. Fertil.,* 31:39–50.

87. Voorhees, J. J., and Duell, E. A. (1975): Imbalanced cyclic AMP-cyclic GMP levels in psoriasis. *Adv. Cyclic Nucleotide Res.,* 5:735–758.

88. YoungLai, E. V. (1972): Effect of mating on follicular fluid steroids in the rabbit. *J. Reprod. Fertil.,* 30:157–165.

Ontogeny of Receptors and Reproductive Hormone Action,
edited by T. H. Hamilton, J. H. Clark, and W. A. Sadler.
Raven Press, New York 1979.

FSH and Calcium as Modulators of Sertoli Cell Differentiation and Function

A. R. Means, J. R. Dedman, M. J. Welsh, M. Marcum, and B. R. Brinkley

Department of Cell Biology, Baylor College of Medicine, Houston, Texas 77030

The primary target cell for follicle-stimulating hormone (FSH) within the testis is the Sertoli cell. Studies using the Sertoli cell-enriched animal model and Sertoli cells in culture have established a sequence of biochemical events initiated on interaction of FSH with the plasma membrane (19,21,23, 29,51,52). It is not the intent of this chapter to review the extensive literature in this field. For this purpose the reader is referred to recent review articles by Means (47) and Fritz (22). Rather, our intent is to discuss new observations from our own laboratory that bear directly on the problem of acute FSH action and offer some possibilities for fruitful new avenues of approach.

FSH binds to receptors located on the plasma membrane of Sertoli cells (21,51,56,65). In immature animals this interaction results in an activation of adenylyl cyclase and a decreased activity of cyclic AMP-phosphodiesterase (cAMP-PDE) (46,48). Together these altered enzyme activities lead to an elevation in the intracellular concentration of cAMP (18,19,21,37,47,51, 52) that, in turn, activates cAMP-dependent protein kinase (20,21,51–53). Many proteins present in every subcellular compartment are then phosphorylated, but, to date, no single protein substrate has been identified that shows altered activity following phosphorylation. All of the events outlined above occur within 1 to 3 min of exposure to FSH and reach maximal degrees of stimulation by 30 min. Rates of RNA and protein synthesis are subsequently elevated, but no convincing data are available to directly link these processes with altered cyclic nucleotide metabolism (19,21,44,49,50,52). However, levels of at least four specific proteins are specifically stimulated by FSH. These proteins include two secretory proteins—androgen binding protein (ABP) (21,23,27,52,53,55,61,70,71) and plasminogen activator (38)—as well as two proteins that remain within the Sertoli cell—protein kinase inhibitor (2) and γ-glutamyl transpeptidase (31,41). With respect to ABP, it is now apparent that FSH is the primary hormone responsible for regulation of both synthesis and secretion *in vivo* (68,69).

One of the enigmas concerning FSH action in the male rat is the fact that all acute responses to this gonadotropin are age dependent. Effects of

FSH can be observed by 2 days postnatally, maximal responsiveness occurs between 10 and 12 days, and by 25 days the Sertoli cell is largely refractory to hormone. Hypophysectomy can restore sensitivity to FSH when performed at any age. In attempting to determine the mechanism for the altered responsiveness, it seemed logical to first determine whether receptor numbers decreased with age. Results of these studies show that the number of FSH receptors per testis increases for the first 15 days of age in a linear fashion (21,35,51,52,67). This is predictable since Sertoli cells only divide for the first 15 days (8,66). Between 15 and 40 days of age, a further increase has been noted but at a much slower rate (35,67). These data demonstrate that the loss of Sertoli cell sensitivity to FSH cannot be due to decreased numbers of hormone receptor sites.

The second step in the sequence of events initiated by FSH is the stimulation of adenylyl cyclase. Figure 1 shows the changes in Sertoli cell FSH sensitivity of this enzyme with respect to age. Between 5 and 15 days, a three to fourfold increase is noted with maximal stimulation at 12 days of age. Thereafter, the effect of FSH decreases until by 30 days only a slight

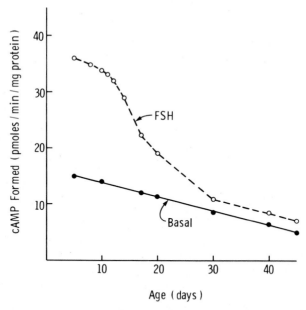

FIG. 1. Developmental pattern of Sertoli cell adenylyl cyclase responsiveness to FSH. Sertoli cell-enriched testis homogenates were prepared by homogenizing detunicated tissue 1:20 (w/v) in 27% sucrose, 10 mM Tris-HCl, pH 7.5, and 1 mM EDTA. Aliquots were assayed for adenylyl cyclase activity in a final volume of 50 µl containing 0.4 mM adenosine 5′-triphosphate (ATP), 5.0 mM $MgCl_2$, 1.0 mM EDTA, 1.0 mM cAMP, 20 mM creatine phosphate, 0.2 mg/ml creatine kinase, 0.1 mg/ml myokinase, and 25 mM bis-tris-propane (BTP)-HCl buffer, pH 8.0. Incubations were carried out at 37°C for 10 min in the absence (●——●) or presence (○----○) of 5.0 µg/ml NIH FSH-S10.

stimulation is observed. Basal catalytic activity declines only slightly during the same time period. These data suggest that at least a part of the loss of responsiveness to FSH concerns hormone-stimulated cyclase. Moreover, since receptor number increases and basal catalytic activity declines minimally, it is likely that the defect lies in the coupling of receptor and cyclase.

Increasing intracellular cAMP in response to FSH promotes the activation of cAMP-dependent protein kinase. This effect is also age-dependent with maximal stimulation observed between 12 and 16 days (21,48,51–54). By 30 days of age, the gonadotropin causes no significant protein kinase activation. The Sertoli cell contains both Type I and Type II enzymes, and the ratio of the forms changes with age (20,48,53,56). However, since FSH activates both enzymes in the immature testis (20), it is unlikely that the changes in isoforms are a contributing factor to the decreased effect of hormone. Indeed, the lack of response in mature rats can be reversed by incubating Sertoli cells in the presence of the potent PDE inhibitor 1 methyl, 3-isobutyl xanthine (MIX) (21,45,51,52,54). Whereas FSH alone causes a threefold activation of protein kinase in cells from 16-day-old rats, no response is noted in 70-day-old Sertoli cells. On the other hand, MIX has a small but significant effect in immature testis, but produces a twofold increase in activity in mature testis. Addition of FSH and MIX to the 70-day-old cells promotes a synergistic stimulation of protein kinase. Determination of cAMP levels revealed that MIX does enhance the intracellular concentration of this nucleotide. Two interesting pieces of information emerge from these experiments. The first is that adenylyl cyclase can still adequately respond to FSH and elevate cAMP levels in mature cells if PDE is inhibited. This finding demonstrates that the loss of FSH effect cannot be solely attributed to a coupling defect between receptor and the catalytic moiety of cyclase. Secondly, these observations suggest that PDE also plays an important role in mediating the effects of FSH on cyclic nucleotide metabolism and contributes to the age-dependent tachyphylaxis.

The importance of PDE as an integral component in hormone action is often neglected since PDE provides a negative control mechanism. However, emphasis must be placed on this enzyme when considering the control of the steady-state levels of cAMP. The steady-state concentration is determined by the activities of two enzymes—one unique for cAMP synthesis, adenylyl cyclase, and the other specific for cAMP degradation, PDE. These reactions are illustrated by the equation:

$$ATP \xrightarrow{\quad 1 \quad} cAMP \xrightarrow{\quad 2 \quad} 5'AMP \qquad [1]$$

where 1 represents adenylyl cyclase and 2 PDE. This reaction sequence is a unique kinetic system in that both reactions are irreversible and that cAMP metabolism is controlled exclusively by the two enzymes. The reactions are also distinct since cyclase functions in a zero-order state with respect to the substrate adenosine 5′triphosphate (ATP). Intracellular levels of ATP exceed 1 mM, whereas the K_m of cyclase for ATP is 10 to 50 μM; therefore, the substrate is not rate limiting. In contrast, concentrations of cAMP (10^{-7} M) are generally well below the K_m of PDE (10^{-5} to 10^{-6} M), which results in a first-order reaction with respect to cAMP. Thus, when a hormone combines with its receptor stimulating adenylyl cyclase and increasing intracellular cAMP, the kinetic rate of PDE also increases. This establishes a dynamic steady-state condition with regard to cAMP levels. Davies and Williams (10) and Arch and Newsholme (1) were the first to derive a kinetic equation describing this situation. We have modified the equilibrium reaction into the following form:

$$[\text{cAMP}]_{ss} = \frac{V_1 \times K_{m2}}{V_2 - V_1} \qquad [2]$$

where V_1 and V_2 are the maximal velocities of adenylyl cyclase and PDE, respectively, and K_{m2} is the K_m of PDE.

It is well established that PDE V_{max} is 50 to 100 times greater than V_{max} adenylyl cyclase, and the low K_m PDE is approximately 10^{-6} M. It is evident from Eq. [2] that when $V_2 \gg V_1$, a change in V_1 does not significantly alter the denominator. Consequently, a change in V_{max} of adenylyl cyclase would be equivalent to a similar change in K_m PDE, and these changes are inversely related to an alteration in V_{max} PDE. These relationships can be summarized as follows:

$$V_1 = K_{m2} = V_2^{-1} \qquad [3]$$

The equality of the terms in Eq. [3] illustrates the importance of considering both changes in cyclase and PDE in response to hormones.

As mentioned previously FSH results in a dose-dependent decrease in cAMP-PDE activity in immature Sertoli cells (48). The time course of this effect is identical to that of activation of protein kinase. When such studies were repeated in mature cells, no inhibition was observed (48). These data, together with the theoretical argument concerning regulation of the steady-state cAMP concentration, suggested that PDE was involved in the mechanism of action of FSH. Initial experiments established that total cAMP-PDE activities are not markedly different between 5 and 55 days of age (21, 47,48). Therefore, the decreased sensitivity of Sertoli cells to FSH cannot be a consequence of gross increases in total soluble PDE activity. Thus, the most reasonable solution to the PDE problem would be if changes in the isoforms of PDE occurred during differentiation.

Most tissues contain at least two forms of cAMP-PDE—one that is calcium sensitive and one that is not (72). Since total PDE levels were similar in 14- and 55-day-old cells, we chose to investigate calcium sensitivity. Indeed, calcium stimulated PDE activity in 14-day testes twofold, whereas no effect was observed in cells from the 55-day-old animals (21,47,48). The reason for this difference was determined by separating PDEs into the various isoforms by ion-exchange chromatography. Two cAMP-PDEs exist in the soluble fraction of the Sertoli cell of the immature rat (13,48). The first to elute is a calcium-dependent enzyme form that will hydrolyze both cGMP and cAMP (form I), whereas the second (form II) does not require calcium activity and only utilizes cAMP at physiological concentrations (form II; $K_m \sim 10^{-6}$ M). The distribution of the enzyme forms is different in immature and mature animals even though the total amount of PDE activity is the same. In the mature animal a previously undetected isozyme is demonstrated to elute from a diethylaminoethyl (DEAE)-cellulose column between the two forms previously discussed (form III) (13,48). Moreover, this PDE only utilizes cGMP and exhibits a marked requirement for Ca^{2+} and the heat-stable calcium-dependent regulator protein (CDR), which is discussed later (13,48). Finally, form I PDE as characterized in the immature cell is apparently absent in mature Sertoli cells. Instead, an isozyme is eluted from DEAE at somewhat higher salt concentrations. This enzyme form (I_M) utilizes predominantly cGMP and is Ca^{2+} dependent. More detailed kinetic analyses of the calcium-dependent cAMP form have revealed that in the immature testis calcium increases the affinity of PDE (form I) for cAMP (K_m changes from 10^{-6} to 10^{-7} M) and also results in a twofold increase in the maximal velocity of this enzyme. On the other hand, calcium has little or no effect on cAMP-PDE activity in the adult testis. Therefore, changes in the calcium concentration within the Sertoli cell (10^{-7} to 10^{-5} M) would alter the activity of cAMP-PDE in immature animals, whereas a similar change in the adult would result in no net difference in cAMP-PDE activity. Even if adenylyl cyclase was stimulated to some extent in both aged animals (Fig. 1), the steady-state level of cAMP might not increase sufficiently to allow subsequent cyclic nucleotide-mediated events to ensue.

If the lack of response in the adult animal was primarily due to the absence of the Ca^{2+}-sensitive isoform of PDE, an effect of FSH on calcium flux need not be age dependent. Therefore, we wished to determine whether other Sertoli cell properties were calcium sensitive. For the initial experiments, we utilized Sertoli cells in culture and asked whether FSH would alter the shape of the cells. Dishes of cells were observed by phase-interference microscopy so that the same population of cells could be followed with time. Under our usual culture conditions, the cells form a monolayer and are very flat (Fig. 2). Addition of FSH (or cAMP or EGTA) results in a time-dependent shape change—the cells round, thereby revealing a decrease in the two-dimensional surface area. This alteration is cell specific since testis fibroblasts co-cultured with the Sertoli cells show no demonstrable decrease in surface area. Quantitation of the change was accomplished by photographing the cells at various times, tracing the images on paper of

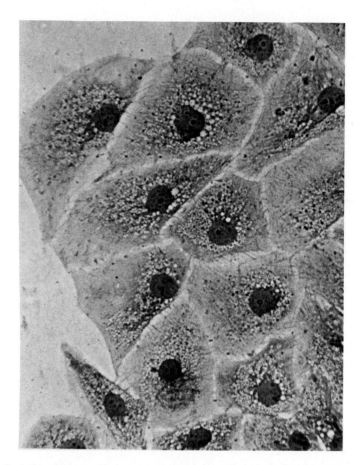

FIG. 2. Sertoli cells from 27-day-old rats, in monolayer culture after 7 days *in vitro.* The cells were isolated from germ cell-free testes produced by irradiating pregnant females with 125 rads on day 20 of gestation (51). The Sertoli cells were purified by first mincing the tissue, then treating the minced tissue for 1 hr with collagenase, then for 20 min with pancreatin, as described by Welsh and Wiebe (74). The resulting suspension of Sertoli cell aggregates was washed free of enzymes, then dispensed into Corning culture flasks in a modified minimal essential medium (74) with 5% fetal calf serum. Cells were incubated at 34°C with air atmosphere, and medium was changed every third day. ×1,000.

uniform density, cutting out the paper, and weighing the pieces. Figure 3 shows a plot of such data. It can be seen that Sertoli cells lose 60% of their initial area within 30 min, whereas no significant change is noted for testis fibroblasts. The shape change was also reversible; removing the cAMP by changing the culture media or by addition of excess calcium resulted in reformation of a uniform flattened cell monolayer.

Specificity was indicated by the observation that neither luteinizing hormone nor testosterone mimicked the effect of FSH on cell shape. However, addition of EGTA, a potent chelator of calcium, did cause identical changes to those observed in the presence of FSH. Moreover, these changes were also reversible since addition of excess calcium to the media resulted in a return to a flattened confluent monolayer. Again, the calcium effect was cell specific and also specific with respect to divalent cations. More importantly, however, the cell shape effects were observed in Sertoli cells isolated from animals of any age. Thus, the effects of Ca^{2+} on cell shape, unlike PDE regulation, was not age dependent.

In a variety of cultured fibroblasts, cell shape has been shown to be a function of the integrity of the cytoplasmic microtubule complex (59). Our next task was to determine whether the cytoskeleton was responsible for the Sertoli cell shape alteration in response to FSH or calcium. In order to approach this problem, we utilized the technique of indirect immunofluorescence (4,39). For this procedure, cells were grown on cover slips and fixed in 3% formalin. Subsequently, the cells were treated with cold acetone in order to promote increased permeability of the plasma membrane. The cells were incubated with a monospecific antibody that was specific for an intracellular protein. After an incubation time sufficient to allow antigen–antibody interaction, the cells were washed free of excess unbound antibody. Next, cells were incubated with a fluorescein-tagged immunoglobulin G (IgG) that would recognize the first antibody. Again, the cells were washed to remove excess second antibody. Preparations were then observed using fluorescence microscopy. When antitubulin is employed as the primary antibody, this procedure results in the specific decoration of polymerized microtubules.

Application of this technique to cultured Sertoli cells in interphase revealed an extensive cytoskeletal network of microtubules radiating from a single intensively fluorescent organizing center associated with the centrioles. Preincubation with the specific microtubule disrupting agent colchicine resulted in a complete dissolution of the microtubule network, but did not reveal significant changes in cell shape. On the other hand, cells incubated with FSH or EGTA showed a distinct change in shape but no apparent diminution in the number of cytoplasmic microtubules. It is entirely likely, however, that the microtubules could have shortened or undergone extensive rearrangement. At any rate, the gross shape of the Sertoli cell is not primarily controlled by the integrity of the cytoplasmic microtubule network.

We next turned our attention to the other major component of the cytoskeleton—the microfilaments. Isolation and chemical characterization of this cytoplasmic component revealed it to be composed primarily of actin and myosin. Actin was chosen as the microfilament marker due to availability of a purified monospecific antibody. When the distribution of microfilaments was examined by fluorescence microscopy, numerous tightly packed actin cables were observed aligned in parallel arrays throughout the cell. Addition of cytochalasin, a microfilament-disrupting agent, resulted in the disappear-

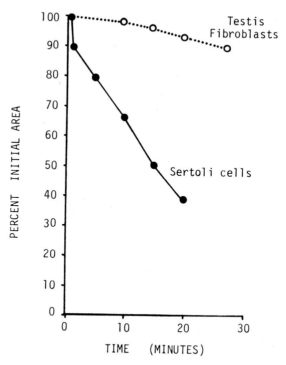

FIG. 3. Sertoli cell shape change in response to dibutyryl cAMP. Sertoli cells were prepared as described in the legend to Fig. 1 and co-cultured on glass coverslips with testis fibroblasts from the same age rats. A coverslip was inverted over a drop of culture medium on a microscope slide. The inverted coverslip was supported on two opposite sides by narrow strips of a #2 thickness coverslip, allowing medium to be applied at one end of the inverted coverslip and withdrawn at the other end. Thus, solutions could be applied or changed while the cells were under observation on the microscope stage. Photographs were taken, and change in shape could be measured by tracing the cell outlines onto uniform weight paper, cutting out the cell tracings, and weighing the cut outs. The results above show the change of Sertoli cell shape when incubated in the presence of 10^{-4} M db-cAMP. Similar changes occurred in response to FSH or by removal of external Ca^{2+} by addition of EGTA.

ance of all actin stress fibers. Furthermore, this drug caused a marked change in the shape of the Sertoli cells. Cells incubated with FSH or EGTA showed an identical shape change and were also characterized by a diminished and reorganized microfilament system. Taken together these data reveal that the shape of interphase Sertoli cells in culture is largely determined by the micro-filament network. Moreover, addition of FSH (or removal of calcium) leads to an altered, more rounded shape by altering the integrity or distribution of the microfilament network.

At this point, it was clear that calcium is important in the control of at least two Sertoli cell events—regulation of PDE in the immature cell and maintenance of the shape of cells isolated from young or old animals. These observations strengthened our supposition that FSH affects the Sertoli cell both via cyclic nucleotide metabolism and by promoting a change in calcium flux. An attempt was made to confirm the latter possibility by preincubating Sertoli cells with $^{45}Ca^{2+}$. Subsequently, FSH was added to half the cells, and the release of $^{45}Ca^{2+}$ into the medium was monitored with time. Figure 4 illustrates that a rapid efflux of $^{45}Ca^{2+}$ occurred from both control and FSH-

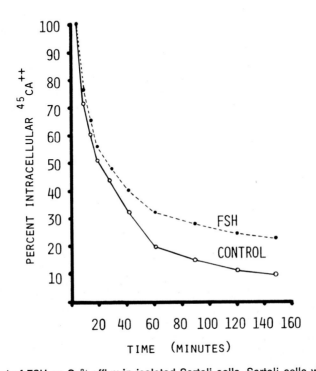

FIG. 4. Effect of FSH on Ca^{2+} efflux in isolated Sertoli cells. Sertoli cells were purified from testes of 12-day-old male rats that had been irradiated *in utero* with 125 rads on day 20 of gestation. Cells were purified by sequential enzyme disaggregation of the minced testis tissue. First, the tissue was treated for 1 hr with collagenase, then washed and treated for 15 min with pancreatin (74). The resulting suspension of Sertoli cell aggregates was washed twice in Ca^{2+}, Mg^{2+}-free phosphate buffered saline (PBS) containing 5% fetal calf serum and twice with the saline minus the serum. After incubating the cell suspension in 2.5 mM EGTA in PBS at 37°C for 20 min, the cells were incubated in 5 ml 1:3 Hank's salts-PBS with 1 μCi $^{45}Ca^{2+}$ for 45 min at 37°C. The cells were centrifuged at 100 × g for 2 min, resuspended in Hank's salt solution, divided into two equal samples, and NIH-FSH-S11 was added to one sample at a final concentration of 5 μg/ml. After 5 min, the two suspensions of cells were centrifuged for 1 min and resuspended in fresh solution, with or without hormone, as before. At intervals, the cells were centrifuged and resuspended in fresh medium with or without hormone. Supernatant fluids containing released ^{45}Ca were counted. After 150 min, the cell pellet was solubilized in sodium dodecyl sulfate and also counted to determine $^{45}Ca^{2+}$ remaining in the cells. A portion of each cell pellet was also used for estimation of cell protein. The results are expressed as the percent of ^{45}Ca remaining with the pellet after each time interval. Total (i.e., 100%) is defined as the number of counts in the supernatants plus the pellet, not including the first 5-min interval. Results were calculated as counts/mg/protein in the pellets at 150 min.

treated cells. However, after equilibrium was achieved, it was noted that FSH did not accelerate efflux of calcium. Rather, the hormone seemed to promote a redistribution of intracellular calcium from the free-exchangeable pool to the bound nonexchangeable pool. The interpretation of these data must be considered preliminary since calcium measurements are fraught with technical difficulties. At any rate, FSH did affect either the levels or distribution of calcium in the Sertoli cell.

It was now necessary to reconcile two effects of FSH acting through calcium—on the one hand, a decrease in PDE activity and, on the other, an increase in nonexchangeable calcium. These events must also be compatible with the effects caused by FSH and decreased free calcium levels with respect to the microfilaments and cell shape. Only one solution seemed to meet the requirements of all these seemingly disparate results. That solution lay in the multifunctional role of a single protein—the CDR. CDR was first described as an activator of calcium-sensitive PDE (7,34). It is a 17,000 M_r, acidic, thermostable protein that specifically binds calcium at physiological concentrations. We have recently isolated and characterized rat testis CDR and determined the primary sequence of the protein (14,15). CDR is a ubiquitous intracellular protein present at high concentrations (1 to 20 μM) that is 50% homologous to the calcium binding protein of skeletal muscle troponin C (TnC). In fact, TnC and CDR cross-react in their respective biological systems (16). Thus, CDR forms a soluble complex with troponin I and troponin T. This complex can then regulate the actin-dependent ATPase activity of skeletal muscle actomyosin in a Ca^{2+}-dependent fashion. Likewise, TnC can activate CDR-deficient PDE, albeit at a 600-fold M excess over the requirement for CDR. CDR has also been implicated in a spectrum of Ca^{2+}-dependent processes including the Ca^{2+}-ATPase (Ca^{2+}-pump) of erythrocyte plasma membrane (26,33,42), regulation of smooth muscle and nonmuscle actomyosin ATPase (11,12), and microfilament organization in isolated postsynaptic densities of dog cerebral cortex (3). Thus, CDR has been implicated in other tissues in three processes affected by FSH and Ca^{2+} in Sertoli cells: (a) regulation of Ca^{2+}-dependent PDE; (b) regulation of intracellular Ca^{2+} levels; and (c) control of microfilaments.

In order to elucidate the role(s) CDR might play in the Sertoli cell, it was desirable to obtain a monospecific antibody. This was accomplished by injecting the pure protein into goats and purifying the nonprecipitating antibody by CDR-affinity chromatography (17). The anti-CDR, when preincubated with its antigen, abolished the ability of CDR to stimulate regulator-deficient PDE from rat brain. However, when the same experiment was attempted using Sertoli cell PDE, no inhibition was observed. Yet, this enzyme was still Ca^{2+} sensitive. These data suggested that whereas the Sertoli cell PDE requires Ca^{2+}, it is not dependent on CDR. Indeed, chromatography of Sertoli cell cytosol on an ion exchange column resulted in complete separation of Ca^{2+}-sensitive PDE from CDR. Even when the entire

PDE peak was treated with EGTA and heat, followed by lyophilization, no CDR could be detected when the extract was added to CDR-deficient brain PDE. Further confirmation of the lack of requirement for CDR can be seen in the Ca^{2+}-activation curve. Maximal Ca^{2+} activation of Sertoli cell PDE occurs at 10^{-7} M Ca^{2+}, whereas the K_D of Ca^{2+} binding to CDR is 10^{-6} M. A final documentation was obtained by employing the antipsychotic drug stelazine. This compound has been shown to bind specifically to CDR with a K_D of 1 μM (40). Thus, incubation of the PDE assay mixture with stelazine should abolish the effect of CDR. This postulate was shown to be true for rat brain PDE, but again no effect of the drug was observed with the Sertoli cell enzyme. It was clear then, that the testis PDE required Ca^{2+} but not CDR.

At first glance, it appeared that this dealt a severe blow to our original hypothesis that CDR mediates the FSH effect on the Ca^{2+}-sensitive PDE. Further scrutiny of the problem yielded a logical explanation that could be tested experimentally. FSH treatment resulted in a decrease in Ca^{2+}-regulated PDE and an increase in intracellular nonexchangeable calcium. The latter effect is most easily explained by a decrease in the free calcium concentration. If CDR was acting as a Ca^{2+} buffer system, then a redistribution of the CDR–Ca^{2+} complex to a particulate locale could lower free Ca^{2+} and inhibit the PDE. Alternatively, CDR may regulate Ca^{2+} redistribution intracellularly by affecting the Ca^{2+} pump of the smooth endoplasmic reticulum. Again, this would increase the nonexchangeable pools and lower free Ca^{2+} levels. To test these possibilities, we chose to use indirect immunofluorescence microscopy to localize CDR within the Sertoli cell at interphase. In most cells examined, CDR was found to be localized in a pattern that closely resembled that of actin stress fibers. Indeed, the distribution of CDR-specific fluorescence in prominent parallel arrays was shown to be identical to the pattern of stress fibers observed in cells by phase-contrast or interference-contrast microscopy (17,73).

We had previously shown that the distribution of actin-specific fluorescence was altered by FSH or EGTA (to remove free Ca^{2+}). It is also known that the actomyosin of smooth muscle and nonmuscle microfilaments (stress fibers) is regulated by Ca^{2+} and CDR (9,11,12,64). CDR mediates the Ca^{2+} regulation of the kinase that phosphorylates the light chain of myosin and thus regulates ATPase activity (11,12). Taken together, these observations suggest that FSH regulates Sertoli cell actomyosin by promoting a redistribution of CDR-Ca^{2+} and thus a decrease in free Ca^{2+}. This would allow the activation of ATPase and the observed shape change. Moreover, in immature cells the decreased free calcium would inactivate the Ca^{2+}-PDE, thereby promoting the events in the Sertoli cell that require increased intracellular concentrations of cAMP. Studies are presently underway to evaluate these postulates.

The cytoskeleton becomes disorganized as cells enter mitosis and undergo the characteristic shape change as reflected by actin (5,39) and tubulin (4) immunofluorescence. As cells enter prophase, the distinct cytoplasmic distribution of CDR appears commensurate with the dissolution of the cytoskeleton, and this protein is randomly dispersed throughout the cell (Fig. 5). During progression of the mitotic cycle, CDR becomes intensely concentrated at the half spindles of the mitotic apparatus, but is conspicuously absent from the interzonal region. This pattern is maintained until late anaphase when CDR-specific immunofluorescence is transiently found in the interzone before rapidly condensing into two small regions, one on either side of the midbody that separates the daughter cells. Thus, CDR placement during mitosis differs significantly from that of actin and tubulin.

The role of Ca^{2+} in mitosis is unclear. It is known, however, that increasing the intracellular Ca^{2+} concentration by microinjection (60) results in the disassembly of the mitotic spindle. Moreover, cytoplasmic microtubules undergo disassembly in culture media containing elevated Ca^{2+} plus ionophore A23187 (25). Since it is known that the chromosome to pole microtubules must shorten during chromosome movement to the poles (32), we questioned whether a CDR–Ca^{2+} complex might serve as a physiological regulator of microtubule assembly-disassembly. It is exceedingly difficult to specifically isolate the microtubules of the mitotic apparatus. Fortunately, however, the *in vitro* polymerization of cytoplasmic microtubules is also a Ca^{2+}-sensitive process. Therefore, we isolated microtubule protein from rat brain by three assembly-disassembly cycles in order to determine whether the Ca^{2+} effects on polymerization are mediated via CDR.

In vitro microtubule protein polymerization was initiated by the addition of guanosine 5'-triphosphate at 37°C and monitored by the change in absorbancy at 320 nm. In the absence of added Ca^{2+}, the polymerization reaction was initiated after a very short lag period, rapidly increased, and reached a stable plateau after about 10 min. Addition of CDR (after removal of bound Ca^{2+} by EGTA) or 0.4 μM free Ca^{2+} did not alter either rate or extent of microtubule polymerization. Increasing the free Ca^{2+} to 11 μM only caused a 10% reduction in the extent of polymerization. However, addition of 11 μM Ca^{2+} bound to CDR resulted in total inhibition of microtubule assembly. Thus, CDR does prevent the assembly of microtubules *in vitro* in a Ca^{2+}-dependent manner. If CDR plays a role in the movement of chromosomes to the poles, it may do so by promoting the disassembly of microtubules. To test this hypothesis, microtubules were fully polymerized *in vitro* before the addition of Ca^{2+} or CDR. Again, Ca^{2+}-free CDR or 11 μM Ca^{2+} caused no significant change in the assembled microtubule population. However, complexing CDR and Ca^{2+} prior to addition resulted in a precipitous decrease in absorbancy at 320 nm followed by a slower rate of change until complete disassembly was achieved. These data offer the possibility that CDR plays a physiological role in the control of microtubule polymerization.

At least two other possibilities must be considered regarding the role of

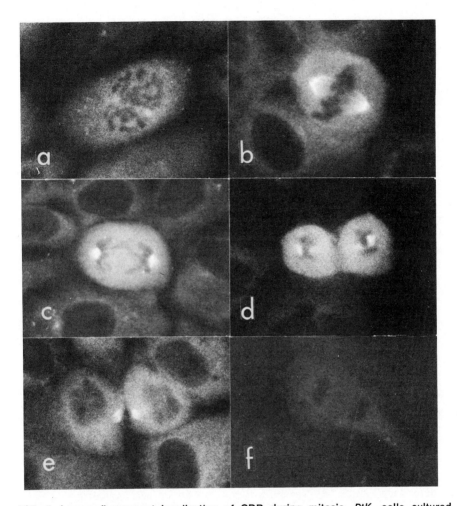

FIG. 5. Immunofluorescent localization of CDR during mitosis. *PtK₁* cells cultured on glass coverslips and prepared for indirect immunofluorescent localization of CDR. Cells were fixed in 3% formalin in Dulbecco's PBS at room temperature for 30 min, then treated with absolute acetone at −20°C for 7 min to permeabilize the cell membrane. The fixed cells were then incubated in a small amount of monospecific goat antibody to CDR (100 μg/ml) for 30 min at 37°C. Cells were washed twice for 10 min in PBS, then incubated in fluorescein conjugated to rabbit antigoat IgG for 30 min at 37°C. Cells were again washed twice for 10 min, rinsed in distilled water, and mounted in pH 9.0 PBS:glycerol (1:9). Cells were observed with a Leitz microscope adapted for fluorescence microscopy using epi-illumination. Images were recorded on Kodak Tri-X film. **a:** Prophase. **b:** Metaphase. **c:** Anaphase. **d:** Lata anaphase. **e:** Telophase. **f:** Control preparation using immune goat IgG (200 μg/ml) not retained on CDR-Sepharose affinity column. **A–F:** ×750.

CDR in mitosis. Both of these take precedent in previously established cytoplasmic functions. The first relates to the similarity to TnC and the fact that CDR can regulate actomyosin ATPase activity in nonmuscle cells. There is evidence for the presence of actin and myosin in the mitotic apparatus (6,24,30,62,63). Results to date suggest the most likely role of these contractile proteins is in cytokinesis (36). This raises the possibility that CDR also regulates the ATPase of actomyosin present in the mitotic cell. The second alternate possibility stems from the finding that CDR regulates the Ca^{2+} pump of erythrocyte plasma membranes. A Ca^{2+}-sensitive ATPase has been shown to be associated with the mitotic apparatus (28,43,57,58). This activity occurs in conjunction with the smooth endoplasmic reticulum of mitotic cells and peaks during metaphase in many cell types. It is possible, therefore, that this ATPase-containing vesicular system could serve a Ca^{2+}-sequestering function similar to the sarcoplasmic reticulum of skeletal and cardiac muscle. CDR, then, might play a functional role due to its Ca^{2+}-binding properties of regulating Ca^{2+} levels in the microenvironment of the spindle. At any rate, the association of CDR with the mitotic apparatus and its widespread distribution in a variety of eucaryotic cells suggest a fundamental role for this protein in motile systems of nonmuscle cells.

Whether CDR plays a regulatory role in the action of FSH remains to be elucidated. This gonadotropin does not appear to regulate the total Sertoli cell concentration of CDR. Rather, Ca^{2+} levels are altered by FSH in a manner that regulates PDE and CDR by independent mechanisms. There is no question that Ca^{2+} and cyclic nucleotides are intimately related in the hormonal control of the Sertoli cells. Available data suggest that, functionally, cyclic nucleotide metabolism is only important during Sertoli cell differentiation. On the other hand, effects of FSH on Ca^{2+} distribution and cell shape are independent of age. Considering the known ultrastructure of the Sertoli cell in testes undergoing cyclic spermatogenesis, it seems possible that FSH, via Ca^{2+} and CDR, regulates the movement of germ cells from basal to adluminal compartments. The release of mature spermatids from the Sertoli cell might also require Ca^{2+} and CDR. Formation and maintenance of the tight junctions between adjacent Sertoli cells could be dependent on the CDR–Ca^{2+} system since microfilament bundles and isolated patches of smooth endoplasmic reticulum are intimately associated with these membrane specializations. Finally, the Sertoli cell is known to secrete both proteins and other organic molecules into the tubular fluid under the control of FSH. Secretion is a contractile process, and such events require Ca^{2+}. Thus, this also could provide a role for CDR during spermatogenesis regardless of the age of the animals. The realization that CDR and Ca^{2+} are important in the control of such fundamental processes as cyclic nucleotide metabolism, motility, and secretion offers new and fertile avenues of approach toward a better understanding of the hormonal control of Sertoli cell function.

REFERENCES

1. Arch, J. R., and Newsholme, E. A. (1976): Activities and some properties of adenylate cyclase and phosphodiesterase in muscle, liver, and nervous tissue from vertebrates and invertebrates in relation to the control of the concentration of adenosine 3′:5′-cyclic monophosphate. *Biochem. J.,* 158:603–622.
2. Beale, E. G., Dedman, J. R., and Means, A. R. (1977): Isolation and regulation of the protein kinase inhibitor and the calcium-dependent cyclic nucleotide phosphodiesterase regulator in the Sertoli cell-enriched testis. *Endocrinology,* 101:1621–1634.
3. Blomberg, F., Cohen, R. S., and Siekevitz, P. (1977): The structure of postsynaptic densities isolated from dog cerebral cortex. II. Characterization and arrangement of some of the major proteins within the structure. *J. Cell Biol.,* 74:204–225.
4. Brinkley, B. R., Fuller, G. M., and Highfield, D. P. (1975): Cytoplasmic microtubules in normal and transformed cells: Analysis by tubulin antibody immunofluorescence. *Proc. Natl. Acad. Sci. USA,* 72:4981–4958.
5. Brinkley, B. R., Miller, C. L., Fuseler, J. W., Pepper, D. A., and Wible, L. J. (1978): Cytoskeletal changes in cell transformation to malignancy. In: *Cell Differentiation and Neoplasia,* edited by G. Saunders. William Wilkins, Baltimore. (*In press.*)
6. Cande, W. Z., Lazarides, E., and McIntosh, J. R. (1977): A comparison of the distribution of actin and tubulin in the mammalian mitotic spindle as seen by indirect immunofluorescence. *J. Cell Biol.,* 72:552–567.
7. Cheung, W. Y. (1970): Cyclic 3′,5′-nucleotide phosphodiesterases: Demonstration of an activator. *Biochem. Biophys. Res. Commun.,* 38:533–538.
8. Clermont, Y., and Perey, B. (1957): Quantitative study of the cell population of the seminiferous tubules in immature rats. *Am. J. Anat.,* 100:241–268.
9. Cohen, I., and DeVries, A. (1973): Platelet contractile regulation in an isometric system. *Nature,* 246:36–37.
10. Davies, J. I., and Williams, P. A. (1975): Quantitative aspects of the regulation of cellular cyclic AMP levels. I. Structure and kinetics of a model system. *J. Theor. Biol.,* 53:1–30.
11. Dabrowska, R., Aromatorio, D., Sherry, J. M. F., and Hartshorne, D. J. (1977): Composition of the myosin light chain kinase from chicken gizzard. *Biochem. Biophys. Res. Commun.,* 78:1263–1272.
12. Dabrowska, R., Sherry, J. M. F., Aromatorio, D. K., and Hartshorne, D. J. (1978): Modulator protein as a component of the myosin light chain kinase from chicken gizzard. *Biochemistry,* 17:253–258.
13. Dedman, J. R., Fakunding, J. L., and Means, A. R. (1977): Protein kinase and phosphodiesterase. In: *Hormone Action and Molecular Endocrinology Workshop Syllabus,* edited by B. W. O'Malley, and W. T. Schrader, pp. 901–946. The Endocrine Society, Bethesda, Md.
14. Dedman, J. R., Jackson, R. L., Schreiber, W. E., and Means, A. R. (1978): Sequence homology of the Ca^{2+}-dependent regulator of cyclic nucleotide phosphodiesterase from rat testis with other Ca^{2+}-binding proteins. *J. Biol. Chem.,* 253:343–346.
15. Dedman, J. R., Potter, J. D., Jackson, R. L., Johnson, J. D., and Means, A. R. (1977): Physicochemical properties of rat testis Ca^{2+}-dependent regulator protein of cyclic nucleotide phosphodiesterase: Relationship of Ca^{2+}-binding, conformational changes and phosphodiesterase activity. *J. Biol. Chem.,* 252:8415–8422.
16. Dedman, J. R., Potter, J. D., and Means, A. R. (1977): Biological cross-reactivity of rat testis phosphodiesterase activator protein and rabbit skeletal muscle troponin-C. *J. Biol. Chem.,* 252:2437–2440.
17. Dedman, J. R., Welsh, M. J., and Means, A. R. (1978): Ca^{2+}-dependent regulator: Production and characterization of a monospecific antibody. *J. Biol. Chem.,* 253: 7515–7521.
18. Dorrington, J. H., and Fritz, I. B. (1974): Effects of gonadotropins on cyclic AMP production by isolated seminiferous tubules and interstitial cell preparations. *Endocrinology,* 94:395–403.

19. Dorrington, J. H., Roller, N. F., and Fritz, I. B. (1975): Effects of follicle stimulating hormone on cultures of Sertoli cell preparations. *Mol. Cell. Endocrinol.,* 3:59–70.
20. Fakunding, J. L., and Means, A. R. (1977): Characterization and follicle stimulating hormone activation of Sertoli cell cyclic-AMP-dependent protein kinase. *Endocrinology,* 101:1358–1368.
21. Fakunding, J. L., Tindall, D. J., Dedman, J. R., Mena, C. R., and Means, A. R. (1975): Biochemical actions of follicle stimulating hormone in the Sertoli cell of the rat testis. *Endocrinology,* 98:392–402.
22. Fritz, I. B. (1978): Sites of action of androgens and follicle stimulating hormone on cells of the seminiferous tubules. In: *Biochemical Actions of Hormones,* Vol. 5, edited by G. Litwack. Academic Press, New York. (*In press.*)
23. Fritz, I. B., Rommerts, F. G., Louis, B. G., and Dorrington, J. H. (1976): Regulation by FSH and dibutyryl cyclic AMP of the formation of androgen-binding protein in Sertoli cell-enriched cultures. *J. Reprod. Fertil.,* 46:17–24.
24. Fujiwara, K., and Pollard, T. D. (1976): Fluorescent antibody localization of myosin in the cytoplasm, cleavage furrow, and mitotic spindle of human cells. *J. Cell Biol.,* 71:848–875.
25. Fuller, G. M., and Brinkley, B. R. (1976): Structure and control of assembly of cytoplasmic microtubules in normal and transformed cells. *J. Supramol. Struct.,* 5:497–514.
26. Gopinath, R. M., and Vincenzi, F. F. (1977): Phosphodiesterase protein activator mimics red blood cell cytoplasmic activator of $(Ca^{2+}-Mg^{2+})$ ATPase. *Biochem. Biophys. Res. Commun.,* 77:1203–1209.
27. Hansson, V., Reusch, E., Trygstad, O., Torgersen, O., Ritzen, E. M., and French, F. S. (1973): FSH stimulation of testicular androgen binding protein. *Nature,* 246:56–58.
28. Harris, P. (1975): The role of membranes in the organization of the mitotic apparatus. *Exp. Cell Res.,* 94:409–425.
29. Heindel, J. J., Rothenberg, R., Robinson, G. A., and Steinberger, A. (1975): LH and FSH stimulation of cyclic AMP in specific cell types isolated from the testes. *J. Cyc. Nucl. Res.,* 1:69–78.
30. Hinkley, R., and Telser, A. (1974): Heavy meiomyosin-binding filaments in the mitotic apparatus of mammalian cells. *Exp. Cell Res.,* 86:161–164.
31. Hodgen, G. D., and Sherins, R. J. (1973): Enzymes as markers of testicular growth and development in the rat. *Endocrinology,* 93:985–989.
32. Inoue, S., and Ritter, H. (1975): Dynamics of mitotic spindle organization and function. In: *Molecules and Cell Movement,* edited by S. Inoue and R. Stephens, pp. 3–30. Raven Press, New York.
33. Jarrett, H. W., and Penniston, J. T. (1977): Partial purification of the $Ca^{2+}-Mg^{2+}$ ATPase activator from human erythrocytes: Its similarity to the activator of 3':5'-cyclic nucleotide phosphodiesterase. *Biochem. Biophys. Res. Commun.,* 77:1210–1216.
34. Kakiuchi, S., Yamazaki, R., and Nakajima, H. (1970): Properties of a heat-stable phosphodiesterase activating factor isolated from brain extract. *Proc. Jpn. Acad.,* 46:587–592.
35. Kettleslegers, J. M., Hetzel, W. D., Sherins, R. J., and Catt, K. J. (1978): Developmental changes in testicular gonadotropin receptors, plasma gonadotropins and plasma testosterone in the rat. *Endocrinology,* 103:212–222.
36. Kiehart, P., Inoue, S., and Mabuchi, I. (1977): Evidence that the force production in chromosome movement does not involve myosin. *J. Cell Biol.,* 75:258a.
37. Kuehl, F., Patanelli, D. J., Tarnoff, J., and Humes, J. L. (1970): Testicular adenyl cyclase: Stimulation by the pituitary gonadotropin. *Biol. Reprod.,* 2:153–163.
38. Lacroix, M., Smith, F. E., and Fritz, I. B. (1977): Secretion of plasminogen activator by Sertoli cell-enriched cultures. *Mol. Cell. Endocrinol.,* 9:227–236.
39. Lazarides, E., and Weber, K. (1974): Actin antibody: The specific visualization of actin filaments in non-muscle cells. *Proc. Natl. Acad. Sci. USA,* 71:2268–2272.
40. Levin, R. M., and Weiss, B. (1977): Binding of trifluoperazine to the calcium-

dependent activator of cyclic nucleotide phosphodiesterase. *Mol. Pharmacol.*, 13:690–697.

41. Lu, C., and Steinberger, A. (1977): Gamma-glutamyl transpeptidase activity in the developing rat testis. Enzyme localization in isolated cell types. *Biol. Reprod.*, 17:84–88.

42. Luthra, M. G., Au, K. S., and Hanahan, D. J. (1977): Purification of an activation of human erythrocyte membrane ($Ca^{2+} + Mg^{2+}$) ATPase. *Biochem. Biophys. Res. Commun.*, 77:678–687.

43. Mazia, D., Petzelt, C., Williams, R. O., and Meza, I. (1972): A Ca^{2+}-activated ATPase in the mitotic apparatus of the sea urchin egg (isolated by a new method). *Exp. Cell Res.*, 70:325–332.

44. Means, A. R. (1971): Concerning the mechanism of FSH action: Rapid stimulation of testicular synthesis of nuclear RNA. *Endocrinology*, 89:981–989.

45. Means, A. R. (1973): Specific interaction of ^{3}H-FSH with rat testis binding sites. *Adv. Exp. Med. Biol.*, 36:431–448.

46. Means, A. R. (1976): Mechanisms of gonadotropic action in the testis. In: *Regulatory Mechanisms of Male Reproductive Physiology*, edited by C. H. Spilman, T. J. Lobl, and K. T. Kirton, pp. 87–96. American Elsevier, New York.

47. Means, A. R. (1977): Mechanisms of action of follicle-stimulating hormone (FSH). In: *The Testis*, Vol. 4, edited by A. D. Johnson, and W. R. Gomes, pp. 163–188. Academic Press, New York.

48. Means, A. R., Dedman, J. R., Fakunding, J. L., and Tindall, D. J. (1978): Mechanism of action of FSH in the male rat. In: *Hormone Receptors*, Vol. 3, edited by L. Birnbaumer, and B. W. O'Malley, pp. 363–393. Academic Press, New York.

49. Means, A. R., and Hall, P. F. (1967): Effect of FSH on protein biosynthesis in testis of the immature rat. *Endocrinology*, 81:1151–1160.

50. Means, A. R., and Hall, P. F. (1969): Protein biosynthesis in the testis: Concerning the nature of stimulation by follicle stimulating hormone. *Biochemistry*, 8:4293–4298.

51. Means, A. R., and Huckins, C. (1974): Coupled events in the early biochemical actions of FSH on the Sertoli cells in the testis. In: *Hormone Binding and Target Cell Activation in Testis*, edited by M. L. Dufau and A. R. Means, pp. 145–165. Plenum Press, New York.

52. Means, A. R., Fakunding, J. L., Huckins, C., Tindall, D. J., and Vitale, R. (1976): Follicle stimulating hormone, the Sertoli cell, and spermatogenesis. *Recent Prog. Horm. Res.*, 32:477–527.

53. Means, A. R., Fakunding, J. L., and Tindall, D. J. (1976): Follicle stimulating hormone regulation of protein kinase activity and protein synthesis in testis. *Biol. Reprod.*, 14:54–63.

54. Means, A. R., MacDougall, E., Soderling, T., and Corbin, J. D. (1974): Testicular adenosine 3',5'-monophosphate-dependent protein kinase: Regulation by follicle stimulating hormone. *J. Biol. Chem.*, 249:1231–1238.

55. Means, A. R., and Tindall, D. J. (1975): FSH-induction of androgen binding protein in testis on Sertoli cell only rats. In: *Hormonal Regulation of Spermatogenesis*, edited by F. French, V. Hansson, E. M. Ritzen, and S. N. Neyfeh. pp. 383–398. Plenum Press, New York.

56. Orth, J., and Christensen, A. K. (1977): Localization of ^{125}I-labeled FSH in the testes of hypophysectomized rats by autoradiography at the light and electron microscope levels. *Endocrinology*, 101:262–278.

57. Petzelt, C., and Auel, D. (1977): Synthesis and activation of mitotic Ca^{2+}-adenosinetriphosphatase during the cell cycle of mouse mastocytoma cells. *Proc. Natl. Acad. Sci. USA*, 77:1610–1613.

58. Petzelt, C., and vonLedenbur-Villiger, M. (1973): Ca^{2+}-stimulated ATPase during the early development of parthenogenetically activated eggs of the sea urchin. *Paracentrotus lividus. Exp. Cell Res.*, 81:87–94.

59. Porter, K. R. (1966): Cytoplasmic microtubules and their function. In: *Principles of Biomolecular Organization*, edited by G. E. Wolstenholm and M. O'Conner, pp. 308–315. Little, Brown, Boston.

60. Salmon, E. D., and Jenkins, R. (1977): Isolated mitotic spindles are depolymerized by μM calcium and show evidence of dynein. *J. Cell Biol.*, 75:295a.
61. Sanborn, B. M., Elkington, J. S. H., Chowdhury, M., Tcholakian, R. K., and Steinberger, E. (1975): Hormonal influences on the level of testicular androgen binding activity: Effect of FSH following hypophysectomy. *Endocrinology*, 96:304–312.
62. Sanger, J. W. (1975): Presence of actin during chromosomal movement. *Proc. Natl. Acad. Sci. USA*, 72:2451–2455.
63. Schloss, J. A., Milstead, A., and Goldman, A. D. (1977): Myosin subfragment binding for the localization of actin-like microfilaments in cultured cells. *J. Cell Biol.*, 74:794–815.
64. Sobieszek, A. (1977): Vertebrate smooth muscle myosin: Enzymatic and structural properties. In: *The Biochemistry of Smooth Muscle*, edited by N. L. Stephens, pp. 413–443. Univ. Park Press, Baltimore.
65. Steinberger, A., Heindel, J. J., Lindsey, J. N., Elkington, J. S. H., Sanborn, B. M., and Steinberger, E. (1975): Isolation and culture of FSH responsive Sertoli cells. *Endocr. Res. Commun.*, 2:261–272.
66. Steinberger, E., Steinberger, A., and Ficher, A. (1970): Study of spermatogenesis and steroid metabolism in cultures of mammalian testis. *Recent Prog. Horm. Res.*, 26:547–588.
67. Thanki, K. H., and Steinberger, A. (1977): Effect of age and hypophysectomy on FSH binding by rat testis. *Andrologia*, 9:307–312.
68. Tindall, D. J., Cunningham, G. R., and Means, A. R. (1978): 5α-dihydrotestosterone binding to androgen binding protein: Effects of testosterone and other compounds *in vivo* and *in vitro* on the number of binding sites measured. *J. Biol. Chem.*, 253:166–169.
69. Tindall, D. J., Mena, C. R., and Means, A. R. (1978): Hormonal regulation of androgen binding protein in hypophysectomized rats. *Endocrinology*, 103:584–594.
70. Vernon, R. G., Kopec, B., and Fritz, I. B. (1974): Observations on the binding of androgens by rat testis seminiferous tubules and testis extracts. *Mol. Cell. Endocrinol.*, 1:167–184.
71. Weddington, S. C., Hansson, V., Ritzen, E. M., Hagenas, L., French, F. S., and Nayfeh, S. N. (1975): Sertoli cell secretory function after hypophysectomy. *Nature*, 254:145–147.
72. Wells, J. N., and Hardman, J. G. (1977): Cyclic nucleotide phosphodiesterase. *Adv. Cyclic Nucleotide Res.*, 8:119–143.
73. Welsh, M. J., Dedman, J. R., Brinkley, B. R., and Means, A. R. (1978): Calcium dependent regulator protein: Localization in the mitotic apparatus of eucaryotic cells. *Proc. Natl. Acad. Sci. USA*, 75:1867–1871.
74. Welsh, M. J., and Wiebe, J. P. (1975): Rat Sertoli cells: A rapid method for obtaining viable cells. *Endocrinology*, 96:618–624.

Ontogeny of Receptors and Reproductive Hormone Action,
edited by T. H. Hamilton, J. H. Clark, and W. A. Sadler.
Raven Press, New York 1979.

Mouse Mammary Tissue Estrogen Receptors: Ontogeny and Molecular Heterogeneity

Thomas G. Muldoon

Medical College of Georgia, Department of Endocrinology, Augusta, Georgia 30902

The development and maintenance of a variety of strains of mice having wide diversity in their probability of mammary tumor incidence has led to the use of mammary tissue as a model system for a great number of studies aimed at elucidating the nature and etiology of neoplastic transformation in the mammary gland; a number of detailed reviews of this subject are available (4,43,47). The literature is replete with investigations into the effects of a vast array of synthetic and naturally occurring agents as tumorigenic determinants in the mouse mammary gland, with somewhat less regard for the molecular mechanisms by which such agents act. In recent years, the hormonal requirements for normal mammary tissue growth and development have been clarified (2) and simultaneously implicated as vital in maintenance of mammary cell harmony. It has also become evident that the responsiveness of mammary tissue to steroid hormones bears a positive relationship to the functional presence of intracellular receptor proteins, and the clinical relevance of this observation forms the basis for a diagnostic tool in treatment of human breast cancer (36).

Little emphasis has been placed on the means by which steroid hormone receptor activity is regulated within mammary cells. Any understanding of the process of neoplastic transformation of cells, with its characteristic loss of dependency on controlled hormonal stimulation, demands precognition of the factors involved in normal modulation of hormonal responsivity. This latter aspect of hormonal action, with specific reference to estrogen action in normal mouse tissue, is the subject of the studies presented here.

MATERIALS AND METHODS

Female mice of the C3H[+] or A[+] strain were grown in our own colony (Kirschbaum Memorial Laboratory). Details of most of the materials and procedures employed have been described previously (21), with the exception of the following points.

Quantification of cytosol receptor levels by saturation binding analysis required partial purification of the receptor by ammonium sulfate fractionation (0 to 40% saturation) to remove excessive amounts of lipid, which interfered with the assay. This procedure did not result in conversion of receptor to the nuclear form (17), as assessed by sucrose gradient analysis, probably because of the inclusion of ethylenediaminetetraacetate (EDTA) in the buffer (0.01 M Tris, 0.0015 M Na$_2$EDTA, 0.005 M dithiothreitol, pH 8.0). However, it was also observed that the presence of EDTA in the buffer contributed to significant amounts of receptor aggregation on sucrose gradients, in agreement with the report of Shyamala (57). Therefore, samples of cytosol prepared in EDTA-containing buffer were dialyzed prior to sucrose gradient analysis.

^{14}C-labeled bovine serum albumin (BSA, 4.6S) was prepared by methylation with [^{14}C]formaldehyde, as described by Stancel and Gorski (60), for use as an internal marker in sucrose gradient experiments.

Receptor complex formation was assayed by protamine sulfate precipitation (29). Levels of steroid-complexed receptor were measured in cytosols and nuclear fractions by the exchange assays of Katzenellenbogen et al. (26) and Anderson et al. (1), respectively.

RESULTS

Nature and Specificity of Estrogen Binding

Sucrose density gradient ultracentrifugation analysis of adult virgin mouse mammary tissue cytosol was approached with the preconceived notion that, in accord with most other receptor systems characterized in this manner, a complex would be readily detectable in the 8S region of the gradient. As shown in Fig. 1, very little binding of [^{3}H]estradiol was located in any portion of the gradient other than the 4S region, as determined by reference to a 4.6S BSA marker. In the unlikely event that 8S binding was being masked by the large 4S component, fractions 9 to 12 from a series of six separate identical cytosol gradients were pooled, concentrated, equilibrated with [^{3}H]estradiol (10^{-9} M), and applied to a sucrose gradient; no enhancement of 8S binding could be induced by this procedure.

The 4S estrogen complexes were not disrupted by the presence of testosterone or progesterone in 200-fold molar excess amounts. Unlabeled estradiol (100-fold) and the nonsteroidal antiestrogenic CI-628 (5,000-fold) each reduced the 4S binding by approximately 50%, demonstrating that half of this binding manifested steroidal specificity consistent with a low-capacity estrogen binding protein. The proteinaceous nature of the 4S complexes was indicated by their susceptibility to proteolytic digestion by pronase.

In all ensuing sucrose gradient patterns presented in this discussion, it should be noted that: (a) all samples are treated with dextran-coated charcoal immediately prior to centrifugation to eliminate the accumulation of

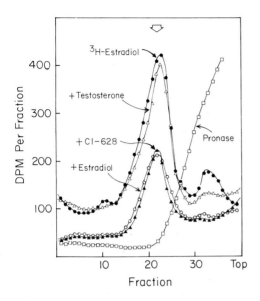

FIG. 1. Steroidal specificity of mammary tissue estrogen binding. Samples of mammary tissue cytosol from mature virgin mice (8.6 mg protein/ml) were incubated for 1 hr at 4°C with [³H]estradiol alone (10⁻⁹ M, ●) or in the presence of testosterone (2 × 10⁻⁷ M, △), CI-628 (5 × 10⁻⁵ M, ▲), or unlabeled estradiol (10⁻⁷ M, ○). A separate aliquot was treated with pronase (100 μg, □) for 30 min at 13°C subsequent to addition of [³H]estradiol. Centrifugation through a linear 5 to 20% sucrose gradient for 15.5 hr at 225,000 × *g* yielded the patterns depicted. The position of simultaneously centrifuged ¹⁴C-labeled BSA is indicated by the arrow. All samples were subjected to dextrancoated charcoal adsorption prior to application onto the gradient.

unbound steroid at the top of the gradient, and (b) the data are represented as "specific binding," indicating that the contribution of nonspecific interactions to the total pattern has been subtracted (i.e., the resultant of ([³H]estradiol alone) − ([³H]estradiol + 100-fold excess unlabeled estradiol) is plotted for each fraction).

Because of the heterogeneity of estrogen binding proteins in plasma and cells, it was necessary to demonstrate that the specific estrogen binding component in the 4S region fulfilled several other criteria for classification as a receptor. In Fig. 2 are presented data that establish protein specificity. In Fig. 2A are shown saturation binding curves for ammonium sulfate-fractionated cytosol following equilibration with progressively increasing amounts of [³H]estradiol. When corrections in total binding were made for the contribution of nonspecific binding elements, the specific saturation level was approximately 30 fmoles/mg cytosol protein. Addition of diluted mouse plasma to the cytosol increased only the nonspecific component of the total binding, yielding the same specific saturation curve as seen in the absence of the plasma. Thus, high-affinity limited-capacity estrogen-specific plasma binding proteins do not interfere with the specific cytosol binding being measured. This result is consistent with the reported observation (33) that protamine sulfate does not precipitate plasma sex steroid binding globulin.

The presence of α-fetoprotein in young rodents is well established (39,53) and may be distinguished from estrogen receptor by its lack of affinity for diethylstilbestrol (51,53). In the 21-day-old mouse (Fig. 2B), all specific cytosol binding in the 4S region was eliminated by diethylstilbestrol, as as-

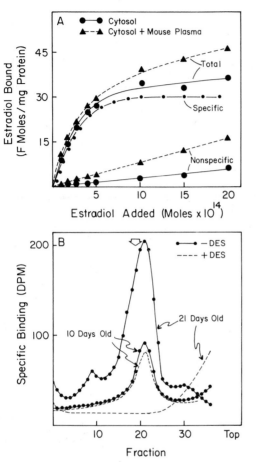

FIG. 2. Protein specificity of mammary tissue estrogen binding. **A:** Saturation binding characteristics of ammonium sulfate-fractionated cytosol from virgin adult mice were determined in the presence or absence of a supplemental addition of 1:30 diluted plasma from 7-day-old female mice. Constant amounts of cytosol were incubated with increasing concentration of [³H]estradiol either with (nonspecific binding) or without (total binding) concomitant addition of excess unlabeled estradiol. Incubations were for 18 hr at 4°C, and the bound species were precipitated with protamine sulfate. Specific binding (●—●—●) is the result of total minus nonspecific. **B:** Specific estradiol binding (———) (i.e., binding in the presence of [³H]estradiol minus that in the presence of [³H]estradiol + unlabeled estradiol) was determined in cytosol of 10- or 21-day-old mice by sucrose gradient analysis and is plotted as the solid line. The estrogenic specificity was then further delineated by introduction of diethylstilbestrol (DES) and analysis of its effect on the estradiol-specific binding.

sessed under conditions where competition for sites is optimal. Under the same conditions, diethylstilbestrol had no significant effect on the level of specific 4S binding in cytosol from 10-day-old animals. Attardi and Ohno (3) have previously noted that the fetoneonatal estradiol binding protein is present in large amounts in cytosol from 3- to 5-day-old mice, less in 9- to 11-day-old, and undetectable at 3 weeks. These findings further substantiate the receptor-like nature of the specific 4S binding in the adult virgin mouse.

Equilibrium binding parameters were calculated from saturation binding data obtained from cytosol of 2- or 4-week-old mice. The data were plotted according to our adaptation (65) of the direct linear plot method (18) (Fig. 3). In this procedure, each point of the saturation curve becomes a straight line in K_D,B_{max} space, characterized by its intersection with the abscissa at the value of unbound steroid, and its ordinate intersection at the value of specifically bound steroid. When the lines are extrapolated into the first quadrant, their median intersection point affords an accurate (66)

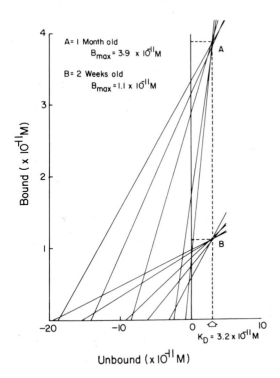

FIG. 3. Direct linear plot analysis of mammary estrogen receptor equilibrium binding parameters. Ammonium sulfate-fractionated cytosol from animals either 2 or 4 weeks of age was incubated to equilibrium with varying amounts of [³H]estradiol. Specific binding was assessed, and molar quantities of unbound and bound steroid at each point were plotted in K_D, B_{max} space.

estimate of the values of the binding site concentration and the equilibrium dissociation constant, when projected onto the ordinate and abscissa, respectively. Between the ages of 2 weeks and 1 month, the high-affinity specific estrogen binding capacity of mouse mammary tissue may be seen to increase some three- to fourfold. The very low K_D, consistent with estrogen receptor values in other systems, was invariant in these groups.

Ontogeny of Specific Estrogen Binding

Receptor concentration of mammary tissue cytosol, measured as diethylstilbestrol-sensitive specific binding, was determined by saturation binding analysis, using cytosol from groups of animals of different ages (Fig. 4). Prior to 2 weeks of age, no receptor activity was demonstrable. Between 2 and 4 weeks, the time of puberty in the mouse, receptor levels rose to a maximal concentration of 48 fmoles/mg protein. This is somewhat later than the development of receptor systems in the brain (50) and uterus (16) of the rat, both of which have attained a full receptor complement well before the onset of puberty. Since mammary glands in the mouse begin to grow only at about 25 to 30 days of age, it is reasonable physiologically to expect that estrogen receptor maturation will be relatively late in this tissue. Beyond 4 weeks, there was a gradual fall in receptor levels to a plateau;

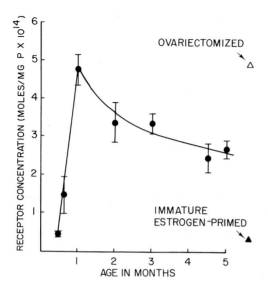

FIG. 4. The ontogeny of specific estrogen receptor binding in mouse mammary tissue. The concentration of diethylstilbestrol-sensitive binding of estradiol was assessed in cytosol from mice of different ages (●). Values are the means ± SE of at least three separate determinations. Receptor content of 2-weeks' ovariectomized adult animals is also shown (Δ). A group of 10-day-old mice was administered estradiol (0.02 μg) daily for 5 days and killed 2 days later (▲). (From Hunt and Muldoon, ref. 21.)

this decline was concomitant with, and presumably the result of, rising endogenous estrogen levels during this period, as substantiated by the observation that chronically ovariectomized adult mice had receptor levels equivalent to those of the 4-week-old animal.

It is also pertinent to note that administration of estradiol in priming dosages to an immature mouse did not result in induction of estrogen receptor activity, as it does in the adult animal (see section immediately following). In an early study, Flux (19) showed that mammary tissue does not grow in response to estrogen or progesterone between 10 and 15 days of age, but exhibits a slight responsiveness to an unfractionated anterior pituitary suspension. It would, therefore, appear that some features of the complex hormonal pattern of mammary tissue growth and differentiation are prerequisites to the development of an estrogen-responsive receptor system in this tissue. In a similar vein, Plapinger and McEwen (50), studying estrogen receptors in rat brain, observed that treatment with estrogen could effect acceleration of puberty only if the estrogen were administered subsequent to day 22 to 23 of age.

Receptor Transformation and Translocation to the Nucleus

The ability of the mammary tissue receptor of the adult virgin mouse to undergo transformation and nuclear translocation similar to other estrogen receptors (23) was examined under *in vitro* and *in vivo* conditions. The data in Fig. 5A are derived from experiments in which cytosol and purified (8) nuclei were prepared. The cytosol was charged with estradiol at elevated

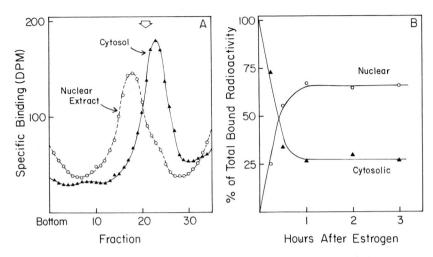

FIG. 5. Transformation and nuclear translocation of the 4S mammary tissue estrogen receptor. **A:** Cytosol was prepared from adult virgin female mice and incubated with [³H]estradiol for 1 hr at 4°C. One portion was further incubated for 95 min at 4°C (▲). A second portion of the same cytosol was incubated for 45 min at 25°C and then used as suspension medium for purified mammary tissue nuclei. After 60 min at 25°C, the nuclei were sedimented (10 min at 1,000 × g, 4°C), washed with 0.32 M sucrose, and extracted with 0.4 M KCl (○). Sucrose gradient centrifugation of both cytosol and nuclear extract was performed in the presence of 0.4 M KCl. **B:** Mature virgin mice were injected with 0.1 μg of [³H]estradiol and killed in groups at the designated intervals thereafter. The tissue homogenate was initially centrifuged at 1,000 × g for 10 min, and the resulting pellet was washed three times with buffer and extracted with 0.4 M KCl in buffer. The extract was isolated by centrifugation for 30 min at 15,000 × g and subjected to sucrose gradient centrifugation in high-salt buffer for quantification of 5.5S estrogen binding (○). The 105,000 × g, 60-min supernatant fraction of the homogenate was applied to a low-salt gradient, and the 4S radioactivity was measured (▲).

temperature to effect transformation and recombined with nuclei, also at increased temperature, to permit translocation of the complex. Samples of the high-salt nuclear extract and the unheated cytosol were applied to a sucrose gradient prepared in 0.4 M KCl. The details of these procedures were exactly as described by Jensen et al. (22). The cytosolic form of the receptor sedimented at 4S, identical with its behavior in a low-salt gradient. The nuclear-bound material migrated principally as a 5.5S component with a trailing 4S shoulder. The nuclear form of this receptor is virtually identical to that encountered in every other estrogen receptor system studied to date (28).

Following administration of [³H]estradiol to mice, the subcellular distribution follows the pattern shown in Fig. 5B. Within 1 hr, the nuclear fraction has risen to a maximal level of bound steroid at the expense of the cytosol compartment. For the next 2 hr, the relative proportions within the two fractions remain constant, reminiscent of the situation described by Giannopoulos and Gorski (20), which was found applicable to a variety of estrogen receptor systems, including the lactating mouse mammary gland (58).

A functional aspect of estrogen receptor activity is the so-called depletion-replenishment response (9,24,55). The injection of estradiol into an animal causes a rapid depletion of cytoplasmic receptors by translocation into the nucleus, followed by a gradual replenishment of these receptors, through a complex dynamic system involving receptor synthesis, reutilization, and recycling (10,14). In Fig. 6, it may be seen that receptor can be depleted and replenished by estradiol in mouse mammary tissue very effectively. The lack of a defect at this stage of the mechanism of estrogen action is an important point to establish because: (a) it has been shown in one instance that mouse mammary tumor tissue may lack the capability to translocate estrogen–receptor complexes into the nucleus (56), and (b) the antagonistic actions of some antiestrogens appear to be related to their inability to effect cytosol receptor replenishment (12,15,25).

Also shown in Fig. 6 is the effect of cycloheximide when administered concomitantly with estradiol. Depletion was unaffected, but replenishment was markedly reduced. A more rigorous analysis of the cycloheximide effect is presented in Table 1. The protein synthesis-blocking action of the drug was analyzed as a function of the time of its administration relative to that of estradiol and was assessed by its effect on the uninhibited 5-hr receptor

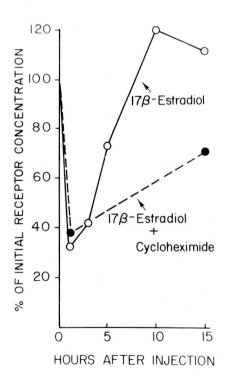

FIG. 6. Estrogen-induced depletion and replenishment of cytoplasmic receptors. Female mice were injected with 0.1 μg of estradiol with (●) or without (○) 0.1 mg of cycloheximide at time zero. At the prescribed intervals, groups of animals were killed and total mammary tissue cytosol receptor content was measured by an exchange assay. Levels are expressed as a function of control concentrations from vehicle-injected animals. (From Hunt and Muldoon, ref. 21.)

TABLE 1. *Cycloheximide effect on 5-hr receptor replenishment levels*

Time of administration (hr)		% Saline-injected 5-hr receptor level	
17β-Estradiol	Cycloheximide	Intact	Castrate
—	0	98.0 ± 1.2	57.7 ± 5.7
0	—	72.5 ± 5.5	36.7 ± 4.8
0	+0.5	50.8 ± 2.8	24.0 ± 4.2
0	+1.0	54.7 ± 5.7	21.5 ± 5.6
0	+3.0	54.7 ± 5.8	15.0 ± 2.3

Groups of intact or 2-weeks' ovariectomized adult animals were injected with 0.1 μg of estradiol at time zero, followed by 0.1 mg of cycloheximide at times indicated. All animals were killed at 5 hr, and cytosol estrogen receptor content was measured. Values were calculated as the means ± SE of the moles of estradiol bound specifically per mg of cytosol protein in three separate determinations and are expressed as percentages of the replenishment values observed in a vehicle-injected group.

replenishment level. (This interval was chosen because it represents a period during which the rate of replenishment is maximal and during which cycloheximide alone has no deleterious effects on the receptor activity.) The presence of cycloheximide decreased the amount of replenishment by 20 to 25% in the intact animal, regardless of the interval, up to 3 hr, between its administration and that of estradiol. This result is distinctly different from the situation in the rat uterus and anterior pituitary (10,14,55), wherein the cycloheximide effect becomes less pronounced with increasing intervals of time of injection after estradiol. The mammary tissue results indicate a persistent requirement for protein synthesis during the initial phases of the replenishment process, in contrast to a transient requirement for early protein synthesis in replenishment in the uterus and anterior pituitary. The physiological action of estrogen on mammary tissue involves a prolonged gradual stimulation of epithelial growth, and the relatively slow process of constantly stimulated synthesis and turnover of receptor should suffice for this type of protracted action. The uterus and anterior pituitary, on the other hand, respond acutely to an estrogen stimulus, and a rapidly-induced activation of receptor would be a more appropriate mechanism of response in these tissues.

Caution must be exercised in extrapolating the receptor-related effects of cycloheximide to different animal models, as exemplified by the data presented in Table 1 for ovariectomized mice. In these animals, cycloheximide alone resulted in the loss of almost half of the control receptor activity. Whereas the final temporal dependency pattern of inhibition was analogous to that of the intact animals, the replenishment levels were greatly reduced. The difference in sensitivity between the intact and castrate mice could easily represent a protective effect of estradiol on the stability of its receptor. We have previously observed (13) that cycloheximide has an adverse effect on rat uterine receptors in ovariectomized animals after about 10 hr.

Effects of Pregnancy and Lactation on Nature and Amount of Receptor

In Table 2, the levels of estrogen receptor in the mammary glands during pregnancy, lactation, and subsequent glandular involution are compared with nonpregnant values. The dry weight of the gland increased constantly throughout pregnancy and lactation, falling rapidly thereafter (lactation ceases at about 21 days postpartum). The total DNA content of the tissue followed the same pattern and the DNA concentration (μg/mg tissue) remained relatively constant. These results were as expected, since epithelial cell proliferation accounts for virtually the entirety of the increase in mammary gland weight during this period. The rigorous DNA analysis performed by Nicoll and Tucker in 1965 (48) showed nicely that isolated mammary parenchymal tissue displays an enormous increase in DNA content during lactation and that the DNA level of the fat pad remains constant. Thus, the enhancement of mammary gland DNA level during this interval is a realistic estimate of the increase in the parenchymal cell population.

The specific binding activity of estradiol is expressed in the last three columns of the table as a function of a unit weight of cytosol protein, DNA, or tissue. In all these instances, the changes throughout pregnancy and

TABLE 2. *Receptor levels during and following pregnancy*

				Specific estrogen binding		
Group	Dry wt tissue per mouse (mg)	Total DNA (mg)	DNA (μg/mg tissue)	moles/mg protein ($\times 10^{-14}$)	moles/mg DNA ($\times 10^{-13}$)	moles/g tissue ($\times 10^{-13}$)
Nonpregnant	42.3	1.75	41.4	4.45	1.77	7.33
Pregnant						
Early (5–9 days)	58.4	2.76	47.3	5.48	1.83	8.65
Mid (10–15 days)	86.2	4.23	49.1	1.02	1.34	6.59
Late (16–19 days)	121.1	6.50	53.7	2.46	1.69	9.09
Postpartum lactating						
3 days	202.0	9.01	44.6	5.82	1.01	4.49
6 days	251.4	9.77	38.9	5.32	0.71	2.76
15 days	335.7	10.20	30.4	3.04	0.49	1.50
30 days	159.9	5.12	32.0	2.28	0.53	1.70
Involuted						
24 days	62.0	2.31	37.3	4.11	1.96	7.32

Animals were killed in groups of 10–15. The body weight of the animals (corrected for weight of fetuses) did not vary by more than 5 g (range of 22 to 27 g). For studies with postpartum lactating mice, the number of pups was adjusted to six on the day of parturition. Lactation normally ceased 21 days postpartum, and the glands began to involute.

lactation were minor, and it is clear that total receptor was constantly increasing only in proportion to the increased total weight of the mammary tissue. In other words, the number of cells containing receptor underwent a striking increase, whereas the number of receptors per cell did not change appreciably. Previous comparisons of virgin and lactating glands, based on DNA content, have yielded similar conclusions (52,58).

In spite of the relative constancy of the specific receptor binding values, a trend was observed that appeared worthy of further investigation. There was a slight rise in specific binding activity in early pregnancy, followed by a fall during midpregnancy and a subsequent second rise in late pregnancy. Serum prolactin levels fluctuate during pregnancy in the mouse in much this same fashion (46), and since prolactin has been reported to enhance estrogen receptor levels in rat dimethylbenz(a)anthracene-induced tumors (32), an analysis of prolactin's effectiveness as a regulator of estrogen receptor content was undertaken. The results are presented in the following section.

A surprising change in the physical nature of the receptor–estradiol complex was encountered when cytosol samples from different stages of pregnancy were examined by sucrose gradient analysis (Fig. 7). In early pregnancy, most of the receptor sedimented in the 4S region as in the virgin animal, but a discrete binding peak was also present at 8S. By midpregnancy, both forms were also present, but the 8S form was clearly the predominant of the two. In late pregnancy, the receptor existed almost completely in the 8S form.

The midpregnancy pattern showed a shift in the presence of 0.4 M KCl

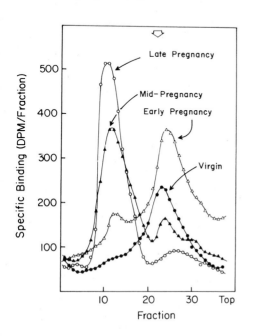

FIG. 7. Sedimentation patterns of specific estrogen binding components in mammary glands throughout pregnancy. Pregnant mice were sacrificed in groups on either day 6 (△), day 12 (▲), or day 18 (○) of pregnancy. Cytosol preparations containing very similar protein concentrations among groups were incubated with [³H]estradiol and applied to low-salt sucrose gradients and centrifuged. Cytosol from virgin animal glands was analyzed simultaneously (●).

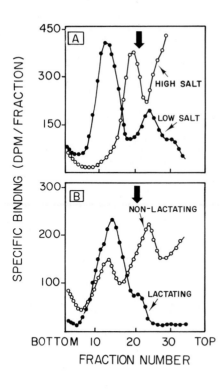

FIG. 8. Behavior of 8S receptor form in high salt and effects of lactation. **A:** Midpregnant (10 to 15 days) mice were sacrificed, and cytosol was prepared. Samples were incubated with [³H]estradiol for 4 hr and applied to a sucrose gradient (●). Identical cytosol samples were extracted with buffer containing 0.4 M KCl, and the extracts were applied to the gradient (○). **B:** Sucrose gradient patterns of cytosol from 7-day postpartum lactating (●) or 24-day postpartum nonlactating involuted (○) mammary glands. Conditions as in **A.** (From Hunt and Muldoon, ref. 21.)

to a 5.3S form (Fig. 8A), which was readily distinguishable from the 4S form of the virgin cytosol. The 8S to 5.3S conversion would appear to be the same octameric-tetrameric shift recently described by Sica et al. (59) for the calf uterine estrogen receptor system.

If postpartum animals were allowed to nurse their young, the receptor complex retained its 8S character (Fig. 8B). If, on the other hand, the pups were removed from the mother at birth and the mammary glands consequently involuted, the receptor largely reverted to the virgin-like 4S form.

When depletion-replenishment experiments of the type exemplified in Fig. 6 were performed in late pregnant animals, the replenished receptor at 5 hr was exclusively an 8S component; when the experiments were conducted with virgin mice, the receptor replenished as a 4S form. This would indicate that the hormonal environment of the cell dictates the physical nature of the receptor produced under estrogen stimulation.

Direct Effect of Prolactin on the Estrogen Receptor

All the results presented in the previous section are compatible with an influence of some stimulatory factor(s) capable of producing an enhanced level of estrogen receptor and an alteration in its molecular form. The likely involvement of prolactin was explored by injecting virgin mice with the protein hormone and examining the cytosol receptors by sucrose gradient studies (Fig. 9). A single injection of prolactin and sacrifice 18 hr later resulted in an increased amount of receptor and a shift into the 8S region. The short-term nature of this effect renders unlikely the possibility that prolactin was inducing a separate cell type that produced a different form of the receptor; similar rapid enhancement of receptor by prolactin has been found by Vignon and Rochefort (62) in rat mammary tumor cytosol. Extended treatment with prolactin, entailing three successive daily injections, further augmented the levels of receptor, almost exclusively of the 8S variety.

Mittra (41) has reported that the mammotropic effect of prolactin in rats is depressed by endogenous thyroid hormones. We, therefore, thyroidectomized groups of animals and looked at the response of the mammary cytosol receptors to prolactin, using the same conditions as described above for intact animals. It may be seen in Fig. 9 that enhanced sensitivity to prolactin did exist in the thyroidectomized mice, as evidenced by a marked elevation in the receptor levels. Since thyroidectomy does not result in increased prolactin or growth hormone secretion (49), it is likely that a direct action on the mammary gland is being affected. It was of peripheral interest in this regard to note that the 4S virgin receptor levels were significantly lower in

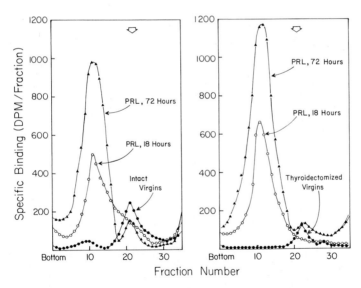

FIG. 9. Effect of prolactin (PRL) on mammary tissue estrogen receptor nature and content in the virgin mouse. Virgin animals, either intact (*left panel*) or 2-weeks' thyroidectomized (*right panel*), were administered rat prolactin in saline solution in a single 0.2 mg dosage i.p. and sacrificed 18 hr later (○) or in three daily injections of 0.2 mg each and killed 1 hr after the final injection (▲). Cytosol was prepared from each group, including untreated animals (●), and incubated with [³H]estradiol (10^{-9} M) with or without unlabeled estradiol (10^{-7} M). Sucrose gradient analysis was performed in low-salt medium. The patterns have been quantitatively normalized to an equivalent concentration of cytosol protein.

the thyroidectomized mice than in the intact animals. A direct effect of thyroid hormones on specific tissue receptor levels has only been demonstrated previously in the rat anterior pituitary gland (11).

Concentration as a Determinant of Receptor Form

In their analysis of the molecular forms of the guinea pig uterine cytosol progesterone receptors, Milgrom et al. (40) found that the 6.7S proestrus form was partially an aggregate of the 4.5S diestrus form and that cytosol dilution could reverse the aggregation to a limited degree. In the righthand panel of Fig. 10 is shown the effect of dilution on the mammary tissue cytosol receptor from the late pregnant mouse. No tendency toward a shift to smaller size particles was observed.

We have repeatedly noted that MTV⁺ strains of mice produce predominantly 8S mammary tissue estrogen receptors beyond the age of 10 to 11 months (using only normal tissue with no signs of spontaneous tumor formation). In Fig. 10 (*left panel*), the effects of dilution on these receptors is seen to be different from the pregnancy 8S forms, in that, on dilution, a progressively greater proportion of the total receptor binding activity appeared in the 4S region. This effect allows differentiation between these dif-

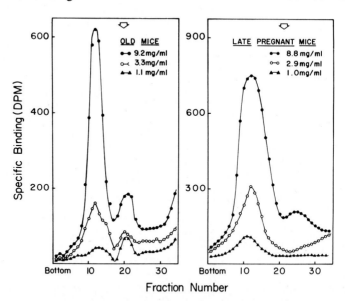

FIG. 10. Protein concentration as a determining factor in sedimentation of 8S forms of receptor. Normal mammary tissue from aging female virgin mice (11 months old) (*left panel*) and from 18-day pregnant animals (*right panel*) was excised, and cytosol was prepared at the highest protein concentration shown. Dilutions to the designated concentrations were then prepared, and all samples were equilibrated with [³H]estradiol ± unlabeled estradiol. Following dextran-charcoal treatment, the samples were applied to low-salt sucrose gradients and centrifuged.

ferent 8S receptors and indicates that the late pregnancy receptors arise from a more complex phenomenon than simple aggregation of 4S forms.

In another series of experiments, we performed sucrose gradient fractionation on samples of virgin and late pregnancy cytosol and equilibrated the fractions individually with [³H]estradiol. Specific binding was limited to the 4S and 8S regions of virgin and pregnancy samples, respectively, although there was an appreciable amount of aggregation to the bottom of the gradient. These results indicate that the 4S component does not arise from a native 8S species in virgin cytosol during incubation in the presence of estradiol, a mechanism suggested by Leclercq and Heuson (31) to explain the appearance of both 8 to 9S and 4 to 5S specific binding moieties in tumor cytosol.

Kinetics of the Interaction between Estradiol and the Different Mammary Tissue Receptor Forms

In the course of a number of experiments, we were unable to detect significant differences between the equilibrium binding constants of 4S estrogen and 8S estrogen complexes from virgin and late pregnant animals, respectively. However, the 4S receptor always appeared to have a slightly higher affinity than the 8S form. Studies of the kinetic parameters of these interactions were therefore undertaken in an attempt to delineate more precisely any differences that do exist. Association rate kinetics were investigated by previously established methods (30), and the data were analyzed as second-order rate functions (Fig. 11). Linearity for at least 10 min was suggestive

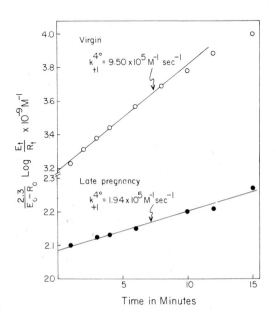

FIG. 11. Association kinetics of estrogen-receptor interactions at 4°C in virgin and pregnancy cytosol. Cytosol was prepared from adult virgin and 18-day pregnant mice, and the receptor concentration (R_0) was determined. [³H]Estradiol was added at a concentration (E_0) of 0.5 nM. At intervals up to 15 min, aliquots were taken and plunged into tubes containing excess unlabeled estradiol. Receptor complexes were precipitated with protamine sulfate, and the radioactivity was measured for determination of estradiol and receptor concentration remaining at each interval (E_t and R_t, respectively). The association rate constant, k_{+1}, was determined as the slope of the lines shown for a second-order kinetic reaction.

of a simple bimolecular association reaction. The rate of formation of the virgin 4S receptor–estradiol complex was five times as great as that of the 8S receptor–estradiol complex, and this difference was highly significant and readily reproducible.

The dissociation rate kinetics of the 8S complex and the 4S complex are presented graphically in Figs. 12 and 13, respectively. The figures (*upper panels*) show the arithmetic data obtained when the preformed complexes were allowed to dissociate in the presence of excessive molar quantities of unlabeled estradiol at either 4 or 25°C. In all cases, the best mathematical fits to these curves were single exponential functions with the addition of a constant term representing unsaturable nonspecific binding. Therefore, the horizontal asymptote of each curve was estimated, and the value thus obtained was subtracted at each point. The resultant curves were plotted as a first-order dissociation function (*shown in the lower panel of each figure*), and the dissociation rate constants were calculated as the slopes of these lines.

At 4°C, the dissociation rate was linear first order in both instances, with the 4S complex dissociating about three times as fast as the 8S complex. At 25°C, the initial rate of dissociation of the 4S complex was faster than that of the 8S complex, but the 4S curve was biphasic, rapidly braking to a

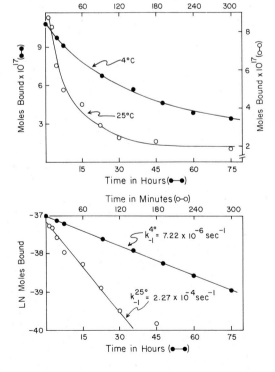

FIG. 12. Dissociation kinetics of 8S receptor–estradiol complexes. Cytosol was prepared from 18-day pregnant animals and equilibrated at 4°C with excess [^{3}H]estradiol, then reacted with dextran-coated charcoal to remove unbound steroid. A 100-fold molar excess amount of unlabeled estradiol was then added, and dissociation of the complex was allowed to proceed at 4 or 25°C, with sampling and evaluation of receptor binding at various time periods. The arithmetic plots (*upper panel*) were corrected for nonspecific binding as the asymptote to infinite time, and the corrected curves were then treated as a semilogarithmic function (*lower panel*) for determination of the first-order dissociation rate constant, k_{-1}.

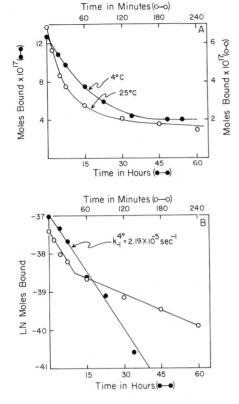

FIG. 13. Dissociation kinetics of 4S receptor–estradiol complexes. Cytosol from mature virgin mice mammary glands was prepared and handled as described in the legend to Fig. 12.

slower rate at longer intervals. It is likely that this was simply a rate limitation occasioned by the small amount of radiolabeled receptor complex still present beyond 45 min at 25°C.

Table 3 summarizes the kinetic data for the two receptor complexes. At 4°C, the 8S complex formed less readily than did the 4S complex, but once formed, dissociated less readily, with a half-life three times as long as the

TABLE 3. *Kinetic parameters of 4S and 8S receptor binding*

Animal source	Temperature (C)	k_{+1} (M$^-$ sec^{-1})	k_{-1} (sec^{-1})	$t_{\frac{1}{2}}$ (hr)	K_A (M^{-1} sec^{-1})
Virgin (4S)	4°	9.5×10^5	2.2×10^{-5}	9.0	4.3×10^{10}
	25°	—	4.5×10^{-4}	0.45	—
Late pregnant (8S)	4°	1.9×10^5	7.2×10^{-6}	26.7	2.7×10^{10}
	25°	—	2.3×10^{-4}	0.85	—

Values for the association and dissociation rate constants are summarized from Figs. 11, 12, and 13. The equilibrium association constants, K_A, were calculated as the ratio k_{+1}/k_{-1}. The sedimentation coefficients of each sample were verified by sucrose gradient centrifugation analysis.

4S complex. These differences are not obvious from a comparison of the equilibrium association constants, calculated in the table as the ratio of the association and dissociation rate constants. The differences in these constants tended to cancel each other out when expressed as a ratio, attesting to the necessity for performing kinetic, as well as equilibrium, analyses of steroid-receptor dynamics as a means of gaining insight into the molecular mechanisms of such interactions.

SUMMARY AND DISCUSSION

These studies have shown that estrogen receptors mature in mouse mammary tissue at about the time of puberty and are demonstrable as 4S-sedimenting moieties in the unstimulated adult gland. During pregnancy and lactation, the receptor levels rise in proportion with the increasing weight of the tissue. The 4S form converts progressively through pregnancy to an 8S species, which is maintained throughout lactation and subsequently reverts to the 4S form during glandular involution. Estradiol can induce the synthesis of its receptor in the adult, but not in the immature animal. In the virgin, the induced receptor sediments at 4S; in the late pregnant animal, at 8S.

The pregnancy 8S form transforms to a 5.3S moiety in high salt, but does not dissociate into smaller units on simple dilution of the cytosol. *In vivo* administration of prolactin to virgin mice causes a dramatic acute increase in receptor number and a shift from the 4S to the 8S form.

Both the 4S and 8S receptor species are able to bind estradiol with equivalent affinity, transforming to 5 to 6S nuclear forms and translocating to binding sites in the nucleus. Although the equilibrium binding parameters are similar for the 4S and 8S forms, the latter can be distinguished by their differing association and dissociation rate kinetic behavior. The 8S complex forms more slowly than does the 4S, but once formed, also dissociates less readily.

The observation made in this study that estrogen receptors are demonstrable as different molecular forms in the presence of differing hormonal surroundings is not unique to the mammary gland. Toft and O'Malley (61) found that the chick oviduct progesterone receptor exists as a 4 to 5S form in the unstimulated tissue and as a 6 to 7S component under estrogen stimulation. In a similar vein, Milgrom et al. (40) reported that the guinea pig uterine progesterone receptor oscillates between a 4.5S and a 6.7S form from diestrus to proestrus in the normal cycle. The ability to induce such a shift in mammary tissue cytosol simply by administration of prolactin, coupled with the fact that the physical nature of the receptor acutely induced by estrogen is dependent on the state of the tissue, presents a model system in which it should be possible to study the molecular characteristics and kinetics of interconversion between these receptor forms. In this regard, Vignon and Rochefort (62) have recently reported the intriguing observation that estrogen cannot induce its own receptors in mammary tumor tissue from rats that were treated with a prolactin-blocking agent.

In human breast tumor tissue, both 4S and 8S forms of the estrogen receptors are commonly encountered, and there is some evidence to suggest that only one of these forms is functional (63,64). The well-known roles of pregnancy (6,42) and prolactin (44) as positive determinants of tumorigenesis in the mouse suggest that our findings are related to changes in receptor activity that take place during neoplastic transformation. Although it has been postulated that development of mammary tumors following estrogen treatment is mediated by the pituitary and results from increased production of prolactin (44), there is good evidence that estrogen has an independent action (45), probably mediated through receptor interactions at the level of the mammary gland (37). The presumed role of estrogen during late pregnancy as an inhibitor of lactation (5,38) would appear to represent an analogous situation. Indeed, tumorigenesis in the adult mouse mammary gland is similar in many respects to a pregnancy-type state wherein excessive hormonal stimulation has become uncontrollable. On the other hand, tumors induced in rudimentary mammary glands of newborn mice, or those produced in aged female rats, are generally autonomous (27). Thus, the hormonal character of a primary tumor, just like the physical nature of the estrogen receptor, appears to reflect the endocrine state of the animal.

The battery of hormones that is known to play a role in mammary tissue growth and development presents an extraordinarily complex melange from which to extract a distinct element and define its nature and control. As a single example of this complexity, the documented role of progesterone in mammary tissue development in the postpartum rat (7,54) and in carcinogen-induced tumors during lactation (34,35) clearly implicates this hormone as a prime candidate for analysis of its interplay with other hormones and their receptors. The approach taken in this study, therefore, has led to results that barely scratch the surface of the interactions occurring within mammary tissue. It is evident that a great deal remains to be learned about the normal regulatory mechanisms in this gland before any conclusions can be drawn about the abnormalities accompanying transformation to a neoplastic state.

ACKNOWLEDGMENTS

The excellent technical assistance of Mildred E. Hunt in all phases of this study is gratefully acknowledged. CI-628 (α-[4-pyrrolidinoethoxy] phenyl-4-methoxy-α-nitrostilbene) was a gift from Dr. Jerry Reel, Parke-Davis Co. This work was supported by U.S. Public Health Service Grants CA 17059 from the National Cancer Institute and AM 17650 from the National Institute of Arthritis, Metabolic and Digestive Diseases, National Institutes of Health.

REFERENCES

1. Anderson, J., Clark, J. H., and Peck, E. J. (1972): Oestrogen and nuclear binding sites. Determination of specific sites by [³H]oestradiol exchange. *Biochem. J.,* 126:561–567.
2. Anderson, R. R. (1974). In: *Lactation,* Vol. 1, edited by B. L. Larson and V. R. Smith, pp. 97–142. Academic Press, New York.
3. Attardi, B., and Ohno, S. (1976): Androgen and estrogen receptors in the developing mouse brain. *Endocrinology,* 99:1279–1290.
4. Banerjee, M. R. (1976): Responses of mammary cells to hormones. *Int. Rev. Cytol.,* 47:1–97.
5. Bruce, J. O., and Ramirez, V. D. (1970): Site of action of the inhibitory effect of estrogen upon lactation. *Neuroendocrinology,* 6:19–29.
6. Bruni, J. E., and Montemurro, D. G. (1971): Effect of pregnancy, lactation, and pituitary isografts on the genesis of spontaneous mammary gland tumors in the mouse. *Cancer Res.,* 31:1903–1907.
7. Caligaris, L., Astrada, J. J., and Taleisnik, S. (1974): Oestrogen and progesterone influence on the release of prolactin in ovariectomized rats. *J. Endocrinol.,* 60:205–215.
8. Chauveau, J., Moule, Y., and Rouiller, C. (1956): Isolation of pure and unaltered liver nuclei. Morphology and biochemical composition. *Exp. Cell Res.,* 11:317–321.
9. Cidlowski, J. A., and Muldoon, T. G. (1974): Estrogenic regulation of cytoplasmic receptor populations in estrogen-responsive tissues of the rat. *Endocrinology,* 95: 1621–1629.
10. Cidlowski, J. A., and Muldoon, T. G. (1975): Cytoplasmic estrogen receptor bio-dynamics: A biphasic response. *Program, 57th Annual Meeting, The Endocrine Society,* Lippincott, Philadelphia. p. 211.
11. Cidlowski, J. A., and Muldoon, T. G. (1975): Modulation by thyroid hormones of cytoplasmic estrogen receptor concentrations in reproductive tissues of the rat. *Endocrinology,* 97:59–67.
12. Cidlowski, J. A., and Muldoon, T. G. (1976): Dissimilar effects of antiestrogens upon estrogen receptors in reproductive tissues of male and female rats. *Biol. Reprod.,* 15:381–389.
13. Cidlowski, J. A., and Muldoon, T. G. (1976): Sex-related differences in the regulation of cytoplasmic estrogen receptor levels in responsive tissues of the rat. *Endocrinology,* 98:833–841.
14. Cidlowski, J. A., and Muldoon, T. G. (1978): The dynamics of intracellular estrogen receptor regulation as influenced by 17β-estradiol. *Biol. Reprod.,* 18:234–246.
15. Clark, J. H., Anderson, J. N., and Peck, E. J. (1973): Estrogen receptor · antiestrogen complex: Atypical binding by uterine nuclei and effects on uterine growth. *Steroids,* 22:707–718.
16. Clark, J. H., and Gorski, J. (1970): Ontogeny of the estrogen receptor during early uterine development. *Science,* 169:76–78.
17. DeSombre, E. R., Mohla, S., and Jensen, E. V. (1972): Estrogen-independent activation of the receptor protein of calf uterine cytosol. *Biochem. Biophys. Res. Commun.,* 48:1601–1608.
18. Eisenthal, R., and Cornish-Bowden, A. (1974): The direct linear plot: A new graphical procedure for estimating enzyme kinetic parameters. *Biochem. J.,* 139:715–720.
19. Flux, D. S. (1954): Growth of the mammary duct system in intact and ovariectomized mice of the CHI strain. *J. Endocrinol.,* 11:223–237.
20. Giannopoulos, G., and Gorski, J. (1971): Estrogen receptors. Quantitative studies on transfer of estradiol from cytoplasmic to nuclear binding sites. *J. Biol. Chem.,* 246:2524–2529.
21. Hunt, M. E., and Muldoon, T. G. (1977): Factors controlling estrogen receptor levels in normal mouse mammary tissue. *J. Steroid Biochem.,* 8:181–186.
22. Jensen, E. V., Brecher, P. I., Numata, M., Smith, S., and DeSombre, E. R. (1975):

Estrogen interaction with target tissues; two-step transfer of receptor to the nucleus. In: *Methods in Enzymology,* Vol. XXXVI, edited by B. W. O'Malley, and J. G. Hardman, pp. 267–275. Academic Press, New York.

23. Jensen, E. V., and DeSombre, E. R. (1972): Mechanism of action of the female sex hormones. *Annu. Rev. Biochem.,* 41:203–230.

24. Jensen, E. V., Suzuki, T., Numata, M., Smith, S., and DeSombre, E. R. (1969): Estrogen-binding substances of target tissues. *Steroids,* 13:417–427.

25. Katzenellenbogen, B. S., and Ferguson, E. R. (1975): Antiestrogen action in the uterus: Biological ineffectiveness of nuclear bound estradiol after antiestrogen. *Endocrinology,* 97:1–12.

26. Katzenellenbogen, J. A., Johnson, H. J., and Carlson, K. E. (1973): Studies on the uterine, cytoplasmic estrogen binding protein. Thermal stability and ligand dissociation rate. An assay of empty and filled sites by exchange. *Biochemistry,* 12:4092–4099.

27. Kim, U., and Depowski, M. J. (1975): Progression from hormone dependence to autonomy in mammary tumors as an *in vivo* manifestation of sequential clonal selection. *Cancer Res.,* 35:2068–2077.

28. King, R. J. B., and Mainwaring, W. I. P. (1974): *Steroid-Cell Interactions.* Univ. Park Press, Baltimore.

29. Korach, K. S., and Muldoon, T. G. (1974): Studies on the nature of the hypothalamic estradiol-concentrating mechanism in the male and female rat. *Endocrinology,* 94:785–793.

30. Korach, K. S., and Muldoon, T. G. (1974): Characterization of the interaction between 17β-estradiol and its cytoplasmic receptor in the rat anterior pituitary gland. *Biochemistry,* 13:1932–1938.

31. Leclercq, G., and Heuson, J. C. (1973): Specific estrogen receptor of the DMBA-induced mammary carcinoma of the rat and its estrogen-requiring molecular transformation. *Eur. J. Cancer,* 9:675–680.

32. Leung, B. S., and Sasaki, G. H. (1975): On the mechanism of prolactin and estrogen action in 7,12-dimethylbenz(a)anthracene-induced mammary carcinoma in the rat. II. *In vivo* tumor responses and estrogen receptor. *Endocrinology,* 97:564–572.

33. Lippman, M., and Huff, K. (1976): A demonstration of androgen and estrogen receptors in a human breast cancer using a new protamine sulfate assay. *Cancer,* 38:868–874.

34. Matsuzawa, A., and Yamamoto, T. (1975): Response of a pregnancy-dependent tumor to hormones. *J. Natl. Cancer Inst.,* 55:447–453.

35. McCormick, G. M., and Moon, R. C. (1967): Hormones influencing postpartum growth of 7,12-dimethylbenz(a)anthracene-induced rat mammary tumors. *Cancer Res.,* 27:626–631.

36. McGuire, W. L., Carbone, P. P., and Vollmer, E. P. (editors) (1975): *Estrogen Receptors in Human Breast Cancer.* Raven Press, New York.

37. Meites, J. (1972): Relation of prolactin and estrogen to mammary tumorigenesis in the rat. *J. Natl. Cancer Inst.,* 48:1217–1224.

38. Meites, J., and Sgouris, J. T. (1953): Can the ovarian hormones inhibit the mammary response to prolactin? *Endocrinology,* 53:17–23.

39. Michel, G., Jung, I., Baulieu, E-E., Aussel, C., and Uriel, J. (1974): Two high affinity estrogen binding proteins of different specificity in the immature rat uterus cytosol. *Steroids,* 24:437–449.

40. Milgrom, E., Atger, M., Perrot, M., and Baulieu, E-E. (1972): Progesterone in uterus and plasma: VI. Uterine progesterone receptors during the estrus cycle and implantation in the guinea pig. *Endocrinology,* 90:1071–1078.

41. Mittra, I. (1974): Mammotropic effect of prolactin enhanced by thyroidectomy. *Nature,* 248:525–526.

42. Muhlbock, O. (1950): Note on the influence of the number of litters upon the incidence of mammary tumors in mice. *J. Natl. Cancer Inst.,* 10:1259–1262.

43. Muhlbock, O. (1956): The hormonal genesis of mammary cancer. *Adv. Cancer Res.,* 4:371–391.

44. Muhlbock, O. (1972): Role of hormones in the etiology of breast cancer. *J. Natl. Cancer Inst.*, 48:1213–1216.
45. Murota, S-I., and Hollander, V. P. (1971): Role of ovarian hormones in the growth of transplantable mammary carcinoma. *Endocrinology*, 89:560–564.
46. Murr, S. M., Bradford, G. E., and Geschwind, I. I. (1974): Plasma luteinizing hormone, follicle-stimulating hormone and prolactin during pregnancy in the mouse. *Endocrinology*, 94:112–116.
47. Nandi, S., and McGrath, C. M. (1973): Mammary neoplasia in mice. *Adv. Cancer Res.*, 17:353–414.
48. Nicoll, C. S., and Tucker, H. A. (1965): Estimates of parenchymal, stromal, and lymph node deoxyribonucleic acid in mammary glands of C3H/Crgl/2 mice. *Life Sci.*, 4:993–1001.
49. Peake, G. T., Birge, C. A., and Daughaday, W. H. (1973): Alterations of radioimmunoassayable growth hormone and prolactin during hypothyroidism. *Endocrinology*, 92:487–493.
50. Plapinger, L., and McEwen, B. S. (1973): Ontogeny of estradiol-binding sites in rat brain. I. Appearance of presumptive adult receptors in cytosol and nuclei. *Endocrinology*, 93:1119–1128.
51. Plapinger, L., McEwen, B. S., and Clemens, L. E. (1973): Ontogeny of estradiol-binding sites in rat brain. II. Characteristics of a neonatal binding macromolecule. *Endocrinology*, 93:1129–1139.
52. Puca, G. A., and Bresciani, F. (1969): Interactions of 6,7-^{3}H-17β-estradiol with mammary gland and other organs of the C3H mouse *in vivo*. *Endocrinology*, 85:1–10.
53. Raynaud, J. P., Mercier-Bodard, C., and Baulieu, E-E. (1971): Rat estradiol binding plasma protein (EBP). *Steroids*, 18:767–788.
54. Rothchild, I. (1965): Interrelations between progesterone and the ovary, pituitary, and central nervous system in the control of ovulation and the regulation of progesterone secretion. *Vitam. Horm.*, 23:209–327.
55. Sarff, M., and Gorski, J. (1971): Control of estrogen binding protein concentration under basal conditions and after estrogen administration. *Biochemistry*, 10:2557–2563.
56. Shyamala, G. (1972): Estradiol receptors in mouse mammary tumors: Absence of the transfer of bound estradiol from the cytoplasm to the nucleus. *Biochem. Biophys. Res. Commun.*, 46:1623–1630.
57. Shyamala, G. (1975): Is the estrogen receptor of mammary glands a metalloprotein? *Biochem. Biophys. Res. Commun.*, 64:408–415.
58. Shyamala, G., and Nandi, S. (1972): Interactions of 6,7-^{3}H-17β-estradiol with the mouse lactating mammary tissue *in vivo* and *in vitro*. *Endocrinology*, 91:861–867.
59. Sica, V., Nola, E., Puca, G. A., and Bresciani, F. (1976): Estrogen binding proteins of calf uterus. Inhibition of aggregation and dissociation of receptor by chemical perturbation with NaSCN. *Biochemistry*, 15:1915–1923.
60. Stancel, G. M., and Gorski, J. (1975): Analysis of cytoplasmic and nuclear estrogen-receptor proteins by sucrose density gradient centrifugation. In: *Methods in Enzymology*, Vol. XXXVI, edited by B. W. O'Malley, and J. G. Hardman, pp. 166–176. Academic Press, New York.
61. Toft, D. O., and O'Malley, B. W. (1972): Target tissue receptors for progesterone: The influence of estrogen treatment. *Endocrinology*, 90:1041–1045.
62. Vignon, F., and Rochefort, H. (1976): Regulation of estrogen receptors in ovarian-dependent rat mammary tumors. I. Effects of castration and prolactin. *Endocrinology*, 98:722–729.
63. Wittliff, J. L., Mehta, R. G., Boyd, P. A., and Goral, J. E. (1976): Steroid-binding proteins of the mammary gland and their clinical significance in breast cancer. *J. Toxicol. Environ. Health (Suppl.)*, 1:231–256.
64. Wittliff, J. L., and Savlov, E. D. (1975): Estrogen-binding capacity of cytoplasmic forms of the estrogen receptors in human breast cancer. In: *Estrogen Receptors in Human Breast Cancer*, edited by W. L. McGuire, P. P. Carbone, and E. P. Vollmer, pp. 73–105. Raven Press, New York.

65. Woosley, J. T., and Muldoon, T. G. (1976): Use of the direct linear plot to estimate binding constants for protein-ligand interactions. *Biochem. Biophys. Res. Commun.,* 71:155–160.
66. Woosley, J. T., and Muldoon, T. G. (1977): Comparison of the accuracy of the Scatchard, Lineweaver-Burk and direct linear plots for the analysis of steroid-protein interactions. *J. Steroid Biochem.,* 8:625–629.

Ontogeny of Receptors and Reproductive Hormone Action,
edited by T. H. Hamilton, J. H. Clark, and W. A. Sadler.
Raven Press, New York 1979.

Hormonal Regulation of Casein Gene Expression in the Mammary Gland

Jeffrey M. Rosen, William A. Guyette, and Robert J. Matusik

Department of Cell Biology, Baylor College of Medicine, Houston, Texas 77030

Steroid Hormones and Gene Expression

The availability of several hormone-responsive model systems, in which steroid hormone-inducible gene products comprise a substantial percentage of the total mRNA and protein synthesized, has greatly facilitated the study of the molecular mechanisms of steroid hormone action. For example, steroid hormones have been shown to regulate the accumulation of numerous mRNAs, including ovalbumin mRNA (28) and conalbumin mRNA (59) in the chick oviduct, tryptophan oxygenase mRNA (74) and $\alpha2$-μ-globulin mRNA (80) in rat liver, vitellogenin mRNA in Xenopus and Avian liver (5,16), pregrowth hormone mRNA in pituitary cells (42), and mouse mammary tumor virus RNA in several cloned mammary tumor cells lines (65). In most of these model systems administration of steroids *in vivo* is required in order to induce mRNA accumulation. However, hormonal induction of specific mRNAs has also been demonstrated in a number of *in vitro* systems (42,65). Therefore, in these cases pulse-labeling studies are possible and the rates of both mRNA synthesis and degradation can be determined (91). Such experiments are necessary because mRNA accumulation has been shown to be a function not only of the rate of synthesis, but also of the rate of turnover of the specific mRNA (33). For example, differential mRNA turnover has been demonstrated to be an important factor regulating the accumulation of histone mRNA during the cell cycle (48), of globin mRNA during erythroid development (4), and ovalbumin mRNA following estrogen treatment in the chick oviduct (14).

Using cDNA probes synthesized from these steroid-inducible mRNAs it has been possible to demonstrate that the primary effect of steroid hormones is to rapidly increase the rate of specific gene transcription (64). These hormones appear to regulate gene expression through a mechanism involving translocation of the steroid-cytoplasmic receptor complex to the cell nucleus, followed by an interaction of the receptor complex with specific chromosomal proteins and DNA sequences (57). However, because of the enormous complexity of eucaryotic DNA it has not been possible to elucidate the precise mechanism of steroid hormone action using total genomic DNA and nuclear protein fractions. High levels of nonspecific binding of regulatory proteins to DNA sequences may mask the more limited specific binding of these proteins to promoter regions.

An alternative approach to this problem is to isolate specific gene sequences and study steroid hormone action using reconstituted mini-chromosomes. The synthesis and amplification of the structural gene sequence for ovalbumin mRNA has recently been accomplished using recombinant DNA technology (31,47). Furthermore, the cloned ovalbumin DNA has been used to identify larger fragments of genomic DNA, which contain within them the ovalbumin structural gene sequence (9,36). However, nontranslated insertion sequences within larger genomic DNA fragments specifying several viral and eucaryotic mRNAs, including ovalbumin mRNA, have recently been detected (9,36). Thus, it may be necessary to isolate these larger DNA fragments in order to elucidate the mechanisms by which steroid hormones regulate specific gene transcription. The isolation of the genomic globin gene which has recently been accomplished using recombinant DNA technology (81) provides a direction for future studies of steroid hormone action.

Prolactin Action in the Mammary Gland: a Model System for Studying Peptide Hormone Action

In contrast to steroid hormones, the mechanism by which peptide hormones regulate gene expression remains a complete enigma. This is especially true for prolactin which does not appear to be acting through the classical peptide hormone pathway, involving the initial binding of the hormone to a membrane receptor followed by the activation of adenylate cyclase and a subsequent activation of protein kinase (11). The binding of prolactin to a membrane receptor has been shown by the studies of Friesen and colleagues (77) to be a necessary prerequisite for the induction of casein synthesis (Fig. 1). This was demonstrated by blocking the induction of casein synthesis with an antibody prepared against the solubilized prolactin receptor. However, prolactin activation of adenylate cyclase and the increased accumulation of intracellular cyclic AMP has not been demonstrated in the mammary gland (41,54). A similar failure to demonstrate prolactin activation of adenylate cyclase has been reported in the corpus luteum which contains prolactin receptors and a peptide hormone (LH)-sensitive adenylate cyclase system (8). Finally, dibutyryl cyclic AMP when added to mammary explants is unable to induce casein synthesis (41,43).

Following the binding of prolactin to specific membrane receptors a rapid induction of cyclic AMP-stimulated protein kinase and cyclic AMP-binding protein has been reported with half-maximal levels observed within 30 min and maximal levels observed by 4 hr (41). These early effects of prolactin were, however, prevented by both actinomycin D and cycloheximide sug-

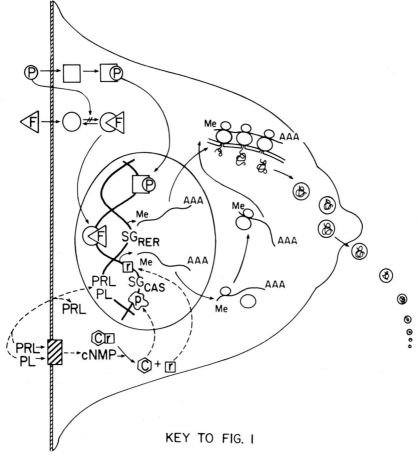

KEY TO FIG. I

Ⓟ PROGESTERONE

△F HYDROCORTISONE

PRL: PROLACTIN

PL: PLACENTAL LACTOGEN

cNMP: CYCLIC NUCLEOTIDE MONOPHOSPHATE

ⓒⓡ CATALYTIC (C) & REGULATORY (r)

        SUBUNITS OF PROTEIN KINASE

SG_{CAS}: STRUCTURAL GENE FOR CASEIN mRNA

SG_{RER}: STRUCTURAL GENE OR GENES FOR

        ROUGH ENDOPLASMIC RETICULUM

Ⓟ NONHISTONE PHOSPHOPROTEIN

Me—AAA: CAPPED & POLYADENYLATED mRNA

8 80S RIBOSOME

⑧ CASEIN MICELLE

☐ & ○ STEROID RECEPTORS

FIG. 1. Multiple hormonal regulation of casein gene expression in the mammary gland. A diagrammatic representation of a secretory mammary cell is shown.

gesting that the concomitant synthesis of RNA and protein was required for induction. Thus, the induction of cyclic AMP-dependent protein kinase probably did not result from the usual mechanisms operable during kinase activation, although conclusions drawn from these inhibitor experiments must be viewed with caution. Furthermore, it was not clear whether increased kinase activity was a cause or an effect of prolactin-regulated gene expression. Moreover, this result has not been repeated in other independent studies (46). Presumably, translocation of the catalytic subunit of protein kinase to the nucleus may result in nuclear protein phosphorylation and gene activation (Fig. 1). Alternatively, translocation of cyclic AMP bound to the regulatory subunit of protein kinase might result in positive regulation of casein gene transcription in an analogous fashion to the action of the CAP protein in bacteria (3).

Recent studies have also indicated that certain peptide hormones, including prolactin (52), insulin (26), and epidermal growth factor (10) may enter the cell (Fig. 1). Furthermore, the recovery of intact internalized prolactin from rat hepatocytes has been reported (60). Thus, regulation of specific gene expression by a peptide hormone may be mediated through a surface membrane site via an indirect mechanism or via the direct interaction of the internalized peptide hormone, or a fragment thereof, with the genetic constituents (Fig. 1). In the past few years both of these pathways have been studied in greater detail. Chomczynski and Topper (13) have reported that prolactin and placental lactogen can directly stimulate the rate of UTP incorporation into RNA in nuclei isolated from mammary epithelial cells. However, the significance of these results is difficult to assess in the absence of experiments to analyze the newly synthesized RNA or to delineate effects of these hormones at the level of initiation or elongation of RNA synthesis. Furthermore, prolactin unexpectedly stimulated UTP incorporation in nuclei isolated from both pregnant and lactating tissue, and in the latter case the rates of RNA synthesis should already have been maximal in the presence of high levels of endogenous prolactin. Because of the multiplicity of artifacts that are known to occur in these *in vitro* transcriptional assays, it is necessary to reassess these results using specific cDNA probes and more sophisticated techniques to determine the fidelity of transcription.

On the other hand, studies from several laboratories including those of Oka (34,54,55) and Rillema (61–63) have been directed toward the identification of possible second messengers that may mediate prolactin action via an indirect mechanism. Rillema's studies have suggested that prostaglandins, especially $PGF_{2\alpha}$, may lead to an increase in cyclic GMP levels and that cyclic GMP and polyamines then act in a coordinated fashion to increase both RNA and casein synthesis in mammary explants. Individually no effect of either polyamines or prostaglandins on casein synthesis was evident (61), although spermidine was able to maintain maximal levels of casein synthesis in the presence of insulin and prolactin, and, therefore, was able to replace hydrocortisone in the culture medium (54). The evidence to support the role of these agents is twofold. Firstly, inhibitors of polyamine (methylglyoxal *bis*-guanyl hydrazone, methyl GAG) and prostaglandin (indomethacin) synthesis prevented the stimulation of casein synthesis by prolactin (62). Secondly, prolactin in the presence of insulin has been reported to stimulate both polyamine biosynthesis and uptake, as reflected by both increased ornithine decarboxylase activity and increases in intracellular spermidine (34,54). In addition, prolactin has been reported to stimulate the release of arachidonic acid from phosphatidyl choline as mediated possibly by a direct effect of the hormone on membrane phospholipase A activity (63).

Once again the significance of these results with respect to prolactin-regulated gene expression is unclear. In these studies casein synthesis was usually measured by the incorporation of ^{3}H-leucine into rennin-Ca^{2+}-precipitable proteins, and as much as 70% of the radioactivity precipitable from cell homogenates may represent proteins other than casein. In many cases the effects of these putative mediators of prolactin action on radioactive precursor uptake and the specific activity of precursor pools were not determined. Thus, conclusions concerning increases in the rates of RNA or protein synthesis may not be warranted. Conclusions based on the use of inhibitors such as methyl GAG and indomethacin may also not be valid, since the nonspecific toxic effects of these compounds were not determined. Finally, both the dose of prolactin required to elicit some of these effects, and the kinetics of the response, are inconsistent with the rapid stimulation (30 min to 1 hr) of nuclear RNA synthesis (82) and casein mRNA accumulation by low doses of prolactin (5 ng to 5 μg/ml). Thus, the roles of polyamines, prostaglandins and cyclic GMP, and protein kinase need to be reexamined using a specific marker for prolactin-mediated gene expression.

In order to elucidate the mechanism by which peptide hormones regulate gene transcription a system is required in which the rapid induction of a specific mRNA can be accurately measured following the addition of the hormone. As previously mentioned, an ideal model system should permit pulse-labeling studies of the specific mRNA, so that the rates of mRNA synthesis and degradation can be determined. In addition, quantitative measurements will require the availability of a specific cDNA hybridization probe synthesized from a well characterized, purified mRNA. This cDNA probe can also be used for the synthesis and amplification of the structural gene sequence, and eventually a larger genomic DNA fragment.

The mammary gland fulfills these criteria: casein mRNA comprises between 50 to 60% of the total mRNA activity and is therefore readily purified in large quantities (67). Furthermore, a full-length complementary DNA probe has been synthesized using purified casein mRNA as a template, and well characterized (69). Utilizing this cDNA probe, the rapid induction of casein mRNA accumulation by prolactin in organ culture of midpregnant rat mammary tissue has been demonstrated (44). Finally, mammary gland organ culture is performed in a serum-free, chemically defined medium, thereby allowing a detailed examination of the mechanism by which prolactin regulates the synthesis of casein mRNA.

Multiple Hormonal Interactions in the Mammary Gland

The regulation of growth and differentiated function in the mammary gland is a complex process involving the multiple interaction of several peptide and steroid hormones (83). It is well established that insulin, hydrocortisone, and prolactin are required for the initiation of milk protein synthesis and secretion in mammary gland organ culture (83,84). The lactogenic effects of prolactin are also modulated by its interaction with a second steroid hormone, progesterone (30,70). In addition, during pregnancy an additional peptide hormone, placental lactogen, appears to play a critical role in the regulation of casein mRNA and casein synthesis (35,84).

The possible sites of action and interaction of these hormones are illustrated in Fig. 1. The ability of the steroid hormones to modulate the effect of prolactin on casein synthesis may be mediated by unique, cytoplasmic receptors. For example, the glucocorticoid receptor complex may potentiate the action of prolactin by promoting the synthesis of rough endoplasmic reticulum (91), thereby stabilizing casein mRNA on membrane-bound polysomes. Progesterone, on the other hand, may antagonize the effect of prolactin or placental lactogen either by competing for binding to the hydrocortisone receptor and preventing nuclear translocation, or by the action of its own unique hormone receptor complex on the transcription or processing of an undetermined gene product. For example, progesterone has been reported to prevent the prolactin-regulated synthesis of its own receptor (20).

Clearly the interaction of these hormones in regulating casein synthesis and secretion is a complex process potentially involving transcriptional, posttranscriptional, translational, and even posttranslational sites of action. Thus, following the transcription of casein mRNA, the primary gene product must undergo polyadenylation, "capping," and possibly cleavage and RNA splicing. The processed mRNA must then associate with free polysomes and, following translation of a signal peptide sequence, associate with endoplasmic reticulum. This results in internalization of the newly synthesized casein peptides and eventual glycosylation, phosphorylation, and association with calcium to form micelles. Finally, under the appropriate hormonal stimuli, casein secretion is elicited. The regulation of casein synthesis and secretion is, therefore, not an "all or none" process dependent solely on the induction of casein mRNA. Furthermore, the action of these hormones to regulate casein synthesis need not be only at the transcriptional level.

In this chapter we will illustrate some of the recent studies from our laboratory that have been directed toward elucidating the hormonal regulation of casein gene expression in the mammary gland. There are obviously still many gaps in the rather simplified and speculative model presented in Fig. 1. In most cases the results of the studies to be described do not permit a definitive evaluation of the mechanism of prolactin action, or its interaction with other hormones, in regulating casein gene expression. However, they present a direction for future studies and illustrate the kind of techniques that will be used subsequently to elucidate the mechanism of prolactin action on casein gene expression.

MOLECULAR PROBES: CHARACTERIZATION OF CASEIN MESSENGER RNAs AND COMPLEMENTARY DNAs

Both the purification and properties of the rat and mouse casein mRNAs have been extensively described in our earlier publications (67,68,72). However, in order to provide a basis for our subsequent discussion of casein gene regulation it is necessary to first describe the rat and mouse casein mRNAs used in these studies. The characterization of several partially purified rat and mouse casein mRNA fractions on 1.5% agarose-urea gels stained with ethidium bromide is shown in Fig. 2. These fractions were purified from total cellular nucleic acid extracts isolated from either 8- to 12-day lactating rats or mice (67). The total nucleic acid extracts were initially treated with proteinase K and reprecipitated several times from 3 M sodium acetate. The resulting nuclease-free RNA did not contain DNA or low molecular weight RNAs (Fig. 2, gel #1). Removal of the majority of 28S and 18S ribosomal RNA was then accomplished by two passages over oligo (dT)-cellulose. Fractionation of the poly (A)-containing mRNA was next performed by chromatography on Sepharose 4B. Three principal mRNA bands were observed during Sepharose 4B chromatography which contained casein mRNA activity when translated in heterologous cell-free translation systems. An additional smaller mRNA, which has been reported to contain α-lactalbumin mRNA (12) is depicted in Fig. 2, gel #2, in addition to ovalbumin mRNA, which was added as an internal molecular weight marker during gel electrophoresis. Two of the rat casein mRNAs migrated slightly faster than 18S rRNA, as a 15S doublet, while the third casein mRNA migrated as a single band at 12S. The α-lactalbumin mRNA migrated at approximately 10S under these conditions. These three casein mRNAs have been partially resolved from each other and α-lactalbumin mRNA by subsequent fractionation using either sucrose gradient ultracentrifugation or by preparative gel electrophoresis (67,72). Although it is relatively easy by these methods to isolate milligram quantities of casein mRNA of greater

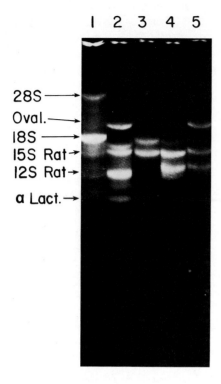

FIG. 2. Characterization of rat and mouse casein mRNAs by agarose-urea gel electrophoresis. Electrophoresis was performed on 1.5% agarose gels containing 6 M urea and 0.025 M sodium citrate, pH 4.2 as previously described (69). Gels were stained for 30 min with 5 μg/ml ethidium bromide in 0.05 M Tris-HCl, pH 8.0, rinsed in H₂O twice and photographed by transillumination using shortwave UV. Gel #1, 5 μg total RNA isolated from 8-day lactating rat after 3 M sodium acetate extraction; gel #2, 2.5 μg partially purified rat milk protein mRNA fraction containing 1 μg purified ovalbumin mRNA as an internal standard; gel #3, 2.5 μg Sepharose 4B purified mouse casein 15S mRNA fraction; gel #4, 2.5 μg mouse casein 15S + 12S casein mRNA fractions; gel #5, same as #4 + 1 μg ovalbumin mRNA.

than 90% purity as assessed by physical, chemical, and biological criteria, the resolution of the individual casein mRNAs into chemically unique species has proven to be more difficult. Partial resolution of the 15S rat casein mRNA doublet from the 12S casein mRNA has been accomplished by preparative gel electrophoresis (67). However, even the most precise sizing technique may not remove fragments of a larger abundant mRNA in a smaller mRNA fraction as detected by molecular hybridization (69). Generation of chemically distinct casein DNA probes will, however, be possible using recombinant DNA technology.

The characterization of the three partially purified mouse casein mRNAs is also shown in Fig. 2 (gels 3–5). Two mRNA fractions were obtained by chromatography on Sepharose 4B followed by an additional passage over oligo (dT)-cellulose to remove a small amount of residual 18S rRNA. Both of these fractions contained greater than 90% casein mRNA activity when translated in the wheat germ cell-free translation system (67). The largest mouse casein mRNA band (Fig. 2, gel #3) migrated slightly slower than the largest rat casein mRNA permitting better resolution of the 15S doublet. In addition, the smallest mouse casein mRNA also migrated slightly slower than the 12S rat casein mRNA (Fig. 2, gel #4).

Using the mobility of the three rat casein mRNAs and ovalbumin mRNA it was possible to calculate the approximate molecular weights of the three mouse casein mRNAs. The sizes of the rat casein and ovalbumin mRNA standards had previously been determined by several independent methods including direct length measurement under the electron microscope (67,68, 89). The molecular weights of the individual rat and mouse casein mRNAs are summarized in Table 1 and compared to the apparent molecular weights of the three rat and mouse caseins. Interestingly, two of the mouse casein mRNAs are larger than the corresponding rat casein mRNAs, and in addition, two of the mouse caseins have apparent molecular weights larger than those of their rat counterparts. The amount of additional noncoding information in the rat and mouse casein mRNAs is also compared in Table 1. The percentage of noncoding bases in each mRNA appears to be highly conserved in both species, suggesting a possible structural role of these sequences in the function of the three casein mRNAs. However, determination of the absolute percentage of noncoding information present in each of these mRNAs will await further sequencing of these nucleic acids and the proteins they specify. Anomalous migration of glycoproteins during SDS-polyacrylamide gel electrophoresis may result in an overestimation of the molecular weight of the glycosylated and phosphorylated caseins. Furthermore, the length of the mouse casein signal peptides has yet to be determined. Finally, both the secondary structure and protonation of nucleic acids on acid-urea agarose gels may affect the migration of the casein mRNAs during electrophoresis (37). However, despite these reservations the apparent conservation of the sizes of the rat and mouse casein mRNAs and their respective protein products is an interesting example of molecular evolution.

The availability of sufficient quantities of the purified rat and mouse casein mRNAs has greatly facilitated their further characterization and use as templates for the synthesis of cDNA hybridization probes. In most of

TABLE 1. *Molecular weights of rat and mouse caseins and casein mRNAs*

Band	Molecular weight mRNA[a]	Molecular weight protein[b]	Additional non-coding information[c]
Mouse 1	500,000 (1462)	46,000	177/1200 = 15%
Rat 1	450,000 (1316)	42,000	135/1096 = 12%
Mouse 2	415,000 (1213)	29,500	358/770 = 47%
Rat 2	415,000 (1213)	29,500	358/770 = 47%
Mouse 3	325,000 (950)	26,500	174/691 = 25%
Rat 3	300,000 (877)	24,500	153/639 = 24%

[a] Average molecular weight of a nucleotide: 342.
[b] Average molecular weight of an amino acid. 115.
[c] Including a signal peptide sequence of approximately 15 amino acids (45 nucleotides) for each precasein (75), and a poly(A) tail of 40 adenosines (67).

the studies to be described the 15S rat or mouse casein mRNA fractions were used to synthesize cDNA. The resulting full-length cDNA probes were shown to be complete copies of the two casein mRNAs employed as templates by both sizing and mRNA-nuclease protection experiments (68,69). These cDNA probes have been utilized to quantitate the levels of casein mRNA in mammary gland RNA extracts, isolated at various stages of normal development (69), isolated from hormone-dependent and independent mammary carcinomas (71), and following hormonal administration both *in vivo* (70) and in mammary gland organ culture (44). Some of the results of these studies will be described in the following sections.

Although the cDNA probe synthesized from 15S casein mRNAs is extremely useful for measuring the levels of both of these casein mRNAs under a variety of conditions, more precise studies of casein gene regulation and organization require chemically unique hybridization probes. This can be accomplished by cloning a mixture of the casein mRNAs and then screening the individual clones for the three casein structural gene sequences. Several screening methods are available, which may involve the use of either specific restriction enzymes and DNA sequencing (86) or cell-free translation systems (59b). Analogous studies have been performed to separate the alpha and beta globin structural genes (22). The separated structural gene probes can then be used to study the induction of the casein genes, and their organization in genomic DNA.

As a first step for cloning and amplifying the rat casein structural gene sequences we have synthesized double-stranded DNA copies of the 15S rat casein mRNAs using avian myeloblastosis virus RNA-directed DNA polymerase. The double-stranded ^{32}P-labeled casein DNA was characterized by electrophoresis on nondenaturing agarose gels and identified by autoradiography (Fig. 3). Several dark bands were observed, the largest of which were 1,250 and 1,000 nucleotides respectively. Preliminary restriction mapping experiments were performed in order to facilitate the further cloning and identification of the individual rat casein structural genes (Fig. 3). By quantitative scanning of the autoradiogram it was possible to determine which of the casein DNAs was cleaved by a given enzyme. In addition, the individual full-length bands could be eluted from the gel and digested individually with each of these restriction enzymes. As shown in Fig. 3, the enzymes HAE III, HHA I, and PST cut the casein DNAs, whereas HIND III, ECO RI, and BAM HI did not appear to digest these DNAs. Further examination of the individual DNAs revealed that only the largest band (1250 N.T.) appeared to be cleaved by HHA I. In addition, a minor fragment was observed using the isolated DNAs with HIND III. Thus, it should be possible to use synthetic linkers, containing, for example, a BAM cleavage site, to insert the casein DNAs into a recombinant plasmid such as pBR322. This would allow the subsequent isolation of an inserted sequence following cloning and amplification. Furthermore, the enzyme HHA I may be useful for dis-

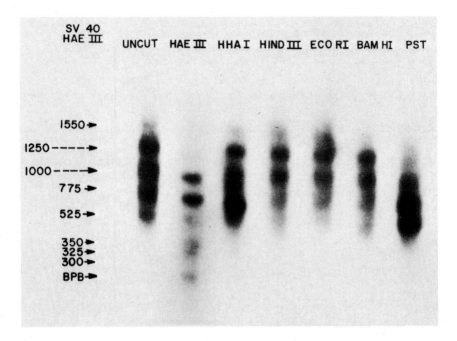

FIG. 3. Restriction mapping of double-stranded casein structural genes. Approximately 23,000 cpm of ^{32}P-labeled double-stranded casein DNA, synthesized from a 15S rat casein mRNA preparation, was digested along with 1.6 μg of SV 40 DNA as an internal standard. Each limit restriction digest was performed for 2 hr with at least 3 units of enzyme added. The products were analyzed by electrophoresis on a 1.5% agarose gel containing 40 mM Tris-HOAC, pH 8.2, 2 mM EDTA, 20 mM NaOAc, 18 mM NaCl, and 500 μg/ml EtBr. Electrophoresis was performed at 75mA, 60 V at room temperature for 4.5 hr and the ^{32}P-DNA detected by autoradiography at −80° using a Dupont enhancing screen.

tinguishing between the different 15S casein DNA clones. These preliminary experiments illustrate the kind of approach that is being utilized in order to characterize the individual casein structural genes. Construction of detailed maps of the three rat casein genes will require the availability of the isolated, cloned DNA fragments. However, using these techniques considerable information about the sequence and organization of these peptide hormone regulated genes should be obtained.

CASEIN GENE EXPRESSION DURING MAMMARY GLAND DEVELOPMENT

Development of the mammary gland during pregnancy is characterized by an increased synthesis of rRNA, polysomes, and tRNA (6,25), including changes in the ratios of specific isoaccepting species of tRNA (23). This increase in RNA synthesis is accompanied by the extensive development of endoplasmic reticulum and the appearance of membrane-bound polysomes

(56,85,90). Furthermore, an increased DNA content (2) and the proliferation of alveolar cells occurs (51), resulting in a highly sophisticated protein-synthetic machinery capable of secreting several grams of casein per day during lactation (32). The regulation of casein synthesis and secretion would, therefore, not be expected to be an "all or none" phenomenon dependent solely upon the induction of casein mRNA, but rather reflects coordinated changes in many components of the protein synthetic and secretory machinery.

Two different methods have been employed in our laboratory in order to determine the levels of casein mRNA present during normal mammary gland development: cell-free translation (72) and molecular hybridization (69). In our initial studies the variations in both casein mRNA activity and total mRNA activity during mammary gland differentiation were determined by assaying total RNA preparations obtained from pregnant, lactating, and regressed (7 days after weaning) mammary tissue in the wheat germ translation system (72). Casein mRNA specific activity increased 20-fold, whereas total mRNA specific activity increased only twofold between day 5 of pregnancy and day 2 of lactation. Accordingly, casein mRNA activities increased from approximately 7% of the total mRNA activity to 60% during this period. After mammary gland regression following weaning there was a rapid loss in casein mRNA activity to a barely detectable level; whereas the total mRNA activity was reduced to a value comparable to that found in a 5-day pregnant mammary gland. When the selective increase in casein mRNA activity was corrected for a 2- to 3-fold increase in RNA recovery, reflecting the increased content of RNA present during mammary gland development, a 60-fold overall increase in the total casein mRNA activity resulted between early pregnancy and lactation. However, during mammary gland development the percentage of fat cells is markedly reduced and glandular tissue may represent as much as 75% of the total gland during lactation (51). When the casein mRNA activity was, therefore, expressed relative to the percentage glandular tissue, the increase in total casein mRNA activity now represented only an 8- to 11-fold change from day 5 of pregnancy until lactation. Thus, the selective induction of casein mRNA activity which occurred during pregnancy and early lactation was accompanied by a proliferation of alveolar cells and resulted in a marked increase in the total concentration of casein mRNA in the lactating rat mammary gland. Moreover, a selective loss in casein mRNA activity was observed during mammary gland involution.

Because of the possibility that different mRNAs may be translated with different efficiencies in cell-free translation systems, the more quantitative hybridization assay was also used to measure the levels of casein mRNA sequences during normal mammary development. A specific casein $cDNA_{15S}$ probe was hybridized with an excess of RNA isolated from mammary tissue obtained from a 6-month-old virgin animal, at different stages of pregnancy, during lactation, or following regression of the gland after weaning (69). This extremely sensitive and quantitative cDNA probe was able to detect a limited amount of casein mRNA even in the virgin mammary gland (70). A series of parallel hybridization curves were generated which displayed progressively faster rates of hybridization. The slowest rate was observed with RNA extracted from virgin mammary tissue, whereas a maximal rate was found using RNA obtained from 8-day lactating tissue. Using these hybridization data and the equivalent $R_{o}t_{1/2}$ for the highly purified 15S casein mRNA, the percentage of casein mRNA in each total RNA extract was determined. As shown in Fig. 4, casein mRNA sequences represented 0.52% of the total cellular RNA in the 8-day lactating tissue, a 19-fold increase over the amount present at 5 days of pregnancy and an overall 300-fold increase relative to the virgin mammary gland. Since the RNA content of the lactating gland is considerably greater than that of the virgin this resulted in almost a 4,000-fold increase in the total number of casein mRNA molecules/g tissue during this developmental period (70). Following regression of the gland, the level of casein mRNA sequences decreased markedly until casein mRNA comprised only 0.014% of the total RNA.

The amount of 15S casein mRNA sequences can also be expressed as

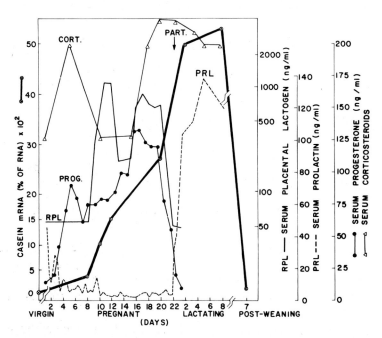

FIG. 4. Changes in casein mRNA levels during normal mammary gland development. Correlation with serum hormone levels. Casein mRNA levels were measured by cDNA hybridization (69,70). Prolactin and placental lactogen levels were obtained from Shiu et al. (78), progesterone levels were taken from Morishige et al. (49), and hydrocortisone serum levels were reported by Simpson et al. (79).

the number of molecules of casein mRNA per alveolar cell (70). A 12-fold increase in the number of molecules per alveolar cell was observed between 5 days of pregnancy and 8 days of lactation, reaching a maximal level of 79,000 molecules of the 15S casein mRNA per cell. Thus, in highly specialized tissues, which are producing large amounts of a given protein, it is not unusual to observe mRNA levels as high as 80,000 to 100,000 specific mRNA molecules per cell.

By comparing the changes in casein mRNA during mammary development with the known changes in several serum hormone levels, it may be possible to determine which hormones are the principal regulators of casein mRNA accumulation *in vivo*. For example, the observed alterations in casein mRNA levels (Fig. 4) may be correlated with the serum levels of prolactin and placental lactogen which undergo marked changes in the rat during pregnancy, lactation, and after weaning (49,78). Thus, the small amount of casein mRNA activity observed in the early pregnant (0 to 7 days) mammary gland may result from the increase in serum prolactin to 50 ng/ml that occurs in the rat after coitus, with levels remaining elevated (20 to 30 ng/ml) for the first 3 days of pregnancy (49). The additional increase in casein mRNA that occurs at approximately day 8 of pregnancy may then be attributable to the vast increase in rat placental lactogen that occurs at this time, reaching levels as high as 1200 ng/ml by day 12 of pregnancy (35). This conclusion is given additional support by the observation that a 90-fold greater difference in the amount of casein mRNA is present in the rat versus the rabbit during midpregnancy (35). This difference, therefore, may reflect the high concentrations of placental lactogen in the rat, compared to very low levels observed in the rabbit; e.g., a maximal level of only 25 ng/ml was found at day 30 of pregnancy (35). Thus, placental lactogen may be of primary importance in initiating rat mammary gland development during pregnancy.

Although serum placental lactogen levels decrease markedly at parturition, prolactin levels are known to increase to approximately 30 ng/ml just prior to parturition. A further increase to levels as high as 300 ng/ml is observed within 9 to 10 hr postpartum (1). This may account for the additional rise in casein mRNA reported during early lactation. Following weaning prolactin levels fall to a basal level of only 10 ng/ml and, accordingly, casein mRNA activity is barely detectable. Thus, prolactin and placental lactogen are performing a dual role in the mammary gland; they initiate alveolar differentiation and proliferation, as well as selectively induce casein synthesis (84).

The effects of these two peptide hormones are also modulated by several steroid hormones. Thus, in the rat the plasma level of progesterone increases as early as day 4 of pregnancy to levels of 80 ng/ml and reaches levels of 120 ng/ml during midpregnancy (49). A dramatic fall in serum progesterone then occurs at parturition which may be the signal for the onset of lactation.

Finally, plasma corticosteroid levels during midpregnancy in the rat have been reported to be comparable to serum progesterone levels, approaching levels of 200 ng/ml (79). Thus, the development of secretory capacity may depend upon the relationship between the stimulatory effects of corticosteroids and the inhibitory effect of progesterone during pregnancy (90).

Because of the complex hormonal interrelationships which occur in the mammary gland it is extremely difficult to elucidate the exact mechanisms by which these hormones control casein gene expression *in vivo*. The preceding discussion was intended to be speculative and many of the above relationships must still be established. Clearly, hormonal effects on casein synthesis may be mediated at several levels within the cell. Furthermore, an increase in casein mRNA observed following the administration of a given hormone *in vivo* should not always be construed to mean that this hormone is acting to stimulate gene transcription. In order to better understand these complex relationships we have utilized a well-defined *in vitro* model system, as follows.

PROLACTIN INDUCTION OF CASEIN mRNA IN ORGAN CULTURE

Kinetics of mRNA Induction

Some progress has recently been made in obtaining hormonally responsive primary epithelial cell cultures from pregnant mammary tissue in which the maintenance and induction of differentiated function has been possible (24). However, since no cloned, prolactin-responsive mammary epithelial cell line is presently available for study, we have chosen mammary gland organ culture as a well characterized and operable system in which to study peptide hormone regulation of a specific mRNA. One advantage of mammary gland organ culture is that it is performed in a serum-free, chemically defined medium, in which the effective concentration of both peptide and steroid hormones can be carefully controlled.

In the experiments to be described, midpregnant mammary gland explants were employed rather than explants derived from virgin tissue in order to study the early effects of prolactin in preexisting, differential alveolar cells. Because of the high levels of casein mRNA which existed in the midpregnant rat, explants were initially exposed for 48 hr to a medium containing only insulin and hydrocortisone. Following the first 24 hr in culture only 10% of the original amount of casein mRNA remained and after the next 24 hr a further decrease to 4% of the original level was observed (45). At this time the medium was changed and supplemented with either insulin (I) and hydrocortisone (F), or insulin, hydrocortisone, and prolactin (M).

The response to prolactin was also found to be dependent upon the day of pregnancy at which the tissue was removed and placed in organ culture. Thus, a 3.5-, 6.5-, and 12.6-fold induction of casein mRNA was usually observed

24 hr following prolactin addition when organ explants were obtained from 7-, 13-, and 15-day pregnant rats, respectively. Within 48 hr after hormone addition a 25-fold induction of casein mRNA was observed in explants obtained from 15-day pregnant rats (44). Therefore, in order to study the early kinetics of casein mRNA accumulation following prolactin addition experiments were performed using 13- to 15-day pregnant animals.

The techniques of both RNA excess and cDNA excess hybridization have been utilized to quantitate casein mRNA levels during the early time period after prolactin addition. The latter technique is especially useful when only small quantities of RNA are available for hybridization analysis. The results of one such experiment obtained by RNA excess hybridization are summarized in Table 2. A 1.3-fold induction of casein mRNA was observed within 1 hr. Within 4 hr after prolactin addition a 2.1-fold induction was detected. Casein mRNA sequences continued to accumulate for 48 hr, reaching a maximal level of 13.4-fold greater than the controls. Using these data, it was possible to estimate the number of molecules of casein mRNA per alveolar cell (Table 2). An initial value of 478 molecules per cell was observed in the untreated control, reaching a level of 6420 molecules per cell at 48 hr after prolactin addition. Thus, on a per cell basis there was an observed accumulation of casein mRNA.

Although these mRNA accumulation data suggested that prolactin must be at least partially acting to increase the rate of casein synthesis, these types of experiments could not distinguish between a direct effect of prolactin on mRNA transcription or an indirect action on mRNA turnover. Analysis of the rate of casein mRNA accumulation suggested that casein mRNA had a long half-life of greater than 24 hr in the presence of prolactin (68). Thus, in order to explain the rapid rate of casein mRNA accumulation in the absence of an increased rate of transcription, its half-life would have to have changed markedly from a half-life of an hour or less to greater than 24 hr. Although this appears unlikely, an alternative method is available in which

TABLE 2. *Prolactin induction of casein mRNA* in vitro

Hormonal milieu	Equivalent $R_0 t_{\frac{1}{2}}$	Fold increase compared to IF	No. molecules/ alveolar cell
IF, 49 and 96 hr	470	—	478
IFM, 1 hr	350	1.3	641
IFM, 4 hr	220	2.1	1022
IFM, 8 hr	135	3.5	1665
IFM, 24 hr	68	6.9	3308
IFM, 48 hr	35	13.4	6428

For a detailed description of methods and calculations see Matusik and Rosen (44). Cultures were performed using 13-day pregnant rats as described in the text.

the rates of casein mRNA synthesis and degradation can be measured directly in the presence or absence of prolactin (see below). The preceding studies have demonstrated that a peptide hormone can rapidly induce the accumulation of a specific gene product *in vitro.* The mammary gland organ culture system, therefore, provides an excellent model for future studies of the mechanism of peptide hormone action, as well as the study of eucaryotic mRNA metabolism in general.

Modulation by Steroid Hormones

If mammary gland explants were incubated for 48 hr in the presence of insulin and hydrocortisone, and then the medium changed to one containing insulin and prolactin, induction of casein mRNA was also observed (44,68). However, hydrocortisone did appear to play at least a permissive role in the accumulation of casein mRNA. For example, insulin and hydrocortisone–treated explants had higher levels of casein mRNA than insulin alone, and insulin, hydrocortisone, and prolactin–treated tissue had higher levels of casein mRNA than insulin and prolactin–treated explants. Thus, hydrocortisone may modulate the rate of prolactin-induced mRNA synthesis and/or the degradation of casein mRNA. Glucocorticoids have also been shown to amplify the capacity of prolactin to increase the concentration of casein mRNA available for translation *in vivo,* but were totally ineffective when administered alone (18). Thus, both *in vitro* and *in vivo,* this steroid hormone has been demonstrated to potentiate the action of prolactin on casein gene expression. It remains to be determined if the initial requirement for hydrocortisone is to promote rough endoplasmic reticulum biosynthesis (90) and the stabilization of casein mRNA on membrane-bound polysomes. Recent studies by Houdebine and his colleagues have also demonstrated the induction of casein mRNA (19) and casein synthesis (17) in organ culture using a reduced O_2 atmosphere (57% versus 95%) with prolactin alone, i.e., in the absence of both insulin and hydrocortisone. Thus, prolactin appears to be the critical hormone regulating casein mRNA synthesis.

Since progesterone has been reported to inhibit casein mRNA accumulation and casein synthesis and secretion *in vivo* (30,70), the possibility that progesterone might directly prevent prolactin-induced casein mRNA synthesis *in vitro* was tested. Progesterone added simultaneously with prolactin reduced the level of casein mRNA, relative to the prolactin-induced control, in a concentration-dependent manner, i.e., by 30, 50, and 100% at concentrations of 0.5, 1.0, and 5 μg/ml, respectively (Fig. 5). Thus, progesterone, in a dose-related fashion, inhibited prolactin induction of casein mRNA in agreement with its suggested role during pregnancy in the rat (70). The inhibitory effect of progesterone, at the lower doses of 0.5 and 1.0 μg/ml, does not appear to be due to a nonspecific toxic effect of the steroid *in vitro* since insulin, hydrocortisone, and progesterone (1 μg/ml)

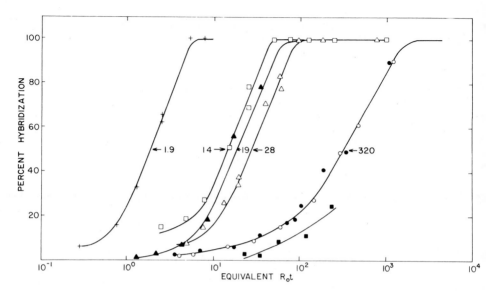

FIG. 5. Progesterone antagonism of casein mRNA induction. Mammary glands from a 15-day pregnant animal were placed in organ culture for 48 hr in the presence of IF (insulin plus hydrocortisone). At this point, the medium was changed and different hormonal combinations were added for an additional 24 hr as follows: controls, IF (○) or IF plus progesterone (P), 1 μg/ml (●); IFM (IF plus prolactin) plus P, 0.5 μg/ml (▲); IFM plus P, 1.0 μg/ml (△); IFM plus P, 5 μg/ml (■); IMF alone, (□). An 8-day lactating rat mammary gland RNA extract (±) served as a standard curve. The arrows indicate the $R_0t_{1/2}$ values for each hybridization curve. (From ref. 44, with permission.)

controls had identical amounts of casein mRNA as did insulin and hydro-cortisone controls. However, alterations in the insulin-hydrocortisone controls did appear when progesterone was added at 5 μg/ml (Fig. 5).

Thus, the ability of two steroid hormones, hydrocortisone and progester-one, to modulate the response to a peptide hormone, prolactin, on casein gene expression has been demonstrated both *in vivo* and *in vitro*. Our ability to study the effects of these hormones in an *in vitro* system should now greatly facilitate the elucidation of their respective mechanisms of action on casein gene expression. For example, the effects of hydrocortisone and pro-gesterone on the rates of synthesis and processing of casein mRNA can now be determined by the pulse-chase and cDNA-cellulose affinity chroma-tography techniques described below.

Possible Mediators of Prolactin Action

Since polyamines and cyclic GMP have been reported to mimic some of the effects of prolactin on casein synthesis in organ culture, we decided to investigate their possible role as mediators of prolactin action on casein gene expression. Addition of spermidine (500 μM) to insulin and hydro-cortisone–treated cultures did not increase the level of casein mRNA (45). However, spermidine did appear to maintain the basal level of casein mRNA observed in the IF-treated cultures when added with insulin alone. This result is consistent with previous data concerning the ability of polyamines to replace glucocorticoids in organ culture (54). The addition of an inhibitor of spermidine synthesis, methyl GAG, at a concentration of 5 μM, was also unable to prevent the usual increase in induction of casein mRNA observed with prolactin alone. These results suggest that polyamines are not critical mediators of prolactin action on casein mRNA induction.

Experiments were then performed to determine if a variety of cyclic nucleotides were able to mimic prolactin action. Dibutyryl cyclic GMP when added to IF-treated cultures resulted in a maximal 2-fold elevation of casein mRNA levels when added at a concentration between 0.01 and 0.07 mM (Fig. 6). Under comparable conditions an 11- to 12-fold increase in casein mRNA levels was routinely observed with prolactin. However, when higher concentrations of dibutyryl cyclic GMP (0.1 to 0.3 mM) were employed no increase in casein mRNA levels was detected. The addition of dibutyryl cyclic AMP, butyric acid, cyclic CMP, or cyclic UMP to the IF-treated cultures also did not result in an increase in casein mRNA levels. Thus, this 1.5- to 2-fold response appeared to be selective for cyclic GMP and occurred over a limited concentration range. Moreover, a combination of spermidine and dibutyryl cyclic GMP was not able to increase further the level of casein mRNA observed with dibutyryl cyclic GMP alone. Therefore, prolactin regulation of casein gene expression appears to be a complex process and may not be directly mimicked by the addition of exogenous cyclic nucleotides or polyamines.

Since cyclic AMP levels have previously been reported to peak just prior to parturition and decrease prior to lactation (73), the possible inhibitory effects of cyclic AMP on prolactin action were also examined. When dibutyryl cyclic AMP was added at a concentration of 0.01 mM (Fig. 6), it was unable to prevent the 11- to 12-fold induction of casein mRNA observed in the presence of prolactin. A complete dose response curve has recently been established (0.01–0.1 mM), and no inhibitory or stimulating effect of cyclic AMP was observed.

A fundamental problem with the above experiments is the ability of added cyclic nucleotides to maintain a critical intracellular level sufficient to evoke the desired biochemical response. Higher concentrations of exogenous cyclic nucleotides may be toxic to the cultures and actually inhibit casein mRNA induction. In the mixed cell populations, containing fat cells, connective tissue, and epithelial cells, present in these cultures of midpregnant tissue, it is difficult to analyze the rapid effects of prolactin on cyclic nucleotide metabolism using conventional biochemical techniques. However, it may be possible to use immunohistochemical methods to localize and even quantitate

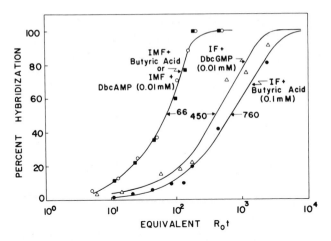

FIG. 6. Effect of cyclic GMP on the induction of casein mRNA. Cultures were initially performed as described in the legend to Fig. 5. After 48 hr in the presence of insulin and hydrocortisone, the medium was changed and supplemented with dibutyryl cyclic GMP (Dbc GMP) or butyric acid in the presence of insulin (I) and hydrocortisone (F) alone, or supplemented with dibutyryl cyclic AMP (Dbc AMP) or butyric acid in the presence of I, F, and prolactin (M). Following a 24-hr incubation, casein mRNA levels were determined in total RNA extracts by cDNA hybridization (44).

changes in cyclic GMP, cyclic AMP, and their respective protein kinases in the prolactin-responsive epithelial cells (21). These experiments may also help define a cellular compartment in which alterations in cyclic nucleotide levels or kinase activity occur in response to prolactin. Such changes may exist in the absence of a net increase in the levels of intracellular cyclic GMP or protein kinase, i.e., as a result of translocation. Therefore, at present there are still no definitive identifiable mediators of prolactin action in the mammary gland. The ability to rapidly quantitate changes in casein mRNA levels following prolactin addition should, however, provide a useful assay for detecting potential mediators in the future.

Use of cDNA Cellulose Affinity Chromatography to Measure the Rates of Specific mRNA Synthesis and Turnover

In order to measure the rates of synthesis and turnover of a specific mRNA it is necessary to be able to analyze pulse-labeled RNA sequences. This is usually accomplished by hybridization with an excess of unlabeled cDNA. At steady state the level of an abundant mRNA species may be greater than 50% of the total cellular mRNA or as much as 1% of total RNA mass. However, a newly synthesized mRNA may represent only 0.1% of the initial RNA transcripts and, therefore, will be difficult to quantitate by hybridization in solution. For example, the background level of radioactivity due to nuclease-resistant secondary structure in the radioactive RNA and trapping during Cl$_3$CCOOH-precipitation may exceed 1% of the radioactivity added. In order to overcome these problems cDNA can be covalently attached to an inert matrix and several cycles of hybridization performed. Such cDNA affinity columns are useful for isolating and analyzing primary transcripts and for studying the kinetics of hormone action on specific mRNA synthesis and processing.

Several chemical methods are available for covalently coupling nucleic acids to cellulose (53), phosphocellulose (76), or agarose (66) resins. Alternatively, a poly(dC)-tail can be attached to cDNA enzymatically using terminal deoxynucleotidyl transferase and, following hybridization in solution, the cDNA-mRNA hybrid isolated on a poly(I)-Sephadex column (15). We have utilized another enzymatic method originally described by Venetianer and Leder (87) in which the cDNA copy is synthesized using an oligo(dT)-cellulose primer. The methodology employed in these studies is illustrated in Fig. 7. A polyadenylated-15S casein mRNA fraction was hybridized to oligo(dT)-cellulose and cDNA synthesis performed using avian myeloblastosis virus reverse transcriptase. Since the oligo(dT)-cellulose was employed

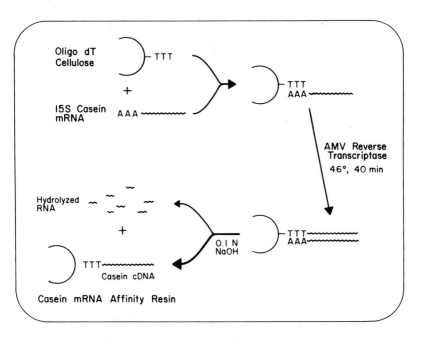

FIG. 7. Synthesis of affinity resin for casein mRNA. Casein cDNA cellulose was synthesized using a purified 15S casein mRNA fraction. Synthesis was performed using AMV reverse transcriptase at 46° C, for 40 min. Synthesis conditions were as follows: Tris-HCl, pH 8.3, 50 mM; DTT, 20 mM; KCL, 150 mM; MgAc, 15 mM; dATP, dCTP, dGTP, dTTP, 0.5 mM; Na$_4$P$_2$O$_7$, 4 mM; actinomycin D, 50 μg/ml; casein mRNA, 100 μg/ml; oligo(dT)-cellulose type 7 (PL Biochemical), 50 mg/ml; ^{32}P-dGTP, 50 μCi/ml; RT, 1,000 units/ml. After completion of synthesis, the RNA was hydrolyzed with 0.1 N NaOH.

as a primer the resulting newly synthesized cDNA is covalently attached to the cellulose matrix. The mRNA bound to the column can then be selectively hydrolyzed with 0.1 N NaOH. The final cDNA affinity resin can now be used to isolate newly synthesized mRNA transcripts. An affinity column prepared in this manner may be less stable than columns in which a chemical attachment method was employed, since a single linkage of each cDNA to the matrix is present rather than the multiple attachment sites present using the latter method. However, as expected, the rate of hybridization observed using the enzymatically synthesized cDNA-cellulose is more rapid than that obtained using chemically immobilized DNA, and approaches the rate of cDNA hybridization in solution (39).

Characterization of the cDNA-cellulose affinity column was performed by measuring its ability to selectively bind casein mRNA (Fig. 8). In this experiment 1 mg RNA isolated from an 8-day lactating rat was allowed to hybridize to the column. Following extensive washing the bound mRNA was eluted by "melting" the cDNA-mRNA hybrid. The initial bound fraction was purified 18-fold as assessed by hybridization to the specific casein ^{3}H-cDNA probe and almost 80% of the casein mRNA sequences were removed from the nonbound RNA. After a second passage over the column an additional 19-fold purification was observed and the resulting mRNA

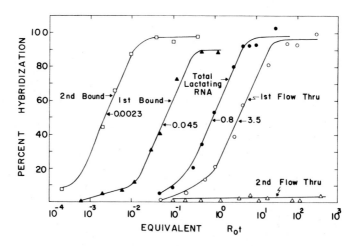

FIG. 8. Selectivity of cDNA-cellulose affinity chromatography. A total of 1 mg of RNA extracted from lactating tissue was hybridized to 50 mg of cDNA-cellulose in 0.6 M NaCl at 68 C for 30 min. The nonbinding fraction was eluted with 0.6 M salt (first flowthrough), and the hybrid was melted off with water at 68 C (first bound). The first bound and first flowthrough were rehybridized under identical conditions to yield the second bound and second flowthrough fractions, respectively. When tested for casein mRNA content by RNA excess hybridization, the second bound fraction reacted with kinetics identical to purified 15S casein mRNA, denoting a 348-fold purification over total lactating RNA. The second flowthrough fraction contained no detectable casein mRNA sequences.

hybridized with the ^{3}H-cDNA with kinetics identical to the purified 15S casein mRNA. A second passage of the nonbound RNA over the column also resulted in the complete removal of detectable casein sequences. The cDNA affinity column was also shown to specifically bind ^{125}I-labeled casein mRNA, but not ^{125}I-labeled ovalbumin mRNA, ^{3}H-labeled *E. coli* RNA, or ^{14}C-labeled KB-cell RNA. Furthermore, selective competition of ^{3}H-casein mRNA binding was detected with as little as 100 ng rat casein mRNA, but no competition was observed with a vast excess of poly(A). Thus, the cDNA-cellulose column provides both a sensitive and selective method for measuring newly synthesized casein RNA sequences.

Total ^{3}H-labeled RNA extracted from mammary gland organ culture exposed to different hormonal combinations was then analyzed by two passages over the cDNA-cellulose affinity column (Table 3). Following 48 hr in the presence of insulin and hydrocortisone the media was changed and labeling performed for an additional 8 hr in the presence of insulin, hydrocortisone, and prolactin or insulin and hydrocortisone alone. Under these conditions prolactin resulted in a 6-fold increase in the incorporation of ^{3}H-uridine into casein mRNA, in close agreement with the previously determined effect of prolactin on the accumulation of casein mRNA in organ cultures of 15-day pregnant mammary tissue. Pulse-labeling studies (15-min incorporation) have also been performed at various times after prolactin addition and revealed a dramatic increase in the rate of casein transcription following the first 30 min after peptide hormone addition (27). Thus, prolactin, following its initial interaction with a membrane receptor, may rapidly increase the rate of transcription of casein mRNA. These types of kinetic studies, therefore, may provide a valuable clue concerning the mechanism by which prolactin regulates casein gene expression. In addition, the methodology employed in these experiments can be applied to the detection of *de novo* synthesized casein mRNA sequences in reconstituted *in vitro* systems, such as isolated nuclei and chromatin. This will permit a distinction between endogenous casein mRNA contaminants and newly synthesized sequences, a major problem in the interpretation of *in vitro* transcription experiments.

Recently, it has also become possible to perform pulse-chase studies in eucaryotic cells without the use of RNA synthesis inhibitors, such as

TABLE 3. *Prolactin induction of ^{3}H-casein mRNA as assayed by cDNA-cellulose affinity chromatography*

	Total RNA (cpm/μg)	First passage (cpm bound)	Second passage (cmp bound)	Casein mRNA (%)
IF	136,600	4650 (3.4%)[a]	619 (13%)	0.45
IMF	142,800	9940 (7%)	3940 (40%)	2.76 (6.1×)

[a] Percentage radioactivity applied at each step that selectively bound to the column.

actinomycin D, which alter the processing and half-life of newly synthesized mRNAs (38). Pretreatment of cells with glucosamine is followed by a pulse of incorporation of radioactive uridine. Labeling is then halted almost instantaneously by the addition of unlabeled uridine and cytidine. Therefore, considerable information concerning the size of putative casein mRNA precursors and the rates of total casein mRNA synthesis and processing may be obtained using a combination of this pulse-chase technique and cDNA-cellulose affinity chromatography. Such an approach may be useful in elucidating the roles of progesterone and hydrocortisone in modulating the accumulation of casein mRNA.

RELATIONSHIP OF DNA SYNTHESIS AND MAMMARY GLAND "DIFFERENTIATION" TO THE INDUCTION OF CASEIN mRNA AND CASEIN SYNTHESIS

An interesting concept in developmental biology concerns the necessary coupling of differentiation to a prior round of cell division (29). It has been previously suggested that hormone-dependent cell differentiation and regulation of specific function in the mammary gland involved, firstly, multiplication and development of a mammary epithelial cell stem cell population into secretory alveolar cells and, secondly, induction of mammary gland specific milk proteins in these daughter cells (58,88). Thus, induction of these milk proteins required prior cell division of a stem cell population, which is accompanied by increased DNA synthesis and DNA polymerase activity. Terminal differentiation of mammary epithelial cells appeared to involve hormone-dependent critical events, which occurred only during a limited portion of the G_1 phase of the cell cycle (88). This concept of a "cell cycle traversal" was based upon the use of inhibitors of DNA synthesis, such as cytosine arabinoside (ara C) and 5-fluorodeoxyuridine (FUdR), which prevented the induction of milk protein synthesis in epithelial cells obtained from mammary glands of mature virgin mice (58). Neither of these inhibitors were shown to have a postmitotic inhibitory effect on casein or α-lactalbumin synthesis. Mitosis was induced by insulin at a pharmacological concentration in organ culture, but similar changes could be induced by epithelial growth factor and growth hormone at physiological concentrations (83). Recent studies have suggested that prolactin may actually be the primary mitogenic hormone in organ culture of mouse mammary tissue (50).

In collaboration with B. Vonderhaar, G. Smith, and Y. Topper at the National Institutes of Health, we have recently reassessed these results using improved techniques developed in our laboratory designed to distinguish effects of prolactin on casein mRNA levels from its effect on casein synthesis. In these experiments casein mRNA was measured by cDNA excess hybridization and casein synthesis was quantitated by a specific immunoprecipitation assay. This latter assay was both more sensitive and more specific than the previously employed precipitation of ^{32}P-labeled casein with rennin and Ca^{2+}. Quite unexpectedly a similar induction of casein mRNA by prolactin was found in virgin explants either in the presence or absence of two inhibitors of DNA synthesis (Fig. 9). Thus, a 3.7-fold induction was observed after 72 hr in the presence of insulin, hydrocortisone, and prolactin compared to insulin and hydrocortisone alone in the presence or absence of ara C (Fig. 9B). A 2.5-fold induction was also detected in the presence of FUdR compared to a 3.5-fold induction in the absence of the inhibitor (Fig. 9A). However, this slight inhibitory effect can be explained by the non-selective toxic action of this inhibitor on RNA and protein synthesis. Thus, DNA synthesis did not appear to be required for the normal induction of casein mRNA in the virgin mammary gland. When companion experiments were performed to measure the level of casein synthesis under the same conditions, a significant, albeit limited, induction was also observed. However, the level of casein synthesis was markedly reduced in the presence of inhibitors of DNA synthesis compared to the prolactin-treated controls. Thus, DNA synthesis may be required for the elaboration of the complex protein synthetic and secretory machinery required in the "fully differentiated"

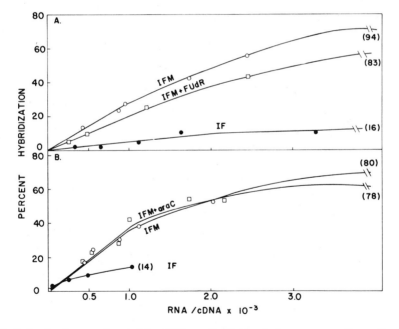

FIG. 9. Lack of a requirement for DNA synthesis in virgin mammary tissue for the induction of casein mRNA. Cultures of virgin mouse mammary tissue were performed for 72 hr as described by Owens et al. (58). The percentage of casein mRNA in each total RNA extract was determined by cDNA excess hybridization using a least-squares analysis to calculate the initial slope of the hybridization. A purified mouse 15S casein mRNA fraction was used as a standard in these experiments. The abbreviations used are the same as those described in Fig. 5.

mammary cell. Furthermore, this may explain the failure to detect an increase in casein synthesis in earlier experiments in which the incorporation of ^{32}P into rennin and Ca^{2+}-precipitable casein was measured. This assay measures predominantly posttranslational modifications of newly synthesized casein, which may have been altered in the presence of ara C and FUdR. Thus, under these conditions the translation of the induced casein mRNA does occur, but probably with a decreased efficiency. Similar results concerning the mechanism of erythropoietin-induced differentiation in cultured marrow cells have also been reported (7). Hormonally induced differentiation in these systems, therefore, appears not to require DNA synthesis prior to the induction of these specific mRNAs, but instead may be a more complex process involving the development of a specialized cell type capable of synthesizing and/or secreting large amounts of a given protein. Clearly, proliferation of alveolar cells is required in the mammary gland prior to lactation in order to permit the synthesis and secretion of large quantities of casein.

CONCLUSIONS

In this chapter we have attempted to summarize the recent developments in our laboratory concerning the mechanisms of hormonally induced casein gene expression. In the past few years considerable progress has been made in the application of the latest techniques of molecular biology to the study of hormone action in the mammary gland. At present, mammary gland organ culture provides both a useful and unique model for studying peptide hormone regulation of gene expression. With the anticipated development of cloned, hormonally responsive cell lines, however, it should be possible in the future to extend these studies and utilize techniques of somatic cell genetics and cell biology. In addition, in the near future considerable information should be obtained concerning the organization of the casein genes in rat DNA and, hopefully, concerning the nature of their primary transcription products. Such information should be invaluable in elucidating the mechanism of prolactin action, i.e., in evaluating reconstituted *in vitro* transcriptional systems, in which the detection of potential mediators of prolactin action may be possible. However, at present the precise mechanisms involved in the hormonal control of casein gene expression remain a complete, but fascinating, enigma.

ACKNOWLEDGMENTS

The authors wish to thank Barbara Vonderhaar and Yale Topper for the inclusion of some of their data prior to publication. The excellent technical assistance of Sherry Barker, Dudley L. O'Neal, and Carol Waugh in the performance of some of these experiments was greatly appreciated. This research was supported by grants CA 16303 and CA 05844 and a Research Career Development Award CA 00154 to Jeffrey Rosen from the National Institutes of Health.

REFERENCES

1. Amenomori, Y., Chen, C. L., and Meites, J. (1970): Serum prolactin levels in rats during different reproductive states. *Endocrinology,* 86:506–510.
2. Anderson, R. R., and Turner, C. W. (1968): Mammary gland growth during pseudo-pregnancy and pregnancy in the rat. *Proc. Soc. Exp. Biol. Med.,* 128:210–214.
3. Anderson, W. B., Gottesman, M. E., and Pastan, I. (1974): Studies with the cyclic adenosine morphosphate receptor and stimulation of *in vitro* transcription of the gal operon. *J. Biol. Chem.,* 249:3592–3596.
4. Aviv, H., Voloch, Z., Bastos, R., and Levy, S. (1976): Biosynthesis and stability of globin mRNA in cultured erythroleukenic Friend cells. *Cell,* 8:495–503.
5. Baker, H. J., and Shapiro, D. J. (1977): Kinetics of estrogen induction of *Xenopus laevis* vitellogenin messenger RNA as measured by hybridization to complementary DNA. *J. Biol. Chem.,* 252:8428–8434.
6. Banerjee, D. N., and Banerjee, M. R. (1973): Rapidly-labeled RNA in the mouse mammary gland before and during lactation. *J. Endocrinol.,* 56:145–152.
7. Bedard, D. L., and Goldwasser, E. (1976): On the mechanism of erythropoietin-induced differentiation. XV. Induced transcription restricted by cytosine arabinoside. *Exp. Cell. Res.,* 102:376–384.
8. Birnbaumer, L., Yang, P.-C., Hunzicker-Dunn, M., Bockaert, J., and Duran, J. M. (1976): Adenylyl cyclase activities in ovarian tissues. I. Homogenization and conditions of assay in Graafian follicles and corpora lutea of rabbits, rats, and pigs. Regulation of ATP and some comparative properties. *Endocrinology,* 99:163–184.
9. Breathnach, R., Mandel, J. L., and Chambon, P. (1977): Ovalbumin gene is split in chicken DNA. *Nature,* 270:314–319.
10. Carpenter, G., and Cohen S. (1976): ^{125}I-labeled human epidermal growth factor. Binding, internalization, and degradation in human fibroblasts. *J. Cell Biol.,* 71:159–171.
11. Catt, K. J., and Dufau, M. L. (1976): Basic concepts of the mechanism of action of peptide hormones. *Biol. Reprod.,* 14:1–15.
12. Chakrabartty, P. K., and Qasba, P. K. (1977): Partial purification of rat α-lactal-bumin mRNA. *Nucleic Acids Res.,* 4:2065–2074.
13. Chomczynski, P., and Topper, Y. J. (1974): A direct effect of prolactin and placental lactogen on mammary epithelial cell nuclei. *Biochem. Biophys. Res. Commun.,* 60:56–63.
14. Cox, R. F. (1977): Estrogen withdrawal in chick oviduct. Selective loss of high abundance classes of polyadenylated messenger RNA. *Biochemistry,* 16:3433–3443.
15. Curtis, P. J., and Weissman, C. (1976): Purification of globin messenger RNA from dimethylsulfoxide-induced Friend cells and detection of a putative globin messenger RNA precursor. *J. Mol. Biol.,* 106:1061–1075.
16. Deeley, R. G., Udell, D. S., Burns, A. T. H., Gordon, J. I., and Goldberger, R. F. (1977): Kinetics of avian vitellogenin messenger RNA induction. *J. Biol. Chem.,* 252:7913–7915.
17. Delouis, C., and Combaud, M.-L. (1977): Lack of mitotic effects of insulin during synthesis of casein induced by prolactin in pseudopregnant rabbit mammary gland organ cultures. *J. Endocrinol.,* 72:393–394.
18. Devinoy, E., and Houdebine, L.-M. (1977): Effects of glucocorticoids on casein gene expression in the rabbit. *Eur. J. Biochem.,* 75:411–416.
19. Devinoy, E., Houdebine, L.-M., and Delouis, C. (1978): Role of prolactin and glucocorticoids on casein genes in rabbit mammary gland organ culture. Quantification of casein mRNA. *Biochim. Biophys. Acta,* 517:360–366.

20. Djiane, J., and Durand, P. (1977): Prolactin-progesterone antagonism in self-regulation of prolactin receptors in the mammary gland. *Nature,* 226:641–643.
21. Dousa, T. P., Barnes, L. D., Ong, S.-H., and Steiner, A. L. (1977): Immunohistochemical localization of 3':5'-cyclic AMP and 3':5'cyclic GMP in rat renal cortex. Effect of parathyroid hormone. *Proc. Natl. Acad. Sci. USA,* 74:3569–3573.
22. Efstratiadis, A., Kafatos, F. C., and Maniatis, T. (1977): The primary structure of rabbit-globin mRNA as determined from cloned DNA. *Cell,* 10:571–585.
23. Elska, A., Matsuka, G., Matiash, U., Nasarenko, I., and Jemenova, N. (1971): tRNA and aminoacyl-tRNA synthetases during differentiation and various functional states of the mammary gland. *Biochim. Biophys. Acta,* 247:430–440.
24. Emerman, J. T., Enami, J., Pitelka, D. R., and Nandi, S. (1977): Hormonal effects on intracellular and secreted casein in cultures of mouse mammary epithelial cells on floating collagen membranes. *Proc. Natl. Acad. Sci. USA,* 74:4466–4470.
25. Gaye, P., Houdebine, L.-M., Petrissant, G., and Denamur, R. (1973): Protein synthesis in the mammary gland. In: *Gene Transcription in Reproductive Tissue, Vol. 5,* edited by E. Diczfalusy, pp. 426–448. Karolinska Institute, Stockholm.
26. Goldfine, I. D., Smith, G. J., Wong, K. Y., and Jones, A. L. (1977): Cellular uptake and nuclear binding of insulin in human cultured lymphocytes. Evidence for potential intracellular sites of insulin action. *Proc. Natl. Acad. Sci. USA,* 74:1368–1372.
27. Guyette, W. A., and Rosen, J. M. (1978): Prolactin induction of casein specific transcripts in mammary gland organ culture. *Fed. Proc.,* 37:1271.
28. Harris, S. E., Rosen, J. M., Means, A. R., and O'Malley, B. W. (1975): Use of a specific probe for ovalbumin messenger RNA to quantitative estrogen-induced gene transcripts. *Biochemistry,* 14:2072–2080.
29. Holtzer, H. (1963): Mitosis and cell transformations. In: *General Physiology of Cell Specialization,* edited by D. Mazia and A. Tyler, pp. 80–90. McGraw-Hill, New York.
30. Houdebine, L.-M. (1976): Effects of prolactin and progesterone on expression of casein genes. *Eur. J. Biochem.,* 68:219–225.
31. Humphries, P., Cochet, M., Krust, A., Gerlinger, P., Kourilsky, P., and Chambon, P. (1977): Molecular cloning of extensive sequences of the *in vitro* synthesized chicken ovalbumin structural gene. *Nucleic Acids Res.,* 4:2389–2406.
32. Jenness, R. (1974): Biosynthesis and composition of milk. *J. Invest. Dermatol.,* 63:109–118.
33. Kafatos, F. C. (1972): mRNA stability and cellular differentiation. In: *Gene Transcription in Reproductive Tissue, Vol. 5,* edited by E. Diczfalusy, pp. 219–345. Karolinska Institute, Stockholm.
34. Kano, K., and Oka, T. (1976): Polyamine transport and metabolism in mouse mammary gland. General properties and hormonal regulation. *J. Biol. Chem.,* 251:2795–2800.
35. Kelly, P. A., Tsushima, T., Shiu, R. P. C., and Friesen, H. G. (1976): Lactogenic and growth hormone-like activities in pregnancy determined by radioreceptor assays. *Endocrinology,* 99:765–774.
36. Lai, E. C., Woo, S. L. C., Dugaiczyk, A., Catterall, J. F., and O'Malley, B. W. (1978): The ovalbumin gene. Structural sequences in native chick DNA are not contiguous. *Proc. Natl. Acad. Sci. USA,* 75:2205–2209.
37. Lehrach, H., Diamond, D., Worney, J. M., and Boedtker, H. (1977): RNA molecular weight determinations by gel electrophoresis under denaturing conditions, a critical reexamination. *Biochemistry,* 16:4743–4751.
38. Levis, R., and Penman, S. (1977): The metabolism of poly(A)+ and poly(A)-hnRNA in cultured Droophila cells studied with a rapid uridine pulse-chase. *Cell,* 11:105–113.
39. Levy, S., and Aviv, H. (1976): Quantitation of labeled globin messenger RNA by hybridization with excess complementary DNA covalently bound to cellulose. *Biochemistry,* 15:1844–1847.
40. Lewis, J. B., Anderson, C. W., and Atkins, J. F. (1977): Further mapping of late

adenovirus genes by cell-free translation of RNA selected by hybridization to specific DNA fragments. *Cell,* 12:37–44.

41. Majumder, G. C., Turkington, R. W. (1971): Hormonal regulation of protein kinase and adenosine 3′,5′-monophosphate-binding protein in developing mammary gland. *J. Biol. Chem.,* 246:5545–5554.

42. Martial, J. A., Baxter, J. D., Goodman, H. M., and Seeburg, P. H. (1977): Regulation of growth hormone messenger RNA by thyroid and glucocorticoid hormones. *Proc. Natl. Acad. Sci. USA,* 74:1816–1820.

43. Matusik, R. J., and Rosen, J. M., unpublished data.

44. Matusik, R. J., and Rosen, J. M. (1978): Prolactin induction of casein mRNA in organ culture. A model system for studying peptide hormone action. *J. Biol. Chem.,* 253:2343–2347.

45. Matusik, R. J., and Rosen, J. M. (1978): Prolactin regulation of casein gene expression. Possible mediators. *Proc. 60th Endocrine Soc.,* Miami, June, 1978, p. 287.

46. McGuire, W. L., personal communication.

47. McReynolds, L. A., Monahan, J. J., Bendure, D. W., Woo, S. L. C., Paddock, G. V., Salser, W., Dorson, J., Moses, R. E. and O'Malley, B. W. (1977): The ovalbumin gene. Insertion of ovalbumin gene sequences in chimeric bacterial plasmids. *J. Biol. Chem.,* 252:1840–1843.

48. Melli, M., Spinelli, G., and Arnold, E. (1977): Synthesis of histone messenger RNA of HeLa cells during the cell cycle. *Cell,* 12:167–174.

49. Morishige, W. K., Pepe, G. J., and Rothchild, I. (1973): Serum luteinizing hormone, prolactin, and progesterone levels during pregnancy in the rat. *Endocrinology,* 92:1527–1530.

50. Mukherjee, A. S., Washburn, L. L., and Banerjee, M. R. (1973): Role of insulin as a "permissive" hormone in mammary gland development. *Nature,* 246:159–160.

51. Munford, R. E. (1963): Changes in the mammary gland of rats and mice during pregnancy, lactation, and involution. I. Histological structure. *J. Endocrinol.,* 28:1–15.

52. Nolin, J. M., and Witorsch, R. J. (1976): Detection of endogenous immunoreactive prolactin in rat mammary cells during lactation. *Endocrinology,* 99:949–958.

53. Noyes, B. E., and Stark, G. R. (1975): Nucleic acid hybridization using DNA covalently coupled to cellulose. *Cell,* 5:301–310.

54. Oka, T., and Perry, J. W. (1976): Studies on regulatory factors of ornithine decarboxylase activity during development of mouse mammary epithelium *in vitro. J. Biol. Chem.,* 251:1738–1744.

55. Oka, T., Perry, J. W., and Kano, K. (1977): Hormonal regulation of spermidine synthetase during the development of mouse mammary epithelium *in vitro. Biochem. Biophys. Res. Commun.,* 79:979–986.

56. Oka, T., and Topper, Y. J. (1971): Hormone-dependent accumulation of rough endoplasmic reticulum in mouse mammary cells *in vitro. J. Biol. Chem.,* 246:7701–7709.

57. O'Malley, B. W., and Means, A. R. (1974): Female steroid hormones and target cell nuclei. *Science,* 183:610–620.

58. Owens, I. S., Vonderhaar, B. K., and Topper, Y. J. (1973): Concerning the necessary coupling of development to proliferation of mouse mammary epithelial cells. *J. Biol. Chem.,* 248:472–477.

59. Palmiter, R. D., Moore, P. B., Mulvihill, E. R., and Emtage, S. (1976): A significant lag in the induction of ovalbumin messenger RNA by steroid hormones. A receptor translocation hypothesis. *Cell,* 8:557–572.

59a. Paterson, B. M., Roberts, P. E., and Kuff, E. L. (1977): Structural gene identification and mapping by DNA: mRNA hybrid arrested cell free translation. *Proc. Natl. Acad. Sci. USA,* 74:4370–4374.

60. Posner, B. I., Josefsberg, Z., Patel, B. A. and Bergeron, J. J. M. (1977): Uptake of prolactin into hepatocytes of female rat liver. *J. Cell Biol.,* 75:191a.

61. Rillema, J. A. (1976): Activation of casein synthesis by prostaglandins plus spermidine in mammary gland explants of mice. *Biochem. Biophys. Res. Commun.,* 75:45–49.

62. Rillema, J. A. (1976): Effects of prostaglandins on RNA and casein synthesis in mammary gland explants of mice. *Endocrinology,* 99:490–495.
63. Rillema, J. A., and Wild, E. A. (1977): Prolactin activation of phospholipase A activity in membrane preparations from mammary glands. *Endocrinology,* 100:1219–1221.
64. Ringold, G. M., Yamamoto, K. R., Bishop, J. M., and Varmus, H. E. (1977): Glucocorticoid-stimulated accumulation of mouse mammary tumor virus RNA. Increased rate of synthesis of viral RNA. *Proc. Natl. Acad. Sci. USA,* 74:2879–2883.
65. Ringold, G. M., Yamamoto, K. R., Tomkins, G. M., Bishop, J. M., and Varmus, H. E. (1975): Dexamethasone-mediated induction of mouse mammary tumor virus RNA. A system for studying glucocorticoid action. *Cell,* 6:299–305.
66. Robberson, D. L., and Davidson, N. (1972): Covalent coupling of ribonucleic acid to agarose. *Biochemistry,* 11:533–537.
67. Rosen, J. M. (1976): Isolation and characterization of purified rat casein messenger ribonucleic acids. *Biochemistry,* 15:5263–5271.
68. Rosen, J. M. (1978): Gene expression in normal and neoplastic breast tissue. In: *Breast Cancer: Experimental Breast Cancer, Vol. 2,* edited by W. L. McGuire, Chapter 6. Plenum Press, New York.
69. Rosen, J. M., and Barker, S. W. (1976): Quantitation of casein messenger ribonucleic acid sequences using a specific complementary DNA hybridization probe. *Biochemistry,* 15:5272–5280.
70. Rosen, J. M., O'Neal, D. L., McHugh, J. E., and Comstock, J. P. (1978): Progesterone-mediated inhibition of casein mRNA and polysomal casein synthesis in the rat mammary gland during pregnancy. *Biochemistry,* in press.
71. Rosen, J. M., and Socher, S. H. (1977): Detection of casein messenger RNA in hormone-dependent mammary cancer by molecular hybridization. *Nature,* 269:83–86.
72. Rosen, J. M., Woo, S. L. C., and Comstock, J. P. (1975): Regulation of casein messenger RNA during the development of the rat mammary gland. *Biochemistry,* 14:2895–2902.
73. Sapag-Hagar, M., and Greenbaum, A. L. (1974): Adenosine 3′,5′-monophosphate and hormone interrelationships in the mammary gland of the rat during pregnancy and lactation. *Eur. J. Biochem.,* 47:303–312.
74. Schutz, G., Beato, M., and Feigelson, P. (1973): Messenger RNA for hepatic tryptophan oxygenase. Its partial purification, its translation in a heterologous cell free system, and its control by glucocorticoid hormones. *Proc. Natl. Acad. Sci. USA,* 70:1218–1221.
75. Shields, D., Rosen, J. M., and Blobel, G., unpublished observations.
76. Shih, T. Y., and Martin, M. A. (1974): Chemical linkage of nucleic acids to neutral and phosphorylated cellulose powders and isolation of specific sequences by affinity chromatography. *Biochemistry,* 13:3411–3418.
77. Shiu, R. P. C., and Friesen, H. G. (1976): Blockage of prolactin action by an antiserum to its receptors. *Science,* 192:259–261.
78. Shiu, R. P. C., Kelly, P. A., and Friesen, H. G. (1973): Radioreceptor assay for prolactin and other lactogenic hormones. *Science,* 180:968–971.
79. Simpson, A. A., Simpson, M. H. W., Sinha, Y. N., and Schmidt, G. H. (1973): Changes in the concentrations of prolactin and adrenal corticosteroids in rat plasma during pregnancy and lactation. *J. Endocrinol.,* 58:675–676.
80. Sippel, A. E., Feigelson, P., and Roy, A. K. (1975): Hormonal regulation of the hepatic messenger RNA levels for $\alpha 2_\mu$-globulin. *Biochemistry,* 14:825–829.
81. Tilghman, S. M., Tiemeier, D. C., Polsky, F., Edgell, M. H., Seidman, J. G., Leder, A., Enquist, L. W., Norman, B., and Leder, P. (1977): Cloning specific segments of the mammalian genome. Bacteriophage containing mouse globin and surrounding gene sequences. *Proc. Natl. Acad. Sci. USA,* 74:4406–4410.
82. Turkington, R. W. (1970): Hormonal regulation of rapidly labeled ribonucleic acid in mammary cells *in vitro. J. Biol. Chem.,* 245:6690–6697.

83. Turkington, R. W. (1972): Multiple hormonal interactions. The mammary gland. In: *Biochemical Actions of Hormones, Vol. II,* edited by G. Litwack, pp. 55–79. Academic Press, New York.
84. Turkington, R. W. (1968): Induction of milk protein synthesis by placental lactogen and prolactin *in vitro. Endocrinology,* 82:575–583.
85. Turkington, R. W., and Riddle, M. (1970): Hormone-dependent formation of polysomes in mammary cells *in vitro. J. Biol. Chem.,* 245:5145–5152.
86. Ullrich, A., Shine, J., Chirgwin, J., Pictet, R., Tischer, E., Rutter, W. J., and Goodman, H. M. (1977): Rat insulin genes. Construction of plasmids containing the coding sequences. *Science,* 196:1313–1319.
87. Venetianer, P., and Leder, P. (1974): Enzymatic synthesis of solid phase-bound DNA sequences corresponding to specific mammalian genes. *Proc. Natl. Acad. Sci. USA,* 71:3892–3895.
88. Vonderhaar, B. K., and Topper, Y. J. (1974): A role of the cell cycle in hormone-dependent differentiation. *J. Cell Biol.,* 63:707–712.
89. Woo, S. L. C., Rosen, J. M., Liarakos, C. D., Choi, Y. C., Busch, H., Means, A. R., O'Malley, B. W., and Robberson, D. L. (1975): Physical and chemical characterization of purified ovalbumin messenger RNA. *J. Biol. Chem.,* 250:7027–7039.
90. Wynn, R. M., Harris, J. A., and Chatterton, R. T. (1976): Interaction of progesterone and adrenocorticoids in ultrastructural development of the mammary gland of the rat. *Am. J. Obstet. Gynecol.,* 126:920–930.
91. Young, H. A., Shih, T. Y., Scolnick, E. M., and Parks, W. P. (1977): Steroid induction of mouse mammary tumor virus. Effect upon synthesis and degradation of viral RNA. *J. Virol.,* 21:139–146.

Ontogeny of Receptors and Reproductive Hormone Action,
edited by T. H. Hamilton, J. H. Clark, and W. A. Sadler.
Raven Press, New York 1979.

Steroid Hormone Receptors in Breast Cancer

William L. McGuire, David T. Zava, Kathryn B. Horwitz,
Roberto E. Garola, and Gary C. Chamness

*Department of Medicine, University of Texas Health Science Center,
San Antonio, Texas 78284*

ESTROGEN RECEPTOR IN BREAST TUMOR CELL NUCLEI

According to the current operational hypothesis, estrogen enters a dependent breast tumor cell and binds to receptor molecules, which then undergo a transformation and enter the nucleus to promote tumor growth. If the source of estrogen is removed by ovariectomy and adrenalectomy, or if its access to estrogen receptor (ER) is blocked by antiestrogens, the chain of action is broken and the tumor regresses. Thus a favorable response to any of these treatments could be expected if the tumor before treatment has available both ER itself and sufficient estrogen to translocate at least some ER to cell nuclei.

It is already well established that tumors lacking cytoplasmic ER are extremely unlikely to respond to such treatments (10). Tumors with ER often respond but frequently do not, perhaps in some cases because insufficient estrogen was present initially to sustain hormone-dependent growth or because the tumor ER failed to translocate to nuclei. If so, the presence of nuclear ER in a tumor specimen would be evidence not only that a functional receptor was present in the cytoplasm at the time of biopsy, but also that estrogen in the environment was entering the cell, binding the receptor, and probably stimulating cell growth. Thus nuclear ER bound to estrogen should be a prerequisite finding in tumors that will respond to ablative procedures designed to reduce endogenous estrogens. We will show here that the situation regarding nuclear receptors is even more complex, as illustrated by our studies on breast tumors in tissue culture.

We first approached the question of nuclear receptor using the MCF-7 human breast cancer cell line, in which we have previously demonstrated receptors for all the major classes of steroid hormones (5). We were surprised to find that the majority of the cellular ER was in the nucleus even in the absence of estrogen (13). This Rn comprises about 75% of the cell's total population of estrogen-binding sites, a subcellular distribution in striking contrast to that found in normal target tissues. (Although, to our knowledge, such high levels of Rn have never been reported, we and others have shown that rat uteri quite consistently contain a small amount of Rn, usually about 10% of the total ER.) The presence of Rn in MCF-7 cells was first suggested by the experiment of Brooks et al. (2), who found that after incubation of cells at 0°C for 1 hr, ^{3}H-estradiol (E*) was located in the nucleus. They felt, however, that E* had translocated Rc at 0°C, while our data suggest that E* is bound directly to pre-existing free nuclear sites (Table 1).

Rn has many of the same properties ascribed to Rc, although we have detected several minor differences. As others have reported, Rc appears to be extremely unstable at elevated temperatures unless saturating doses of estrogen are present, suggesting either intrinsic instability of Rc or presence of cytoplasmic contaminants that degrade uncharged Rc. We find, in contrast, that Rn is stable at 37°C and even slightly increases its affinity for estrogen. In addition, Rc yields different discrete sedimentation values in various concentrations of salt in contrast to Rn, which like Rc sediments at 4S in high salt but partially aggregates when diluted to low salt concentrations.

We have attempted to exclude the possibility that the apparent abnormal accumulation of Rn in MCF-7 nuclei is an artifact of tissue preparation. Clean whole nuclei prepared with Triton X-100 bind E* in a similar fashion to salt-extracted, protamine-precipitated Rn, and the measured level of Rn is similar by both procedures. There is no specific E* binding in control nuclei from other tissues even though Rc is present. Nor do we find Rc contaminating the nuclear-myofibrillar pellet of the rat uterus even after vigorous homogenization procedures that completely disrupt nuclei.

In normal target tissue such as the rat uterus, Rc gains entry into the nucleus only after first binding estrogen. Furthermore, nuclear retention of RnE in the rat uterus appears to require the bound ligand. We find, however, that in MCF-7 cells, Rn does not appear to be bound to any endogenous ligand that could have facilitated Rn transport into the nucleus. If Rn

TABLE 1. *ER distribution in MCF-7 cells*

ER[a]	Control	Estradiol treated
Rc	400	0
RcE	0	0
Rn	1,200	0
RnE	0	1,500
Total	1,600	1,500

[a] Rc, unoccupied cytoplasmic receptor; RcE, estradiol-occupied cytoplasmic receptor; Rn, unoccupied nuclear receptor; RnE, estradiol-occupied nuclear receptor. Methods are described in Horwitz et al., ref. 5. Control conditions are cells growing on charcoal-stripped serum to provide an estradiol-free environment. Estradiol-treated cells are exposed to 10^{-8} M estradiol for 1 hr. Values are expressed as fmoles/mg DNA.

TABLE 2. *ER in MCF-7 cells*

Cell type	Cytosol	Nuclei
Solid tumor[a]	800	400
Tissue culture	500	1,500

[a] Solid tumor: MCF-7 cells were transplanted into a nude athymic mouse. When the tumor reached 1 cm in diameter, ER was measured by methods described in ref. 5. The values obtained from the nuclei represent unoccupied nuclear receptor measured at 4°C. Values are expressed as fmoles/mg DNA.

was originally transported to the nucleus via endogenous ligands; the ligands are no longer present. It also seems unlikely that very low affinity ligands cause translocation of Rn, since it would require enormous concentrations of such ligands with affinities much weaker than estradiol to translocate 75% of the ER into the nucleus.

Our final evidence that Rn is not unique to cells in tissue culture is derived from studies where MCF-7 cells are transplanted into athymic nude mice. We find that Rn is present in the intact cycling MCF-7 tumor-bearing mouse (Table 2). The absolute levels of Rn are lower than in MCF-7 cells in culture and may be due to partial "processing" of Rn (see below).

When the MCF-7 cells are rapidly dividing in log phase growth, cytoplasmic proteins per cell are stimulated almost 3-fold. By phase microscopy, the ratio of cytoplasmic to nuclear mass is also increased. In concert with increased cytoplasmic protein, we find an almost 2-fold rise in Rc and 4-fold enhancement of Rn. The significance of this 4-fold increase in Rn is not yet known, although one could speculate that since its level is higher during log phase growth, it might play some role in the process of cell growth and division.

Further investigating such a suggestion, growth studies have shown that MCF-7 cells are not dependent upon estrogen for growth, but are nonetheless paradoxically inhibited by the antiestrogen tamoxifen (9). To confirm the generality of this antiestrogen action, we treated growing MCF-7 cells with the antiestrogen nafoxidine (Upjohn U-11,100A) for 48 hr in medium with 2% fetal calf serum that had been stripped of steroids by treatment with charcoal. We found that growth of MCF-7 cells was markedly reduced by nafoxidine (Fig. 1). Despite the absence of estrogen in the medium, estradiol alone showed only slight stimulation of growth, demonstrating again that the hormone is not required for growth. Estradiol did, however, completely reverse the inhibition of growth caused by nafoxidine, as it had also reversed the effect of tamoxifen. It seemed likely, therefore, that antiestrogens were acting through the ER, thus explaining why estradiol countered their inhibitory effects. But why did the cells not appear to require estrogen otherwise?

We suggest the possibility that the uncharged nuclear ER is capable of

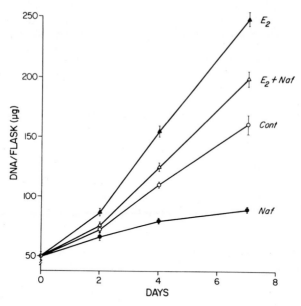

FIG. 1. The effect of estrogen and antiestrogen on MCF-7 cell growth: estradiol (E_2), 10^{-8} M; nafoxidine (NAF), 10^{-7} M. Culture conditions are described in ref. 9.

stimulating the growth of MCF-7 cells (11), as diagrammed in Fig. 2. In the upper left panel MCF-7 cells are growing in the absence of estrogen because biologically active Rn is present. The further stimulation of growth by estrogen would then be due to the resulting small increase in receptors translocated from the cytoplasm (Fig. 2, upper right). If this interpretation is

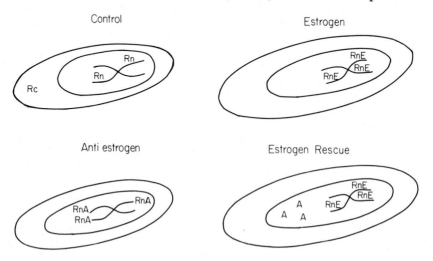

FIG. 2. Schematic representation of ER binding by estrogen and antiestrogen in MCF-7 cells.

correct, then the binding of antiestrogen molecules must inactivate the receptor (Fig. 2, lower left), whereas the rescue from the antiestrogenic effect by estradiol results from the reversal of the antiestrogen binding and subsequent reactivation of the receptor molecule (Fig. 2, lower right).

In order to see if the unusual distribution of ER found in MCF-7 cells might also occur in some human breast tumors *in vivo,* we studied biopsy specimens of solid human breast cancers. We had successfully used the protamine exchange assay to measure cytoplasmic and nuclear ER in normal reproductive tissue (12) and in human breast cancer cells in culture (13), but we found that in solid human tumor biopsies, nuclear ER was rapidly degraded by proteolytic enzymes in the nuclear extract (3). This problem could be avoided by adsorbing nuclear ER onto hydroxylapatite prior to

TABLE 3. *Cytosol and nuclear ER in human breast cancer*[a]

Specimen	Cytosol	Nuclear extract	
		Free sites	Occupied sites
1	20,670	286	206
2	5,469	195	208
3	4,156	513	70
4	3,526	284	177
5	2,761	411	870
6	2,160	87	14
7	1,781	364	158
8	1,493	450	141
9	1,481	243	174
10	1,161	309	17
11	1,030	533	0
12	899	0	225
13	847	501	242
14	547	0	0
15	294	165	256
16	233	0	0
17	45	0	0
18	36	29	29
19	25	0	35
20	22	81	88
21	14	9	2
22	0	0	0
23	0	0	0
24	0	0	0
25	0	0	0
26	0	0	0
27	0	0	0
28	0	0	0

[a] Values expressed fmoles/mg DNA. Cytosol assayed as in ref. 3. Free nuclear sites measured at 4°C for 6 hr. Occupied nuclear sites measured by the difference in total sites (30°C for 5 hr) and free sites.

performing the exchange assay (4). Table 3 demonstrates that certain tumors contain only Rc, while some tumors also contain RnE, as would be expected in tumors from patients having circulating estrogens. However, there are also tumors with considerable Rn, as we had previously seen in breast tumor cells in culture.

We would anticipate that tumors with RnE should respond to ablative endocrine therapy, but the situation is less obvious if a tumor contains appreciable Rn, either alone or along with RnE. If both Rn and RnE were active in stimulating replication and endocrine ablation reduced only RnE, tumor regression might not occur or might be of brief duration, since the Rn stimulus would remain. However, since antiestrogens bind and render inoperative both Rc and Rn, they might cause a more substantial decrease in DNA synthesis and lead to more measurable tumor regression. It must be emphasized that there is no direct evidence that Rn is capable of stimulating replication in the absence of hormone. Additional studies are required to clarify its function.

PROGESTERONE RECEPTORS IN HUMAN BREAST CANCER

In the preceeding section we described our approach to evaluating the function of a tumor's estrogen response system as far as the localization of RnE and Rn in tumor cell nuclei. Here we carry the analysis further by examining the tumor for a specific response to estrogen.

We have demonstrated progesterone receptor (PgR) in human breast tumors (6) and have proposed that this receptor, whose synthesis is known to be controlled by estrogen in the uterus, might serve as a marker of estrogen action in breast cancer (7,8). Thus the presence of PgR in a tumor would indicate that the entire sequence involving estrogen binding to cytoplasmic receptor, movement of the receptor complex into the nucleus, and stimulation of a specific end product can be achieved in the tumor cell. This would rule out the existence of a defect beyond the binding step.

Estrogen Dependence

Though this proposal assumes that PgR is under the control of estrogen acting through ER, this inductive effect had not previously been demonstrated in human breast cancer cells. We therefore examined these points in MCF-7 cells. MCF-7 cells have persistently low, but nevertheless measurable, quantities of PgR, ranging from 50 to 100 fmoles/mg cytosol protein (300 to 700 fmoles/mg DNA). Figure 3 shows that this basal level is increased threefold by 4-day treatment of log phase cells with 10 nM estradiol. The extent of induction varies from 2-to 6-fold among experiments, depending on cell density and growth rate.

The role of estradiol and ER in PgR induction is indicated by studies in

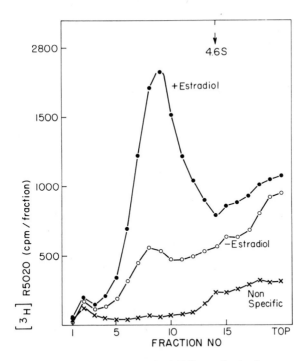

FIG. 3. Sucrose gradient pattern of PgR in MCF-7 cells in the presence (+) and absence (−) of estradiol.

which the levels of the two receptors are contrasted after treatment with and withdrawal from estradiol. Within minutes after exposure to estradiol, all ER is in the nuclei in bound form (RnE). Then begins a rapid turnover or processing of RnE, so that 70% of RnE has disappeared after 5 hr. This processing step precedes PgR induction, and RnE remains low as long as estrogen is present in the medium. If estrogen is removed, processing stops and ER is restored while PgR falls to basal levels. This relationship between PgR levels and ER processing suggests that processing may be a necessary intermediate step following RnE binding and preceding synthesis of a specific estrogen-regulated protein. Table 4 illustrates the effect of increasing doses of estradiol in the processing of ER and on PgR induction. If processing is measured as the reduction of total ER, then it is clear that physiological doses of estradiol are sufficient to maximize PgR, processing, and RnE levels. Higher doses have no further effect. There is a strong correlation between ER processing and PgR induction and also a correlation between RnE and PgR induction, so that it is not possible to say with certainty which is more directly related to the mechanism of induction.

The nature of processing is unclear. Apparently, only RnE is processed, no matter how much estradiol is added. Processing may represent an active state in which a new equilibrium between receptor degradation and syn-

TABLE 4. *Effect of estradiol dose on ER distribution and PgR synthesis*

Estradiol (E_2) dose	ER (pmoles/mg DNA)					PgR
	Rc	Rn	RnE	Total	Processed[a]	
Control	1.79	1.80	0.22	3.81	—	0.25
10^{-12} M	1.61	1.57	0.09	3.27	0.54	0.56
10^{-11} M	0.99	1.44	0.18	2.61	1.20	0.69
10^{-10} M	0.28	0.49	0.69	1.46	2.35	1.17
10^{-9} M	0.02	0.21	0.97	1.20	2.61	1.25
10^{-8} M	0	0.17	0.98	1.15	2.66	1.10
10^{-7} M	0	0	1.13	1.13	2.68	1.09

[a] Processed = total receptor in control — total remaining at each estradiol dose.

thesis is achieved; the degradation step is presumably saturable, since some RnE remains. It is equally possible that processing represents a redistribution of receptor within nuclear binding sites of differing affinities or specificities, or sequestration of receptor to sites inaccessible to salt extraction. Clearly, more investigation on the processing of receptor is needed.

Clinical Correlations

The clinical relevance of estrogen-dependent PgR induction in human breast cancer tissue is currently under investigation in many laboratories and clinics. A few preliminary reports have appeared (1,10), which indicate that patients whose tumors contain both ER and PgR respond to endocrine therapies at nearly double the rate of tumors known to contain ER but not PgR. These encouraging results need confirmation in larger clinical trials, but they emphasize the fact that basic scientists investigating the mechanisms of hormone action in model systems are capable of making important contributions to improved therapeutic strategies in breast cancer.

ACKNOWLEDGMENTS

These studies were supported in part from the National Cancer Institute (CA 11378, CB 23862), the American Cancer Society, and the Robert A. Welch Foundation. We thank J. P. Raynaud of Roussel Uclaf for providing ^{3}H-R5020 and Lois Trench of ICI Chemicals for providing the tamoxifen.

REFERENCES

1. Bloom, N., Tobin, E., and Degenshein, G. A. (1977): Clinical correlations of endocrine ablation with estrogen and progesterone receptors in advanced breast cancer. In: *Progesterone Receptors in Normal and Neoplastic Tissue,* edited by W. L. McGuire, J. P. Raynaud, and E. E. Baulieu, pp. 125–139. Raven Press, New York.

2. Brooks, S. C., Locke, E. R., and Soule, H. D. (1973): Estrogen receptor in a human cell line (MCF-7) from breast carcinoma. *J. Biol. Chem.,* 248:6251–6253.
3. Garola, R. E., and McGuire, W. L. (1977): Estrogen receptor and proteolytic activity in human breast tumor nuclei. *Cancer Res.,* 37:3329–3332.
4. Garola, R. E., and McGuire, W. L. (1977): An improved assay for nuclear estrogen receptor in experimental and human breast cancer. *Cancer Res.,* 37:3333–3337.
5. Horwitz, K. B., Costlow, M. E., and McGuire, W. L. (1975): MCF-7: A human breast cancer cell line with estrogen, progesterone and glucocorticoid receptors. *Steroids,* 26:785–795.
6. Horwitz, K. B., and McGuire, W. L. (1975): Specific progesterone receptor in human breast cancer. *Steroids,* 25:497–505.
7. Horwitz, K. B., and McGuire, W. L. (1977): Estrogen and progesterone: their relationship in hormone-dependent breast cancer. In: *Progesterone Receptors in Normal and Neoplastic Tissue,* edited by W. L. McGuire, J. P. Raynaud, and E. E. Baulieu, pp. 102–124. Raven Press, New York.
8. Horwitz, K. B., McGuire, W. L., Pearson, O. H., and Segaloff, A. (1975): Predicting response to endocrine therapy in human breast cancer: A hypothesis. *Science,* 189:726–727.
9. Lippman, M. E., Bolan, G., and Huff, K. (1976): The effects of estrogens and antiestrogens on hormone-responsive human breast cancer in long-term tissue culture. *Cancer Res.,* 36:4595–4601.
10. McGuire, W. L., Horwitz, K. B., Pearson, O. H., and Segaloff, A. (1977): Current status of estrogen and progesterone receptors in breast cancer. *Cancer,* 39:2934–2947.
11. Zava, D. T., Chamness, G. C., Horwitz, K. B., and McGuire, W. L. (1977): Human breast cancer: biologically active estrogen receptor in the absence of estrogen? *Science,* 196:663–664.
12. Zava, D. T., Harrington, N.Y., and McGuire, W. L. (1976): A new exchange assay for nuclear bound estrogen receptor. *Biochemistry,* 15:4292–4297.
13. Zava, D. T., and McGuire, W. L. (1977): Estrogen receptor: unoccupied sites in nuclei of a breast tumor cell line. *J. Biol. Chem.,* 252:3703–3708.

Ontogeny of Receptors and Reproductive Hormone Action,
edited by T. H. Hamilton, J. H. Clark, and W. A. Sadler.
Raven Press, New York 1979.

Regulation of Expression of the Vitellogenin Gene in Avian Liver

Roger G. Deeley and Robert F. Goldberger

Laboratory of Biochemistry, National Cancer Institute, National Institutes of Health, Bethesda, Maryland 20014

The term *"vitellogenesis"* has been used for many years as a general term to describe formation of egg yolk, a process known to be hormonally regulated. During the past few years, however, the term has evolved so that the molecular biologist now uses the word *"vitellogenin"* to refer to a specific gene product. This evolution reflects not only a shift from physiological to biochemical approaches in studying vitellogenesis, but also a change in our understanding of what this specific gene product is.

Historically, avian egg yolk proteins were divided into two groups—the low-density lipoproteins and the high-density lipoproteins—all of which were known to be synthesized in the liver. The major egg yolk *phospho*proteins are found among the high-density lipoproteins; they are known as phosvitin (1,21,25,26) and lipovitellin (4,11,23). The more distinctive of these proteins is phosvitin, which has a molecular weight of around 30,000 and is composed of more than 50% serine, all the residues of which are phosphorylated (9). Since phosvitin is small and extremely unusual in composition and constitutes a major part of the egg yolk and since it is known that its synthesis in the liver is responsive to estrogen (10,17,18,22,24), it seemed to us to be an ideal subject for studies on regulation of gene expression. Another feature of the phosvitin system that was appealing was that it is possible to induce the synthesis of large amounts of the protein in the male chicken (which normally does not make detectable amounts of it) by simply injecting estrogen. Furthermore, the response occurs in a fully differentiated, highly active tissue—the liver—and does not require DNA synthesis (14).

The first (and rather embarrassing) problem we encountered when we began work on the system was that we could not find any phosvitin in the liver or the serum—not even in the hen! We soon recognized that the reason for this problem was that phosvitin is synthesized in the liver only as part of a much larger polypeptide that also includes lipovitellin (3,13). The egg yolk phosphoproteins never exist as such until this large precursor protein is cleaved within the oocyte. It is this large precursor—known as vitellogenin—that is induced in the liver by estrogen.

Figure 1 shows a schematic representation of avian vitellogenesis as we now picture it. Vitellogenin is synthesized in the liver of the female as a polypeptide with a molecular weight of 240,000 (13). This polypeptide is subsequently phosphorylated, glycosylated, and associated with lipid. It is carried in the blood as a dimer (with a molecular weight of about 500,000) to the ovary, where it is taken up by the developing oocyte and cleaved specifically to form the two egg yolk phosphoproteins known as lipovitellin and phosvitin. The function of vitellogenin has never been elucidated, although most authors state (without any apparent hesitation) that it serves as a phosphate storage protein for the developing embryo or as a metal-ion transport protein. Whatever the true function of vitellogenin may be, it is a molecule to be reckoned with. Each egg contains the equivalent of approximately one gram vitellogenin, which represents approximately 150 mg high-energy phosphate.

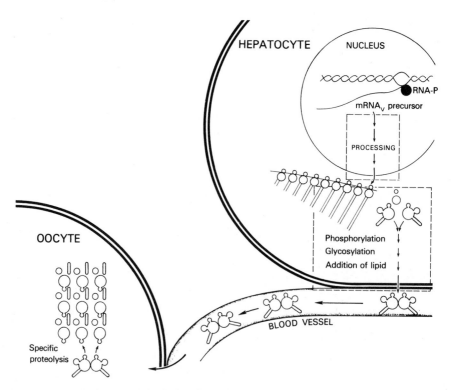

FIG. 1. Avian vitellogenesis. Vitellogenin is synthesized in the hepatocyte on membrane-bound polysomes. After posttranslational modifications, it is immediately secreted into the blood, where it exists as a dimer associated with lipid. It is taken up by the oocyte, possibly by pinocytosis, and cleaved specifically to form the egg yolk phosphoproteins—phosvitin and lipovitellin.

A tabulation of the component parts of vitellogenin in the chicken is shown in Fig. 2. On the basis of the molecular weights of its component parts and their phosphate contents, it is possible to assign two different phosvitins, each of which can be purified from the egg yolk (9), and one lipovitellin to each subunit of the precursor vitellogenin (13). This model is also supported by immunological (13) and amino acid sequence (8) data.

The rooster does not synthesize detectable quantities of vitellogenin. Following a single injection of estrogen, however, as much as 15% of the biosynthetic activity of the liver becomes committed to the synthesis of this one protein (6). It is secreted into the blood, where, having no place to go, it accumulates to quite high concentrations. Eventually, however, it is degraded and disappears from the blood. Figure 3 shows the kinetics of appearance of vitellogenin in chicken plasma from the work of Goldstein and Hasty (15). Depending on what technique is used, vitellogenin can first be detected in the plasma of a rooster between a few hours and 1 day after estrogen stimulation; it reaches a maximum level in about 5 days and disappears during the following 7 days (5,19). It is important to recognize that the relatively insensitive techniques that are used for assaying changes in the plasma concentration of a protein are not suitable methods with which to monitor changes in gene expression. Even if we had very sensitive techniques, the appearance of vitellogenin in the plasma is many steps removed from transcription of the vitellogenin gene. Nonetheless, one of the very interesting features of the vitellogenic response emerges clearly from the data shown in Fig. 3—the greater rapidity and greater magnitude of the response

PROTEIN	MOLECULAR WEIGHT	PHOSPHATE (moles/mole polypeptide)	
Phosvitin 1	28,000	100	
Phosvitin 2	34,000	100	
Lipovitellin	180,000	25	
Vitellogenin Subunit	240,000	250	
Vitellogenin	500,000	500	

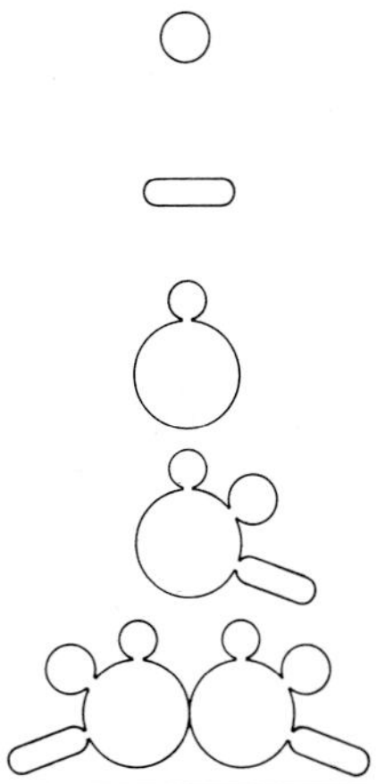

FIG. 2. The component parts of avian vitellogenin.

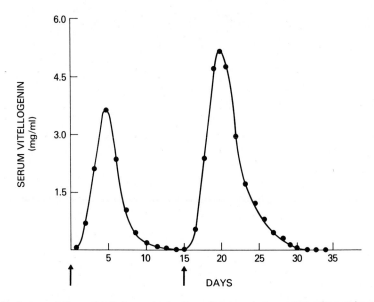

FIG. 3. Accumulation of vitellogenin in the plasma of a rooster after primary and secondary administration of 17β-estradiol. (This figure is redrawn, with permission of the authors from the data of Goldstein and Hastey, ref. 15. The data have been re-calculated to give the weight of vitellogenin rather than phosvitin, although the original measurements were expressed as μg of protein-bound phosphate, a measurement that subsequent work has shown to provide a valid estimate of the vitellogenin content of plasma.)

when the same animal is again stimulated with estrogen, weeks or even months after the first response has died away (5,15,19,20). This effect, which you can see in the second peak in Fig. 3, is reminiscent of the anamnestic response of the immune system and has been the subject of much speculation. Although many aspects of the vitellogenic response are transient, there are some long-term effects of the hormone that are reflected in this anamnestic response. Thus, vitellogenesis is a system in which we can study not only those mechanisms involved in modulating gene expression, but also the less readily reversible mechanisms by which a hormone renders available for transcription a previously dormant gene.

In recent years, one of the most successful approaches for studying the molecular mechanisms by which hormones regulate gene expression in eukaryotic organisms has involved the purification of a specific messenger RNA (mRNA) and its use as a template for the synthesis of complementary DNA (cDNA). The cDNA may then be used as a highly specific and sensitive probe for transcripts from the relevant gene. Studies successfully utilizing this approach have been limited to specialized tissues in which the potential for gene expression is severely restricted and only one (or a very few) mRNA(s) is made in large amounts. Because vitellogenin mRNA is so abundant under certain conditions and because this mRNA is so unusually large, we have been able to apply this strategy to a system in a tissue with diverse genetic expression and myriad biochemical functions—the liver—in which we also have access to a system that is not regulated by estrogen—serum albumin synthesis.

Purification of vitellogenin mRNA from chicken liver has been fraught with difficulties resulting from the extremely high levels of ribonucleases present in this tissue. In order to avoid these difficulties, we dissolve the tissue in 8 M guanidine at −20°C in the presence of a reducing agent (12). It is then possible to obtain RNA, free of protein and DNA, by a series of ethanol precipitations and guanidine extractions during which the concentration of guanidine never drops below 5 M. It is a relatively simple matter to prepare a gram of total RNA in 1 day. This procedure yields RNA that contains no detectable DNA, allowing the RNA to be used directly for even the most demanding hybridization studies.

Figure 4 shows the results of electrophoresis of RNA in a 1.5% agarose gel containing 5 mM methyl mercury—a gel system, developed by Bailey and Davidson (2), that completely denatures both the RNA and DNA. Various amounts of total RNA from the livers of estrogen-treated roosters have been applied to the gel. As expected, there are two major species, 28S and 18S ribosomal RNA; but even before any purification it is possible to

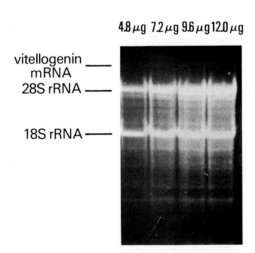

FIG. 4. Electrophoresis of total RNA from the livers of estrogen-treated roosters in 1.5% agarose gels containing methylmercuric hydroxide (5 mM). RNA was extracted with guanidine · HCl from the livers of estrogen-treated roosters and was denatured in sodium borate buffer, pH 8.19, containing methylmercuric hydroxide (5 mM). The figure shows a photograph of the gel stained with ethidium bromide and examined under ultraviolet light. The amounts of RNA applied to the gel are indicated in the figure, as are the positions of vitellogenin mRNA and 18S and 28S ribosomal RNA. (The figure is reprinted from Deeley et al., ref. 12, with permission of the authors and publisher.)

detect vitellogenin mRNA, which, on this gel system, is well resolved from 28S ribosomal RNA.

These RNA preparations can be translated efficiently in either the wheat germ or rabbit reticulocyte cell-free system. We have purified vitellogenin mRNA by a series of isokinetic sucrose gradients run under dissociating conditions and by two cycles of chromatography on poly(U)-Sephadex G-10. This matrix is very similar to poly(U)-Sepharose except that long tracts of poly(U) are attached exclusively to the surface of the resin, and it is more tolerant of heat and formamide than is Sepharose. We have found that poly(U)-Sephadex G-10 is particularly effective for isolating very large mRNAs (2).

A methyl mercury agarose gel is again shown in Fig. 5. On one track are 2.1 μg of purified vitellogenin mRNA and on the other are 1.25 μg each of 18S and 28S ribosomal RNA. The limit of detection of this particular gel system is approximately 20 ng per species, or a little less than 1% of the load applied. The RNA has been sized using this gel system; Fig. 6 shows the results obtained with both 1.5 and 2% agarose gels using a variety of ribosomal and viral RNA preparations as standards. As seen in the figure, vitellogenin mRNA has a molecular weight of 2.3 to 2.35 million. This estimate agrees very well with contour length measurements of the RNA. Figure 7 shows an electron micrograph of four typical vitellogenin mRNA molecules, each of which is approximately 2 μm long. On the basis of the molecular weights estimated on methyl mercury agarose gels and on the basis of the contour lengths of many molecules of vitellogenin mRNA visualized by electron microscopy, we estimate that vitellogenin mRNA is about 7,000 nucleotides—approximately 10% longer than would be required to specify the vitellogenin polypeptide (2).

The purified vitellogenin mRNA can be translated in a wheat germ cell-free system, as shown in Fig. 8 (16). This figure shows a fluorograph of a

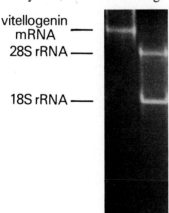

vitellogenin
mRNA —

28S rRNA —

18S rRNA —

FIG. 5. Electrophoresis of purified vitellogenin mRNA in a 1.5% agarose gel containing methylmercuric hydroxide (5 mM). The gel was stained with ethidium bromide and photographed under ultraviolet light. The figure shows a photograph of the gel to which had been applied 2.1 μg of purified vitellogenin mRNA and 2.5 μg of 18S and 28S ribosomal RNA from chicken liver. (Reprinted from Deeley et al., ref. 12, with permission of the authors and publisher.)

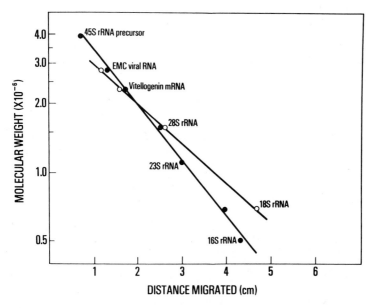

FIG. 6. Molecular weight determination of vitellogenin mRNA in agarose gel containing methylmercuric hydroxide (5 mM). Vitellogenin mRNA was partially purified by centrifugation through a 5 to 30% linear sucrose gradient, chromatography on poly(U)-Sephadex G-10, and centrifugation in an isokinetic 10–23% sucrose gradient. This RNA was subjected to electrophoresis in 1.5 and 2% agarose gels containing methylmercuric hydroxide (5 mM). Gels were stained with ethidium bromide and photographed under ultraviolet light. The mobility of vitellogenin mRNA was compared to that of 45S ribosomal RNA (rRNA) precursor, encephalomyocarditis (EMC) viral RNA, and 28S, 23S, 18S, and 16S rRNA. The figure shows a semilogarithmic plot of the molecular weight as a function of mobility for two concentrations of agarose, 2% (○) and 1.5% (●). (Reprinted from Deeley et al., ref. 12, with permission of the authors and publisher.)

sodium dodecyl sulfate-polyacrylamide gel in which the *total* polypeptides synthesized in the cell-free system have been separated. Track 4 contains authentic ^{14}C-labeled vitellogenin. In track 3 are the ^{35}S-methionine-labeled polypeptides synthesized from purified vitellogenin mRNA. For comparison, the polypeptides synthesized from total RNA are shown in track 1—also labeled with ^{35}S-methionine.

It can be seen in track 3 that the polypeptides synthesized from purified vitellogenin mRNA include one major protein, which is the same size as intact vitellogenin and which constitutes about 75% of the material synthesized, plus small amounts of other polypeptides with molecular weights greater than 200,000. We have been able to demonstrate that these minor polypeptides represent incomplete vitellogenin molecules, since they all react with purified antivitellogenin antibody and they are all preferentially labeled with serine, which (as described earlier) constitutes more than 50% of the phosvitin portions of the vitellogenin polypeptide. In tracks 1 and 2 of Fig. 8

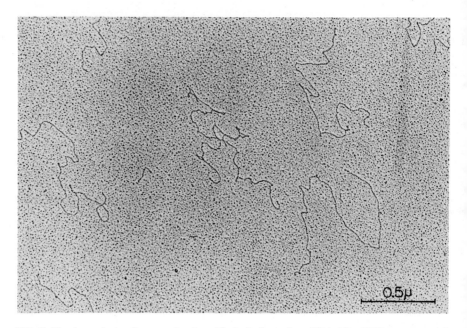

FIG. 7. Electron photomicrograph of purified vitellogenin mRNA. The RNA was spread in formamide (75%, v/v) and urea (6 M). Its length was determined, by measurement of many such molecules, to be 2 μm. (Reprinted from Deeley et al., ref. 12, with permission of the authors and publisher.)

are displayed the polypeptides synthesized from the same preparation of total RNA, the only difference being that the polypeptides displayed in track 1 were labeled with ^{35}S-methionine, whereas those displayed in track 2 were labeled with ^{3}H-serine. It is clear that the serine-rich polypeptides are the very ones that appear together with full-sized vitellogenin made from the purified mRNA. We were intrigued by the fact that all of these peptides are so large—around a molecular weight of 200,000—and wondered what the reason could be that enrichment with serine was seen only when the vitellogenin polypeptide was almost complete. Investigation of this question led us to a further understanding of the structure of vitellogenin (16).

As described earlier, both phosvitins are more than 50% phosphoserine. Furthermore, they contain essentially no methionine. When vitellogenin is cleaved with cyanogen bromide, we obtain a highly unusual peptide with a molecular weight of 90,000 that contains both of the phosvitins. This characteristic peptide is obtained by cyanogen bromide cleavage of not only authentic vitellogenin, but also the 240,000 molecular weight polypeptide synthesized by the wheat germ system *in vitro*. The abundance of serine in this peptide and the absence of methionine led us to believe that we might be able to localize it within the vitellogenin molecule by comparing the kinetics of incorporation of serine and methionine during translation of vitellogenin mRNA in the wheat germ system. We found that as the polypeptide chain

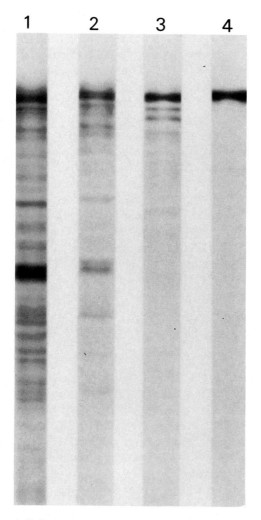

FIG. 8. Translation of vitellogenin mRNA in a wheat germ cell-free system. Total liver RNA from estrogen-treated roosters was centrifuged through a linear 5–30% sucrose gradient, and the fractions containing vitellogenin mRNA were chromatographed on poly(U)-Sephadex G-10. The fraction of RNA that had been bound to the matrix was translated in the wheat germ cell-free translation system, using either ^{35}S-methionine or ^{3}H-serine as the labeled amino acid. Purified vitellogenin mRNA was also translated using ^{35}S-methionine as the labeled amino acid. The polypeptides synthesized were separated in a 7.5% polyacrylamide gel containing sodium dodecyl sulfate (0.1%). The figure shows a fluorograph of the gel. Position **1:** ^{35}S-methionine-labeled polypeptides synthesized from poly(A)$^+$ RNA; **2:** ^{3}H-serine-labeled polypeptides synthesized from poly(A)$^+$ RNA; **3:** ^{35}S-methionine-labeled polypeptides synthesized from purified vitellogenin mRNA; and **4:** authentic, ^{14}C-labeled, delipidated vitellogenin. (Reprinted from Deeley et al., ref. 12, with permission of the authors and publisher.)

grows *in vitro* from the amino-terminus, it grows as a peptide with quite ordinary composition until it reaches a molecular weight of about 150,000. Then it suddenly begins to incorporate huge amounts of serine and to stop incorporating methionine. The shift in the ratio of serine to methionine that occurs during translation of the last portion of the messenger is a spectacular 40-fold! From these studies we conclude that both phosvitins are located within the carboxy-terminal portion of the vitellogenin molecule (16).

We have used vitellogenin mRNA as a template for reverse transcriptase in order to synthesize cDNA (2). Although it has been possible to synthesize extremely long cDNA, the cDNA used for the studies described below was about 1,000 nucleotides long and had a specific activity of about 9×10^7 dpm/μg. As an initial check on the homogeneity of the cDNA, it was hybridized both to purified vitellogenin mRNA and to total mRNA from the liver of an estrogen-treated rooster, under conditions of RNA excess. The results are shown in Fig. 9 plotted as a $R_o t$[1] analysis. In the figure, the curve to the left shows hybridization of vitellogenin cDNA to pure vitellogenin mRNA; the middle curve shows hybridization to total mRNA isolated from the liver of an estrogen-treated rooster. Both reactions proceeded to the same extent, and in both instances the data gave a good fit with the theoretical pseudo-first-order curves expected for the hybridization of a single species. The $R_o t_{1/2}$ determined for the purified messenger was 5.6×10^{-3} moles $\cdot$ s $\cdot$ l^{-1}. The shift in the $R_o t_{1/2}$ observed with total mRNA indicates that it is about 10% by weight vitellogenin mRNA. The cDNA was also hybridized to total mRNA from the livers of normal (*not* estrogenized) roosters, under conditions such that the presence of one molecule of vitellogenin mRNA per cell would provide a 3-fold sequence excess in the hybridization reaction, the minimum required to give pseudo-first-order kinetics. The results are shown in the curve on the right in Fig. 9. Significant hybridization did occur but only at very high $R_o t$ values. However, some decrease in the rate of hybridization at these high $R_o t$ values was observed. The $R_o t_{1/2}$ obtained from this hybridization indicates that vitellogenin mRNA is present in normal rooster liver at a level of somewhat less than 1 molecule/cell. Although these nonideal kinetics make it difficult to estimate the number of vitellogenin molecules accurately, we can say that the level is not zero and is not more than 5 molecules/cell (2).

We have used this approach to examine the rates of accumulation and decay of vitellogenin mRNA following primary and secondary injections of estrogen (7,14). All of these experiments were carried out with total cellular RNA rather than poly(A)$^+$ RNA to avoid the risk of selecting against part of the vitellogenin mRNA population. The disadvantage of doing this is that it is difficult, from a practical standpoint, to remain in adequate RNA excess

---

[1] $R_o t$ is the product of the initial RNA concentration in nucleotides per liter and the time of incubation in seconds.

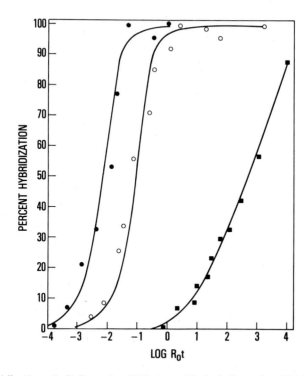

FIG. 9. Hybridization of vitellogenin cDNA to purified vitellogenin mRNA and to total poly(A)$^+$ RNA from the livers of estrogen-treated and normal roosters. Vitellogenin cDNA (0.066 ng) was incubated with various amounts of either purified vitellogenin mRNA (1–250 ng) or total poly(A)$^+$ RNA from the livers of estrogen-treated (10 ng–10 μg) and normal (1–30 μg) roosters. Incubations were carried out at 68°C in 2, 5, or 10 μl 0.01 M Hepes, pH 7.0, containing NaCl (0.6 M) and Na$_2$-ethylene diamine tetra acetic acid (0.002 M), to the $R_o t$ values indicated. The extent of hybridization was assayed with Sl nuclease. The hybridization curves are: **A:** purified vitellogenin mRNA (●); **B:** total poly(A)$^+$ RNA from the livers of estrogen-treated rooster (○); and **C:** total poly(A)$^+$ RNA from the livers of normal roosters. The lines drawn for hybridization of vitellogenin cDNA with purified vitellogenin mRNA and with total poly(A)$^+$ RNA from the livers of estrogen-stimulated roosters are those theoretically expected for an ideal pseudo-first-order reaction. The line drawn for hybridization of vitellogenin cDNA with total poly(A)$^+$ RNA from the livers of normal roosters is drawn through the experimental points and does not represent the curve expected for an ideal pseudo-first-order reaction. (Reprinted from Deeley et al., ref. 12, with permission of the authors and publisher.)

when the level of vitellogenin mRNA falls below about 10 molecules/cell. We reach this limit by hybridizing 250 μg RNA with 60 to 70 pg cDNA. (Hybridization can be detected at levels below this, but it is not possible to quantitate such levels accurately.) Figure 10 shows the levels of vitellogenin mRNA in the liver at various times following a primary injection of estrogen. Figure 10A shows that the level of vitellogenin mRNA increases from less than 10 molecules/cell (the actual estimate is 2) to almost 2,000 during the first 12 hr following administration of the hormone. The maximum level, about 6,000 molecules/cell, is reached in 3 days, after which the level drops sharply until at 17 days there are once again less than 10 molecules/cell. If the assumption is made that vitellogenin mRNA synthesis essentially ceases at 3 days, then the rate at which the message disappears agrees very well with the theoretical decay curve (Fig. 10A, *descending broken line*) expected for a molecule with a half-life of 30 hr. Using this as an initial estimate of the half-life of vitellogenin mRNA and the observed rate of its accumulation, we calculate a transcription rate of about 340 nucleotides/sec/cell. The theoretical accumulation curve for such a transcription rate and half-life is shown by the ascending broken line. It extrapolates to a steady-state level of about 7,000 molecules/cell, which is approximately the steady-state level of vitellogenin mRNA found in the laying hen. Figure 10B shows once again the levels of vitellogenin mRNA, this time determined both by hybridization analysis and by translation of the RNA in a wheat germ system. The results obtained by both methods agree very well.

Figure 10C shows the temporal relationship between the accumulation and disappearance of vitellogenin mRNA in the liver, accumulation and disappearance of vitellogenin protein in the plasma, and the serum levels of 17β-estradiol at various times after injection of the hormone. It is interesting to note that at the time when there is a drastic reduction in the rate of synthesis of vitellogenin mRNA, the serum level of 17β-estradiol is still 10- to 20-fold higher than the level found in the serum of the hen. This level is more than adequate to saturate the high-affinity estrogen binding sites that have been shown to exist in rooster liver nuclei. Whether or not the rooster possesses a mechanism for excluding estrogen from the liver cell or for overriding the effect of estrogen remains to be determined.

The early kinetics of induction of vitellogenin mRNA synthesis following both primary and secondary injections of hormone are shown in Fig. 11. The curves indicate that vitellogenin mRNA begins to accumulate about 30 min after injection of the hormone following both primary and secondary stimulation. However there is a 6- to 7-fold difference in the rates observed. Following a primary injection (Fig. 11, *the lower curve*), the mRNA accumu-

$\longrightarrow$

FIG. 10. Kinetics of accumulation and degradation of vitellogenin mRNA following a primary injection of 17β-estradiol. Total RNA was prepared from the liver of cockerels (55 g) at various times after injection of 17β-estradiol. Vitellogenin mRNA was quantitated by RNA excess hybridization to vitellogenin cDNA, using purified vitellogenin mRNA as a standard, and also by translation in a wheat germ cell-free system. **A:** The levels of vitellogenin mRNA (●—●) determined by $R_o t$ analysis and the theoretical accumulation and decay curves (−−−) for a mRNA with a half-life of 29 hr. **B:** A comparison of the levels of vitellogenin mRNA determined by $R_o t$ analysis (●—●) and by translation in a wheat germ cell-free system (■−−−■). **C:** The temporal correlation between vitellogenin mRNA levels (●—●) and the levels of 17β-estradiol (■−−−■) and vitellogenin (○—○) in plasma. (Reprinted from Burns et al., ref. 7, with permission of the authors and publisher.)

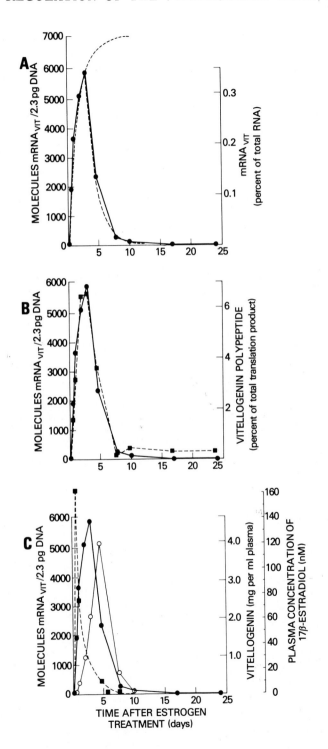

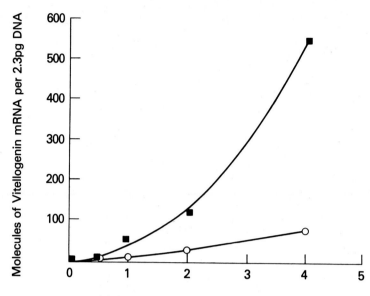

FIG. 11. Initial rates of accumulation of vitellogenin mRNA during primary and secondary stimulation with 17β-estradiol. Cockerels were injected with hormone (20 mg/kg of body weight) and sacrificed at the times indicated. The animals used for the secondary stimulation had received a primary injection of hormone 1 month previously. Total cellular RNA (free of DNA) was extracted from the liver, and hybridization to vitellogenin cDNA under conditions of RNA excess was carried out by the procedures reported by Deeley et al. (12). The amount of vitellogenin mRNA in each sample was determined by comparison of the $R_o t$ value with that obtained using purified vitellogenin mRNA. The number of molecules of vitellogenin mRNA per nuclear equivalent of DNA (2.3 pg) was calculated for each sample from the hybridization data, using the known molecular weight of vitellogenin mRNA (2.35×10^6), the yield of RNA per g of tissue (5 to 6 mg/g), and the DNA content of the tissue (2 to 2.5 mg/g). The data demonstrate that during the first 4 hr the rate of accumulation of vitellogenin mRNA following secondary hormonal stimulation (■—■) was about six to seven times higher than following primary stimulation (○—○). (Reprinted from Deeley et al., ref. 14, with permission of the authors and publisher.)

lates linearly at a rate of only 50 nucleotides/sec/cell for the first 4 hr or so. However, as mentioned above (Fig. 10), the rate of accumulation taken over the first 3 days of the primary response approaches 350 nucleotides/sec/cell. Thus, the primary response is markedly biphasic, with a 6- to 7-fold increase in the rate of accumulation of the mRNA occurring between 4 and 12 hr. In contrast, following a secondary injection of hormone, accumulation begins at the high rate and then increases still further by a factor of about 1.4. Figure 12 shows the overall accumulation of vitellogenin mRNA during the primary and secondary responses. During the primary response (Fig. 12, *lower curve*) vitellogenin mRNA begins to accumulate at a very low rate and then increases greatly after about 4 hr, reaching approximately 6,000 molecules/cell after 3 days. During the secondary response (Fig. 12, *upper curve*) accumulation begins at a high rate and reaches approximately 9,000 molecules/cell after 3 days. In both cases, the mRNA decays with a half-life of about 30 hr. Thus, the anamnestic response of the vitellogenin system, in which secondary stimulation with estrogen causes more rapid and more extensive accumulation of vitellogenin in rooster plasma than does primary stimulation, can be explained on the basis of the difference in the rates and extents of accumulation of vitellogenin mRNA in the liver.

In summary, we have described a system in which estrogen induces a many thousandfold increase in the transcription of a previously dormant gene in a fully differentiated tissue that is already highly active metabolically, without any requirement for cellular proliferation. Thus, a major advantage of the vitellogenin system is that unlike other systems, such as the synthesis of ovalbumin in the highly specialized tubular gland cell of the oviduct and the synthesis of globin in the reticulocyte, the responsive tissue here is available for study even before the gene has been activated. Vitellogenin mRNA has been purified to homogeneity, and cDNA prepared from this mRNA has

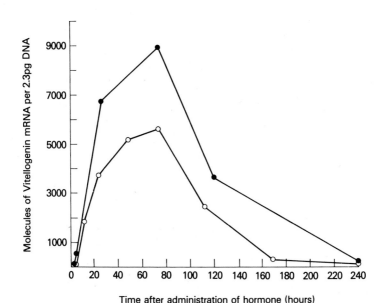

FIG. 12. Rates of accumulation and decay of vitellogenin mRNA after primary and secondary hormonal stimulation. The experiments were carried out and the data analyzed as described in the legend for Fig. 10. The data demonstrate that after primary (O—O) or secondary (O—O) stimulation, the mRNA accumulated for a period of 3 days. Its rate of disappearance during the next 7 days was consistent with that expected for a molecule with a half-life of approximately 30 hr. The maximum level of vitellogenin mRNA reached following secondary stimulation was 1.5 times higher than that reached following primary stimulation. (Reprinted from Deeley et al., ref. 14, with permission of the authors and publisher.)

been used as a probe to monitor with great sensitivity and specificity the levels of vitellogenin mRNA in rooster liver under a variety of conditions. On the basis of our findings, we are considering the intriguing possibility that primary exposure to estrogen not only causes an increased rate of transcription of the previously dormant vitellogenin gene, a phenomenon that lasts for a few days, but also causes a long-term change in the structure of the vitellogenin gene so that this gene becomes more readily available for induction on subsequent exposure to estrogen. Studies designed to probe the structure of the vitellogenin gene before and after primary stimulation with estrogen are now in progress in our laboratory.

REFERENCES

1. Allerton, S. E., and Perlmann, G. E. (1965): Chemical characterization of the phosphoprotein phosvitin. *J. Biol. Chem.*, 240:3892–3898.
2. Bailey, J. M., and Davidson, N. (1976): Methylmercury as a reversible denaturing agent for agarose gel electrophoresis. *Anal. Biochem.*, 70:75–85.
3. Bergink, E. W., Wallace, R. A., Van De Berg, J. A., Bos, E. S., Gruber, M., and Ab, G. (1974): Estrogen-induced synthesis of yolk proteins in roosters. *Am. Zool.*, 14:1175–1195.
4. Bernardi, G., and Cook, W. H. (1960): An electrophoretic and ultracentrifugal study on the proteins and the high density fraction of egg yolk. *Biochim. Biophys. Acta*, 44:86–96.
5. Beuving, G., and Gruber, M. (1971): Induction of phosvitin synthesis in roosters by estradiol injection. *Biochim. Biophys. Acta*, 232:529–536.
6. Bos, E. S., Vonk, R. J., Gruber, M., and Ab, G. (1972): Lipovitellin synthesizing polysomes: specific and quantitative isolation. *FEBS Lett.*, 24:197–200.
7. Burns, A. T. H., Deeley, R. G., Gordon, J. I., Udell, D. S., Mullinix, K. P., and Goldberger, R. F. (1978): Primary induction of vitellogenin mRNA in the rooster by 17β-estradiol. *Proc. Natl. Acad. Sci. USA*, 75:1815–1819.
8. Christmann, J. L., Grayson, M. J., and Huang, R. C. C. (1977): Comparative study of hen yolk phosvitin and plasma vitellogenin. *Biochemistry*, 16:3250–3256.
9. Clark, R. C. (1973): Amino acid sequence of a cyanogen bromide cleavage peptide from hens egg phosvitin. *Biochim. Biophys. Acta*, 310:174–187.
10. Clegg, R. E., Sanford, P. E., Hein, R. E., Andrews, A. C., Hughes, J. S., and Mueller, C. D. (1951): Electrophoretic comparison of the serum proteins of normal and diethylstilbestrol-treated cockerels. *Science*, 114:437–438.
11. Cook, W. H. (1961): Proteins of hen's egg yolk. *Nature*, 190:1173–1176.
12. Deeley, R. G., Gordon, J. I., Burns, A. T. H., Mullinix, K. P., Bina-Stein, M., and Goldberger, R. F. (1977): Primary activation of the vitellogenin gene in the rooster. *J. Biol. Chem.*, 252:8310–8319.
13. Deeley, R. G., Mullinix, K. P., Wetekam, W., Kronenberg, H. M., Meyers, M., Eldridge, J. D., and Goldberger, R. F. (1975): Vitellogenin is the precursor of the egg yolk phosphoproteins. *J. Biol. Chem.*, 250:9060–9066.
14. Deeley, R. G., Udell, D. S., Burns, A. T. H., Gordon, J. I., and Goldberger, R. F. (1977): Kinetics of avian vitellogenin messenger RNA induction. *J. Biol. Chem.*, 252:7913–7915.
15. Goldstein, J. L., and Hasty, N. A. (1973): Phosvitin kinase from the liver of the rooster: purification and partial characterization. *J. Biol. Chem.*, 248:6300–6307.
16. Gordon, J. I., Deeley, R. G., Burns, A. T. H., Paterson, B. N., Christman, J. L., and Goldberger, R. F. (1977): *In vitro* translation of avian vitellogenin messenger RNA. *J. Biol. Chem.*, 252:8320–8327.
17. Greengard, O., Gordon, M., Smith, M. A., and Acs, G. (1964): Studies on the mechanism of diethylstilbestrol-induced formation of phosphoprotein in male chicken. *J. Biol. Chem.*, 239:2079–2082.

18. Hillyard, L. A., Entenman, C., and Chaikoff, I. L. (1956): Concentration and composition of serum lipoproteins of cholesterol-fed and stilbestrol-injected birds. *J. Biol. Chem.*, 223:359–368.
19. Jailkhani, B. L., and Talwar, G. P. (1972): Oestradiol-induced synthesis of phosvitin. *Nature [New Biol.]*, 236:239–240.
20. Jost, J.-P., Keller, R., and Dierks-Ventling, C. (1973): Deoxyribonucleic acid and ribonucleic acid synthesis during phosvitin induction by 17β-estradiol in immature chicks. *J. Biol. Chem.*, 248:5262–5266.
21. Mecham, D. K., and Olcott, H. S. (1949): Phosvitin, the principal phosphoprotein of egg yolk. *J. Am. Chem. Soc.*, 71:3670–3679.
22. Ranney, R. E., and Chaikoff, I. L. (1951): Effect of functional hepatectomy upon estrogen-induced lipemia in fowl. *Am. J. Physiol.*, 165:600–603.
23. Schjeide, O. A., and Urist, M. R. (1959): Proteins and calcium in egg yolk. *Exp. Cell Res.*, 17:84–94.
24. Schjeide, O. A., Wilkens, M., McCandless, R. G., Munn, R., Peterson, M., and Carlsen, E. (1963): Liver synthesis, plasma transport, and structural alterations accompanying passage of yolk proteins. *Am. Zool.*, 3:167–184.
25. Shainkin, R., and Perlmann, G. E. (1971): Phosvitin, a phosphoglycoprotein: composition and partial structure of carbohydrate moiety. *Arch. Biochem. Biophys.*, 145:693–700.
26. Shainkin, R., and Perlmann, G. E. (1971): Phosvitin, a phosphoglycoprotein. I. Isolation and characterization of a glycopeptide from phosvitin. *J. Biol. Chem.*, 246:2278–2284.

Ontogeny of Receptors and Reproductive Hormone Action,
edited by T. H. Hamilton, J. H. Clark, and W. A. Sadler.
Raven Press, New York 1979.

Estrogen Regulation of *Xenopus laevis* Vitellogenin Gene Expression

David J. Shapiro and Hollis J. Baker

Department of Biochemistry, University of Illinois, Urbana, Illinois 61801

Analysis of the molecular mechanisms of reproductive hormone action has been complicated by the existence of a limited number of well-defined biochemical responses to reproductive hormones and by the inability to elicit most hormone responses *in vitro* in a defined environment. These problems are exacerbated by the enormous analytic complexity of most eucaryotic cells, which makes recognition and isolation of structural and regulatory elements in DNA, in specific messenger RNA (mRNA) and mRNA precursors, and in hormone receptors extremely difficult. The general absence of the type of regulatory mutations that proved so useful in procaryotic systems has led to a biochemical approach to the problem of hormone control of gene expression. In spite of intensive investigation of the estrogen induction of ovalbumin synthesis (16,19,21) and of estrogen action in several other systems (14,34), many of the most fundamental questions of gene organization and expression remain unresolved.

Our analysis of the mechanism of estrogen action at the gene level has focused on the estrogen induction of the egg yolk precursor protein vitellogenin in *Xenopus laevis*. The synthesis and secretion of vitellogenin have been investigated by Wallace and his associates (6,28) and by others (10). The formation of egg yolk proteins in oviparous vertebrates, such as *X. laevis,* appears to involve synthesis of a yolk protein precursor by the liver and its secretion into the blood and uptake and processing by the ovary. Synthesis of the *X. laevis* yolk protein precursor vitellogenin can be induced in the livers of both male and female *X. laevis* by the administration of estradiol-17β and related steroids. *X. laevis* vitellogenin is a calcium-binding phospholipoglycoprotein with a monomer molecular weight of approximately 200,000 (6). The vitellogenin monomer is cleaved in the ovary to yield lipovitellin, with a molecular weight of about 120,000, and phosvitin, an extensively phosphorylated protein of molecular weight 35,000 to 40,000 that is 55% serine. Induction of vitellogenin proceeds without cytodifferentiation or DNA synthesis (11) and is, therefore, readily amenable to biochemical analysis.

We have attempted to resolve several questions relevant to an analysis of estrogen action in this system.

1. What is the extent of estrogen induction of vitellogenin synthesis? Is the induction sufficient to permit ready isolation of the mRNA and protein and other components of the RNA and protein-synthesizing systems?
2. Is induction of vitellogenin synthesis achieved by an increase in the level of vitellogenin mRNA or by preferential translation of preexisting mRNA or by some combination of these mechanisms?
3. Are the kinetics of vitellogenin mRNA induction consistent with the hypothesis that an estradiol-17β–receptor complex interacts directly with the vitellogenin gene, or does vitellogenin synthesis result from a kind of developmental cascade, or from the interaction of the hormone–receptor complex with a regulatory gene?
4. Does estrogen stimulation of male *Xenopus laevis* that have been previously induced to synthesize vitellogenin, but are inactive in vitellogenin synthesis at the time of restimulation, result in an altered pattern of vitellogenin gene expression?

ASSAY OF VITELLOGENIN mRNA IN RABBIT RETICULOCYTE LYSATE

We employed an immunologic approach for determination of the relative rate of vitellogenin synthesis and for initial studies on the vitellogenin mRNA content of *Xenopus* liver cells. *Xenopus* vitellogenin was purified from the serum of estrogen-stimulated male *Xenopus laevis* and used to elicit the production of monospecific rabbit antivitellogenin (22). Vitellogenin synthesis in liver cells and in the reticulocyte lysate protein-synthesizing system was quantitated by indirect immunoprecipitation (21–25).

The large size of vitellogenin mRNA and the high level of ribonuclease activity make extraction of undegraded vitellogenin mRNA quite difficult. Analysis of the pH-activity profile of *Xenopus* liver RNase revealed a major peak of acidic lysosomal RNase and little RNase activity at alkaline pH (9). The discrete peaks of neutral and alkaline RNase activity observed in rat liver (5) are absent in *Xenopus* liver. We developed a method for the extraction of undegraded vitellogenin mRNA based on homogenization at alkaline pH and phenol-chloroform extraction (2). A related method has been employed by Berridge et al. (7).

Vitellogenin mRNA was efficiently translated in the rabbit reticulocyte lysate cell-free protein-synthesizing system and biosynthesized vitellogenin isolated by indirect immunoprecipitation (22). Optimization of the reticulocyte lysate system for production of full-length 200,000-dalton vitellogenin monomer was achieved through the use of high levels of Mg^{2+} and K^+ (22).

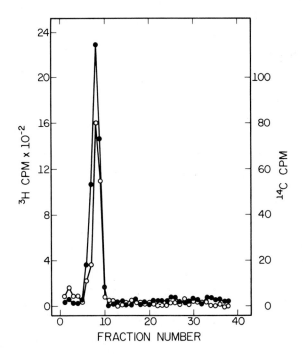

FIG. 1. Immunoprecipitated material from a pooled 300 μl reaction mixture (●—●) was mixed with authentic [^{14}C]vitellogenin and subjected to electrophoresis (○—○). In order to eliminate nascent vitellogenin chains that would complicate analysis, the reaction was continued for 30 min in the presence of a large excess of labeled leucine, and ribosome-associated nascent peptide chains were removed by sedimentation at 140,000 × g. The direction of migration is from left to right. (From Shapiro et al. ref. 22.)

The specificity of the indirect immunoprecipitation reaction and the size of the biosynthesized product were established by analysis of the immuno-precipitated [^{3}H]-labeled reaction product and authentic [^{14}C]vitellogenin on sodium dodecyl sulfate-polyacrylamide gel electrophoresis (Fig. 1; 22). The immunoprecipitated material comigrates with authentic [^{14}C]vitellogenin and has a molecular weight of 190,000 to 200,000. RNA from livers of un-stimulated male *Xenopus laevis* does not code for immunoprecipitable vitellogenin (22). Additional evidence that the immunoprecipitated material is vitellogenin was obtained by comparing the immunologic reactivity of the immunoprecipitated material with that of authentic vitellogenin. The *in vitro* reaction product and vitellogenin exhibit identical precipitin curves, demon-strating the antigenic identity of the two species (22).

The ability of RNA preparations to code for immunoprecipitable vitel-logenin in the reticulocyte lysate system provides a quantitative assay for vitellogenin mRNA. We employed this assay for vitellogenin mRNA and a related immunoprecipitation assay for determination of the intracellular rate

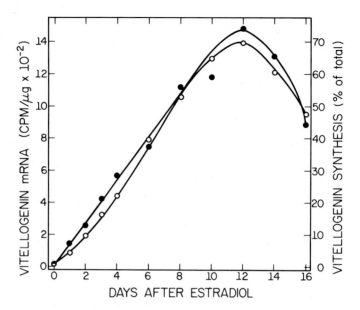

FIG. 2. Male *X. laevis* received injections of 0.2 mg of estradiol-17β/10 g body weight on day 0. The fraction of liver protein synthesis devoted to the synthesis of vitellogenin was determined by comparing the level of immunoprecipitable vitellogenin to the level of trichloroacetic acid-precipitable protein in a pulse-labeled liver cube system (O—O). RNA was extracted and assayed for its ability to code for immunoprecipitable vitellogenin in the rabbit reticulocyte lysate cell-free protein-synthesizing system. Results are plotted as vitellogenin synthesis per μg of RNA (●—●). (From Shapiro et al., ref. 22.)

of vitellogenin synthesis in a study of the induction of vitellogenin synthesis and vitellogenin mRNA by estradiol-17β. Our goals were to establish the extent of induction and determine whether estrogen induction proceeded exclusively through an increased level of vitellogenin mRNA, or whether the preferential translation of vitellogenin mRNA was implicated in the induction process. The data in Fig. 2 demonstrate that the amount of translatable vitellogenin mRNA in *Xenopus* liver is regulated by estrogen administration (22). Vitellogenin synthesis was undetectable in unstimulated male *Xenopus* and was induced to high levels by administration of estradiol-17β, reaching a maximum of nearly 70% of liver protein synthesis 12 days after estrogen administration (Fig. 2). A general correspondence between the rate of vitellogenin synthesis and the level of vitellogenin mRNA was observed (Fig. 2), indicating that selective translation of vitellogenin mRNA is not the major control mechanism in this system. These preliminary investigations established that the estrogen induction of vitellogenin synthesis represents an unusually favorable system for investigation of the control of gene expression by steroid hormones.

PURIFICATION OF VITELLOGENIN mRNA AND SYNTHESIS OF COMPLEMENTARY DNA

The detailed investigation of estrogen action in *Xenopus* liver requires the isolation of vitellogenin mRNA. Purified vitellogenin mRNA can be employed in investigations of mRNA structure and function and as a template for the synthesis of a complementary DNA (cDNA) hybridization probe.

Isolation of vitellogenin mRNA required its separation from ribosomal (rRNA) and from total cell mRNA. Quantitative separation of poly(A)-containing mRNA and rRNA was achieved by two passages through oligo(dT)-cellulose columns.

Vitellogenin mRNA was separated from total poly(A) RNA by virtue of its large size (30S) relative to other mRNAs and the fact that it is the major species of mRNA in the estrogen-stimulated *Xenopus* (22). Vitellogenin mRNA represents a discrete peak of poly(A)-containing RNA sedimenting at 30S, and it was separated from total poly(A) mRNA by fractionation on successive isokinetic sucrose gradients (23). Purification of vitellogenin mRNA results in an increase of approximately 175-fold in the specific vitellogenin-synthesizing activity of the RNA (Table 1; 23). This purified RNA preparation was free of significant contamination by both RNA and other species of poly(A)-containing RNA. The absence of RNA contamination was demonstrated by the fact that isolated vitellogenin mRNA represents a single, sharp symmetrical peak on sucrose density gradient centrifugation (Fig. 3) and polyacrylamide gel electrophoresis (Fig. 4). Discrete peaks of 18S and 28S RNAs were not observed (23). The absence of contamination by other poly(A)-containing RNA was demonstrated by examination of the translation products coded by the purified mRNA in an mRNA-dependent reticulocyte lysate protein-synthesizing system and by the kinetics of its hybridization to cDNA.

A rabbit reticulocyte lysate dependent on added mRNA for activity was prepared by incubating crude lysate with micrococcal nuclease to destroy endogenous mRNA and then inactivating the nuclease by chelating the calcium required for its activity (18). Residual globin synthesis represented about 13% of total protein synthesis during translation of purified vitellogenin mRNA. Gel profiles of the total lysate reaction product and of the material immunoprecipitated with antivitellogenin were identical over the entire 25,000 to 300,000 molecular weight range (Fig. 5), indicating that the purified vitellogenin mRNA was free of significant contamination by other species of translatable mRNA (23).

We have employed the RNA-dependent DNA polymerase of avian myeloblastosis virus to synthesize a nucleic acid sequence complementary to vitellogenin mRNA. The enzyme copies single stranded RNA by extending the DNA in an RNA–DNA primer. Addition of oligo(dT) to the purified

TABLE 1. *Purification of vitellogenin mRNA*

Fraction	RNA (mg)	Specific activity (cpm/µg RNA)	Total activity (cpm × 10^{-6})	Purification (—fold)	Yield (%)
1. Total cell RNA	22.8	399	9.08	1	100
2. First oligo(dT)-cellulose eluate	0.256	21,626	5.54	54.2	61
3. Second oligo(dT)-cellulose eluate	0.184	26,095	4.80	65.4	53
4. First sucrose gradient	0.035	57,057	2.00	143	22
5. Second sucrose gradient	0.017	69,426	1.18	174	13

RNA was isolated from livers of six estrogen-induced *X. laevis* toads. The RNA was fractionated twice on oligo(dT)-cellulose to remove traces of RNA, which nonspecifically adsorb to the cellulose. The specific activity represents the ability of RNA preparations to code for the synthesis of immunoprecipitable vitellogenin in the rabbit reticulocyte system. The total activity is the product of the amount of RNA (µg) and the specific activity of the RNA. Purification and yield were calculated from the specific activity and total activity, respectively. Each assay was carried out at three RNA concentrations and the results averaged. (From Shapiro and Baker, ref. 23).

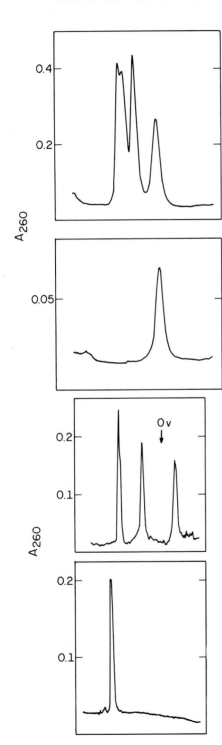

FIG. 3. Sedimentation profile of vitellogenin mRNA on isokinetic sucrose gradients. Purified vitellogenin mRNA (*lower panel*) and purified 16S, 18S, 23S, and 28S rRNA markers (*upper panel*) were sedimented on 5 to 29.9% isokinetic sucrose gradients for 4½ hr at 200,000 × g and 25°. The direction of migration is from left to right. Fractions were collected from the gradient containing purified vitellogenin mRNA and assayed in the reticulocyte lysate system. More than 90% of the vitellogenin mRNA activity was in the fractions coinciding with the center of the absorbance peak, and the remainder was in the immediately adjacent fractions. (From Shapiro and Baker, ref. 23.)

FIG. 4. Polyacrylamide gel electrophoresis of purified vitellogenin mRNA. The upper panel is an absorbance scan of a 2.5% polyacrylamide gel containing purified 28S, 23S, and 18S rRNA markers. Purified ovalbumin mRNA (Ov) was run in a separate gel and migrated to the point indicated by the arrow. The lower panel is an absorbance scan of a gel containing 2.5 μg of purified vitellogenin mRNA. The top of the gels is on the left. (From Shapiro and Baker, ref. 23.)

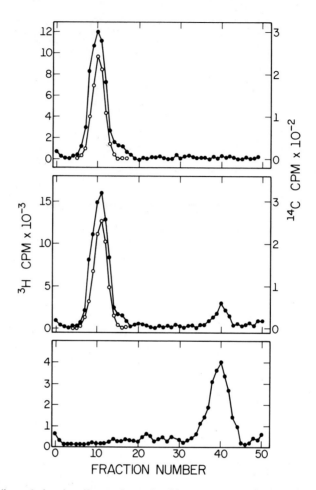

FIG. 5. Sodium dodecyl sulfate-polyacrylamide gel electrophoresis of the reaction products from mRNA-dependent rabbit reticulocyte lysates primed with purified vitellogenin mRNA. Endogenous reticulocyte mRNA was destroyed by preincubation with micrococcal nuclease and the nuclease inactivated by addition of EGTA, which chelates the calcium required for nuclease activity. The products (e.g., globin chains) synthesized in the absence of exogenous mRNA are illustrated in the lower panel. In the upper panel the [³H]-labeled reaction product immunoprecipitated from lysates primed with purified vitellogenin mRNA (●) and authentic [¹⁴C]vitellogenin (○) was subjected to electrophoresis. In the middle panel the total [³H]-reaction product (●) and authentic [¹⁴C]vitellogenin (○) were subjected to electrophoresis, sliced and counted. (From Shapiro and Baker, ref. 23.)

vitellogenin mRNA produces a poly(rA)-oligo(dT) primer at the 3′-OH terminus of the mRNA. Most cDNAs represent partial transcripts of the mRNA templates from their 3′-OH termini and not full length cDNA copies. When long vitellogenin cDNAs are synthesized, we have observed significant radiolytic decomposition on storage (H. J. Baker, *unpublished observations*).

Since the rate of RNA–cDNA hybridization is influenced by the size of the cDNA, we have employed cDNA preparations 1,500 to 3,000 nucleotides in length in these investigations; cDNA in this size range is reasonably stable.

Examination of the kinetics of hybridization of vitellogenin cDNA and vitellogenin mRNA provides us with a useful tool for evaluation of the purity of the isolated vitellogenin mRNA. If the isolated vitellogenin mRNA were contaminated with other poly(A) mRNAs, they should be transcribed into cDNA also. These cDNAs would reassociate with the minor species of mRNA they are transcribed from at higher $C_r t$ (the product of concentration of RNA in moles of nucleotides per liter and time in seconds) values than the major RNA species, resulting in a gradual increase in the extent of hybridization over a wide range of $C_r t$ values. Contaminating RNAs present in livers of unstimulated male *Xenopus laevis* would also reassociate with cDNAs transcribed from these RNAs and would exhibit a gradual increase in the extent of hybridization over a wide range of $C_r t$ values. Vitellogenin cDNA does not undergo significant hybridization with liver RNA from the unstimulated male *Xenopus* even at high $C_r t$ values (Fig. 6; 23), demonstrating that the purified vitellogenin mRNA used as a template for cDNA synthesis

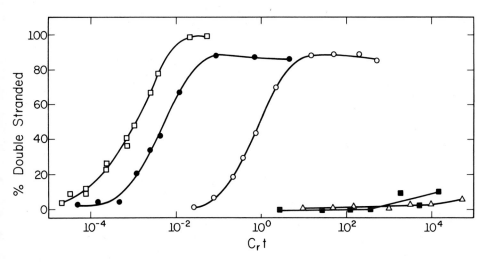

FIG. 6. Hybridization of vitellogenin cDNA to purified mRNA and RNA from induced and control *Xenopus laevis*. Purified vitellogenin mRNA (●) and purified oval-bumin mRNA (□) were hybridized to their respective single-stranded cDNAs and the extent of hybridization quantitated after digestion of unreacted cDNA with S_1 nuclease. The RNA from six induced *Xenopus laevis* (○) that was used to purify vitellogenin mRNA (Table 1), hybridized 163 times more slowly than purified vitellogenin mRNA. RNA from unstimulated male *Xenopus laevis* (■), hen oviduct (△), rabbit reticulocytes and calf thymus and chicken DNA, all hybridized with less than 10% of the vitellogenin cDNA at high $C_r t$ values. For simplicity only the data for control *Xenopus* RNA and oviduct RNA are shown. The percent of the cDNA resistant to digestion by S_1 nuclease (7% for ovalbumin cDNA, 1% for vitellogenin cDNA) was subtracted from all values. (Redrawn from Shapiro and Baker, ref. 23.)

is free of significant contamination by all of the RNA species present in *Xenopus* liver prior to estrogen administration. The extent of hybridization of both purified vitellogenin mRNA and total cell RNA to vitellogenin cDNA is high (approximately 88%) and does not increase when the hybridization is carried to high $C_r t$ values (Fig. 6). In addition, both hybridizations proceed over a 100-fold range in $C_r t$ values (Fig. 6), as would be expected if the vitellogenin cDNA were hybridizing to a single RNA species.

The molecular weight of vitellogenin mRNA when purified ovalbumin mRNA is run as a standard is 2,070,000 on sucrose gradients and 2,000,000 on polyacrylamide gels (equivalent to 6,450 and 6,230 nucleotides, respectively) (23). We have also determined the size of vitellogenin mRNA by nucleic acid hybridization, a method that is independent of secondary structure and does not require an intact RNA molecule. Within broad limits, the rate of hybridization in RNA excess RNA–DNA hybridizations is unaffected by the size of the hybridizing RNA (8). The rate of hybridization of vitellogenin mRNA to its cDNA, therefore, provides an estimate of the number of vitellogenin sequences in a given amount of mRNA. The rate of hybridization of purified vitellogenin mRNA to its cDNA is compared with that of purified ovalbumin mRNA hybridizing to its cDNA in Fig. 6. Purified vitellogenin mRNA hybridized with a $C_r t_{1/2}$ of $4.10 \cdot 10^{-3}$ compared to a $C_r t_{1/2}$ of $1.23 \cdot 10^{-3}$ for purified ovalbumin mRNA. The complexity of vitellogenin mRNA can be calculated by comparison to the $C_r t_{1/2}$ and the complexity of ovalbumin mRNA as follows:

$$\frac{\text{complexity of vitellogenin mRNA}}{\text{complexity ovalbumin mRNA}} = \frac{C_r t_{1/2} \text{ vitellogenin mRNA}}{C_r t_{1/2} \text{ ovalbumin mRNA}}$$

The complexity of vitellogenin mRNA is 6,670 nucleotides, which corresponds to a molecular weight of 2,140,000 (23). These figures are in excellent agreement with the values of 6,450 and 6,230 nucleotides obtained from sucrose gradients and acrylamide gels, respectively. Our data are in reasonable accord with those of Wahli et al. (27), who used contour-length measurements of electron micrographs and found the apparent length of vitellogenin mRNA to be approximately 7,000 nucleotides. *X. laevis* vitellogenin contains approximately 1,700 amino acids (6), and the minimum size of vitellogenin mRNA is, therefore, 5,100 nucleotides. Vitellogenin mRNA, with a size of approximately 6,500 nucleotides, could, therefore, contain more than 1,000 untranslated nucleotides.

QUANTITATION OF VITELLOGENIN mRNA SEQUENCES DURING ESTROGEN INDUCTION BY HYBRIDIZATION TO cDNA

Quantitation of vitellogenin mRNA sequences by hybridization to cDNA offers several advantages over quantitation by *in vitro* translation. Eucaryotic mRNAs may undergo a variety of posttranscriptional modifications including nuclear processing, capping, and polyadenylylation, and estrogen could, in principle, regulate the rate of vitellogenin synthesis by controlling the translational efficiency of vitellogenin mRNA *in vivo* and *in vitro*. A sequence-specific hybridization assay quantitates vitellogenin mRNA molecules independent of their translational efficiency. Hybridization is also far more sensitive than translation and can reliably detect small numbers of RNA sequences. Vitellogenin mRNA sequences need not be intact to hybridize, so that vitellogenin mRNA "nicked" intracellularly or during isolation are still quantitated by hybridization.

The induction of specific mRNAs by steroid hormones has been examined in several systems. These studies have led to four major models that attempt to account for steroid hormone activation of specific gene transcription.

1. "Regulator" genes respond directly and rapidly to steroid–receptor complexes, whereas activation of other genes may require transcription of these regulator genes and synthesis of new gene products. Ashburner has described such a system of "early" and "late" genes based on analysis of puffing patterns in *Drosophila* salivary chromosomes (1).

2. Estrogen activates a type of "developmental cascade" in which the products of earlier genes activate later genes, leading ultimately to the activation of the vitellogenin gene.

3. Palmiter has proposed (17) that the lag in appearance of new ovalbumin mRNA molecules is due to nonspecific binding of receptor to large numbers of low-affinity chromatin-binding sites from which the receptor rapidly dissociates until it finds and binds tightly to the high-affinity regulatory site.

4. Yamamoto and Alberts (34) have suggested that the accumulation of a critical threshold concentration of estrogen receptors at the chromatin binding site is required for gene activation and that this slow accumulation is the rate-limiting step responsible for the lag periods observed for estrogen action.

It was necessary to establish the time at which vitellogenin mRNA and vitellogenin synthesis can first be detected in order to determine which of the above models best accounts for the observed kinetics of vitellogenin mRNA induction. Despite studies in several laboratories (7,29,31,33), the time at which vitellogenin mRNA and vitellogenin synthesis can first be detected had not been clearly established. Wangh and Knowland (29) reported that vitellogenin synthesis was detectable only 1 hr after estrogen administration *in vitro,* whereas Berridge et al. (7) reported a lag of 2 to 3 days in the appearance of vitellogenin in liver and in serum. In contrast, Witliffe and Zelson (33) and Witliffe and Kenney (31) detected vitellogenin in *Xenopus* liver 12 hr after estrogen administration. In our preliminary studies we observed substantial amounts of vitellogenin mRNA and a significant rate of vitellogenin synthesis at our earliest time point—1 day after estrogen administration (22). In these hybridization studies we administered a massive dose of estradiol-17β (0.4 mg/10 g body weight) to saturate the system with estrogen as rapidly as possible. Any lag observed in the induction would therefore be due to some intrinsic feature of the vitellogenin system and not to the absence of a sufficient level of the stimulatory hormone.

Vitellogenin mRNA sequences are undetectable in liver RNA from the unstimulated male *Xenopus* (Figs. 6 and 7) (2,23). Vitellogenin mRNA sequences remain undetectable at 1, 2, and 3 hr after estrogen administration (Fig. 7; Table 2). Four and one-half hours after estrogen administration vitellogenin mRNA sequences are clearly detectable by hybridization (Fig. 7; Table 2; 2). We were able to exclude the possibility that vitellogenin DNA contaminating our RNA preparations made a significant contribution to the hybridization rates observed since vitellogenin is a unique gene present in only two copies per cell (20) and DNA is efficiently solubilized by our sodium acetate washing procedure. The absence of detectable hybridization with RNAs isolated from control *Xenopus laevis* and from *Xenopus laevis* either 1, 2, or 3 hr after estrogen administration provided further evidence for the absence of DNA contamination.

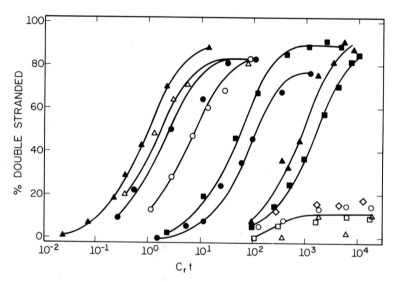

FIG. 7. Kinetics of hybridization of vitellogenin cDNA with *Xenopus* liver RNA during primary estrogen stimulation. RNA isolated from control (uninjected) *Xenopus laevis* (△) and from *Xenopus laevis* either 1 (◇), 2 (○), 3 (□) hr after estrogen stimulation does not contain vitellogenin mRNA and is plotted on a single line. The hybridization curves from right to left are for RNA obtained from animals 4½ ((■), 6 (▲), and 12 (● and ■) hr after estrogen administration and 1 (○), 3 (●), 6 (△), and 12 (▲) days after estrogen stimulation. Duplicate animals were injected, and their livers were pooled for extraction at each time point of induction. RNAs were hybridized to the indicated $C_r t$ values in 10 μl reaction mixtures, that contained 8 to 11 pg single-stranded vitellogenin cDNA and were scored after digestion of single-stranded cDNA with S_1 nuclease. (Redrawn from Shapiro and Baker, ref. 23.)

TABLE 2. *Accumulation of vitellogenin mRNA sequences during primary stimulation with estradiol-17β*

Time after estrogen administration	Number of determinations	$C_r t_{1/2}$	Molecules of vitellogenin mRNA Cell
0 hr	2	$> 10^5$	< 1
1 hr	2	$> 10^5$	< 1
2 hr	2	$> 10^5$	< 1
3 hr	2	$> 10^5$	< 1
4½ hr	2	$1.3 \cdot 10^3$ $(1.3 \cdot 10^3; 1.25 \cdot 10^3)$	16 $(13;19)$
6 hr	2	$8.3 \cdot 10^2$ $(8.0 \cdot 10^2; 8.5 \cdot 10^2)$	25 $(17;34)$
9 hr	2	48 $(42;55)$	335 $(240;430)$
12 hr	2	63 $(41;85)$	265 $(360;170)$
1 day	2	5.0 $(3.6;6.3)$	3,150 $(3,100;3,200)$
3 days	1	2.4	8,900
6 days	1	1.1	23,000
9 days	1	2.2	17,000
12 days	1	1.0	36,000

In most experiments RNA from two male *Xenopus laevis* was pooled and hybridized with cDNA. Typical hybridizations from which these data were calculated are illustrated in Fig. 7. The calculation of the number of molecules of vitellogenin mRNA/cell is as follows. The fraction of an RNA sample that is vitellogenin mRNA is determined by comparison of the $C_r t_{1/2}$ to that of purified vitellogenin mRNA. Since the molecular weight of vitellogenin mRNA is known (2,090,000), we can calculate the number of molecules of vitellogenin mRNA in the sample. The yield of RNA/g of tissue is known from absorbance measurements, and the number of vitellogenin mRNA molecules/g is calculated. The size of the *Xenopus* genome is known, and the amount of DNA/g of liver can be determined by spectrophotometric assay and the number of cells/g calculated. The number of molecules of vitellogenin mRNA/cell is calculated by dividing the number of molecules of vitellogenin mRNA/g by the number of cells/g (see Baker and Shapiro, ref. 2). The data presented at early times after estrogen stimulation represent the average of two separate experiments. The actual values calculated for each experiment are shown in parentheses. All hybridizations for which actual values of $C_r t_{1/2}$ are shown went to at least 70% of completion. (Condensed from Baker and Shapiro, ref. 2, in which methodological details are provided.)

The number of vitellogenin mRNA sequences per cell increases approximately 140-fold between 12 hr and 12 days after estrogen stimulation (Table 2), at which time each liver cell contains approximately 35,000 vitellogenin mRNA sequences and vitellogenin mRNA comprises approximately 0.4% of total cell RNA. The accumulation of vitellogenin mRNA molecules is roughly linear with a rate of approximately 3,000 molecules/cell/day during the 1-day to 12-day time period (2).

The kinetics of vitellogenin mRNA accumulation described in our studies differ substantially from those reported recently by Ryffel et al. (20). In their study vitellogenin mRNA was first detected in *Xenopus* liver RNA slightly less than 12 hr after primary estrogen stimulation and peaked 7 days after estrogen administration. The higher and more concentrated doses of estrogen used in our investigations may be responsible for the earlier appearance of vitellogenin mRNA in our studies and the continued increase in vitellogenin mRNA until 12 days after estrogen stimulation. The appearance of vitellogenin mRNA between 3 and 4½ hr after estrogen stimulation may well represent the ultimate physiologic constraints of the vitellogenin system when a maximal level of hormone is employed. At lower doses of estradiol-17β several hours may be required for the attainment of a sufficient level of estradiol-17β to stimulate liver synthesis of vitellogenin mRNA.

Maintenance of a high level of vitellogenin mRNA requires the continued presence of estradiol-17β. If additional estradiol-17β is not administered, the level of vitellogenin mRNA in the liver of the male *Xenopus* declines approximately 12,000-fold between 12 and 50 days after a single injection of estradiol-17β, when approximately 3 molecules of vitellogenin mRNA/cell are present. Sixty days after administration of estrogen, the level of vitellogenin mRNA is below 1 molecule/cell, and we conclude that there is no significant transcription of the vitellogenin gene at this time.

We next examined the effect of restimulation with estrogen (secondary stimulation) on these male *Xenopus laevis,* which were previously induced to synthesize vitellogenin mRNA but were inactive in vitellogenin mRNA

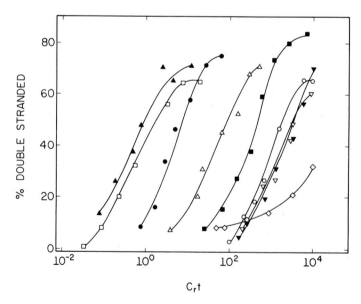

FIG. 8. Kinetics of hybridization of vitellogenin cDNA with liver RNA following secondary estrogen stimulation. Male *X. laevis* that were without estradiol-17β for 60 days were restimulated with 0.4 mg/10 g body weight and RNA extracted at 0 hr (◇), 1 hr (△), 2 hr (▲), 3 hr (○), 4½ hr (■), 6 hr (△), 1 day (●), 6 days (□), and 12 days (▲) after restimulation with estradiol-17β. Hybridization curves are for RNA from individual animals. (From Baker and Shapiro, ref. 3.)

synthesis at the time of restimulation. Our objective was to ascertain whether prior stimulation of vitellogenin mRNA synthesis with estrogen alters the pattern of vitellogenin gene expression on restimulation with estrogen. Sixty days after primary estrogen stimulation male *Xenopus laevis* were restimulated with estradiol-17β and the accumulation of vitellogenin mRNA monitored. New vitellogenin mRNA sequences are detected in some but not all animals as early as 1 hr, and in all animals within 2 hr, after secondary estrogen stimulation (Fig. 8; Table 3). The accumulation of vitellogenin mRNA sequences early in secondary stimulation is compared to that observed early in primary stimulation in Fig. 9. Vitellogenin mRNA sequences accumulate about 10 times more rapidly in early secondary stimulation than in early primary stimulation. Sequence accumulation is best described by a biphasic curve (Fig. 9). Accelerated accumulation of vitellogenin mRNA sequences is maintained throughout secondary induction (Table 3) with a rate of about 19,000 molecules/cell/day or 13 molecules/cell/min (about six times the maximum observed in primary stimulation). The peak level of vitellogenin mRNA, 90,000 molecules/cell, is reached at 6 days and is twice the peak level that is reached 12 days after primary stimulation (2).

Although these studies actually measure the number of vitellogenin mRNA molecules at any given time, they provide important information on the rate of vitellogenin mRNA synthesis and degradation also. The minimum possible rate of vitellogenin mRNA synthesis during secondary stimulation (assuming no degradation of the mRNA during this period) is 13 molecules/cell/min or 7 molecules/gene copy/min, since vitellogenin is a unique gene present

TABLE 3. *Accumulation of vitellogenin mRNA sequences during secondary stimulation with estradiol-17β*

Time after estrogen administration	Number of determinations	$C_r t_{1/2}$	Molecules of vitellogenin mRNA Cell
0 hr	4	$> 4.0 \cdot 10^4$	< 1 (0.4 $\pm$ 0.1)
1 hr	4	$1.1 \cdot 10^4$	2.6 $\pm$ 1.5
2 hr	3	$3.8 \cdot 10^3$	4.4 $\pm$ 0.6
3 hr	4	$2.8 \cdot 10^3$	11 $\pm$ 5.6
4½ hr	4	$5.5 \cdot 10^2$	69 $\pm$ 37
6 hr	3	58	256 $\pm$ 94
12 hr	4	20	1,095 $\pm$ 234
1 day	2	5.1	2,900 (2,700;3,100)
6 days	2	0.43	96,000 (71,000;120,000)
12 days	2	0.56	82,000 (70,000;94,000)

RNA from single animals was extracted and hybridized with vitellogenin cDNA. Typical hybridizations from which these data are calculated are illustrated in Fig. 8. The calculation of values for $C_r t_{1/2}$ and molecules/cell is as described in the legend to Table 2 and Baker and Shapiro (2,3). (Condensed from Baker and Shapiro, ref. 3, in which methodological details are provided.)

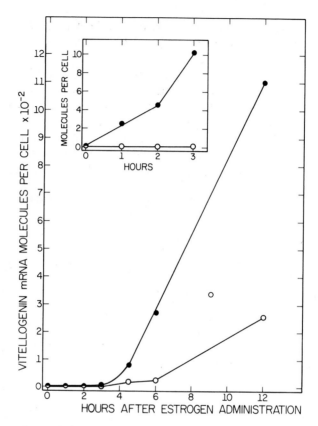

FIG. 9. Comparison of vitellogenin mRNA sequence accumulation early in primary and secondary stimulation. RNA was isolated at the indicated times during primary (○) and secondary estrogen stimulation (●) and hybridized to vitellogenin cDNA. The number of molecules of vitellogenin mRNA was calculated as outlined in the legend to Table 2. (Redrawn from Baker and Shapiro, ref. 3.)

in two copies per cell (20). The maximum rate at which a eucaryotic gene may be transcribed was estimated by Kafatos (12) to be 32 to 48 transcript/min at 37°C and 12 to 18 transcripts/min at 25°C; by extension, this would be 10 to 14 transcripts/min at 22°C. The observed minimum rate of vitellogenin gene transcription of 7 molecules/min approaches the maximum rate possible and, given the temperature differential, is comparable to the rates observed in other systems (16). The high rate of vitellogenin mRNA accumulation during secondary stimulation suggests that vitellogenin mRNA must be relatively stable. If vitellogenin mRNA were degraded with a half-life of only a few hours, transcription of the vitellogenin gene at the maximum theoretical rate would be insufficient to permit the observed rate of vitellogenin mRNA accumulation. Preliminary data from the labeling experiments of Wahli et al. (27) also suggest that vitellogenin mRNA is relatively stable during induction.

These studies demonstrate that activation of vitellogenin gene transcription in primary estrogen stimulation results in a stable alteration in the pattern of future expression of the vitellogenin gene as manifested by: (a) activation of gene transcription and mRNA accumulation, as early as 1 hr after estrogen stimulation (compared to 4½ hr in primary stimulation), (b) accelerated accumulation of vitellogenin mRNA molecules at a rate about six times that observed in primary stimulation and approaching the maximum possible rate of gene transcription, and (c) accumulation of 90,000 molecules of vitellogenin mRNA per cell—approximately twice the level observed with the same dose of estradiol-17β following primary stimulation.

INDUCTION OF VITELLOGENIN SYNTHESIS *IN VITRO*

A major obstacle to the investigation of the molecular basis of estrogen action has been the inability to elicit tissue responses to estrogens in a defined *in vitro* environment, although some partial *in vitro* responses such as the synthesis of "induced protein" (13) have been achieved. The induction of vitellogenin synthesis is unique in that vitellogenin synthesis can be efficiently induced *in vitro* by physiologic levels of estradiol-17β (4,11,29). The *in vitro* induction systems are particularly well suited to the study of estrogen and antiestrogen effects and to a variety of experiments involving radioisotope labeling techniques and the use of inhibitors of protein and RNA synthesis.

Although the extent of induction *in vitro* is substantial, the maximum rate of vitellogenin synthesis is far lower than is observed *in vivo* and rarely exceeds 20% of cell protein synthesis (4,11,29). The relative rate of vitellogenin synthesis is similar when diethylstilbesterol or estradiol-17β is used as an *in vitro* inducer (P. Stock and D. J. Shapiro, *unpublished observations*). We routinely use estradiol-17β to induce vitellogenin mRNA *in vitro* to facilitate comparison of our *in vivo* and *in vitro* data.

Our early *in vitro* studies were designed to reexamine the question of control of vitellogenin mRNA synthesis and translation during induction, and to ascertain whether secondary stimulation *in vitro* alters the pattern of vitellogenin gene expression as is observed *in vivo*. The rate of vitellogenin synthesis during secondary stimulation *in vitro* as well as the fraction of the cell RNA that is vitellogenin mRNA is substantially higher than is observed during primary stimulation *in vitro* (Fig. 10; 4). The accumulation of vitellogenin mRNA is slower than in secondary stimulation, and the level of vitellogenin mRNA peaks after about 8 days when vitellogenin synthesis represents approximately 10 to 15% of cell protein synthesis (4).

The limited solubility of estradiol-17β in aqueous media makes it impossible to employ *in vitro* the very high pharmacologic doses of estradiol-17β we have administered *in vivo*. Perhaps as a consequence of the lower concentration of estradiol-17β employed in our *in vitro* studies, the appear-

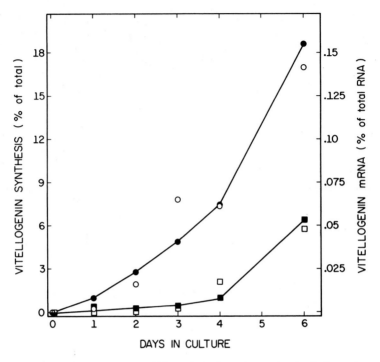

FIG. 10. Induction of vitellogenin mRNA and vitellogenin synthesis during primary and secondary stimulation *in vitro*. Liver cubes were prepared from unstimulated and from withdrawn male *Xenopus laevis* and cultured at 25°C in medium containing 10^{-6} M estradiol-17β. Vitellogenin mRNA is undetectable in control liver cubes cultured in estrogen-free medium. Liver cubes were removed from the culture medium at the indicated times and pulse-labeled in leucine-free medium. The relative rate of vitellogenin synthesis was determined immunochemically. RNA was extracted from liver cubes by standard techniques (2) and hybridized with vitellogenin cDNA. The percentage of the cell RNA represented by vitellogenin mRNA was obtained by comparison of the observed values of $C_r t_{1/2}$ to those of purified vitellogenin mRNA. The data are plotted using only the data for the relative rates of vitellogenin synthesis during primary (■) and secondary (●) stimulation *in vitro*. The data for the vitellogenin mRNA content of cells during primary (□) and secondary (○) stimulation in cultures are accurately represented by these lines also.

ance and early accumulation of vitellogenin mRNA are much slower than are observed *in vivo* (4; Fig. 10). Nevertheless, during secondary stimulation *in vitro,* vitellogenin mRNA appears earlier and accumulates more rapidly than it does during primary stimulation *in vitro* (Fig. 10). Restimulation of vitellogenin mRNA synthesis *in vitro* elicits the same altered pattern of vitellogenin gene expression observed in whole animal studies, indicating that it is the liver cells themselves that have been altered in some way.

Our *in vitro* investigations provide no support for the hypothesis that pronounced variations in the efficiency with which vitellogenin mRNA is translated in hepatocytes make a significant contribution to the observed kinetics of induction of vitellogenin synthesis. The rate of vitellogenin synthesis in both primary and secondary estrogen stimulation *in vitro* is an accurate reflection of the rate of vitellogenin mRNA accumulation in liver cells (Fig. 10; 4). The pronounced lag in the onset of vitellogenin synthesis during primary stimulation *in vitro* is not due to the inability to translate newly synthesized vitellogenin mRNA—it represents a lag in the synthesis of new vitellogenin mRNA (Fig. 10).

DISCUSSION

Administration of estradiol-17β to the male *X. laevis,* which does not ordinarily synthesize vitellogenin, evokes the massive and prolonged synthesis of vitellogenin. Vitellogenin synthesis represents up to 70% of cell protein synthesis in fully induced animals (23,26). The absence of vitellogenin mRNA in unstimulated and in withdrawn male *Xenopus laevis* demonstrates that a large pool of sequestered vitellogenin mRNA is not responsible for induction of vitellogenin mRNA synthesis in either primary or secondary estrogen stimulation. We have also observed a close correspondence between the vitellogenin mRNA content of hepatocytes and the rate of vitellogenin synthesis. We, therefore, conclude that regulation in this system occurs primarily at the level of cell mRNA content and that translation control plays, at best, a secondary role.

Hybridization experiments measure the number of vitellogenin mRNA sequences, not the rate of vitellogenin mRNA synthesis and degradation. The data suggest that selective transcription of the vitellogenin gene plays a major role in the estrogen-induced accumulation of vitellogenin mRNA sequences. Vitellogenin mRNA is undetectable in the livers of control male *Xenopus laevis* and must be present at much less than 1 molecule/cell (Figs. 6 and 7; 2,23). If vitellogenin gene transcription were occurring at rates comparable to those observed during secondary stimulation and failure to detect vitellogenin mRNA were exclusively due to rapid intranuclear breakdown of newly synthesized RNA, the half-life of newly synthesized vitellogenin mRNA would have to be on the order of a few seconds or less (2,3). Regulation based exclusively on changes in rates of mRNA degradation seems inefficient, but the existence of this type of control cannot be completely excluded in this and other hormonally regulated systems (16). Selective stabilization of vitellogenin mRNA against cytoplasmic degradation may well play a significant role in the accumulation of the high levels of vitellogenin mRNA observed in fully induced *Xenopus* liver since preliminary data suggest that vitellogenin mRNA is extremely stable in induced *Xenopus laevis* (3,27).

Vitellogenin mRNA appears between 3 and 4½ hr after primary estrogen stimulation and approximately 1 hr after secondary estrogen stimulation. These data support the hypothesis that induction of vitellogenin mRNA synthesis results from a direct interaction of an estrogen–receptor complex with chromatin structures regulating vitellogenin gene expression. Our view that induction of vitellogenin mRNA represents the liver's initial response to estrogen is consistent with the data of Witliff et al. (32), who observed that *Xenopus* liver RNA synthesis begins to increase between 3 and 6 hr after primary estrogen stimulation.

The appearance of vitellogenin mRNA as early as 1 hr after secondary stimulation with estrogen is particularly striking for several reasons. *X. laevis* toads are maintained at 22°C, and the rate of many physiologic processes is two to five times slower than in 37°C organisms. Consequently, the appearance of vitellogenin mRNA in 1 hr may be roughly equivalent to its appearance in 15 to 20 min in 37°C organisms. The induction of vitellogenin mRNA requires transport of the estradiol-17β from the site of injection to the liver where it must accumulate to a level sufficient to elicit translocation of the estrogen receptor to the nucleus. Translocation of the estrogen receptor into the nucleus and attainment of a steady-state level of nuclear receptors require 15 to 20 min at 37°C in chick oviduct (15) and less than 10 min in rat uterus (30). The attainment of a critical level of nuclear estrogen receptor may well consume most of the 1 hr required for appearance of vitellogenin mRNA. The nuclear receptor must then seek out and interact with the appropriate regulator region of the chromatin and initiate transcription of the vitellogenin gene. Finally, detection of vitellogenin mRNA by hybridization requires synthesis of the complete 6,500 nucleotide vitellogenin mRNA chain since the cDNA is a copy of 2,000 to 2,500 nucleotides at the 3'-OH terminus of the RNA (2,3,23) and transcription presumably begins at the 5'-OH end of the molecule. The 1-hr lag period for appearance of vitellogenin mRNA is much shorter than the 2- to 4-hr lag periods observed for secondary stimulation of ovalbumin synthesis (17) and in several other steroid sensitive systems (16,34). An interval of only 1 hr between the administration of estradiol-17β and the appearance of vitellogenin mRNA seems insufficient to permit the activation of a regulatory gene and the synthesis of a regulatory protein that controls vitellogenin synthesis. Our data are most consistent with the view that activation of vitellogenin gene transcription is an initial effect of estrogen on *Xenopus* liver and does not require prior expression of other estrogen-sensitive genes.

Secondary stimulation of withdrawn *Xenopus* organisms evokes the rapid appearance of new vitellogenin mRNA, accelerated accumulation of vitellogenin mRNA sequences, and an increase in the maximum number of mRNA sequences present. This dramatic alteration in the pattern of vitellogenin gene expression may represent a consequence of a cellular estrogen "memory effect." The duration of this cellular memory effect suggests that it is not due to the synthesis of new proteins (such as the estrogen receptor) since most proteins undergo intracellular degradation within a few days. The nature of the cellular alteration induced by prior administration of estrogen remains to be elucidated.

The induction of massive amounts of *Xenopus* vitellogenin mRNA by estradiol-17β without cytodifferentiation and without production of a new cell type in both primary and secondary estrogen stimulation and the fact that induction can occur *in vivo* and *in vitro* suggest that this is an unusually favorable system for investigation of the molecular mechanisms of reproductive hormone action.

REFERENCES

1. Ashburner, M. (1974): Sequential gene activation by ecdysone in polytene chromosomes of *Drosophila melanogaster* II. The effects of inhibitors of protein synthesis. *Dev. Biol.*, 39:141–157.
2. Baker, H. J., and Shapiro, D. J. (1977): Kinetics of estrogen induction of *Xenopus laevis* vitellogenin messenger RNA determined by hybridization to complementary DNA. *J. Biol. Chem.*, 252:8428–8434.
3. Baker, H. J., and Shapiro, D. J. (1978): Rapid accumulation of vitellogenin messenger RNA during secondary estrogen stimulation of *Xenopus laevis*. *J. Biol. Chem.*, 253:4521–4524.
4. Baker, H. J., and Shapiro, D. J. (1979): Efficient *in vitro* induction of *Xenopus laevis* vitellogenin messenger RNA by estradiol-17β. (*In preparation.*)
5. Bartholeyns, J., Peeters-Joris, C., Reychler, H., and Baudhuin, P. (1975): Hepatic nucleases. I: Methods for the specific determination and characterization in rat liver. *Eur. J. Biochem.*, 57:205–211.
6. Bergink, E. W., and Wallace, R. A. (1974): Precursor-product relationship between amphibian vitellogenin and phosvitin. *J. Biol. Chem.*, 249:2897–2903.
7. Berridge, M. V., Farmer, S. R., Green, C. D., Henshaw, E. D., and Tata, J. R. (1976): Characterization of polysomes from *Xenopus* liver synthesizing vitellogenin and translation of vitellogenin and albumin messenger RNAs *in vitro*. *Eur. J. Biochem.*, 62:161–172.
8. Birnsteil, M. L., Sells, B. H., and Purdom, I. F. (1972): Kinetic complexity of RNA molecules. *J. Mol. Biol.*, 63:21–39.
9. Ellison, J., and Shapiro, D. J. (1978): Analysis and inhibition of the ribonucleases of *Xenopus laevis* liver. (*Undergraduate thesis.*)
10. Follett, B. K., Nicholls, T. J., and Redshaw, M. R. (1968): The vitellogenic response in the South African clawed toad (*Xenopus laevis* Daudin). *J. Cell. Physiol.*, 72(Suppl.):91–105.
11. Green, C. D., and Tata, J. R. (1976): Direct induction by estradiol of vitellogenin synthesis in organ cultures of male *Xenopus laevis* liver. *Cell*, 7:131–139.
12. Kafatos, R. (1972): The cocoonase zymogen cells of silk moths: A model for terminal cell differentiation for specific protein synthesis. *Curr. Top. Dev. Biol.*, 7:125–192.
13. Katzenellenbogen, B. S., and Gorski, J. (1972): Estrogen action *in vitro:* induction of the synthesis of a specific uterine protein. *J. Biol. Chem.*, 247:1299–1305.
14. Katzenellenbogen, B. J., and Gorski, J. (1975): Estrogen actions on syntheses of macromolecules in target cells. In: *Biochemical Actions of Hormones*, Vol. 3, edited by G. Litwack, p. 187. Academic Press, New York.
15. Mulvihill, E., and Palmiter, R. D. (1977): Relationship of nuclear estrogen receptor levels to induction of ovalbumin and conalbumin mRNA in chick oviduct. *J. Biol. Chem.*, 252:2060–2068.
16. Palmiter, R. D. (1975): Quantitation of parameters that determine the rate of ovalbumin synthesis. *Cell*, 4:189–197.
17. Palmiter, R. D., Moore, P. B., Mulvihill, E. R., and Emtage, S. (1976): A significant lag in the induction of ovalbumin messenger RNA by steroid hormones: A receptor translocation hypothesis. *Cell*, 8:557–572.
18. Pelham, H., and Jackson, R. J. (1976): An efficient mRNA-dependent translation system from reticulocyte lysates. *Eur. J. Biochem.*, 67:247–256.

19. Rosen, J., and O'Malley, B. W. (1975): Hormonal regulation of specific gene expression in the chick oviduct. In: *Biochemical Actions of Hormones,* Vol. 3, edited by G. Litwack, p. 271. Academic Press, New York.

20. Ryffel, G. V., Wahli, W., and Weber, R. (1977): Quantitation of vitellogenin messenger RNA in the liver of male *Xenopus* toads during primary and secondary stimulation by estrogen. *Cell,* 11:213–221.

21. Schimke, R. T., McKnight, G. S., and Shapiro, D. J. (1975): Nucleic acid probes and analysis of hormone action in oviduct. In: *Biochemical Actions of Hormones,* Vol. 3, edited by G. Litwack, p. 245. Academic Press, New York.

22. Shapiro, D. J., Baker, H. J., and Stitt, D. T. (1976): *In vitro* translation and estradiol-17β induction of *Xenopus laevis* vitellogenin messenger RNA. *J. Biol. Chem.,* 251:3105–3111.

23. Shapiro, D. J., and Baker, H. J. (1977): Purification and characterization of *Xenopus laevis* vitellogenin messenger RNA. *J. Biol. Chem.,* 252:5244–5250.

24. Shapiro, D. J., and Schimke, R. T. (1975): Immunochemical isolation and characterization of ovalbumin messenger RNA. *J. Biol. Chem.,* 250:1759–1764.

25. Shapiro, D. J., Taylor, J. M., McKnight, G. S., Palacios, R., Gonzales, C., and Schimke, R. T. (1975): Isolation of hen oviduct ovalbumin and rat liver albumin polysomes by indirect immunoprecipitation. *J. Biol. Chem.,* 249:3665–3671.

26. Skipper, J. K., and Hamilton, T. H. (1977): Regulation by estrogen of the vitellogenin gene. *Proc. Natl. Acad. Sci. USA,* 74:2384–2388.

27. Wahli, W., Wyler, T., Weber, R., and Ryffel, G. U. (1976): Size, complexity and abundance of a specific poly(A)-containing RNA of liver from male *Xenopus* induced to vitellogenin synthesis by estrogen. *Eur. J. Biochem.,* 66:457–465.

28. Wallace, R. A. (1972): The role of protein uptake in vertebrate oocyte growth and yolk formation. In: *Oogenesis,* edited by J. D. Biggers and A. W. Schultz, pp. 339–359. University Park Press, Baltimore.

29. Wangh, L. J., and Knowland, J. (1975): Synthesis of vitellogenin in cultures of male and female frog liver regulated by estradiol treatment *in vitro. Proc. Natl. Acad. Sci. USA,* 72:3172–3175.

30. Williams, D., and Gorski, J. (1972): Kinetic and equilibrium analysis of estradiol in uterus: a model of binding site distribution in uterine cells. *Proc. Natl. Acad. Sci. USA,* 69:3464–3468.

31. Witliffe, J. L., and Kenney, F. T. (1972): Regulation of yolk protein synthesis in amphibian liver: I. Induction of lipovitellin synthesis by estrogen. *Biochim. Biophys. Acta,* 269:485–492.

32. Witliffe, J. L., Lee, K. L., and Kenney, F. T. (1972): Regulation of yolk protein synthesis in amphibian liver: II. Elevation of ribonucleic acid synthesis by estrogen. *Biochim. Biophys. Acta,* 269:493–504.

33. Witliffe, J. L., and Zelson, P. R. (1974): Estrogen specificity of the induction of lipovitellin synthesis and evidence for a specific estrogen-binding component. *Endocr. Res. Commun.,* 1:117–131.

34. Yamamoto, K. R., and Alberts, B. M. (1976): Steroid receptors: elements for modulation of eukaryotic transcription. *Annu. Rev. Biochem.,* 45:722–746.

Ontogeny of Receptors and Reproductive Hormone Action,
edited by T. H. Hamilton, J. H. Clark, and W. A. Sadler.
Raven Press, New York 1979.

Hormonal Regulation of Lipoprotein Synthesis in the Cockerel

Lawrence Chan, L. Dale Snow, Richard L. Jackson, and
Anthony R. Means

*Departments of Cell Biology and Medicine, Baylor College of Medicine,
Houston, Texas 77030*

Multiple genetic and environmental factors are known to modulate lipoprotein metabolism (14,28,38). One environmental agent which exerts a profound influence on the plasma levels of lipoproteins is the female steroid hormone estrogen. Women taking oral contraceptives were found to have higher plasma triglycerides as well as very low density lipoproteins (VLDL) than those who were not on the pill (12,15,40,50,51).

The mechanisms by which estrogen elevates plasma triglycerides and VLDL levels are complex (19). Although there is general agreement that triglyceride production is probably increased by estrogen (17,30,31), considerable controversy exists with respect to the effect of the hormone on triglyceride removal. Both normal as well as impaired removal have been reported (17,19,20,21,30,31). Studies on changes in postheparin lipolytic activity (PHLA) which might function in triglyceride removal are not helpful in this respect: estrogen seems to affect the ease of release of the lipolytic activity by heparin (13,20) and studies of PHLA may not reflect true total lipolytic activity *in vivo*. Furthermore, PHLA represents at least two enzymatic activities, hepatic triglyceride lipase and extrahepatic lipoprotein lipase (32). The effects of estrogen on these two enzymatic activities may be discordant (18,19). Progesterone, a common component of oral contraceptives, appears to increase PHLA, and to lower triglyceride levels by accelerating its removal (16,31).

Although a number of studies have indicated that plasma VLDL levels are elevated by estrogen, there is little information on the levels of specific apolipoproteins in women taking estrogen (19,51). The level of apoA-I, a major high density lipoprotein (HDL) apoprotein, appears to be increased in these individuals (19). The changes in VLDL and HDL and probably of their apoproteins produced by estrogen also vary according to the age of the individual (49). To date, there is no information on the effects of the sex steroids on the rate of synthesis of any of the human apolipoproteins.

THE COCKEREL AS A MODEL FOR THE STUDY OF ESTROGEN-INDUCED HYPERLIPOPROTEINEMIA

Over 20 years ago, Hillyard et al. (22) first reported that estrogen injections resulted in markedly elevated levels of plasma triglyceride and VLDL in the avian species. The lipemic response was massive, and the experimental animals developed lipoprotein levels in excess of 10 mg/ml (22,23). The initial report has been confirmed by a number of laboratories (11,33,38). The chick is similar to the human in that estrogen appears to result in an increased production of triglyceride (11,34) as well as an increased synthesis of VLDL (38).

Although differences likely exist between the avian and the mammalian species, the estrogen-treated cockerel is a valuable model for studies on steroid hormone regulation of lipoprotein metabolism. The untreated lipoprotein patterns in this animal are quite similar to those of man (33). The VLDL response in this species is much more marked than that in rats or mice, rendering its detection much easier. Furthermore, the effect of estrogen on VLDL appears to be selective: plasma albumin and HDL are relatively unaffected. We have thus employed the cockerel as a model for our studies on the steroid hormone regulation of lipoprotein metabolism. The same system also serves as a valuable model to investigate the mechanism of action of estrogen in the liver at the molecular level.

PURIFICATION AND CHARACTERIZATION OF ApoVLDL-II

Very low density lipoproteins in the cockerel, similar to those of other animals, are heterogeneous in protein composition. To ensure that we are studying a specific and well-defined end-product of estrogen action, we have purified a major VLDL apoprotein from both the hen and the estrogen-

TABLE 1. *Steps in purification of ApoVLDL-II from hen plasma*

---

1. VLDL isolated by ultracentrifugal flotation[a] at density 1.006 g/ml.
2. Lyophilization of the VLDL followed by delipidation.[b]
3. Chromatography on Sepharose G-150 in buffer A.[c]
4. DEAE-cellulose chromatography.[d]
5. Chromatography on Sephadex G-75 in buffer B.[e]

---

[a] Ultracentrifugal flotation was performed twice in KBr (26).
[b] Delipidation was performed by extraction twice with diethyl ether-ethanol (3:1).
[c] Buffer A: 0.1 M Tris HCl, pH 8.0, 2 mM Na decyl sulfate, 0.01% EDTA.
[d] Column was equilibrated with 0.01 M Tris-HCl, pH 8.0, in 6 M urea (Buffer B). Elution was performed with a 0–0.125 M NaCl gradient.
[e] Usually chromatography on Sephadex G-75 reveals only a single protein peak. It is, however, always included in our purification procedure since occasional preparations will show small amounts of contamination by other proteins which will be removed by this step.

Lys – Ser – Ile – Ile – Asp – Arg – Glu – Arg – Arg – Asp – Trp – Leu – Val – Ile – Pro

5 10 15

Asp – Ala – Ala – Ala – Ala – Tyr – Ile – Tyr – Glu – Ala – Val – Asn – Lys – Val – Ser

20 25 30

Pro – Arg – Arg – Ala – Gly – Glu – Phe – Leu – Leu – Asp – Thr – Val – Ser – Gln – Thr

35 40 45

Val – Val – Ser – Gly – Ile – Arg – Asn – Phe – Leu – Ile – Asn – Thr – Ala – Glu – Arg

50 55 60

Leu – Thr – Lys – Leu – Ala – Glu – Gln – Leu – Met – Glu – Lys – Ile – Lys – Asp – Leu

65 70 75

Cys – Tyr – Thr – Lys – Val – Leu – Gly

80

FIG. 1. Primary amino acid sequence of apoVLDL-II (26).

treated cockerel (8) (Table 1). The apoprotein, designated apoVLDL-II, contains two identical polypeptides of 82 amino acid residues each which are linked by a single disulfide bond at residue 76. The primary amino acid sequence of apoVLDL-II has been determined (26) (Fig. 1). An interesting feature of the primary structure of the peptide is that it contains a stretch of amphipathic helix (residues 53 to 75) typical of many plasma lipoproteins (25,29,44). The amino acid composition of the plasma apoVLDL-II is identical in hen and in the estrogen-treated cockerel.

To characterize apoVLDL-II immunologically, antisera to the purified protein were produced in goats. The antisera produced single precipitin lines of identity against purified apoVLDL-II, reduced S-carboxymethylated apoVLDL-II, hen and cockerel VLDL, LDL, and hen plasma; there was no reaction against HDL, albumin, or apoVLDL-I, the other major VLDL apoprotein. Similarly, antisera produced against chick apoA-I, a major HDL protein, or albumin did not cross-react with apoVLDL-II. Hence, biologically and immunologically apoVLDL-II is a distinct protein and a basic understanding of its primary structure as well as its immunochemical properties has been helpful in our subsequent studies on its biosynthesis.

EFFECTS OF FEMALE STEROID HORMONES ON PLASMA VLDL

Plasma VLDL levels were relatively low in immature cockerels. Following a single injection of estrogen, there was an initial drop in VLDL concentrations which lasted from about 2 hr to 4 or 5 hr after treatment; both the magnitude and the duration of this initial fall in VLDL varied directly with the dose of hormone administered. This fall was immediately followed by a rapid increase in the concentration of VLDL (Fig. 2). Depending on the dose and preparation of estrogen, VLDL levels eventually returned toward baseline (8).

Progesterone is an estrogen antagonist in a number of experimental systems (9,23,36) and possesses a protective effect against estrogen-induced hyperlipoproteinemia in women (16,31). We have studied the effect of this hormone on the VLDL response to estrogen in the cockerel. In this animal, the concomitant administration of progesterone did not significantly alter the plasma VLDL response (Fig. 3) (6). We next studied another estrogen antagonist with a different mode of action, nafoxidine hydrochloride (Table 2). A low dose of nafoxidine-HCl was without any effect on plasma lipids

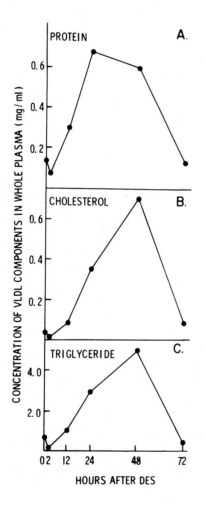

FIG. 2. Effects of estrogen on plasma VLDL protein and lipids. Cockerels in groups of four were given a subcutaneous injection of DES 2.5 mg in sesame oil. At the times indicated the animals were decapitated and blood was collected in 0.1% EDTA, pH 7.5. Erythrocytes were removed by low speed centrifugation. VLDL were prepared by ultracentrifugal flotation as previously described (8). Protein was determined by the methods of Lowry et al. (37). Total cholesterol and triglycerides were determined by autoanalyzer techniques (39). (From Chan et al. (8), with permission.)

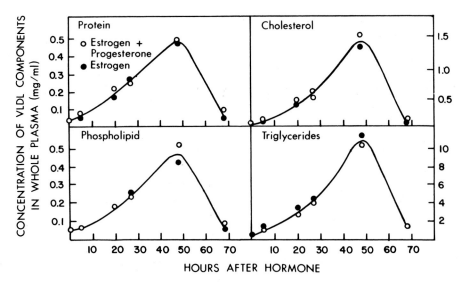

FIG. 3. Effect of estradiol and estradiol + progesterone on plasma VLDL. Groups of four cockerels were given a single subcutaneous injection of estradiol alone (1 mg) or estradiol (1 mg) plus progesterone (2 mg). At various times after hormone treatment, the animals were sacrificed and blood from each group of animals was pooled. VLDL were isolated from plasma by ultracentrifugal flotation. Protein was determined by the method of Lowry et al. (37). Total cholesterol and triglycerides were determined by autoanalyzer techniques (39), and phospholipids by the method of Bartlett (2). (From Chan et al. (6), with permission.)

or lipoproteins. A high dose (25 mg/kg) caused slight elevations of plasma triglyceride and VLDL. The stimulatory effect of nafoxidine-HCl is consistent with the previous observation in the uterus that this compound by itself is a weak estrogen agonist (10). The high dose of nafoxidine-HCl was found to be effective in inhibiting the plasma lipid response to estradiol,

TABLE 2. *Effects of nafoxidine on estrogen-induced hyperlipidemia*

| | VLDL (mg/ml plasma) | | |
Treatment	Cholesterol	Triglycerides	Protein
Oil	0.055	0.48	0.11
Estradiol (1 mg)	0.29	2.05	0.62
Estradiol (1 mg) + nafoxidine-HCl (2 mg)	0.16	1.26	0.40
Estradiol (1 mg) + nafoxidine-HCl (5 mg)	0.08	0.57	0.23
Nafoxidine-HCl (2 mg)	0.09	0.57	0.10
Nafoxidine-HCl (5 mg)	0.04	0.47	0.15

Plasma lipids were determined on pooled plasma from three animals 16 hr following the indicated treatment. VLDL were isolated by ultracentrifugation at plasma density (2×10^8 g-min) as described in Fig. 3. Cholesterol and triglyceride were assayed by autoanalyzer methods (39). Protein was determined by the method of Lowry et al. (37). The recovery of VLDL from total plasma by the ultracentrifugal flotation method was about 60%. (Modified from Chan et al. (6) with permission.)

resulting in a 70% reduction in the net increase in plasma triglyceride and a 77% reduction in the net increase in VLDL proteins 16 hr following the administration of the hormones (Table 2) (6).

EFFECTS OF THE SEX STEROIDS ON VLDL SYNTHESIS

As emphasized in the introduction, the effects of estrogen and progesterone on VLDL metabolism are complex; experimental evidence suggests that both synthesis and removal of the lipoproteins could be affected. Hence, the hyperlipoproteinemic response of the cockerels to estrogen could reflect alterations either in the rate of synthesis of VLDL or in their rate of degradation. We have carried out studies on the effects of the hormone on the synthesis of VLDL in the cockerel liver (8).

Cockerels were given a single injection of estrogen *in vivo*. At various times thereafter, they were sacrificed and their livers removed. Slices of liver were incubated in culture medium *in vitro* in the presence of L-[^{3}H]-lysine for 2 hr, during which time incorporation of the label into protein was linear. The rate of VLDL synthesis was thus measured by the total incorporation of radioactivity into VLDL. As shown in Fig. 4, the rate increased within 2 hr of hormone treatment and reached a maximum at 24 hr, gradually returning to baseline by 72 hr. At the peak of synthesis, VLDL represented about 11% of the total newly synthesized soluble proteins (8). The effects of estrogen on total protein synthesis were much more variable and less marked. When radioactivity incorporated into VLDL was determined in ultracentrifugally prepared VLDL from the liver homogenates after extensive dialysis, the same relative increase in radioactivity after estrogen was observed. This experiment suggests that the immunoprecipitated VLDL did in fact represent VLDL as defined by their flotation properties (8).

When the rate of VLDL synthesis was determined in liver slices after estrogen plus progesterone, no significant difference from the rate after estrogen alone was observed (6). Nafoxidine-HCl, on the other hand, completely inhibited the VLDL biosynthetic response to estrogen (6).

The experimental data on the synthesis of VLDL after the different hormonal manipulations suggest that estrogen has a major action on the synthesis of VLDL. Since nafoxidine-HCl completely inhibits both the stimulation of VLDL synthesis by, and the plasma VLDL response to, estrogen, it is unlikely that alterations in VLDL degradation play a significant role in estrogen-induced hyperlipoproteinemia. Furthermore, as we have pointed out, plasma triglycerides showed a paradoxical initial fall after estrogen. This fall took place when VLDL biosynthetic rate started to increase (see Figs. 2 and 4 and below). This discrepancy in VLDL synthesis and triglyceride kinetics suggests that estrogen acts primarily on VLDL synthesis and the subsequent rise in triglyceride is a late or secondary effect of the hormone.

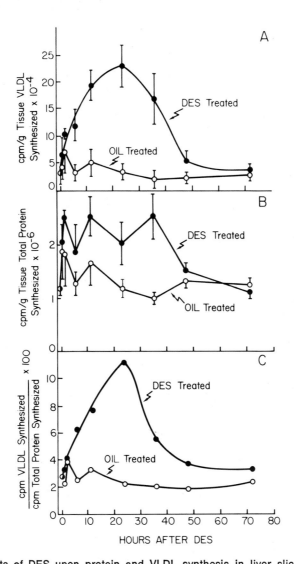

FIG. 4. Effects of DES upon protein and VLDL synthesis in liver slices. Groups of four cockerels were treated with a single subcutaneous injection of either DES (2.5 mg) or sesame oil. At the indicated times after injection, 50-mg liver slices were prepared and incubated *in vitro* (200 to 350 mg/flask) in Medium 199 (Grand Island Biologicals Co., Grand Rapids, N.Y.) containing one-eighth the usual amount of amino acids, 1.2 mg/ml NaHCO$_3$, 1 U/ml penicillin, and 1 mg/ml streptomycin. Twenty-five μC L-4,5-^{3}H-lysine monohydrochloride was added and the incubation was carried out at 37 C for 2 hr. The tissue was homogenized and centrifuged at 105,000 $\times$ g for 1 hr at 4 C. **A:** Immunoprecipitable VLDL synthesized *in vitro* was determined in liver homogenate by precipitation with rabbit anti-VLDL antisera in the presence of carrier VLDL, 5% Triton X-100, and 1% sodium deoxycholate. Incubation was at 23 C for 3 hr or 23 C for 30 min and then overnight at 4 C. Immunoprecipitates were collected by centrifugation at 2000 $\times$ g for 5 min. They were washed once with wash buffer (10 mM sodium phosphate, pH 7.5, 15 mM NaCl, 4 mM L-Lysine, 1% sodium deoxycholate, and 1% Triton X-100) and twice with 0.9% NaCl. The washed pellet was dissolved in 1 ml NCS and counted in spectrofluortoluene. **B:** Total TCA-precipitable counts were determined on the liver homogenates and taken as the total protein synthesized (–●–●–), cpm in liver homogenate of DES-treated animals. (–○–○–), cpm in liver homogenate of oil-treated control animals. **C:** VLDL synthesized as proportion of total protein synthesized was obtained by division of the values in **A** by those from **B**. (–●–●–), ratio in DES-treated animals. (–○–○–), ratio in oil-treated controls. (From Chan et al. (8), with permission.)

EFFECTS OF ESTROGEN ON ApoVLDL-II mRNA ACTIVITY

To understand the level of regulation of VLDL by estrogen, we have studied mRNA activity for a major VLDL apoprotein, apoVLDL-II. For comparison, we have also studied the mRNA for two other proteins synthesized in the liver, albumin and apoA-I, a major HDL protein (27). Following a single injection of estrogen, the plasma level of apoVLDL-II increased rapidly so that in 30 hr, the concentration increased 18-fold. In contrast, the concentrations of plasma albumin and apoA-I were unaltered (Fig. 5) (7). When the respective mRNA activities were determined in RNA isolated from the livers of cockerels similarly treated, apoVLDL-II

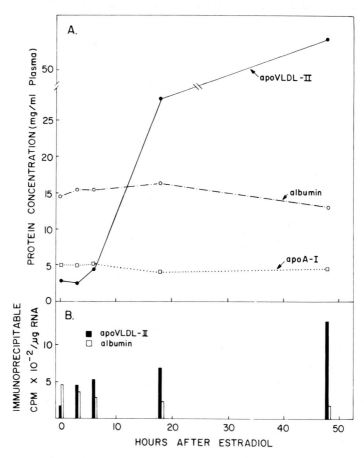

FIG. 5. Effect of estrogen on apo-VLDL II, albumin, and apoA-I. **A:** Concentrations of plasma apoVLDL-II, albumin, and apoA-I at various times after a single intramuscular injection of estradiol (5 mg/kg). ApoVLDL-II was measured by specific immunoprecipitation. Albumin and apoA-I were measured by rocket immunoelectrophoresis (35). **B:** Translational activity of hepatic polysomal RNA at various times after estradiol (5 mg/kg). Translation was carried out in a wheat germ system *in vitro* (8). Specific proteins synthesized *in vitro* were quantified by immunoprecipitation (7).

mRNA activity was found to increase rapidly following estrogen, whereas albumin mRNA activity showed a slight decline and apoA-I activity (data not shown) remained extremely low and not significantly changed (Fig. 5).

The translation data, coupled with the fact that the effect of estrogen on VLDL synthesis is inhibited by actinomycin D (8), suggest that estrogen regulates apoVLDL-II synthesis at a pretranslational level. In fact, the increase in apoVLDL-II mRNA activity precedes increases in plasma apoVLDL-II and occurs much earlier than elevations in plasma triglyceride. This observation supports the conclusion that estrogen acts primarily on the apoprotein of VLDL, and the changes in plasma lipid follow rather than precede the hyperlipoproteinemia.

CHARACTERIZATION OF ApoVLDL-II, ALBUMIN, AND ApoA-I SYNTHESIZED in vitro

In measuring the mRNA activities *in vitro,* the radiolabeled proteins were selectively immunoprecipitated with specific antibodies. As an additional check on the specificity of the immunoprecipitation, and to further characterize the product, the *in vitro* synthesized radiolabeled apoVLDL-II, albumin, and apoA-I were analyzed by SDS slab gel electrophoresis and fluorography. As shown in Fig. 6, in each of these instances the *in vitro* product was found to be considerably larger than the authentic protein isolated from the plasma. The molecular weights of the products were 11,000, 27,800, and 69,000, compared to 9500, 24,400, and 65,000 of plasma apoVLDL-II, apoA-I, and albumin, respectively (7). These products have been designated pro-apoVLDL-II, pro-apoA-I, and pro-albumin. The detection of larger products of translation in wheat germ extracts *in vitro* is compatible with Blobel's signal hypothesis (3,4). According to this hypothesis, intracellular precursors exist for most secretory proteins. Segregation of these proteins in the rough endoplasmic reticulum is accomplished by a metabolically short-lived "signal" sequence in the amino terminal of the nascent polypeptide chain. This sequence is coded for by codons immediately to the 3' end of the initiation codon. Once the signal sequence "leads" the nascent polypeptide chain across the endoplasmic membrane, it is removed by a specific peptidase (24) and is no longer present in the finally secreted molecule. Since apoVLDL-II, apoA-I, and albumin are all secretory proteins, it is quite possible that their primary translation products contain the signal sequence. The size of the *in vitro* translation products is consistent with such an interpretation.

DEMONSTRATION AND CHARACTERIZATION OF THE NUCLEAR ESTROGEN RECEPTOR OF THE COCKEREL LIVER

We have presented evidence that estrogen stimulates VLDL synthesis, probably at a pretranslational level. The mechanism by which estrogen

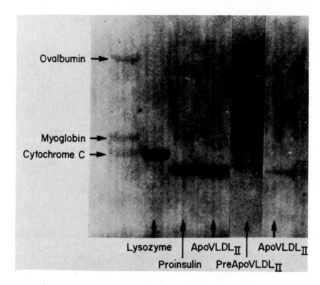

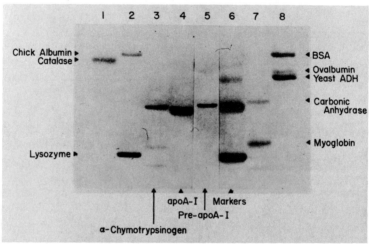

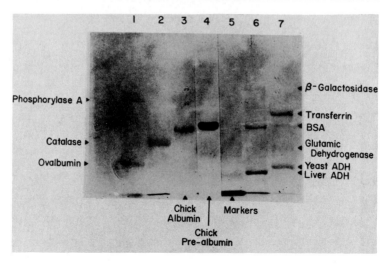

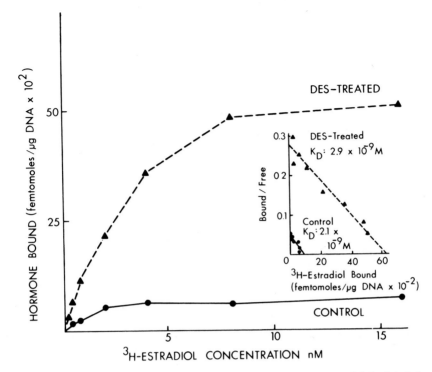

FIG. 7. Binding of ^{3}H-estradiol by estrogen receptor in liver nuclei isolated from chicks treated with 2.5 mg DES (▲—▲) and controls (●—●). Animals were killed 4 hr after injection and nuclei were prepared by homogenization in 50% glycerol, 10 mM NaCl, 1.5 mM MgCl, 10 mM Tris (pH 7.5) buffer at —20 C and centrifugation at 3,900 × g for 15 min. Nuclei were washed twice by centrifugation with homogenization buffer supplemented with 0.1% Triton X-100. The final nuclear pellet was washed with 10 mM Tris (pH 7.4), 1 mM MgCl$_2$, 25% glycerol buffer and used for estradiol exchange assays. Aliquots of nuclei were incubated at 30 C for 30 min with varied concentrations of ^{3}H-estradiol in the presence (series A) and absence (series B) of a 100-fold excess of DES. Following incubation, 1.0 ml of 10 mM Tris (pH 7.4), 1.5 mM EDTA (TE buffer) was added and the samples were centrifuged at 800 × g for 10 min. Pellets were washed three times with cold TE buffer and extracted with 1.0 ml absolute ethanol. After centrifugation at 800 × g for 10 min, the radioactivity in the ethanol extract was determined by liquid scintillation spectrometry. Specific ^{3}H-estradiol binding was determined by subtraction of set A from set B. Inset shows a Scatchard (43) plot of the same data. (From Snow et al. (45), with permission.)

exerts its influence on the selective stimulation of apoVLDL-II synthesis is the subject of our next series of experiments.

Specific nuclear estrogen binding sites were measured by a ^{3}H-estradiol exchange assay (Fig. 7) (45). Nuclei from untreated cockerel liver con-

←

FIG. 6. Slab gel electrophoresis and fluorography of pre-apoVLDL-II, pre-apoA-I, and prealbumin. The putative precursors were synthesized *in vitro* in a wheat germ system using ^{35}S-methionine as a label (8). Acrylamide concentrations were 13.5%, 12% and 7.5% respectively (7).

TABLE 3. *Specificity of ³H-estradiol exchange by isolated liver nuclei from chicks treated with DES*

Competitor	Estradiol binding: % bound radioactivity remaining
Buffer	100
Estradiol	0
DES	0
Estrone	34
Estriol	36
Progesterone	100
Corticosterone	100
Testosterone	100

Sixteen hr after treatment with DES (2.5 mg), animals were sacrificed and liver nuclei were isolated. Nuclei (0.31 mg DNA) were incubated at 30 C for 30 min with 12 nM ³H-estradiol with or without a 100-fold excess of different competitors in the exchange assay as described in Fig. 8. Bound radioactivity of 100% represents 32,000 cpm. Approximately 75% of the ³H-estradiol was specific for estradiol. By definition this represents 100% competition (or 0% bound radioactivity remaining).

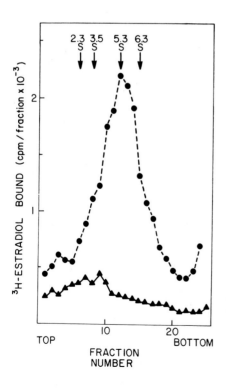

FIG. 8. Sucrose gradient sedimentation profile of KCl-extracted nuclear estrogen receptor. Isolated nuclei from DES-treated chick liver were extracted with 0.4 M KCl-TE buffer at 4 C for 30 min. The suspension was centrifuged at 50,000 × g for 20 min in a Beckman SW56 rotor. The extracted DES-receptor complexes were incubated with 12 nM ³H-estradiol under exchange conditions at 30 C for 30 min prior to centrifugation. The extract was treated with charcoal to remove bound radioactivity. Sucrose gradients (5 to 20%) were fractionated in 0.2-ml aliquots and the ³H-estradiol bound per fraction was determined. ³H-estradiol only (●—●) and ³H-estradiol plus 100-fold DES (▲—▲). ¹⁴C-ovalbumin (3.5S), ¹²⁵I-apoA-I lipoprotein (2.3S), and ¹⁴C-alkaline phosphatase (6.3S) were used as markers in the gradients. (From Snow et al. (45), with permission.)

tained low concentrations (0.08 fmoles/μg DNA) of specific binding sites. Treatment with a single dose of estrogen 4 hr before sacrifice increased the concentration of these binding data by sevenfold (up to 0.6 fmoles/μg DNA). Scatchard (43) analysis of the sites revealed that there was a single class of high affinity sites with an apparent K_d of 2×10^{-9} M. The specific nuclear binding sites were saturated at a hormone concentration of 8 nM and the K_d was unaffected by estrogen treatment.

The binding sites were highly specific for estrogen (Table 3). Substantial competition for the binding was demonstrated only by estrogenic compounds. The presence of a 100-fold excess of either diethylstilbestrol or estradiol obliterated ^{3}H-estradiol binding. The addition of estrone or estriol resulted in a 66 and 64% reduction in binding, respectively. Nonestrogenic steroid hormones, such as progesterone, corticosterone, or testosterone had no apparent effect on the binding (Table 3) (45).

KCl treatment of the nuclear bound receptor complex extracted about 20% of the binding activity. The KCl-resistant sites and KCl-extractable sites had roughly the same K_d (2.9×10^{-9} M and 1.9×10^{-9} M, respectively), which was also similar to that for intact nuclei (2.6×10^{-9} M). Analysis of the KCl-extractable complex on sucrose gradient centrifugation revealed a peak of binding activity in the 5.3S region. The presence of a 100-fold excess of diethylstilbestrol abolished the peak (Fig. 8).

CHARACTERIZATION OF ESTRADIOL BINDING SITES ON CHROMATIN

Since the bulk of the nuclear estrogen receptor complexes was resistant to to extraction with KCl, the effect of estrogen on specific ^{3}H-estradiol binding to liver chromatin *in vitro* was studied. Binding sites were again quantified by the ^{3}H-estradiol exchange assay. As shown in Fig. 9, chromatin from control animals contained low concentrations (0.08 fmoles/μg DNA in the experiment depicted) of binding sites. Sixteen hr after a single injection of estrogen, the number of binding sites increased by four- to fivefold (0.3 fmoles/μg DNA). The K_d was unaltered by estrogen treatment, and remained at 2.6×10^{-9} M (45). Hence, the bulk of ^{3}H-estradiol binding sites present in cockerel liver nuclei appeared to be associated with chromatin. Since the kinetics of binding in both purified nuclei and isolated chromatin were quite similar, it is likely that we were studying the same sites under our experimental conditions.

EFFECT OF ESTROGEN ON ENDOGENOUS NUCLEAR RNA POLYMERASE ACTIVITIES

Since our studies on apoVLDL-II synthesis suggest that estrogen stimulates apoVLDL-II mRNA transcription, it would be of interest to study the effect of the hormone on endogenous nuclear DNA-dependent RNA poly-

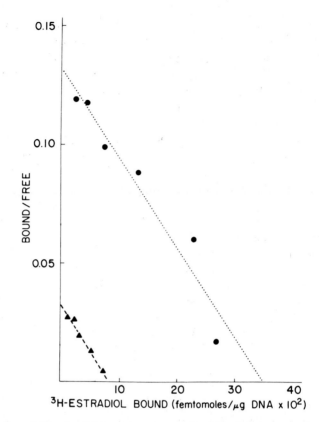

FIG. 9. Binding of ³H-estradiol by estrogen receptor in isolated chromatin. Livers from DES-treated and control chicks were used for chromatin preparation at 16 hr after treatment. To isolate chromatin, purified nuclei were washed three times by resuspensio nin 80 mM NaCl, 20 mM Na EDTA, pH 6.3, and subjected to centrifugation at 2,000 $\times$ *g* for 10 min. The washed nuclei were then suspended in 50 ml 0.01 X SSC (0.15 M NaCl, 15 mM sodium citrate, pH 7.0) and kept at 0 C until at least 90% of the nuclei had lysed. The released chromatin was collected by centrifugation at 10,000 $\times$ *g* for 10 min. Portions of the chromatin from each treatment group were incubated at 30 C for 30 min in a ³H-estradiol exchange assay. ³H-estradiol binding was determined, and the data plotted according to Scatchard (43). DES treatment (●—●) and control (▲—▲).

merase I and II activities in cockerel liver chromatin (45). As shown in Table 4, estrogen caused a 100% increase in RNA polymerase II and a 40 to 50% increase in polymerase I activity within 1 hr. Maximal stimulation of both enzyme activities occurred between 1 and 3 hr. The effects of estrogen were still evident at 24 hr.

EFFECT OF ESTROGEN ON RNA POLYMERASE INITIATION SITES ON LIVER CELL CHROMATIN

Estrogen treatment stimulated total liver soluble protein synthesis in the cockerel by about twofold (Fig. 4) (8). To determine whether estrogen

TABLE 4. *Effect of DES on endogenous RNA polymerase activities in isolated nuclei*

Time (hr)	RNA polymerase I (pmoles ^{3}H-UMP incorporated/ 100 μg DNA)	RNA polymerase II (pmoles ^{3}H-UMP incorporated/ 100 μg DNA)
0	200	223
1	278	424
3	288	437
12	288	386
24	266	326

Three-week-old cockerels were sacrificed at the times indicated after treatment with 2.5 mg DES/0.2 kg. Liver cell nuclei were prepared as described in Fig. 8. Endogenous RNA polymerase activities were measured by adding an aliquot of purified nuclei to an assay mixture containing 2.5 μmole Tris (pH 8.0), 0.05 μmole MnCl$_2$, 0.1 μmole MgCl$_2$, 3 μmole (NH$_4$)$_2$SO$_4$, 0.1 μmole dithiothreitol, 0.03 μmole each of ATP, CTP, and GTP, 0.003 μmole unlabeled UTP, and 2.5 μCi ^{3}H-UTP in a final volume of 50 μliter. 0.005 μg α-amanitin was also added to the appropriate tubes. Reactions were incubated at 30 C for 20 min and terminated by spotting an aliquot on a 2.5 cm DE-81 filter paper disk. Filters were washed and counted as described by Roeder (42).

effects such changes predominantly at the transcriptional level, we quantified the number of RNA polymerase initiation sites on liver chromatin at various times after hormone treatment (45). As shown in Table 5, estrogen causes a doubling effect in the number of initiation sites in 24 hr. The stimulatory effects of the hormone were evident within an hour of treatment.

TABLE 5. *Effect of DES on number of RNA synthesis initiation sites on isolated chromatin*

Hours after treatment	Number of initiation sites/ pg DNA $\times$ 10^{-4}
0	2.6
1	4.9
3	4.2
12	5.0
24	5.3

At the times indicated following treatment of cockerels with DES (2.5 mg), liver chromatin was isolated. Purified nuclei were washed three times with 80 mmole NaCl, 20 mmole Na EDTA, pH 6.3, by resuspension and centrifugation at 2000 $\times$ g for 10 min. Washed nuclei were suspended in 50 ml 0.01 X SSC [0.15 M NaCl, 0.015 M sodium citrate, pH 7.0] and kept at 0 C until at least 90% of the nuclei had lysed. The released chromatin was collected by centrifugation at 10,000 $\times$ g for 10 min and re-suspended in 0.01 X SSC. RNA synthesis on liver chromatin following addition of *E. coli* RNA polymerase was performed in the presence of rifampicin and heparin as described by Tsai et al. (48).

TEMPORAL RELATIONSHIP BETWEEN THE CHANGES IN NUCLEAR ESTROGEN RECEPTOR CONCENTRATION, RNA POLYMERASE I AND II ACTIVITIES, NUMBER OF RNA SYNTHESIS INITIATION SITES, VLDL SYNTHESIS, AND LEVELS OF PLASMA TRIGLYCERIDES

Figure 10 is a composite diagram of the various events taking place in the cockerel liver after a single injection of estrogen. It is apparent that nuclear estrogen receptor concentration increases within 1 hr of hormone treatment. This is accompanied by increases in polymerase I and II activities and in the number of RNA synthesis initiation sites. These changes are followed closely by an increase in the rate of VLDL synthesis. Hypertriglyceridemia, the estrogen-related event which prompted our studies, finally ensued.

DISCUSSION AND FUTURE PROSPECTS

We have summarized studies performed in our laboratory over the past few years on the mechanism of estrogen-induced hyperlipoproteinemia. Our major findings are illustrated in Figs. 5 and 10. Based on these observations,

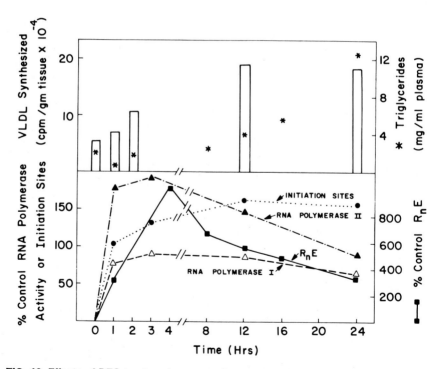

FIG. 10. Effects of DES treatment upon nuclear estrogen receptor concentration, RNA polymerase I and II activities, number of RNA synthesis initiation sites, VLDL synthesis *in vitro,* and level of plasma triglyceride.

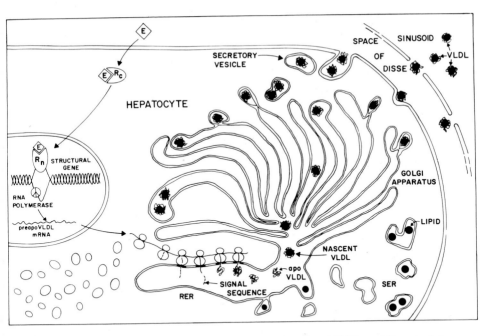

FIG. 11. Schematic model for the synthesis and assembly of VLDL and the regulatory effect of estrogen. E, estrogen; RER, rough endoplasmic reticulum; SER, smooth endoplasmic reticulum; ●, lipids; R_c, cytoplasmic receptor; R_n, nuclear receptor.

we can now construct a model for the estrogenic regulation of lipoprotein synthesis in the liver (Fig. 11). Estrogen appears to bind to specific nuclear receptors in the liver. There is considerable controversy as to the existence of cytoplasmic estrogen receptors in the liver of this species. The association of the estrogen-receptor complex with chromatin somehow leads to a stimulation of RNA polymerase I and II activities, and the derepression of specific gene transcription. As a consequence, the synthesis of apoVLDL-II mRNA and probably of some other specific mRNAs is stimulated. The apoVLDL-II mRNA is exported into the cytoplasm where its translation is initiated on free ribosomes. Translation of the initial "signal" codons results in a unique sequence of amino acid residues on the amino terminal of the nascent polypeptide chain. Emergence of this signal sequence is thought to result in attachment of the ribosomes to the rough endoplasmic membrane. Translation continues on the bound ribosomes. The signal sequence, still attached to the nascent polypeptide, is then cleaved by a specific peptidase. On completion of translation of the mRNA, the apoVLDL-II is released and vectorially discharged. The apoVLDL-II then binds triglycerides which are synthesized mainly in the smooth endoplasmic reticulum (1,46). The completed VLDL are transported to the Golgi apparatus and are concentrated in secretory granules. They are finally secreted by fusion of the vesicular membranes with the plasma membranes of the hepatocyte.

Although the general scheme of the various steps in the effect of estrogen on VLDL synthesis is probably correct, little detailed information on each of these steps is available. The mRNA for apoVLDL-II has been purified to apparent homogeneity (7) and studies are currently in progress on the use of cDNA for the quantification of the specific RNA sequences before and after estrogen treatment.

We have previously demonstrated that nuclear estrogen receptors remain intact for at least 2 weeks in liver slices maintained in organ culture (5). In this system, however, the VLDL response to estrogen is extremely inconsistent. We are currently developing a primary hepatocyte culture system according to the method of Tarlow et al. (47). The hepatocytes are viable and actively synthesize albumin and VLDL for at least 2 weeks. Studies are currently in progress on the response of these cells to estrogen. The development of an *in vitro* hormone-responsive system will allow a much more detailed study on the kinetics of hormone induction and the molecular events involved in such a process.

ACKNOWLEDGMENTS

We wish to acknowledge the collaboration and advice of Drs. Hakan Eriksson and James H. Clark on the studies of estrogen-receptor characterization. We should also like to thank Dr. James Hardin who performed the RNA polymerase and RNA synthesis initiation site assays. Lawrence Chan and Richard L. Jackson are Established Investigators of the American Heart Association. Anthony R. Means is the recipient of a Faculty Research Award from the American Cancer Society. L. Dale Snow is the recipient of a fellowship 1 F32 HL 05534–01 from the National Heart, Lung, and Blood Institute. This work was supported by Health, Education, and Welfare Grant HL 16512 and Grant-in-Aid 75–914 of the American Heart Association.

REFERENCES

1. Alexander, C. A., Hamilton, R. L., and Havel, R. J. (1976): Subcellular localization of B apoprotein of plasma lipoproteins in rat liver. *J. Cell Biol.*, 62:241–263.
2. Bartlett, G. R. (1959): Phosphorus assay in column chromatography. *J. Biol. Chem.*, 234:466–471.
3. Blobel, G., and Dobberstein, B. (1975): Transfer of proteins across membranes. I. Presence of proteolytically processed and unprocessed nascent immunoglobulin light chains on membrane-bound ribosomes of murine myeloma. *J. Cell Biol.*, 67:835–851.
4. Blobel, G., and Dobberstein, B. (1975): Transfer of proteins across membranes. II. Reconstitution of functional rough microsomes from heterologous components. *J. Cell Biol.*, 67:852–862.
5. Chan, L., Eriksson, H., Jackson, R. L., Clark, J. H., and Means, A. R. (1977): Effects of estrogen on very low density lipoprotein (VLDL) synthesis in avian liver slices *in vitro*. Lack of correlation with nuclear estrogen receptors. *J. Steroid Biochem.*, 8:1189–1191.

6. Chan, L., Jackson, R. L., and Means, A. R. (1977): Female steroid hormones and lipoprotein synthesis in the cockerel. Effects of progesterone and nafoxidine on the estrogenic stimulation of very low density lipoproteins (VLDL) synthesis. *Endocrinology*, 100:1636–1643.
7. Chan, L., Jackson, R. L., and Means, A. R. (1978): Regulation of lipoprotein synthesis. Studies on the molecular mechanisms of lipoprotein synthesis and their regulation by estrogen in the cockerel. *Circ. Res.*, 43:207–217.
8. Chan, L., Jackson, R. L., O'Malley, B. W., and Means, A. R. (1976): Synthesis of very low density lipoproteins in the cockerel. Effects of estrogen. *J. Clin. Invest.*, 58:368–379.
9. Chan, L., and O'Malley, B. W. (1976): Mechanism of action of the sex steroid hormones. *N. Engl. J. Med.*, 294:1430–1437.
10. Clark, J. H., Peck, E. J., Jr., Hardin, J. W., and Eriksson, H. (1977): The biology and pharmacology of estrogenic receptor binding. Relationship to uterine growth. In: *Receptors and Hormone Action, Vol. 2*, pp. 1–31, edited by B. W. O'Malley and L. Birnbaumer. Academic Press, New York.
11. Coleman, R., Polokoff, M. A., and Bell, R. M. (1977): Mechanism of estrogen-induced hypertriacylglycerolemia in the chick. Investigations on microsomal enzymes of triacylglycerol and phospholipid synthesis. *Metabolism*, 26:604–655.
12. Davidoff, F., Tishler, S., and Rosoff, C. (1973): Marked hyperlipidemia and pancreatitis associated with oral contraceptive therapy. *N. Engl. J. Med.*, 289:552–555.
13. Ence, T. J., Wilson, D. E., Flowers, C. M., Chen, A. L., Glad, B. W., and Hershgold, E. J. (1976): Heparin metabolism and heparin-released lipase activity during long-term estrogen-progestin treatment. *Metabolism*, 25:139–145.
14. Fredrickson, D. S., Goldstein, J. L., and Brown, M. S. (1978): Familial hyperlipoproteinemia. In: *The Metabolic Basis of Inherited Disease*, 4th edition, edited by J. B. Stanbury, J. B. Wyngaarden, and D. S. Fredrickson, pp. 250–275. McGraw-Hill, New York.
15. Furman, R. H., Howard, R. P., Lakshmi, K., and Norcia, L. N. (1961): The serum lipids and lipoproteins in normal and hyperlipidemic subjects as determined by preparative ultracentrifugation. *Am. J. Clin. Nutr.*, 9:73–102.
16. Glueck, C. J., Brown, W. V., Levy, R. I., Greten, H., and Fredrickson, D. S. (1969): Amelioration of hypertriglyceridemia by progestational activity in familial type-V hyperlipoproteinemia. *Lancet*, 1:1290–1291.
17. Glueck, C. J., Fallat, R. W., and Scheel, D. (1975): Effects of estrogenic compounds on triglyceride kinetics. *Metabolism*, 24:537–545.
18. Glueck, C. J., Gartside, P., Fallat, R. W., and Mendoza, S. (1976): Effects of sex hormones on protamine inactivated and resistant postheparin plasma lipases. *Metabolism*, 25:625–632.
19. Hazzard, W. R., Brunzell, J. D., Applebaum, D. M., Goldberg, A. P., Gagne, C., Albers, J. J., Wah, P. W. and Hoover, J. J. (1977): Steroid contraceptives and human lipoprotein metabolism. Effects and mechanisms. In: *Pharmacology of Steroid Contraceptive Drugs*, edited by S. Garattini and H. W. Berendes, pp. 251–266. Raven Press, New York.
20. Hazzard, W. R., Notter, D. T., Spiger, M. J., and Bierman, E. L. (1972): Oral contraceptives and triglyceride transport. Acquired heparin resistance as the mechanism for impaired post-heparin lipolytic activity. *J. Clin. Endocrinol. Metab.*, 35:425–437.
21. Hazzard, W. R., Spiger, M. J., Bagdade, J. D., and Bierman, E. L. (1969): Studies on the mechanism of increased plasma triglyceride levels induced by oral contraceptives. *N. Engl. J. Med.*, 280:471–474.
22. Hillyard, L. A., Entenman, C., and Charkoff, I. L. (1956): Concentration and composition of serum lipoproteins of cholesterol-fed and stilbestrol-injected birds. *J. Biol. Chem.*, 223:359–368.
23. Hsuch, A. J. W., Peck, E. J., Jr., and Clark, J. H. (1975): Progesterone antagonism of the oestrogen receptor and oestrogen-induced uterine growth. *Nature*, 254:337–339.
24. Jackson, R. L., and Blobel, G. (1977): Post-translational cleavage of presecretory

proteins with an extract of rough microsomes from dog pancreas containing signal peptidase activity. *Proc. Natl. Acad. Sci. USA,* 74:5598–5602.

25. Jackson, R. L., Chan, L., Snow, L. D., and Means, A. R. (1978): Hormonal regulation of lipoprotein synthesis. In: *Disturbances in Lipid and Lipoprotein,* edited by J. M. Dietschy, A. M. Gotto, and J. A. Ontko, pp. 139–154. American Physiological Society, Bethesda.

26. Jackson, R. L., Lin, H.-Y., Chan, L., and Means, A. R. (1977): Amino acid sequence of a major apoprotein from hen plasma very low density lipoproteins. *J. Biol. Chem.,* 252:250–253.

27. Jackson, R. L., Lin, H.-Y., Chan, L., and Means, A. R. (1976): Isolation and characterization of the major apolipoprotein from chick high density lipoproteins. *Biochim. Biophys. Acta,* 420:342–349.

28. Jackson, R. L., Morrisett, J. D., and Gotto, A. M., Jr. (1976): Lipoprotein structure and metabolism. *Physiol. Rev.,* 52:259–316.

29. Jackson, R. L., Morrisett, J. D., Gotto, A. M., Jr., and Segrest, J. P. (1975): The mechanisms of lipid-binding by plasma lipoproteins. *Mol. Cell. Biochem.,* 6:43–50.

30. Kekki, M., and Nikkila, E. A. (1971): Plasma triglyceride kinetics during use of oral contraceptives. *Metabolism,* 20:879–889.

31. Kissebah, A. H., Harrigan, P., and Wynn, V. (1973): Mechanism of hypertriglyceridemia associated with contraceptive steroids. *Horm. Metab. Res.,* 5:184–190.

32. Krauss, R. M., Levy, R. I., and Fredrickson, D. S. (1974): Selective measurement of two lipase activities in postheparin plasma from normal subjects and patients with hyperlipoproteinemia. *J. Clin. Invest.,* 54:1107–1124.

33. Kudzma, D. J., Hegstad, P. M., and Stoll, R. E. (1973): The chick as a laboratory model for the study of estrogen-induced hyperlipidemia. *Metabolism,* 22:423–434.

34. Kudzma, D. J., St. Claire, F., DeLallo, L., and Friedberg, S. J. (1975): Mechanism of avian estrogen-induced hypertriglyceridemia. Evidence for overproduction of triglyceride. *J. Lipid Res.,* 16:123–133.

35. Laurell, C. B. (1966): Quantitative estimation of proteins by electrophoresis in agarose gel containing antibodies. *Anal. Biochem.,* 15:45–52.

36. Lerner, L. J. (1964): Hormone antagonists. Inhibitors of specific activities of estrogen and androgen. *Recent Prog. Horm. Res.,* 20:435–490.

37. Lowry, O. H., Rosebrough, N. J., Farr, A. L., and Randall, R. J. (1951): Protein measurement with the Folin phenol reagent. *J. Biol. Chem.,* 193:265–275.

38. Luskey, K. L., Brown, M. S., and Goldstein, J. L. (1974): Stimulation of the synthesis of very low density lipoproteins in rooster liver by estradiol. *J. Biol. Chem.,* 249:5939–5947.

39. *Manual of Laboratory Methods* (1975): Lipid Research Clinics Program. University of North Carolina, Chapel Hill, North Carolina.

40. Molich, M. E., Oill, P., and Odell, W. D. (1974): Massive hyperlipemia during estrogen therapy. *J.A.M.A.,* 227:522–525.

41. Motulsky, A. G. (1976): The genetics of hyperlipidemias. *N. Engl. J. Med.,* 294:823–827.

42. Roeder, R. G. (1974): Multiple forms of deoxyribonucleic acid–dependent ribonucleic acid polymerase in *Xenopus laevis. J. Biol. Chem.,* 249:241–248.

43. Scatchard, G. (1949): The attractions of proteins for small molecules and ions. *Ann. N.Y. Acad. Sci.,* 51:660–672.

44. Segrest, J. P., Jackson, R. L., Morrisett, J. D., and Gotto, A. M., Jr. (1974): A molecular theory of lipid-protein interactions in the plasma lipoproteins. *FEBS Lett.,* 38:247–253.

45. Snow, L. D., Eriksson, H., Hardin, J. W., Chan, L., Jackson, R. L., Clark, J. H., and Means, A. R. (1978): Nuclear estrogen receptor in the avian liver. Correlation with biologic response. *J. Steroid Biochem.,* 9:1017–1026.

46. Stein, Y., and Stein, O. (1975): Lipoprotein synthesis, intracellular transport, and secretion in liver. In: *Atherosclerosis III,* edited by G. Schettler and A. Weizel, pp. 652–657. Springer-Verlag, Berlin.

47. Tarlow, D. M., Watkins, P. A., Reed, R. E., Miller, R. S., Zwergel, E. E., and Lane, M. D. (1977): Lipogenesis and the synthesis and secretion of very low density lipoprotein by avian liver cells in nonproliferating monolayer culture. *J. Cell Biol.,* 73:332–353.
48. Tsai, M. J., Schwartz, R. J., Tsai, S. Y., and O'Malley, B. W. (1975): Effects of estrogen on gene expression in the chick oviduct. IV. Initiation of RNA synthesis on DNA and chromatin. *J. Biol. Chem.,* 250:5165–5174.
49. Wallace, R. B., Hoover, J., Sandler, O., Rifkind, B. M., and Tyroler, H. A. (1977): Altered plasma-lipids associated with oral contraceptive or oestrogen consumption. *Lancet,* 2:11–13.
50. Wynn, V., Doar, J. W. H., and Mills, G. L. (1966): Some effects of oral contraceptives on serum lipid and lipoprotein levels. *Lancet,* 2:720–723.
51. Wynn, V., Doar, J. W. H., Mills, G. L., and Stokes, T. (1969): Fasting serum triglyceride, cholesterol, and lipoprotein levels during oral contraceptive therapy. *Lancet,* 2:756–760.

Ontogeny of Receptors and Reproductive Hormone Action,
edited by T. H. Hamilton, J. H. Clark, and W. A. Sadler.
Raven Press, New York 1979.

Estrogen-binding Proteins in Avian Liver: Characteristics, Regulation, and Ontogenesis

Catherine B. Lazier

Biochemistry Department, Dalhousie University, Halifax, Nova Scotia, Canada B3H 4H7

Estrogen induction of vitellogenin synthesis in the liver of oviparous vertebrates is an appealing model system for the study of sex steroid control of gene expression in a highly specific primary response (53). The absence of massive cell differentiation or proliferation in response to estradiol in this differentiated tissue presents some advantages over the more widely investigated reproductive tissues. Considerable progress has been made recently in measuring estrogen-induced vitellogenin mRNA accumulation in frogs and roosters (12,46), and the data strongly support the hypothesis that the major site of action of the hormone is at the level of transcription. The receptor system through which estradiol presumably effects the transcriptional changes in avian liver is the subject of this review.

The now classical model for sex steroid receptors was derived mainly from studies with rat uterus and chick oviduct (8,28). In immature or castrated animals the receptor is largely in the cytoplasm. Exposure to hormone leads to rapid transfer of most of the binding sites to the nucleus. After an interval of time, which may vary with the hormone or hormone-derivative, the dose, and the target tissue examined, the nuclear receptor levels fall and cytosol levels rise, possibly due to recycling of receptor as well as to synthesis of new receptor protein (1,29,38). A central feature of this model is the early stoichiometric transfer of cytosol receptor to the nucleus. Chick liver contains nuclear and cytosol estradiol-binding proteins, but it has been difficult to clearly demonstrate their relationship in terms of the classical model for receptor regulation. Whether this is due to technical problems or to qualitative differences in the receptor system in the different target tissues is not yet clear.

CYTOPLASMIC ESTROGEN-BINDING PROTEINS

The Hepatic Steroid-binding Protein

In the first reported study of estradiol-binding proteins in chick liver, Arias and Warren (2) found a cytosol protein which bound estradiol with high affinity (K_d, 2 nM). The estrophile was unlike a classical cytoplasmic receptor, however, in that instead of enhancing estradiol transfer to the nucleus it appeared to retard the process. Subsequently, Mester and Baulieu (39), Ozon and Bellé (42), Gschwendt (19), and Lazier (30) were unable to demonstrate specific high-affinity binding of estradiol in liver cytosol from laying hens, roosters, or immature chicks. Teng and Teng (54) found few high-affinity binding sites in embryonic liver cytosol. Several groups reported, however, that crude liver cytosol contains a high concentration of low-affinity binding sites for estradiol (K_d, 1 to 0.1 μM), which exhibit a sedimentation coefficient variously estimated to range between 2.7 and 4S (14,15,19, 33). This binding protein can be distinguished from the faster-sedimenting plasma estrogen binder and is absent from a variety of other tissues. Particularly distinctive features of the binding protein in crude liver cytosol are that it has a tenfold higher affinity for estrone than for estradiol, and that it does not translocate estrogen to the nucleus under a variety of conditions (14,15,19). The concentration of binding sites in liver of laying hens is 100 to 200 pmole/mg cytosol protein (5 μM, or 5×10^6 sites/cell), but it is reduced in liver of nonlaying hens or roosters, and is considerably lower in cockerel liver cytosol (20 pmole/mg protein) (14). These observations are clearly inconsistent with the classical criteria for specificity, ligand affinity, concentration, and intracellular distribution usually ascribed to a typical estradiol receptor. Furthermore, the protein has recently been purified to 95% homogeneity and has been found to bind pregnenolone and progesterone equally as well as estrone (K_d, 30 to 60 nM). 17β-Estradiol and dihydrotestosterone show tenfold less affinity and 17α-estradiol, estriol, and corticosterone are poor ligands (14). Given this specificity, Dower has recently coined the term "hepatic steroid-binding protein" or HSBP. Purified HSBP is a slightly acidic, globular protein with a molecular weight of 30,000, a Stokes radius of 25Å, and a sedimentation coefficient of 2.7S (14).

The function of HSBP is unknown. It could serve as a hepatic sink or reservoir for estrone and progesterone (14). Although binding studies imply that the two hormones compete for the same site, a more complex model involving separate but interacting sites is also possible. However the intracellular concentration of HSBP is so high that it is unlikely that competition need occur under physiological conditions. As a reservoir for estrogens, HSBP could function to maintain the intracellular estradiol concentration in the face of large fluctuations in plasma levels of estrone and estradiol. When plasma estrogen levels drop HSBP-bound estrone could be released and reduced to estradiol, which could then carry out its intranuclear receptor-linked functions. None of the common steroid-metabolizing functions could be assigned to HSBP (14). Whatever the function of this remarkable protein, its presence in liver should be considered in any model of the physiological regulation of estrogen responses.

High-Affinity Estradiol Binding in Cytosol

The initial failure to reproducibly demonstrate high-affinity binding of estradiol in chick liver cytosol led to speculation that the mechanism of estrogen action in chick liver might be independent of cytosol receptor and hence different from that in other target tissues (33,47). Such speculation was probably premature, since it now appears that high-affinity binding can be detected at certain stages of development and under certain conditions of assay (18). The limited data available thus far preclude confident identification of this binding activity as a receptor, but certain properties are consistent with that designation.

Gschwendt (18) characterized high-affinity binding (K_d, 1 nM) in cytosol from liver of chick embryos at the 19th day of development. The binding site concentration appears to be higher at this stage than at a number of earlier or later stages examined. We have investigated a similar high-affinity estrogen binder (K_d, 0.7 nM) in liver cytosol from 2-week-old cockerels (Fig. 1) (34). In both Gschwendt's and our studies, detection of the high-affinity estradiol-binding protein was facilitated by its precipitation from cytosol by up to 40% saturation with ammonium sulfate. This presumably permits separation of large amounts of the lower-affinity HSBP, which precipitates at much higher concentrations of the salt (14). Furthermore, we found

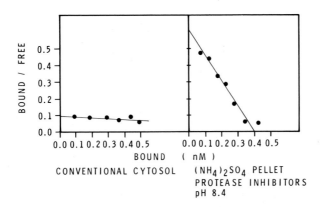

FIG. 1. *Estradiol binding in chick liver cytosol.* Cytosol was prepared from liver of cockerels (Cobbs, 100 g bodyweight), using either the conventional procedure of homogenization in 0.33 M sucrose, 5 mM Tris-HCl, 3 mM MgCl₂, 10 mM thioglycerol, pH 7.4 (2.5 ml/g liver), followed by centrifugation at 100,000 × *g* for 1 hr (30); or by homogenizing the tissue in 0.33 M sucrose, 20 mM Tris-HCl, 3 mM MgCl₂, 10 mM thioglycerol, 0.3 mM phenylmethylsulfonyl fluoride, and 1 mM benzamidine, pH 8.4, followed by centrifugation at 100,000 × *g* for 1 hr, addition of (NH₄)₂SO₄ to the supernatant to give 33% saturation, and dissolution of the pellet in a volume of buffer (10 mM Tris-HCl, 1.5 mM EDTA, 10 mM thioglycerol, 0.5 M KCl, pH 7.4) corresponding to one-half the original liver weight. Scatchard analysis of ³H-estradiol binding was performed as described earlier (31).

some improvement in the yield of high-affinity binding by the inclusion of protease inhibitors in the initial homogenization buffer and by ensuring that the pH of the homogenate was not less than 7.0. Homogenization of chick liver in conventional sucrose-containing buffer at pH 7.4 results in homogenate pH of 5 to 6. In addition, the temperature and time of assay must be chosen carefully. These factors may have contributed to the earlier failures to detect cytosol high-affinity estrogen-binding activity. They also indicate that some caution must be exercised in interpretation of the quantitative data.

In the 19-day embryonic chick liver the concentration of estradiol-binding sites is 1850 fmole/g liver or 1,100 sites/cell (18). After hatching, the concentration appears to be several-fold lower: 200 to 400 fmole/g (Fig. 1) (Table 1) (18). The significance of the peak of binding activity at day 19 is unknown. Mixing experiments with extracts from 19-day and 8-week-old chicks indicate that about 20% of the total binding sites are destroyed or inhibited by a factor present in the extracts from hatched chick liver (18).

The high-affinity binding activity in both 19-day embryonic and cockerel liver cytosol is specific only for estrogens, and progesterone, dihydrotestosterone, and cortisol have no effect. This clearly differentiates the activity from the lower-affinity HSBP described by Dower (14).

Optimal activity of the embryonic high-affinity binder is found when the cytosol precipitate is dissolved in buffer containing 0.5 M KCl (18). Gel filtration in a high-salt medium shows a molecular weight of 50,000 to 60,000 (18). The optimal conditions of incubation require a relatively low temperature. Some instability of the binding sites is apparent at 20°C, and at 37°C marked destruction is observed (18). Similar temperature sensitivity is found for the high-affinity binder in cockerel liver cytosol (C. Lazier, *unpublished results*). This necessitates careful choice of conditions for exchange assays in which elevated temperatures are required.

The effect of injection of estradiol into the egg upon the concentration of the high-affinity binding sites in 19-day embryonic liver cytosol was tested using an exchange assay (18). Liver preparations from control and estradiol-treated embryos were preincubated with charcoal-dextran briefly at 2°C in order to remove unbound estradiol, and then were incubated with ^{3}H-estradiol at 30°C for 40 min in order to effect exchange of bound hormone. Assuming that exchange was complete under the conditions used, the preparations from embryos treated with estradiol for 10 min contained about 40% of the number of high-affinity binding sites present in the untreated controls. After 20 min the level rose to about 60% of control and remained more or less the same for 4 hr.

Depletion of cytosol high-affinity binding sites after estradiol injection can also be demonstrated in cockerel liver (Table 1). In this case an exchange assay involving 16 hr incubation at 2°C followed by 1 hr at 25°C was used. These conditions permit exchange of about 90% of the occupied binding sites (shown by comparison of specific binding in a control cytosol prepa-

TABLE 1. *Effect of estradiol injection on intracellular distribution of high affinity estradiol binding sites in cockerel liver*

| Time after estradiol injection (hr) | [³H]-Estradiol specific binding (fmol/g liver) | |
	Cytosol ammonium sulfate fraction	Nuclear extract
0	386 ± 43	199 ± 43
0.25	101 ± 55	269 ± 70
1.5	36 ± 12	568 ± 130
4.0	192 ± 55	980 ± 141
26	244 ± 14	2540 ± 120
48	451 ± 12	1520 ± 187
120	398 ± 76	407 ± 148

Cockerels (Cobbs, 100 g bodyweight) were injected intra-peritoneally with 2.5 mg 17β-estradiol in 0.1 ml propylene glycol. Specific binding of ³H-estradiol in liver nuclear extracts was measured as detailed earlier (31). The cytosol $(NH_4)_2SO_4$ pellet (Fig. 1) was dissolved in buffer (0.5x cytosol volume). Endogenous free steroids were removed by incubation with charcoal-dextran for 15 min at 2°C (51). Incubation with saturating ³H-estradiol (10 nM) was for 16 hr at 2°C followed by a 1-hr period for exchange at 25°C. Bound ³H-estradiol was separated from free ³H-estradiol by charcoal-dextran treatment (31). Nonspecific binding was determined by parallel incubations including 1 μM diethylstilbestrol as well as the [³H]-estradiol. The results represent the mean ± SEM for duplicate determinations on separate preparations from each of 4 animals.

ration with one which had been presaturated with unlabeled estradiol). Maximal loss of cytosol sites is seen at 90 min, and restoration to control levels is gradual over the next 48 hr.

Both in the embryonic and the cockerel liver, the loss of high-affinity binding sites from cytosol after estradiol injection is accompanied by an increase in high-affinity sites in salt extracts of liver nuclei, but the early increase is less than that which might be expected by stoichiometric translocation of sites to that fraction (18) (Table 1). This might be accounted for by the fact that nuclear receptor is also present in an insoluble form (22, 37). At later times, the increment in the soluble nuclear receptor is far greater than the total of the depleted cytosol binding sites (Table 1). Various lines of evidence, discussed more fully below, lead to the proposal that the early increase in nuclear receptor following estradiol injection reflects translocation of binding sites from cytosol, but that later increases are dependent upon a mechanism involving protein synthesis (44).

The cytosol high-affinity estradiol binder thus bears some resemblance to classic cytosol steroid receptors, but major reservations must be held until further experiments on its properties and role are conducted. Compared to chick oviduct (51) and rat uterus (28), the cockerel liver contains a very low concentration of high-affinity cytoplasmic estradiol binding sites. However, several reports indicate that the high-affinity estradiol-binding site concentration in rat liver cytosol is of the same order of magnitude (5,16, 57). These data assume that receptor is evenly distributed in all cells of the tissue, which of course may not be the case.

NUCLEAR ESTROGEN-BINDING PROTEINS

High-affinity binding of estradiol in avian liver nuclei has been examined in salt extracts of nuclei (30,39,42), in the salt-inextractable nuclear residue and a solubilized preparation thereof (35,36), in isolated chromatin and solubilized fractions (22,23), and in crude and purified suspensions of nuclei (6,44). Each type of binding has been found to have a dissociation constant in the range of 0.1 to 2 nM and specificity for estrogens only. Relatively low-affinity estrogen binding similar to that of the cytosol HSBP has not been characterized in avian liver nuclei. Preparations from crude nuclei do contain an appreciable level of nonsaturable estradiol binding, but this is greatly reduced upon purification of nuclei or of the extract (C. Lazier and W. Murdock, *unpublished results*).

The various liver nuclear preparations from immature chicks, embryos, or roosters contain low concentrations of high-affinity estradiol binding sites. Increased concentrations of the nuclear sites are found in liver of laying hens in which serum estradiol levels are elevated, or in liver of immature animals or roosters given an appropriate injection of estrogen (23,30,37).

The response to exogenous estrogen and the binding characteristics of the liver nuclear high-affinity estrogen-binding sites are such that the use of the term "receptor" has generally been considered to be appropriate. The relationship between the different forms of nuclear receptor is not yet clear, as will be apparent in the following dicussion.

The Soluble Nuclear Receptor

The so-called soluble nuclear receptor in avian liver has usually been assayed by variations of a charcoal-dextran adsorption technique first reported by Mester and Baulieu (39). Nuclei are extracted with a salt-containing buffer (0.5 to 1.0 M KCl), followed by freezing and thawing and high-speed centrifugation. The supernatant contains 30 to 66% of the total nuclear high-affinity estradiol binding sites (37,44). An early report that a significant degree of release of nuclear binding sites could be accomplished with low–ionic strength buffer was not confirmed (32,44).

For quantitation of the estradiol binding sites the salt extracts of nuclei have usually been preincubated with charcoal-dextran suspension for 15 to 30 min at 37 to 45°C in order to remove endogenous bound and free estrogens. The "stripped" extract is then incubated with ^{3}H-estradiol for the desired time, after which free hormone is removed by charcoal-dextran treatment at 2°C. Usually, although not always in some earlier reports, non-saturable binding is determined by parallel incubations containing an excess of radioinert estradiol or diethylstilbestrol as well as the labeled estradiol (25,30,39,42).

The general use of the high-temperature charcoal-dextran preincubation for assay of the soluble nuclear binding sites may be queried on the grounds that it may be unduly harsh treatment for a receptor, but Schneider and Gschwendt (44) have reported that this technique in fact gives slightly higher values for specific binding of estradiol than an exchange assay conducted for 15 min at 37°C. In our experience preincubation of liver nuclear extracts from estrogen-treated chicks with charcoal-dextran is an essential step in the assay. If the preincubation is at 2°C then the same number of binding sites is detectable on an exchange assay at 30 to 37°C as is found when the preincubation is at 37°C and the labeled-hormone incubation at 2°C (C. Lazier and W. Murdock, *unpublished results*). The crude soluble nuclear receptor appears to be relatively temperature-stable (33).

Studies of the binding properties of crude preparations of the soluble nuclear receptor indicate specificity for estrogens but not for progesterone, androgens, or glucocorticoids (33,39). Values reported for the equilibrium dissociation constant vary widely, from 0.1 to 1.8 nM (25,30,33,39,45). Although the Scatchard plots generally appear to be linear, indicating a single class of high-affinity binding sites, the range of ^{3}H-estradiol concentrations used may have been too narrow to permit detection of a second class of sites with intermediate binding affinity.

The soluble nuclear receptor in crude extracts or a fraction partly purified by $(NH_4)_2SO_4$ precipitation sediments in salt-containing sucrose density gradients with a coefficient of 4.1 to 4.5S (33,41). Optimal conditions for centrifugation involve sedimentation of the receptor preparations through a sucrose gradient containing an even distribution of ^{3}H-estradiol, followed by charcoal treatment of each fraction (33). Such a system was used earlier for detection of the estrogen receptor from chick oviduct cytosol (24). Centrifugation of the liver soluble nuclear receptor in the presence of low salt concentrations results in reduction of specifically bound ^{3}H-estradiol in the 4S region of the gradient (33). No binding entities sedimenting in the 8 to 10S region are detected, and it is likely that the receptor has aggregated and sedimented as a pellet under such conditions.

Optimal conditions for determination of the Stokes radius of the soluble nuclear receptor by gel filtration also require the presence of a high salt concentration. Otherwise, pronounced receptor aggregation is found (32). In 0.5 M KCl–containing buffer, specifically bound ^{3}H-estradiol elutes from a calibrated column of Sephacryl-S200 (Pharmacia) in a fraction corresponding to a protein with a Stokes radius of 30.5 Å (W. Murdock and C. Lazier, *unpublished results*). Gel filtration on Agarose A.5M (BioRad) appears to give a significantly lower Stokes radius (41), possibly indicating anomalous interaction of the receptor with the column matrix. Using 4.1S as the value for the sedimentation coefficient (41), 30.5 Å for the Stokes radius, and assumptions for other physical parameters as given by Sherman et al. (48), a molecular weight of 54,400 and a frictional ratio of 1.12 can be calculated for the soluble nuclear receptor. This suggests a compact, globular character for the molecule.

Limited studies on the purification of the soluble nuclear receptor have been reported. An initial step of $(NH_4)_2SO_4$ precipitation (using up to 42% saturation with the salt) gives purification of two- to fivefold, yielding specific activities of 100 to 250 fmole ^{3}H-estradiol bound/mg protein (13, 41). This step appears to remove RNA polymerase, DNase, and DNA from the extract, but RNase contamination is still apparent (13). The $(NH_4)_2SO_4$ preparation of the soluble nuclear receptor stimulates transcriptional activity in isolated nuclear and nucleolar chromatin fractions from immature chicks (4,13). These interesting results underline the necessity for further purification and characterization of receptor fractions. Hydroxylapatite chromatography gives some purification of the receptor (22,41), but it seems most likely that the method of choice for intensive investigation should be estrogen affinity chromatography (11,49,50). In some preliminary studies, we have recently achieved apparently substantial purification of an $(NH_4)_2SO_4$ fraction of the soluble nuclear receptor using this technique.

The crude salt soluble nuclear estrogen receptor from chick oviduct appears to have binding and physical characteristics similar to that from liver (3,26). Eventual comparison of the properties and biological activities of purified preparations from both tissues is of obvious interest.

The Insoluble Nuclear Receptor

Lebeau et al. (35,36) first reported that a significant proportion of ^{3}H-estradiol bound in liver nuclei of estrogen-treated chicks or laying hens appeared to be associated with a salt-insoluble proteinaceous nuclear residue. Solubilization can be effected by a limited trypsin treatment, yielding a protein with a binding affinity and hormone specificity very similar to the soluble nuclear receptor (37). Analysis of the physical properties of the "trypsinized" insoluble receptor shows a sedimentation coefficient of about 4S and a Stokes radius of 35.5 Å (somewhat larger than the soluble nuclear receptor) (37).

Solubilization of chromatin-bound estrogen receptor has also been achieved by treatment of chromatin with 2 M KCl/5 M urea followed by hydroxylapatite chromatography (22). Under these conditions two estradiol-binding activities are resolved, one of which appears to contain residual DNA from chromatin and the other of which is nucleic acid–free. This latter binding protein is a little larger than the soluble nuclear receptor on gel filtration, although after mild trypsin treatment it behaves in a similar manner to a trypsinized soluble nuclear receptor preparation.

The physical and binding studies on the soluble and insoluble (solubilized) nuclear estrogen receptors do not clearly indicate the relationship between these fractions. The soluble receptor could be the form of receptor prior to tight chromatin binding, or it could be a "spent" form of the binder. Further purification and investigation of their physical, chemical, and immunological properties and of their biological activities is essential.

In some studies nuclear estrogen binding sites in unextracted chromatin or in preparations of whole or lysed nuclei have been measured (6,23,44). Theoretically, the concentration should reflect the sum of the soluble plus the insoluble receptor. Schneider and Gschwendt indeed (44) demonstrated that binding by the two types of site is additive when compared to total specific binding by lysed nuclei. However, they also point out that the proportion of nuclear binding sites in the insoluble form depends to a certain extent upon the method of extraction of the nuclei. This important observation requires further study with regard to the relationship between the soluble and insoluble receptors. One view is that salt-insoluble nuclear receptor represents sites bound tightly to functional chromatin acceptor sites (9).

Kinetic Studies on Nuclear Receptor Following Estradiol Injection

The concentration of nuclear estrogen receptor in liver rises rapidly after estradiol injection (23,25,30,39,44). The time course for the increase depends upon the route of injection of the hormone. Intravenous injection of 2 μg estradiol/kg into the portal vein elicits a marked increase in soluble nuclear receptor within one minute (25). Depot injection of larger amounts of estradiol (20 mg/kg) subcutaneously, intramuscularly, or intraperitoneally gives very prolonged substantial increases in nuclear receptor concentration, which are first observed by 5 to 10 min, but which are maximal at 24 to 48 hr and fall off gradually (23,25,30,39). Only the large depot doses of the hormone have been observed to give vitellogenin induction (30,44). The very early increase (< 10 min) in nuclear receptor concentration does not appear to be sensitive to inhibition by cycloheximide or by actinomycin D (25,44). Later increases in receptor concentration are, however, inhibited by the drugs (23,30,44). Schneider and Gschwendt (44) found that estradiol provoked an initial drug-resistant increase in both soluble and insoluble nuclear receptor levels at 10 min, followed by a slight decline and then by a drug-sensitive rise in both forms of receptor.

Lebeau et al. (37), on the other hand, find that there is a differential effect initially, the percent increase in the soluble receptor being greater than in the insoluble receptor, followed by a parallel increase in both. This might indicate that the soluble form of the receptor is that which has not yet been incorporated into chromatin. Regardless of the incongruities with regard to the relative kinetics of the increase in the soluble and insoluble forms, the important point is that nuclear receptor levels rise quickly, in a manner which is in keeping with the notion of initial translocation of receptor from cytosol, as in the classic model of steroid action. The cycloheximide sensitivity of later increases is consistent with a model in which cytosol receptor is turning over rapidly or in which estradiol induces synthesis of the receptor.

The necessity for the large depot doses of estradiol for vitellogenin induction in roosters or immature chicks of either sex has been acknowledged by many workers (12,30,44; and references cited therein). The average serum estradiol concentration in the laying hen is about 0.7 nM (40). Injection of the usual vitellogenin-inducing dose of estradiol into the cockerel (25 mg/kg) produces the remarkable serum estradiol level of 200 nM after 15 min (W. Moger and C. Lazier, unpublished results). This falls off rapidly, but is still at least 20 to 30 times the control levels (0.06 nM) 6 days after hormone treatment. At this time nuclear receptor levels are approaching control values and serum vitellogenin concentration is declining.

We have recently found that a range of doses of estradiol from 1 to 25 mg/kg each give a similar initial stimulation of soluble nuclear receptor levels, but that only the highest doses provoke prolonged elevation of the receptor and accumulation of substantial serum vitellogenin. Even with the lowest dose, however, the serum estradiol at 66 hr is twice that of the laying hen, yet no vitellogenin is detected and nuclear receptor levels have fallen to baseline values. It appears, therefore, that the cockerel liver is less sensitive to estradiol than the laying hen liver. This does not detract from the usefulness of the male bird as a model for estrogen control of gene expression, but in physiological terms it would be most illuminating to understand the molecular basis for this phenomenon. A possible role of HSBP should be considered in this regard (14).

Antiestrogen Action in Chick Liver

The study of the action of nonsteroidal antiestrogens has received widespread attention in mammalian systems and has led to useful insights with regard to estrogen receptor regulation. It is well established that injection of rats with antiestrogens such as nafoxidine (Upjohn 11,100A) or CI-628 (Parke Davis) causes translocation of the uterine cytoplasmic receptor to the nucleus, long-term nuclear retention of receptor, and significant uterotrophic responses. For a certain period after administration of the drug the uterus is refractory to estrogen stimulation. The basis for this antagonistic action appears to be inhibition by the antiestrogen of the normal process of restoration of cytoplasmic estrogen receptor levels (10,17,27).

Administration of a single dose of antiestrogen (nafoxidine or CI-628) to the chick results in an increase in soluble nuclear and chromatin-bound estrogen receptor levels, but the increase is delayed and is quantitatively much smaller than that evoked by the usual doses of estradiol (20,33). The antiestrogens are not themselves estrogenic in terms of vitellogenin induction in the chick (20,33). When given with estradiol, however, nafoxidine exhibits marked antiestrogenic action, 50% inhibition of serum vitellogenin levels being obtained by an antiestrogen/estrogen molar ratio of about 0.6. With this dose ratio soluble nuclear estrogen receptor levels at 48 and 64 hr are strongly reduced, implying that the antiestrogen interferes with the mechanism responsible for sustained maintenance of elevated nuclear receptor levels required for the full vitellogenic response. A molar ratio of antiestrogen/estrogen of 1.2 completely inhibits the estrogen-induced rise in the soluble nuclear receptor and accumulation of serum vitellogenin (33). Similar potent inhibition by nafoxidine of estrogen-induced VLDL synthesis has been found in cockerel liver (7).

In vitro nafoxidine appears to compete for chick liver soluble nuclear estradiol receptor sites, but the affinity of the drug for the receptor is only 4% of that of estradiol itself (33). The potent antiestrogenic action of nafoxidine in the chick liver is therefore probably not by simple blockade of nuclear estrogen binding sites. Other possible considerations include metabolism of the drug to a more potent inhibitor, interference with the receptor system at some point other than the soluble nuclear receptor, and/or non–receptor-linked actions. In view of the recent suggestion that nafoxidine is probably a pro-antiestrogen in rat uterus (17), the question of metabolism of the drug in the chick liver deserves further study.

In the chick oviduct the antiestrogen, tamoxifen (ICI, 46,474), causes depletion of the cytosol estrogen receptor, translocation and long-term retention of nuclear receptor, along with a concomitant block in the normal restoration of cytosol receptor levels (52). In this behavior the drug is similar to antiestrogens in uterus. However, no estrogenic action by tamoxifen alone is seen in terms of oviduct weight, DNA content, or progesterone receptor concentration, and potent antiestrogenic activity is found in a single injection with estradiol. In the predominance of antagonistic activities, antiestrogen action in chick oviduct is more similar to that in chick liver than in rat uterus. It is difficult, however, to draw firm conclusions from comparison of the available studies on antiestrogen action in different tissues and animals. Different doses of drug and/or hormone, different exposure times, and different antiestrogens which might have quite distinctive durations of action have been used. In oviduct and uterus, inhibition of replenishment of cytosol receptor is clearly evident and it has been suggested that the block is in receptor synthesis (27). In chick liver the effect of antiestrogens on the recently discovered high-affinity cytosol estradiol binding protein has not yet been reported, but it is known that the drugs block the estradiol-induced increase in nuclear receptor concentration. This may also possibly reflect a receptor synthesis defect.

ONTOGENY OF THE HEPATIC ESTROGEN-BINDING PROTEINS

The pattern of ontogenesis of steroid receptors is of particular interest when it can be examined in relation to development of complete responsiveness to the hormone. An early study suggested that the developing chick embryo only became able to respond to exogenous estrogen with elevated serum phosphoprotein levels at about days 19 to 20 of egg incubation (43). Since an indirect, and in current terms relatively insensitive, assay was used, we reexamined the question of the development of the capacity to synthesize vitellogenin (31). Livers from control and estradiol-treated embryos at different stages of development were incubated with [³H-leucine and the vitellogenin synthesized was identified by gel electrophoresis of vitellogenin-antivitellogenin immunoprecipitates. Figure 2 shows that injection of estradiol into the yolk sac clearly stimulates vitellogenin production by liver of the 15-day embryo. A slight response is seen in the estrogen-treated 13-day embryo, but no discrete peak corresponding to labeled vitellogenin is detectable before that stage or in control livers.

In terms of the percentage of total protein labeled by equal weights of liver tissue, the vitellogenin synthesis response in the 15- to 21-day embryos appears to be about the same as that seen in a primary response to the usual dose of estradiol in the cockerel (31). To judge whether the embryonic and

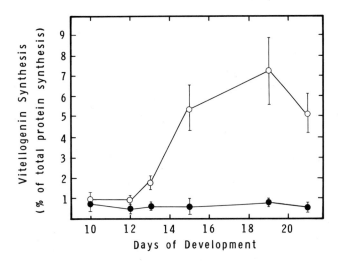

FIG. 2. *Vitellogenin synthesis by livers of estradiol-treated or control embryos at different stages of development.* Incorporation of ³H-leucine into the vitellogenin band on SDS-polyacrylamide gel electrophoresis is expressed as percentage of total protein labeled. The methods for liver incubation, immunoprecipitation of ³H-vitellogenin, and electrophoresis have been described (31). Each point is the mean ± SEM for duplicate determinations on two or three separate preparations. ○, estradiol-treated embryos (1.25 mg, in 0.5 ml propylene glycol, injected into the yolk sac 48 hr before assay); ●, propylene glycol–treated control embryos. (From Lazier ref. 31, with permission.

postembryonic livers are in fact equally responsive to the hormone would require knowledge of the serum estradiol levels reached with the differing routes of hormone injection. This has not been determined.

High affinity estradiol-binding proteins have been characterized in chick embryo liver cytosol (18) and in salt extracts of nuclei (21,31). There have been no studies reported on the relatively low-affinity HSBP in embryonic liver.

As was discussed above, Gschwendt (18) has reported that liver cytoplasmic preparations from 19-day embryos contain a limited number (1,100 sites/cell) of high-affinity binding sites specific only for estrogens and which appear to translocate to the nucleus on estrogen treatment. The properties of the embryonic binder are similar to those of an activity found in cockerel liver cytosol, although the concentration of binding sites is in fact higher in the 19-day embryonic liver. The developmental pattern observed by Gschwendt for the cytosol estradiol binding sites in the embryos of a "fattening chicken" strain has some pronounced features. No change in the K_d of the binding occurs in development, but the binding site concentration increases from about 344 fmole/g liver at day 14 to 1,850 fmole/g at day 19, followed by a decline to 290 fmole/g in the 11-day-old hatched chick. The capacity to synthesize vitellogenin was not investigated in this strain of chicken. However, the ontogeny of the cytosol binder does not correlate well with the ontogeny of the vitellogenic response in white Leghorn chick embryos discussed earlier. There is about a fourfold increase in the concentration of the cytosol binder between day 15 and day 19, yet no such difference in vitellogenin synthetic capacity is found. Similarly, no apparent drop in the ability to produce vitellogenin is found after hatch. The most facile explanation for this dilemma is that it may be accounted for by the different strains of chicken used, which have very different growth rates and may conceivably show distinctive ontogenesis of the vitellogenic response. On the other hand, the apparent peak in cytosol binder concentration at day 19 may reflect a developmental event in liver cells which is independent from control of estrogen responsiveness. Alternatively, the peak may be more apparent than real, perhaps due to loss of binding sites at other stages for technical reasons. Finally, it remains possible that the high-affinity cytosol binder is not a true estrogen receptor.

Liver nuclei from estrogen-treated embryos contain salt-soluble high-affinity binding sites which are very similar to the soluble nuclear estrogen receptor of hatched chicks (21,31). Figure 3 shows the effect of yolk sac injection of estradiol on the concentration of the nuclear sites at various stages of development of white Leghorn embryos. The dose of estradiol used and time of exposure were chosen in preliminary experiments which showed that these conditions were optimal for demonstrating both vitellogenin synthesis and the estrogen-induced increase in nuclear receptor concentration in pooled livers from the same group of 15-day embryos (31). Thus, exposure

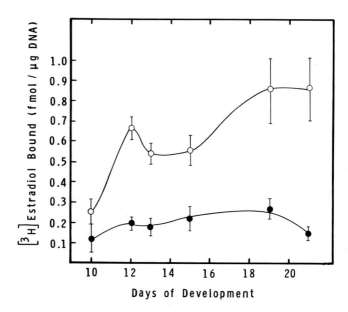

FIG. 3. *Effect of estradiol on the soluble nuclear estrogen receptor in embryonic chick liver.* Embryonated eggs were injected with estradiol (Fig. 2, legend), and after 48 hr the concentration of the soluble nuclear estrogen receptor in liver was determined (31). Each point represents the mean ± SEM for at least three different liver preparations. ○, estradiol-treated embryos; ●, propylene glycol–treated control embryos. (From Lazier, ref. 31, with permission.)

to estradiol for 48 hr marginally stimulates nuclear receptor levels measured on the 10th day of egg incubation, but from the 12th day onward such treatment gives a three- to fourfold increase in the nuclear receptor. The extent of the increase at day 12 is the same as that at day 15, but vitellogenin synthesis proceeds only at the later stage. The nuclear receptor response to exogenous estradiol thus appears to be mature by day 12, but some functional consequences of receptor binding are not yet developed. Some receptor functions may be mature. For example, there is some indication that estrogen treatment increases serum VLDL concentration in very young embryos (43). A direct comparative study of the ontogeny of different specific responses to estradiol could be most informative.

It is interesting that the sequence for receptor ontogeny observed in Müllerian duct of white Leghorn chick embryos is similar to that seen in liver (55,56). Maximal concentrations of cytoplasmic estrogen receptor are found at day 12, and estradiol treatment provokes stoichiometric translocation of these binding sites to the nucleus. The rapid period of Müllerian duct growth, however, is not seen until days 14 to 18 (56).

The appearance of the nuclear receptor response at day 12 in the white Leghorn embryo liver does not seem to be compatible with the peak in the concentration of high-affinity cytoplasmic estradiol binding sites seen at day 19 in the embryos of the "fattening chicken" strain (18). Some possible reasons for this apparent paradox have been given, but further work on this problem is imperative before any conclusions can be drawn. Comparison of the ontogeny of nuclear and cytoplasmic binding sites in the same strain of chicken is obviously the first requirement.

Another aspect of the developmental studies which raises intriguing questions for further study is the observation of the apparent dissociation between nuclear receptor development and attainment of complete hormone responsiveness. Is vitellogenin mRNA synthesized at day 12 but not translated? If the specific mRNA is not transcribed, is it because the putative acceptor sites in chromatin are blocked, absent, or nonfunctional? Is receptor in fact mature? It could have an active hormone-binding site but lack the proper chromatin-binding site (subunit?). Experiments with a complementary DNA probe to vitellogenin mRNA could answer the first question.

The developmental stimulus for the appearance of estrogen receptors in ontogeny is unknown. It could be an autonomous feature of differentiation of the tissue in question, or it could be dependent on an endocrine factor(s).

SUMMARY

This review of estrogen-binding proteins in avian liver emphasizes the properties of the binding proteins and their relation to the regulation of the vitellogenic response to estradiol. The high-affinity estradiol-binding protein in embryonic chick or cockerel liver cytosol is hormone-specific and appears to be depleted upon estrogen treatment. This is accompanied by a large and protracted increase in the concentration of specific high-affinity estradiol binding sites in both salt-soluble and insoluble fractions of the nucleus. Circumstantial evidence gathered from temporal, developmental, and dose-response correlations and from experiments with antiestrogens implies that the increased concentration of nuclear binding sites is essential in the induction of vitellogenin. No direct proof yet exists. Protein synthesis is involved in attainment of the prolonged increase in the nuclear receptor levels.

Avian liver cytosol contains a very high concentration of an unusual steroid-binding protein (HSBP) which preferentially binds estrone, progesterone, and pregnenolone with "intermediate" affinity. HSBP is not a receptor and it may function as a hepatic sink or reservoir for the steroids.

For further understanding of the role of the estrogen-binding proteins in avian liver it is essential that receptor be purified and receptor-chromatin interaction analyzed. The dynamics of receptor regulation under a variety of physiological conditions need to be investigated in more detail. The contribution of HSBP to the overall control of estrogen responsiveness should be assessed.

ACKNOWLEDGMENT

The support of the Medical Research Council of Canada is gratefully acknowledged.

REFERENCES

1. Anderson, J. N., Peck, E. J., Jr., and Clark, J. H. (1975): Estrogen-induced uterine responses and growth: relationship to receptor estrogen binding by uterine nuclei. *Endocrinology,* 96:160–167.
2. Arias, F., and Warren, J. C. (1971): An estrophilic macromolecule in chicken liver cytosol. *Biochim. Biophys. Acta,* 230:550–559.
3. Best-Belpomme, M., Mester, J., Weintraub, H., and Baulieu, E.-E. (1975): Oestrogen receptors in chick oviduct: Characterization and subcellular distribution. *Eur. J. Biochem.,* 57:537–547.
4. Bieri-Bonniot, F., Joss, U., and Dierks-Ventling, C. (1977): Stimulation of RNA polymerase activity by 17β-estradiol-receptor complex on chick liver nucleolar chromatin. *FEBS Lett.,* 81:91–96.
5. Chamness, G. C., Costlow, M. E. and McGuire, W. L. (1975): Estrogen receptor in rat liver and its dependence on prolactin. *Steroids,* 26:363–371.
6. Chan, L., Eriksson, H., Jackson, R. L., Clark, J. H., and Means, A. R. (1977): Effects of estrogen on very low density lipoprotein (VLDL) synthesis in avian liver slices *in vitro:* lack of correlation with nuclear estrogen receptors. *J. Steroid Biochem.,* 8:1189–1191.
7. Chan, L., Jackson, R. L., and Means, A. R. (1977): Female steroid hormones and lipoprotein synthesis in the cockerel: effects of progesterone and nafoxidine on the estrogenic stimulation of very low density lipoproteins (VLDL) synthesis. *Endocrinology,* 100:1636–1643.
8. Chan, L., and O'Malley, B. W. (1976): Mechanism of action of the sex steroid hormones. *N. Engl. J. Med.,* 294:1322–1328.
9. Clark, J. H., Eriksson, H. A., and Hardin, J. W. (1976): Uterine receptor-estradiol complexes and their interaction with nuclear binding sites. *J. Steroid Biochem.,* 7:1039–1043.
10. Clark, J. H., Peck, E. J., Jr., and Anderson, J. N. (1974): Oestrogen receptors and antagonism of steroid hormone action. *Nature,* 251:446–448.
11. Coffer, A. L., Milton, P. J. D., Pryse-Davies, J., and King, R. J. B. (1977): Purification of oestradiol receptor from human uterus by affinity chromatography. *Mol. Cell. Endocrinol.,* 6:231–246.
12. Deeley, R. G., Gordon, J. L., Burns, A. T. H., Mullinix, K. P., Bina-Stein, M., and Goldberger, R. F. (1977): Primary activation of the vitellogenin gene in the rooster. *J. Biol. Chem.,* 252:8310–8319.
13. Dierks-Ventling, C., and Bieri-Bonniot, F. (1977): Stimulation of RNA polymerase I and II activities by 17β-estradiol receptor in chick liver chromatin. *Nucleic Acids Res.,* 4:381–395.
14. Dower, W. J. (1977): The binding of steroids in the avian liver. Ph.D. thesis, University of California, San Diego.
15. Dower, W. J., and Ryan, K. J. (1976): A cytoplasmic estrone-specific binding protein (E_1BP) in hen liver. *Fed. Proc.,* 35:1366, abstract no. 70.
16. Eisenfeld, A. J., Aten, R. F., Haselbacher, G. K., and Halpern, K. (1977): Specific macromolecular binding of estradiol in mammalian liver supernatant. *Biochem. Pharmacol.,* 26:919–922.
17. Ferguson, E. R., and Katzenellenbogen, B. S. (1977): A comparative study of antiestrogen action: temporal patterns of antagonism of estrogen-stimulated uterine growth and effects on estrogen receptor levels. *Endocrinology,* 100:1242–1251.
18. Gschwendt, M. (1977): A cytoplasmic high-affinity estrogen-binding protein in embryonic chicken liver. *Eur. J. Biochem.,* 80:461–468.

19. Gschwendt, M. (1975): A cytoplasmic oestrogen-binding component in chicken liver. *Hoppe Seylers Z. Physiol. Chem.,* 356:157–165.
20. Gschwendt, M. (1975): The effect of antiestrogens on egg yolk protein synthesis and estrogen-binding to chromatin in the rooster liver. *Biochim. Biophys. Acta,* 399: 395–402.
21. Gschwendt, M. (1977): Estrogen binding sites in the embryonic chick liver. *FEBS Lett.,* 75:272–276.
22. Gschwendt, M. (1976): Solubilization of the chromatin-bound estrogen receptor from chicken liver and fractionation on hydroxylapatite. *Eur. J. Biochem.,* 67: 411–419.
23. Gschwendt, M., and Kittstein, W. (1974): Specific binding of estradiol to the liver chromatin of estrogen-treated roosters. *Biochim. Biophys. Acta,* 361:84–96.
24. Harrison, R. W., and Toft, D. O. (1975): Estrogen receptors in the chick oviduct. *Endocrinology,* 96:199–205.
25. Joss, U., Bassand, C., and Dierks-Ventling, C. (1976): Rapid appearance of estrogen receptor in chick liver nuclei: partial inhibition by cycloheximide. *FEBS Lett.,* 66: 293–298.
26. Kalimi, M., Tsai, S. Y., Tsai, M. J., Clark, J. H., and O'Malley, B. W. (1976): Effect of estrogen on gene expression in the chick oviduct: correlation between nuclear-bound estrogen receptor and chromatin initiation sites for transcription. *J. Biol. Chem.,* 251:516–523.
27. Katzenellenbogen, B. S., Ferguson, E. K., and Lan, N.C. (1977): Fundamental differences in the action of estrogens and antiestrogens in uterus: comparison between compounds with similar duration of action. *Endocrinology,* 100:1252–1259.
28. King, R. J. B., and Mainwaring, W. I. P. (1974): *Steroid-Cell Interactions.* University Park Press, Baltimore.
29. Lan, N., and Katzenellenbogen, B. S. (1976): Temporal relationships between hormone receptor binding and biological responses in the uterus. Studies with short- and long-acting derivatives of estriol. *Endocrinology,* 98:220–228.
30. Lazier, C. B. (1975): [³H]-Estradiol binding by chick liver nuclear extracts: mechanism of increase in binding following estradiol injection. *Steroids,* 26:281–298.
31. Lazier, C. B. (1978): Ontogeny of the vitellogenic response to oestradiol and of the soluble nuclear oestrogen receptor in embryonic chick liver. *Biochem. J.,* 174:143–152.
32. Lazier, C. B., and Alford, W. S. (1976): Estradiol-binding proteins in chick liver: effects of salt. *Proc. Can. Fed. Biol. Soc.,* 19:590 (abstract).
33. Lazier, C. B., and Alford, W. S. (1977): Interaction of the antioestrogen, nafoxidine-hydrochloride, with the soluble nuclear oestradiol-binding protein in chick liver. *Biochem. J.,* 164:659–667.
34. Lazier, C. B., Haggarty, A. J., and Comeau, T. L. (1978): A high-affinity estradiol-binding protein in chick liver cytosol. *Proc. Can. Fed. Biol. Soc.,* 21: 397 (abstract).
35. Lebeau, M.-C., Massol, N., and Baulieu, E.-E. (1974): Extraction, partial purification, and characterization of "the insoluble estrogen receptor" from chick liver nuclei. *FEBS Lett.,* 43:107–111.
36. Lebeau, M.-C., Massol, N., and Baulieu, E.-E. (1973): An insoluble receptor for oestrogens in the residual nuclear proteins of chick liver. *Eur. J. Biochem.,* 36:294–300.
37. Lebeau, M.-C., Massol, N., Lemonnier, M., Schmelck, P.-H., Mester, J., and Baulieu, E.-E. (1977): Estrogen-binding proteins in chick liver. Aspects of DNA polymerase activity. In: *Hormonal Receptors in Digestive Tract Physiology,* edited by S. Bonfils et al., pp. 183–195. North Holland, Amsterdam.
38. Mester, J., and Baulieu, E.-E. (1975): Dynamics of oestrogen-receptor distribution between the cytosol and nuclear fractions of immature rat uterus after oestradiol administration. *Biochem. J.,* 146:617–623.
39. Mester, J., and Baulieu, E.-E. (1972): Nuclear estrogen receptor of chick liver. *Biochim. Biophys. Acta,* 261:236–244.
40. Mulvihill, E. R., and Palmiter, R. D. (1977): Relationship of nuclear estrogen re-

ceptor levels to induction of ovalbumin and conalbumin mRNA in chick oviduct. *J. Biol. Chem.*, 252:2060–2068.

41. Murdock, W., and Lazier, C. B. (1977): Partial purification and characterization of the soluble nuclear estrogen receptor from chick liver. *Proc. Can. Fed. Biol. Soc.*, 20:87 (abstract).
42. Ozon, R. and Bellé, R. (1973): Récepteurs de l'oestradiol-17β dans le foie de poule et de l'amphibian *Discoglossus pictus. Biochim. Biophys. Acta*, 297:155–163.
43. Schjeide, O. A., Binz, S., and Ragan, N. (1960): Estrogen-induced serum protein synthesis in the liver of the chicken embryo. *Growth*, 24:401–410.
44. Schneider, W., and Gschwendt, M. (1977): Kinetics of the appearance of nuclear estrogen binding sites in chick liver. *Hoppe Seylers Z. Physiol. Chem.*, 358:1583–1589.
45. Sen, K. K., Gupta, P. D., and Talwar, G. P. (1975): Intracellular localization of estrogens in chick liver. Increase in binding sites for the hormone on repeated treatment of the birds with the hormone. *J. Steroid Biochem.*, 6:1223–1227.
46. Shapiro, D. J., and Baker, H. J. (1977): Purification and characterization of *Xenopus laevis* vitellogenin messenger RNA. *J. Biol. Chem.*, 252:5244–5250.
47. Sheridan, P. (1975): Is there an alternative to the cytoplasmic receptor model for the mechanism of action of steroids? *Life Sci.*, 17:497–502.
48. Sherman, M. R., Tuazon, F. B., Diaz, S. C., and Miller, L. K. (1976): Multiple forms of oviduct progesterone receptors analyzed by ion exchange filtration and gel electrophoresis. *Biochemistry*, 15:980–989.
49. Sica, V., Nola, E., and Puca, G. A. (1976): Purification of estrogen receptor by affinity chromatography. In: *Receptors and the Mechanism of Action of Steroid Hormones, Part I*, edited by J. R. Pasqualini, p. 85. Marcel Dekker, New York.
50. Smith, R. G., Schwartz, R. J., and O'Malley, B. W. (1976): Purification of a biologically active estradiol receptor from hen nuclei. *V Internatl. Congr. Endocrinol.*, p. 36 (abstract).
51. Sutherland, R. L., and Baulieu, E.-E. (1976): Quantitative estimates of cytoplasmic and nuclear oestrogen receptors in chick oviduct. Effect of oestrogen on receptor concentration and subcellular distribution. *Eur. J. Biochem.*, 70:531–541.
52. Sutherland, R. L., Mester, J., and Baulieu, E.-E. (1977): Tamoxifen is a potent "pure" anti-oestrogen in chick oviduct. *Nature*, 267:434–435.
53. Tata, J. R. (1976): The expression of the vitellogenin gene. *Cell*, 9:1–14.
54. Teng, C. S., and Teng, C. T. (1975): Studies on sex-organ development. Isolation and characterization of an oestrogen receptor from chick müllerian duct. *Biochem. J.*, 150:183–190.
55. Teng, C. S., and Teng, C. T. (1976): Studies on sex-organ development. Oestrogen-receptor translocation in the developing chick müllerian duct. *Biochem. J.*, 154:1–9.
56. Teng, C. S., and Teng, C. T. (1975): Studies on sex-organ development. Ontogeny of cytoplasmic oestrogen receptor in chick müllerian duct. *Biochem. J.*, 150:191–194.
57. Viladin, P., Delgado, C., Pensky, J., and Pearson, O. H. (1975): Estrogen binding protein in rat liver. *Endocr. Res. Commun.*, 2:273–280.

Ontogeny of Receptors and Reproductive Hormone Action,
edited by T. H. Hamilton, J. H. Clark, and W. A. Sadler.
Raven Press, New York 1979.

Nature of Hormone Action in the Brain

Roger A. Gorski

*Department of Anatomy and Brain Research Institute, UCLA School of Medicine,
Los Angeles, California 90024*

Investigation of the molecular mechanisms of action of steroid hormones on neural tissue has lagged behind study of hormone action on peripheral organs such as the uterus or oviduct. Nevertheless, the central nervous system, particularly the brain, is clearly a target organ for reproductive hormones and is, in fact, an integral and critical component of the reproductive system. The purpose of the present contribution is to review certain aspects of the nature of hormone action in the brain in the adult and during differentiation in order to provide a background for the more sophisticated studies presented in this volume. We consider the problems inherent in brain research, the conceptual nature of steroid hormone action in the brain, the possible mechanisms of that action in the adult, and finally, possible mechanisms of steroid action during development.

There are a number of factors that contribute to the difficulty of the task of elucidating the action of steroid hormones on the brain. Although there clearly is a functional heterogeneity among the cells of an organ such as the uterus, anatomical complexity probably reaches its highest degree within the brain. Neurons and glia of different structure and function are mixed in the same brain area, probably adjacent one to the other and perhaps interconnected. Moreover, there appears to be considerable functional and perhaps morphological plasticity even in the adult organism. To complicate matters further, it is likely that an individual neuron participates in more than one, perhaps many, functions, often unknown or unrecognized; or equally likely, one specific function may involve a complex constellation of widely dispersed neurons. Finally, it appears to be a reasonable generalization to state that there is a multiplicity or redundancy of regulatory components for various functions. There is a rich opportunity for the interaction of several neural systems and the modulation of particular functions. Unfortunately, it is difficult to determine whether any one neural system is critical for a function or only exerts a modulatory influence on that function.

Perhaps some of these obvious problems could be overcome or circumvented technically if it were possible to study individual neurons. Although under certain conditions, e.g., electrophysiologically, histochemically, or anatomically, it is possible to study one neuron, currently it is not possible to relate one cell to a particular function, nor is it possible to measure the chemical changes in one cell or, in many cases, even in one population of highly similar cells. Even given these abilities, it would still appear to be very difficult conceptually to relate a specific and local hormone-induced chemical change to an alteration in the behavior of the whole animal. Perhaps in response to the probability that each technical approach to the study of the brain has its serious limitations, multiple approaches have been utilized, often simultaneously. Thus, the use of radiolabeled steroids has identified regions of the brain containing high concentrations of neurons that take up and retain steroids (71,88,92). Moreover, the existence in the brain of classic steroid-binding (receptor) proteins has been established in similar neural populations (20,47,63,66,96). The regional activity of the steroid metabolic enzymes has been characterized (26,58). These chemical approaches have been combined with hormone implantation, lesioning, deafferentation, stimulation, and recording studies, as well as with the regional assay of metabolic or synthetic activity of neurotransmitter systems or protein synthesis. Although functional alterations, e.g., gonadotropin secretion or sexual behavior, have been concurrently analyzed in many of these studies, no one study has included all of the possible functions with which the neuronal region under study could be implicated. Although a thorough analysis of the above-mentioned data is not warranted here, at least at certain levels there are obvious consistencies, for example, the comparable localization of steroid uptake mechanisms and the feedback regulation of gonadotropin secretion (31,66). Whether or not these consistencies are powerful enough to elucidate molecular mechanisms remains to be seen.

NATURE OF STEROID ACTION IN THE BRAIN

With respect to the conceptual nature of steroid action on the brain, it is still useful to adopt a general model first proposed in 1959. Working with the guinea pig, Phoenix et al. (75) described a permanent action of exogenous testosterone on the eventual reproductive behavior of the female fetus. At the same time, Barraclough (11,12) was independently pursuing a similar problem in the rat, although in this species it was possible to obtain a permanent effect following postnatal hormone exposure. Phoenix et al. (75) proposed that steroids could have two actions—activational and organizational.

Throughout most of the life of the animal, steroid hormones appear to "activate" (or inhibit) *existing* functional neuronal circuits. In contrast to this transitory effect, during an apparently specific period of development, exposure to steroids actually "organizes" or *determines* the functional neurocircuitry of the adult. In general, research in the area of steroid action on the brain has focused on the former or activational effects, both in the adult and more recently in the developing animal.

Even in the case of the proposed organizational action of steroids, most studies have centered on the adult animal, that is, a long time after the actual organizing action of steroids has occurred (27,29,36). Although it certainly makes sense to identify probable steroid-determined functional differences in the adult brain before one studies the possible mechanisms of their development, it should be stressed that purported studies of the organizational action of steroids that are actually carried out in the adult will be complicated by the activational effects of these hormones, as well as by the experiences of the animal prior to experimentation. Concurrent studies of the molecular action of steroids during the perinatal period are sorely needed.

ACTIVATIONAL ACTION OF STEROIDS IN THE ADULT

In considering the possible mechanisms of the activational action of steroid hormones on the adult brain, I focus on the role of the ovarian hormones in facilitating female sexual behavior. The sexually receptive female rat, when mounted by a male, will exhibit a characteristic posture called lordosis. Thus, it is possible to quantitate female sexual behavior in terms of the lordosis quotient (LQ), defined as the percentage of mounts by the male that are accompanied by lordosis in the female.

Numerous brain regions have been implicated in the control of lordosis behavior, including the preoptic area, the mediobasal hypothalamus, septum, amygdala, the mesencephalic reticular formation, and the cortex (32,33). Many of these same areas, particularly the components of the hypothalamic-limbic system, contain neurons with steroid-binding proteins with biochemical characteristics comparable to those of the peripheral "steroid-receptor" (20,47,63,66,96). This subject, including important developmental aspects, is considered by MacLusky et al. (*this volume*).

I will now review several studies performed in our laboratory a number of years ago. Although these are relatively crude compared to the sophistication of present biochemical techniques, hopefully they reveal something about the nature of steroid hormone action in the adult brain, and perhaps, moreover, point up certain problems. In the first experiment we attempted to characterize the temporal pattern of the action of estrogen in facilitating female sex behavior.

Lordosis behavior can be facilitated in the ovariectomized animal by appropriate hormone treatment. Although it is possible to facilitate high levels of lordosis response with estrogen alone, a more common procedure is to take advantage of the synergism between estrogen and progesterone (32). Although the use of two separate ovarian hormones must complicate interpretations, we chose this regime because of the well-timed response of the female rat. Thus, if the ovariectomized rat was given a single injection of $5\mu g/kg$ estradiol benzoate (EB) at time 0 (1,000 hr) and was tested for behavior 55 hr later, lordosis responses were absent. However, if similarly estrogen-primed females received as well an injection of 0.5 mg progesterone at 48 hr, the LQ at 55 hr approximated 100 (see Arai and Gorski, ref. 4, and Fig. 1).

For the experiment, the antiestrogen CI-628 (then called CN 55,945–27) was administered in two injections of 2 mg/kg, separated by 24 hr. The first of the two injections of CI-628 was given simultaneously with or 6,12,24, or 48 hr after the injection of EB, and the animals were tested for behavior at 55 hr, e.g., 7 hr after progesterone. As shown in Fig. 1, the antiestrogen significantly suppressed lordosis behavior when administered as late as 12 hr after EB. We interpreted these data to indicate that estrogen had to be present in the system for a prolonged period of 12 to 24 hr. Although the original interpretation was based on the assumption that the antiestrogen would block estradiol receptors, perhaps a more current view would involve the phenomenon of cytoplasmic receptor replenishment (2,41). In any case, these data suggest that the action of estrogen needed to facilitate lordosis behavior in

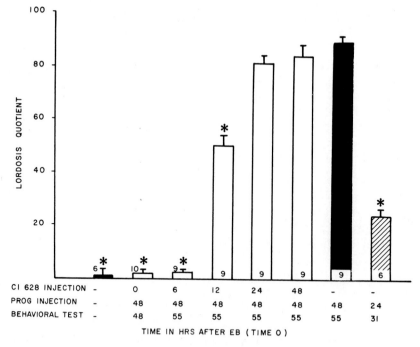

FIG. 1. The time course of EB facilitation of lordosis behavior as determined by the inhibition of the behavioral response produced by the simultaneous or delayed administration of the antiestrogen CI-628 with respect to the injection of 5 μg/kg EB at 1000 hr (time 0). In this study all animals were given EB. The bars indicate the mean LQ for the number of animals shown at the base of the bar. Standard errors of the mean are indicated by the thin vertical lines. Asterisk indicates a result significantly different from control EB–progesterone-treated group (*solid bar*). (Unpublished data from Arai and Gorski which confirms and extends a previous report.)

the rat is complete after 24 hr. If this is indeed true, we argued, the animal should be able to respond behaviorally to progesterone 24 hr after EB injection. However, when rats were given progesterone 1 day after EB and tested 7 hr later, lordosis behavior was only minimally facilitated (Fig. 1, *shaded bar*). Therefore, we proposed that there was a second phase of estrogen action independent of antiestrogen and, therefore, perhaps totally secondary to receptor–steroid interaction.

Armed with this tentative information about the time course of estrogen action, we attempted to study its mechanism of action by applying the antibiotic actinomycin D directly to the preoptic area, the region of the brain thought to be involved in the estrogen facilitation of lordosis behavior subsequent to steroid implantation studies (81). As shown in Fig. 2, implants of actinomycin D were able to inhibit lordosis behavior when administered 12 hr after EB injection. However, there were two major problems with this experiment. First, since this was an acute study, animals had to be anesthetized and operated on during this presumed critical phase of estrogen action. Perhaps because of this, the sham procedures themselves disrupted lordosis

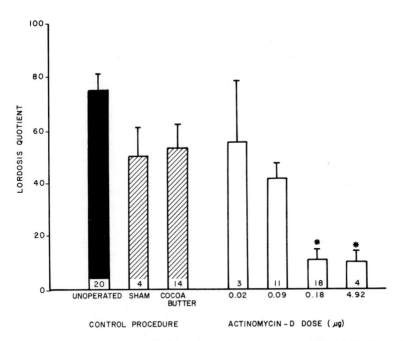

FIG. 2. The inhibition of the facilitation of lordosis behavior produced by the acute implantation of varying doses of actinomycin D in cocoa butter pellets into the preoptic area 12 hr after the injection of 3 μg EB. The influence of control procedures is also shown. All animals (numbers indicated at the base of each bar) were tested for lordosis (50-mount test) 5 hr after 0.5 mg progesterone, which was given 36 hr after estrogen. Asterisk indicates a result significantly different from the cocoa butter controls. (From Quadagno et al., ref 81.)

behavior, so that only the higher doses of actinomycin D produced a significant suppression in lordosis behavior relative to that shown by the shams. The second problem was potentially the most significant. At effective concentrations, actinomycin D implants frequently produced lesions surrounding the implant. Therefore, the question whether the suppression of lordosis behavior was a specific response to the inhibition of RNA–protein synthesis or a nonspecific response to the cytotoxic action of actinomycin D remained unanswered.

We next utilized a system that allowed us to infuse a solution of actinomycin D into the preoptic area in animals that had permanently indwelling cannulae. Not only did this procedure avoid our having to anesthetize the animal during the course of estrogen action under study, but it also became possible to retest the animals at a later date and assess the degree of reversibility of any effects of actinomycin D. Thus, for this experiment ovariectomized rats were tested for lordosis behavior under three conditions: (a) the animals were pretested after saline infusion was performed at different times following the injection of EB; (b) 1 week later the experimental test was performed following the infusion of actinomycin D at the same times; and (c) 1 week later a posttest was conducted in which lordosis behavior was evaluated again following saline infusion at these times. The infusion of actinomycin D into the preoptic area 6 hr after EB injection significantly and reversibly suppressed lordosis behavior (98). Unfortunately, such infusions of actinomycin D also induced lesions, and it was still not possible to separate completely a cytotoxic from a possible specific effect of the antibiotic. As shown in Fig. 3, however, intraventricular injections of 0.11 μg actinomycin D produced a clear and reversible suppression of lordosis behavior, and no lesions were detected. Although the intraventricular site of infusion meant that the identity of the specific locus of actinomycin D action was unknown, it seemed reasonable to conclude that actinomycin D blocks estrogen-induced receptivity by its action on RNA–protein synthesis rather than cytotoxicity. Subsequent experiments by Quadagno and his colleagues (44,79,106) appear to support this hypothesis.

Although complicated by possible nonspecific cytotoxic effects, these data on the effectiveness of intracranial actinomycin D in inhibiting steroid-induced lordosis behavior do suggest that estrogen exerts its behavioral action via an alteration of RNA or protein synthesis, or both. A comparable conclusion has been made with respect to the feedback action of estradiol with the use of systemic antibiotics (45,46), and estrogen-induced alterations in hypothalamic protein synthesis have been detected (69).

In this context it is important to indicate that Beinfeld and Packman (15) have reported that the intraperitoneal injection of immature female rats with estradiol induces a differential incorporation of labeled leucine *in vitro* into a protein(s) of an approximate molecular weight of 18,000. This "hypothalamic-induced protein" appeared to peak approximately 1 hr after estra-

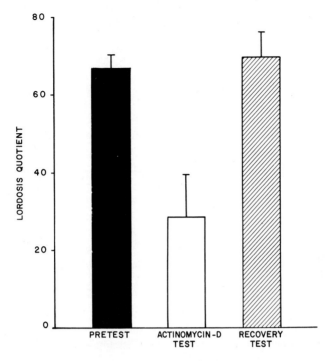

FIG. 3. The reversible suppression of lordosis behavior produced by the intraventricular infusion of 0.11 μg actinomycin D 6 hr after the subcutaneous injection of 3 μg EB. The LQ was significantly lower for the drug test than for the pre- and posttest carried out approximately 2 weeks prior to and after the drug test, respectively. (Modified from Terkel et al., ref. 98.)

diol injection. Although it is difficult to relate this study to the preceding experimental results, it should be noted that Clemens (L. G. Clemens, *unpublished observations*) has reported that a perfusate of the hypothalamus of a sexually receptive rat apparently contains a substance that can facilitate lordosis behavior in a nonresponsive animal when infused intracerebrally. Collectively these studies certainly support the likelihood that the ovarian hormones act in the adult brain by transiently modifying neurochemical processes probably analogous to those that occur in peripheral organs.

MECHANISMS OF THE ORGANIZATIONAL ACTION OF STEROIDS

In this final section I will briefly document the concept of the organizational action of gonadal steroids on the developing brain and consider our current understanding of the possible mechanism of this action. It is now generally accepted that in mammalian species there are functional sex differences within the brain but that these differences are not the direct or sole expressions of neuronal genomic activity. The functional dimorphism is at least partially imposed on what appears to be an indifferent or inherently female brain by the hormonal environment during a specific period of development. A number of brain functions, including the pattern of gonadotropin release, the regulation of lordosis and male sexual behavior, and aggressive behavior have been repeatedly shown to undergo the process of sexual differentiation (for reviews see 13,27,29,32,78,84,85). Somewhat less substantially documented is the sexual differentiation of the regulation of food intake and body weight (67,97,103), social, open field, and emergence behavior (22,38,72,80,94), territorial marking (99,100), micturition posturing (13,61), and learning or cognitive functions (14,21,74,84,107). Recently it has been reported that this process also applies to brain peptidase activity that inactivates luteinizing hormone-releasing hormone (40) and, as well, to the enzymatic activity of the tanacytes of the third ventricle (1). Although it is not feasible to review all these observations, it is sufficient to refer to the regulation of gonadotropin release and lordosis behavior to illustrate this process.

At a phenomenological level, it can be stated categorically that the male and female rat differ in their response to gonadal steroids in the pattern of secretion of gonadotropins and of lordosis behavior (13,27,29,32,36,78,84). The intact female exhibits a cyclic pattern of both gonadotropin release and lordosis behavior, but more importantly, steroid treatment of the ovariectomized rat can elicit a surge of luteinizing hormone as well as facilitate lordosis behavior. Neither of these occurs in response to the same hormone regime in the male. However, when exogenous steroids are administered to the neonatal rat, be they androgenic or estrogenic, the female, but not the male, is affected; in fact, the female develops the masculine pattern of gonadotropin acyclicity and of low levels of lordosis display. The response to the perinatal administration of steroids, however, is dose dependent (28,97), and one might argue that the "masculinization" of the female is a pharmacological effect. However, the effect of neonatal gonadectomy is markedly different as well. The female rat appears to be little affected by neonatal gonadectomy, whereas the neonatally orchidectomized male, when he matures to adulthood, exhibits the female pattern of gonadotropin responsiveness to exogenous steroids as well as the female level of lordosis response to steroid treatment. Thus, during the first few days of life in the rat, and presumably before birth in other species, the gonadal steroids appear to have a permanent effect or effects on the brain and establish what is recognized neuroendocrinologically as normal masculine brain function. In the case of many of the other functional parameters that have been suggested to undergo sex differentiation, the alterations induced by the perinatal environment are not necessarily as marked.

Since the activational actions of steroids on the brain are transitory, it may be suggested that the mechanisms of the permanent organizational action within the neonate cannot be identical. However, it could be that the apparent difference in steroid hormone action is due to the state of the neural tissue at the time of exposure and not necessarily to a real difference in the molecular action of androgen or estrogen. In any case, the fact that the effects of hormone treatment early in life are permanent makes this a useful model system for molecular studies. Indeed, the relationship—if any—between this permanent action of steroids and "classic" receptor mechanisms is an important question. If this permanent action of steroid hormones in the neonate is actually independent of classic receptors, it might indicate that receptor development signals an end to a period of unmodulated influence of hormones. In other words, have steroid receptors in the brain (or elsewhere) evolved to permit hormone action or merely to restrict or regulate it?

The relevance of the preceding question was increased by early reports that steroid receptors in the hypothalamus did not develop until approximately the second week of life, well beyond the period of differentiation (49,76). However, these biochemical observations were contradicted by the autoradiographic experimental data of Sheridan et al. (89), and it now appears that the existence of a systemic perinatal estrogen-binding protein in the rat has masked the time of the initial appearance of intraneuronal estradiol-binding proteins (63,64,77). Although steroid receptors do exist at the time of sexual differentiation, it has not yet been proved that these receptors are critical for, or even involved in, the organizational actions of steroids, however likely it may appear (8,9,48,60,64,83,104,105).

As listed in Table 1, however, there is some experimental evidence to suggest that the perinatal action of steroid hormones subsequent to the initial steroid-receptor interaction involves the expected biochemical events. How-

TABLE 1. *Agents reported to attenuate androgenization*

Systemic administration (ref. nos.)	
Barbiturates (5,6,93)	DNA Synthesis inhibitors
Colchicine (87)	Sarkomycin (37)
Thyroxine (52,73)	Hydroxyurea (86)
5-Hydroxytryptamine (90)	
Chlorpromazine (5,51,57)	RNA Synthesis inhibitors
Reserpine (5,50)	Actinomycin D (56)
Thymic cell suspensions (55)	Refampicin (37)
Progesterone (5,54)	α-Amanitin (86)
Antiandrogen (7,68,108)	
Antiestrogens (25,59,65)	Protein synthesis inhibitors
Aromatization inhibitors (16,65,102)	Puromycin (56,86)
	5-Fluoruracil (86)
Intracerebral administration (ref. no.)	
Cycloheximide (37)	

ever, before reviewing this evidence, it is necessary to point out that there are several possible effects of the administration of exogenous steroids to the female rat, the model system usually studied in mechanistic experiments because the experimentor can control both the dose of hormone and the temporal onset of treatment. Thus, if the dosage of exogenous hormone administered is below threshold, there is no apparent effect of the treatment, whereas if an adequate amount of hormone is administered, the females are anovulatory from the time of vaginal opening, the so-called androgenization syndrome (Fig. 4). Of considerable interest are those animals exposed to a low dosage of testosterone propionate (TP) who, after exhibiting apparently normal ovulatory cycles for a period after puberty, spontaneously become anovulatory at a very early age (95). This has been called the delayed anovulation syndrome (DAS; 28). Similarly, there are three possible effects of administering a potential inhibitor of androgen action along with a fully sterilizing dose of the steroid. First, there may be a complete protection, i.e., the females continue to ovulate even as adults; second, there is no protection, the sterilizing action of the administered steroid is expressed, and sterility ensues with vaginal opening; or finally, the attenuation of androgen action, that is, the DAS develops in response to the combined injection of a potential inhibitor and a fully sterilizing dose of steroid. Table 1 indicates that there are a number of agents that protect against or attenuate the action of androgen neonatally.

The effectiveness of certain antibiotics is of particular importance, in that it may provide insight into the mechanism of hormone action in terms of a rather classic biochemical approach. However, there are inconsistencies in this literature. For example, Salaman (86) reported that the DNA-synthesis

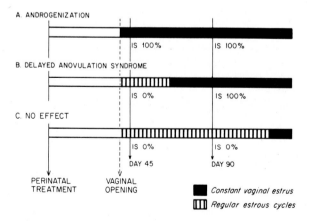

FIG. 4. Schematic representation of three possible effects of the perinatal injection of androgen to the intact female rat on her ovarian function as determined by the pattern of change of the vaginal epithelium and the incidence of anovulatory sterility (IS; number sterile/number treated). (Reprinted from Gorski, ref. 30.)

inhibitor hydroxyurea, at a dose of 4,000 μg/10 g protected against 200μg TP, but in their experiment DNA synthesis was not measured. On the other hand, Barnea and Lindner (10) found that an injection of 1,000 μg/rat pup, although effective in reducing thymidine incorporation by 95% for 6 hr, failed to interfere with the action of 100 μg TP. Although MER-25 has proved effective when given systemically (25), it has not been effective when implanted intracerebrally (43). In our own studies, we found that cycloheximide was effective in attenuating androgenization when applied directly to the preoptic area, but not when given systemically (37); however, in neither experiment did we measure protein synthetic activity. Barnea and Lindner (10) reported a marked decrease (for less than 6 hr) in protein synthesis but no inhibition of androgen action by cycloheximide.

These inconsistencies emphasize the rather obvious fact that there are a number of factors important for the potential success of attempts to dissect the mechanism of the organizing action of gonadal steroids on the brain. These include the relative site(s) of action of both hormone and potential inhibitor and hence the route of administration, the temporal pattern of hormone and inhibitor action including onset and duration, relative dose and rate of delivery to the site of action of the two agents, the age of the animal at treatment relative to the duration of the period of perinatal development and differentiation, the age of the animal at observation, and the specific function used to evaluate androgenization. Considering that successful inhibition of androgenization must require the synchrony of several processes in the right area at the right time, one perhaps should be surprised that protection has been observed, rather than that the inconsistencies described above are reported. The fact that sexual differentiation may be accompanied by discrete morphological changes (see below) may serve as a useful system for further neurochemical studies.

As indicated previously, much of the research into the process of sexual differentiation of the brain has centered on the adult animal, i.e., attempts to identify functional, chemical, and/or anatomical differences in the androgenized adult female and the normal female or differences between the adult male and female. The interpretation of such studies is often complicated by the fact that the different animal preparations have been subjected to different activational actions of hormones, due to different hormonal secretory patterns and possibly to neural substrates of differing hormonal sensitivity. Several investigators, for example, have studied steroid binding in the brain of the adult animal (62,101; and see 27,29). Although certainly a subject germane to this volume, I find the results of these studies unclear. In addition to obvious differences in the hormone environment prior to gonadectomy or sacrifice, or both, I am not sure what parameter should be measured. Should one expect sexual differences in brain function to be manifest in differences in steroid uptake, receptor affinity, receptor synthesis, steroid receptor translocation, or receptor replenishment or in steps further down the line such as protein synthesis, etc.? There are even further complications of experiments of this type. We have behavioral evidence that when female rats are given a low, but sterilizing, dose of TP, some hormone-sensitive systems may actually be *more* sensitive to the activational effects of hormones (17,19,29). Thus, the precise effects of early hormone exposure may well be dependent on the specific functional system under study. Since uptake alone does not elucidate function, carefully controlled functionally and regionally specific experiments are needed to determine the molecular mechanisms of either the organizational or activational action of steroids.

Several studies have suggested that neuronal morphology itself may be altered by the neonatal hormone environment (23,24,34,39,70,82,91). Recently we have discovered in the rat a very marked sex difference in the morphology of the preoptic area (35,36). This observation dramatizes the permanent action of steroids on the brain and also may provide a most useful model for further study, since one may actually be able to see at least one site of the organizational action of steroids.

As illustrated in Fig. 5, an intensely staining component of the medial preoptic nucleus is clearly larger in volume in the male than in the female when both animals are gonadectomized as adults. This fact is also apparent quantitatively when one considers the total volume of this tissue (Fig. 6). This figure also illustrates that this sex dimorphism is not influenced by the activational effects of the gonadal steroids, at least not under hormonal regimes that alter reproductive behavior. Although less convincing and perhaps subject to the activational influences of gonadal hormones, it is possible that the suprachiasmatic nuclei (Fig. 6) are also sexually dimorphic.

The data illustrated in Fig. 7 are perhaps the most relevant to the topic of this book since they demonstrate that this neuroanatomical sex difference is subject to the organizational action of gonadal steroids during the perinatal period. Androgenization of the 4-day-old female increases and castration of the 2-day-old male decreases the volume of this intensely staining cellular region, whereas there is no significant alteration of the suprachiasmatic nuclei (35).

We have not yet quantified, or even identified, the source of this sex difference at the level of individual cells (neurons and glia), nor do we have a clear idea of the specific function of this region of the hypothalamus. Nuclear volume alone does not appear to correlate with the pattern of gonadotropin secretion nor with the potential to exhibit lordosis behavior. Note that nuclear volume is equivalent in the androgenized female (1 mg TP) and the neonatally gonadectomized male, yet these animal preparations differ markedly with respect to these two functions (Fig. 7). Currently we are continuing our efforts to identify the function(s) of this region that may relate to the dimorphism and to study the development of the volume difference with perinatal age. In order to elucidate the molecular mechanisms

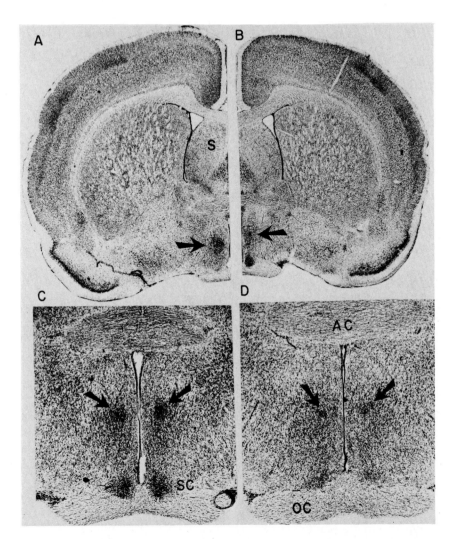

FIG. 5. Thionin-stained histological sections (60 μm) through the point of maximum development of the sexually dimorphic intensely staining component (*arrows*) of the medial preoptic nucleus in two gonadectomized male **(A,C)** and two gonadectomized female **(B,D)** adult rats. AC, anterior commissure; OC, optic chiasm; S, septum; SC, suprachiasmatic nucleus. (Brain tissue from a study reported in Gorski et al., ref. 36. Figure reprinted from Gorski, ref. 34.)

of the organizational action of androgen or estrogen it will be necessary to perform biochemical studies at the time of hormone exposure (18). The fact that this anatomical difference can be clearly recognized and identified readily suggests that it is a particularly useful model system.

Although it is generally accepted that this organizational action of steroid

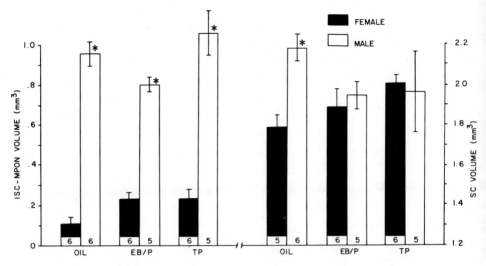

FIG. 6. The influence of hormonal treatment of the adult gonadectomized rat on the mean (+SE) volume of the intensely staining component of the medial preoptic nucleus (ISC-MPON; *left half of figure*) and of the suprachiasmatic nucleus (SC; *to the right*). The number of rats per group is indicated at the base of each bar. EB/P: 2 μg EB/day for 3 days followed by 500 μg progesterone; TP: 500 μg TP/per day for 2 weeks; oil: 0.05 ml sesame oil/day for 4 days. Asterisk indicates that nuclear volume in the male was significantly greater than that of the female under the same conditions. (From a study reported by Gorski et al., ref. 35. Reprinted from Gorski, ref. 34.)

hormones on the developing brain is limited to a so-called critical or competent period, recent evidence appears to challenge this concept and to offer another experimental model that may be of unusual significance. As indicated above, the DAS is the consequence of exposing the neonatal female rat to a low dose of androgen (or to the attenuation of a higher dose). We have been interested in determining what may contribute to ovulatory failure in the DAS rat and, in particular, whether or not the ovary itself is involved in this process. Kikuyama and Kawashima (53) and Arai (3) have suggested that the ovary is involved, since ovariectomy prolongs the period of cyclic gonadotropin release as measured by corpora lutea formation in ovarian grafts. Recently we have confirmed this using the ability of progesterone to facilitate luteinizing hormone secretion after estrogen priming (42). Following progesterone injection there is a significant difference in plasma luteinizing hormone titers between lightly androgenized rats ovariectomized at 30 and those ovariectomized at 95 days of age (Fig. 8). Moreover, the prolonged administration of either EB or TP beginning on day 30 after ovariectomy hastens the onset of the DAS, as indicated by the failure of progesterone to facilitate luteinizing hormone release in these animals (Fig. 8). Thus, in those animals exposed to a low dose of TP neonatally, the ovary

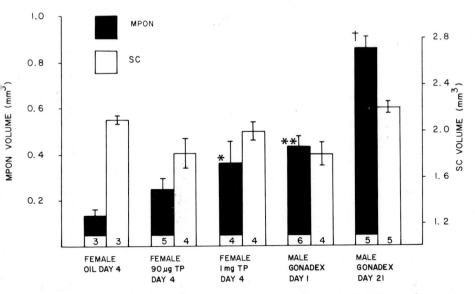

FIG. 7. The influence of neonatal hormone treatment on the mean (+SE) volume of the intensely staining sexually dimorphic component of the medial preoptic nucleus (MPON) and of the suprachiasmatic nucleus (SC). Females were injected with oil, 90 μg, or 1 mg TP on day 4 of life, and ovariectomized 2 weeks prior to sacrifice. Males were gonadectomized on day 1 (fale) or 21 of life. The numbers of animals per group are indicated at the base of each bar. Plus indicates results significantly ($p < 0.01$) larger than any other MPON group. Asterisks indicate statistical significance with respect to the control oil-treated female. (Reprinted from Gorski et al., ref. 35.)

or postpubertal exposure to exogenous steroids apparently leads to the early failure of the cyclic gonadotropin release system.

Figure 9 attempts to summarize the developmental history of the hypothalamus, perhaps of those specific neurons that are morphologically sexually different. Although the mechanism of this process of directing neuronal development is unknown, it may well involve protein synthesis, as appears to be the case for the activational effects of steroids. In the female rat, presumably in the absence of steroid hormone action, there is normal development and maturation of the neural substrate on which the activational action of the gonadal steroids takes place, as indicated by the appearance of a conceptual female "product" of hormone action in these neurons. In the male and the fully androgenized female, however, due to the organizational action of androgen (or estrogen), the neurons are changed, or directed differently during development, such that the activational effects of steroids are different, as represented conceptually by a "male" product. Note that the consequences of hormone action on neurons may vary because of differences in anatomical connectivity, hormone receptor characteristics and interactions, neurotransmitter production and release, specific messenger synthesis and

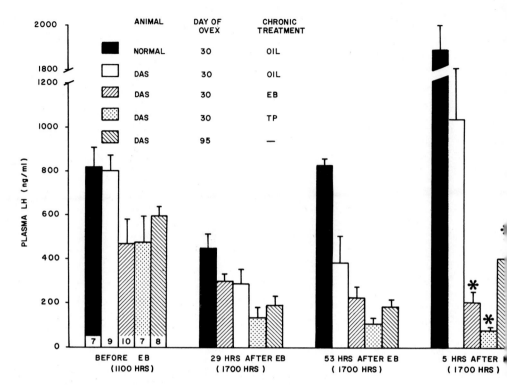

FIG. 8. Evidence that the ovary plays a role in the loss of the ability of progesterone (P) to facilitate luteinizing hormone secretion in lightly androgenized (10 μg TP, day 5; DAS) rats. As indicated, some animals were ovariectomized (Ovex) at day 30 and treated daily for 60 days with oil, EB (0.5 μg), or TP (100 μg). At approximately 125 days of age, EB (8 μg/100 g) was injected at noon, and blood samples were obtained 29 and 53 hr later and again on the third day 5 hr after progesterone. Asterisk indicates results significantly different from rats given chronic oil injections. (Modified from Harlan and Gorski, ref. 42.)

transport, etc. The concept of a male or female neuronal product of Fig. 9 encompasses all such possibilities and does not imply the existence of a specific male or female substance.

When the neonatal animal is exposed to a DAS-inducing dose of TP, there must be some effect of that androgen since in this animal, at least for a period of time including part of its postpubertal life, steroids appear to exert a combined effect. In the case of gonadotropin secretion, for example, ovarian steroids are able to support ovulation, but at the same time lead to the eventual loss of that functional capacity. Whatever the precise mechanisms of the organizational and activational actions of the gonadal steroids on the brain, the interactions of these two processes, which appear to take place in the case of the DAS, may be of considerable conceptual importance in the eventual elucidation of the action of steroid hormones on the brain.

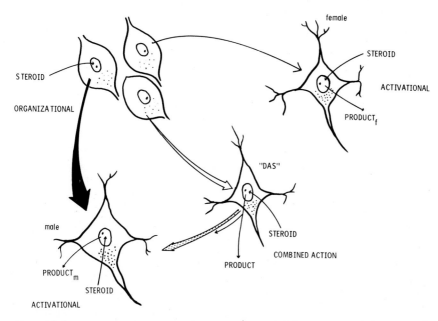

FIG. 9. Highly schematic representation of the sexual differentiation of the brain. See text for explanation. Product$_f$, the "product(s)" of hormone action on a neuron of the adult female; product$_m$, the "product(s)" of hormone action on a neuron of the adult male.

ACKNOWLEDGMENTS

Original work from the author's laboratory supported by National Institutes of Health Grants HD-01182 and AM-18254 and by the Ford Foundation.

REFERENCES

1. Akmayev, I. G., and Fidelina, O. V. (1976): Morphological aspects of the hypothalamic-hypophyseal system. VI. The tanycytes: their relation to the sexual differentiation of the hypothalamus. An enzyme-histochemical study. *Cell Tissue Res.*, 173:407–416.
2. Anderson, J. N., Peck, E. J., Jr., and Clark, J. H. (1975): Estrogen-induced uterine responses and growth: relationship to receptor estrogen binding by uterine nuclei. *Endocrinology*, 96:160–167.
3. Arai, Y. (1971): A possible process of the secondary sterilization: Delayed anovulation syndrome. *Experientia*, 27:463–464.
4. Arai, Y., and Gorski, R. A. (1968): Effect of anti-estrogen on steroid induced sexual receptivity in ovariectomized rats. *Physiol. Behav.*, 3:351–353.
5. Arai, Y., and Gorski, R. A. (1968): Protection against the neural organizing effect of exogenous androgen in the neonatal female rat. *Endocrinology*, 82:1005–1009.
6. Arai, Y., and Gorski, R. A. (1968): The critical exposure time for androgenization of the developing hypothalamus in the female rat. *Endocrinology*, 82:1010–1014.

7. Arai, Y., and Gorski, R. A. (1968): Critical exposure time for androgenization of rat hypothalamus determined by anti-androgen injection. *Proc. Soc. Exp. Biol. Med.,* 127:590–593.
8. Attardi, B., and Ohno, S. (1976): Androgen and estrogen receptors in the developing mouse brain. *Endocrinology,* 99:1279–1290.
9. Barley, J., Ginsburg, M., Greenstein, B. D., MacLusky, N. J., and Thomas, P. J. (1974): A receptor mediating sexual differentiation? *Nature,* 252:259–260.
10. Barnea, A., and Lindner, H. R. (1972): Short-term inhibition of macromolecular synthesis and androgen-induced sexual differentiation of the rat brain. *Brain Res.,* 45:479–487.
11. Barraclough, C. A. (1961): Production of anovulatory, sterile rats by single injections of testosterone propionate. *Endocrinology,* 68:62–67.
12. Barraclough, C. A., and Gorski, R. A. (1961): Evidence that the hypothalamus is responsible for androgen-induced sterility in the female rat. *Endocrinology,* 68:68–79.
13. Beach, F. A. (1975): Hormonal modification of sexually dimorphic behavior. *Psychoneuroendocrinology,* 1:3–23.
14. Beatty, W. W., and Beatty, P. A. (1970): Hormonal determinants of sex differences in avoidance behavior and reactivity to electric shock in the rat. *J. Comp. Physiol. Psychol.,* 73:446–455.
15. Beinfeld, M. C., and Packman, P. M. (1976): Estrogen induction of specific soluble proteins in the hypothalamus of the immature rat. *Biochem. Biophys. Res. Commun.,* 73:646–652.
16. Booth, J. E. (1977): Effects of the aromatization inhibitor, androst-4-ene-3,6,17-trione on sexual differentiation induced by testosterone in the neonatally castrated rat. *J. Endocrinol.,* 72:53P–54P.
17. Christensen, L. W., and Gorski, R. A. (1978): Independent masculinization of neuroendocrine systems by intracerebral implants of testosterone or estradiol in the neonatal female rat. *Brain Res.,* 146:325–340.
18. Clayton, R. B., Kogura, J., and Kraemer, H. C. (1970): Sexual differentiation of the brain: Effects of testosterone on brain RNA metabolism in newborn female rats. *Nature,* 226:810–812.
19. Clemens, L. G., Hiroi, M., and Gorski, R. A. (1969): Induction and facilitation of female mating behavior in rats treated neonatally with low doses of testosterone propionate. *Endocrinology,* 84:1430–1438.
20. Davies, I. J., Naftolin, F., Ryan, K. J., and Siu, J. (1976): Specific binding of steroids by neuroendocrine tissues. In: *Subcellular Mechanisms in Reproductive Neuroendocrinology,* edited by F. Naftolin, K. J. Ryan, and I. J. Davies, pp. 263–275. Elsevier, Amsterdam.
21. Dawson, J. L. M., Cheung, Y. M., and Lau, R. T. S. (1975): Developmental effects of neonatal sex hormones on spatial and activity skills in the white rat. *Biol. Psychol.,* 3:213–229.
22. Denti, A., and Negroni, J. A. (1975): Activity and learning in neonatally hormone treated rats. *Acta Physiol. Lat. Am.,* 25:99–106.
23. Dörner, G., and Staudt, J. (1968): Structural changes in the preoptic anterior hypothalamic area of the male rat, following neonatal castration and androgen substitution. *Neuroendocrinology,* 3:136–140.
24. Dörner, G., and Staudt, J. (1969): Structural changes in the hypothalamic ventromedial nucleus of the male rat, following neonatal castration and androgen treatment. *Neuroendocrinology,* 4:278–281.
25. Doughty, C., and McDonald, P. G. (1974): Hormonal control of sexual differentiation of the hypothalamus in the neonatal female rat. *Differentiation,* 2:275–285.
26. Farquhar, M. N., Namiki, H., and Gorbman, A. (1976): Cytoplasmic and nuclear metabolism of testosterone in the brains of neonatal and prepubertal rats. *Neuroendocrinology,* 20:358–372.
27. Flerkó, B. (1975): Perinatal androgen action and the differentiation of the hypothalamus. In: *Growth and Development of the Brain,* edited by M. A. B. Brazier, pp. 117–137. Raven Press, New York.

28. Gorski, R. A. (1968): Influence of age on the response to perinatal administration of a low dose of androgen. *Endocrinology,* 82:1001–1004.
29. Gorski, R. A. (1971): Gonadal hormones and the perinatal development of neuro-endocrine function. In: *Frontiers in Neuroendocrinology, 1971,* edited by L. Martini and W. F. Ganong, pp. 237–290. Oxford Univ. Press, New York.
30. Gorski, R. A. (1973): Perinatal effects of sex steroids on brain development and function. In: *Drug Effects on Neuroendocrine Regulation,* edited by E. Zimmermann, W. H. Gispen, B. H. Marks, and D. De Wied, *Prog. Brain Res.,* 39:149–162. Elsevier, Amsterdam.
31. Gorski, R. A. (1974): Extrahypothalamic influences on gonadotropin regulation. In: *The Control of the Onset of Puberty,* edited by M. M. Grumbach, G. D. Grave, and F. E. Mayer, pp. 182–207. Wiley, New York.
32. Gorski, R. A. (1974): The neuroendocrine regulation of sexual behavior. In: *Advances in Psychobiology, Vol. II,* edited by G. Newton and A. H. Riesen, pp. 1–58. Wiley, New York.
33. Gorski, R. A. (1976): The possible neural sites of hormonal facilitation of sexual behavior in the female rat. *Psychoneuroendocrinology,* 1:371–387.
34. Gorski, R. A. (1979): Long-term hormonal modulation of neuronal structure and function. In: *The Neurosciences: Fourth Study Program,* edited by F. O. Schmitt and F. G. Worden. MIT Press, Cambridge (*in press*).
35. Gorski, R. A., Gordon, J. H., Shryne, J. E., and Southam, A. M. (1978): Evidence for a morphological sex difference within the medial preoptic area of the rat brain. *Brain Res.,* 148:333–346.
36. Gorski, R. A., Harlan, R. E., and Christensen, L. W. (1977): Perinatal hormonal exposure and the development of neuroendocrine regulatory processes. *J. Toxicol. Environ. Health,* 3:97–121.
37. Gorski, R. A., and Shryne, J. (1972): Intracerebral antibiotics and androgenization of the neonatal female rat. *Neuroendocrinology,* 10:109–120.
38. Goy, R. W., and Resko, J. A. (1972): Gonadal hormones and behavior of normal and pseudohermaphroditic nonhuman female primates. *Recent Prog. Horm. Res.,* 28:707–733.
39. Greenough, W. T., Carter, C. S., Steerman, C., and DeVoogd, T. J. (1977): Sex differences in dendritic patterns in hamster preoptic area. *Brain Res.,* 126:63–72.
40. Griffiths, E. C., Hooper, K. C., Jeffcoate, S. L., and Holland, D. T. (1976): Effect of neonatal androgen on the activity of peptidases in the rat brain inactivating luteinizing hormone-releasing hormone. *Horm. Res.,* 7:218–226.
41. Hardin, J. W., Clark, J. H., Glasser, S. R., and Peck, E. J., Jr. (1976): RNA polymerase activity and uterine growth: Differential stimulation by estradiol, estriol, and nafoxidine. *Biochemistry,* 15:1370–1374.
42. Harlan, R. E., and Gorski, R. A. (1978): Effects of postpubertal ovarian steroids on reproductive function and sexual differentiation of lightly androgenized rats. *Endocrinology,* 102:1716–1724.
43. Hayashi, S. (1974): Failure of intrahypothalamic implants of antiestrogen, MER-25, to inhibit androgen-sterilization in female rats. *Endocrinol. Jpn.,* 21:453–457.
44. Ho, G. K., Quadagno, D. M., Cooke, P. H., and Gorski, R. A. (1974): Intracranial implants of actinomycin-D: Effects on sexual behavior and nucleolar ultrastructure in the rat. *Neuroendocrinology,* 13:47–55.
45. Jackson, G. L. (1972): Effect of actinomycin-D on estrogen-induced release of luteinizing hormone in ovariectomized rats. *Endocrinology,* 91:1284–1287.
46. Jackson, G. L. (1973): Time interval between injection of estradiol benzoate and LH release in the rat and effect of actinomycin-D or cycloheximide. *Endocrinology,* 93:887–891.
47. Kato, J. (1976): Characterization and function of steroid receptors in the hypothalamus and hypophysis. In: *Endocrinology, Vol. I,* edited by V. H. T. James, pp. 12–17. Exc. Med. Int. Congr. Ser. No. 402. Excerpta Medica, Amsterdam.
48. Kato, J. (1976): Cytosol and nuclear receptors for 5α-dihydrotestosterone and

testosterone in the hypothalamus and hypophysis, and testosterone receptors isolated from neonatal female rat hypothalamus. *J. Steroid Biochem.*, 7:1179–1187.

49. Kato, J., Sugimara, N., and Kobayashi, T. (1971): Changing pattern of the uptake of estradiol by the anterior hypothalamus, the medial eminence, and the hypophysis in the developing female rat. In: *Hormones in Development*, edited by M. Hamburgh and E. J. W. Barrington, pp. 689–703. Appleton, New York.

50. Kikuyama, S. (1961): Inhibitory effect of reserpine on the induction of persistent estrus by sex steroids in the rat. *Annot. Zool. Jpn.*, 34:111–116.

51. Kikuyama, S. (1962): Inhibition of induction of persistent estrus by chlorpromazine in the rat. *Annot. Zool. Jpn.*, 35:6–11.

52. Kikuyama, S. (1966): Influence of thyroid hormone on the induction of persistent estrus by androgen in the rat. *Sci. Papers Coll. Gen. Ed. Univ. Tokyo*, 16:265–270.

53. Kikuyama, S., and Kawashima, S. (1966): Formation of corpora lutea in ovarian grafts in ovariectomized adult rat subjected to early postnatal treatment with androgen. *Sci. Papers Coll. Gen. Ed. Univ. Tokyo*, 16:69–74.

54. Kincl, F. A., and Maqueo, M. (1965): Prevention by progesterone of steroid-induced sterility in neonatal male and female rats. *Endocrinology*, 77:859–862.

55. Kincl, F. A., Oriol, A., Folch, Pi, A., and Maqueo, M. (1965): Prevention of steroid-induced sterility in neonatal rats with thymic cell suspension. *Proc. Soc. Exp. Biol. Med.*, 120:252–255.

56. Kobayashi, F., and Gorski, R. A. (1970): Effects of antibiotics on androgenization of the neonatal female rat. *Endocrinology*, 86:285–289.

57. Ladosky, W., Kesikowski, W. M., and Gaziri, I. F. (1970): Effect of a single injection of chlorpromazine into infant male rats on subsequent gonadotrophin secretion. *J. Endocrinol.*, 48:151–156.

58. Lieberburg, I., and McEwen, B. S. (1977): Brain cell nuclear retention of testosterone metabolites, 5α-dihydrotestosterone and estradiol-17β, in adult rats. *Endocrinology*, 100:588–597.

59. Lieberburg, I., Wallach, G., and McEwen, B. S. (1977): The effects of an inhibitor of aromatization (1,4,6-androstatriene-3,17-dione) and an anti-estrogen (CI-628) on *in vivo* formed testosterone metabolites recovered from neonatal rat brain tissues and purified cell nuclei. Implications for sexual differentiation of the rat brain. *Brain Res.*, 128:176–181.

60. MacLusky, N. J., Chaptal, C., Lieberburg, I., and McEwen, B. S. (1976): Properties and subcellular inter-relationships of presumptive estrogen receptor macromolecules in the brains of neonatal and prepubertal female rats. *Brain Res.*, 114:158–165.

61. Martins, T., and Valle, J. R. (1967): Micturition behaviour of neonatally testosterone treated female dogs. *Experientia*, 23:921–922.

62. Maurer, R. A., and Woolley, D. E. (1975): ³H-Estradiol distribution in female, androgenized female, and male rats at 100 and 200 days of age. *Endocrinology*, 96:755–765.

63. McEwen, B. S. (1976): Steroid receptors in neuroendocrine tissues: Topography, subcellular distribution, and functional implications. In: *Subcellular Mechanisms in Reproductive Neuroendocrinology*, edited by F. Naftolin, K. J. Ryan, and J. Davies, pp. 277–304. Elsevier, Amsterdam.

64. McEwen, B. S., Lieberburg, I., MacLusky, N., and Plapinger, L. (1977): Do estrogen receptors play a role in the sexual differentiation of the rat brain? *J. Steroid Biochem.*, 8:593–598.

65. McEwen, B. S., Lieberburg, I., Chaptal, C., and Krey, L. C. (1977): Aromatization: Important for sexual differentiation of the neonatal rat brain. *Horm. Behav.*, 9:249–263.

66. McEwen, B. S., and Pfaff, D. W. (1973): Chemical and physiological approaches to neuroendocrine mechanisms: attempts at integration. In: *Frontiers in Neuroendocrinology, 1973*, edited by W. F. Ganong and L. Martini, pp. 267–335. Oxford Univ. Press, New York.

67. Nance, D. M., and Gorski, R. A. (1975): Neurohormonal determinants of sex differences in the hypothalamic regulation of feeding behavior and body weight in

the rat. In: *Central Neural Control of Eating and Obesity, Pharmacol. Biochem. Behav.*, 3(Suppl. 1):155–162.

68. Neumann, F., and Kramer, M. (1967): Female brain differentiation of male rats as a result of early treatment with an androgen antagonist. In: *Hormonal Steroids,* edited by L. Martini, F. Fraschini, and M. Motta, pp. 932–941. Excerpta Medica, Amsterdam.

69. Ozaki, M., and Kurachi, K. (1977): Changes in synthesis of protein and nucleic acids in hypothalamo-pituitary gonadal system after administration of large dose of estradiol-17β. *Acta Obstet. Gynaecol. Jpn.*, 29:277–284.

70. Pfaff, D. W. (1966): Morphological changes in the brains of adult male rats after neonatal castration. *J. Endocrinol.*, 36:415–416.

71. Pfaff, D., and Keiner, M. (1973): Atlas of estradiol-concentrating cells in the central nervous system of the female rat. *J. Comp. Neurol.*, 151:121–158.

72. Pfaff, D. W., and Zigmond, R. E. (1971): Neonatal androgen effects on sexual and non-sexual behavior of adult rats tested under various hormone regimes. *Neuroendocrinology*, 7:129–145.

73. Phelps, C. P., and Sawyer, C. H. (1976): Postnatal thyroxine modifies effects of early androgen on lordosis. *Horm. Behav.*, 7:331–340.

74. Phillips, A. G., and Deol, G. S. (1977): Neonatal androgen levels and avoidance learning in prepubescent and adult male rats. *Horm. Behav.*, 8:22–29.

75. Phoenix, C. H., Goy, R. W., Gerall, A. A., and Young, W. C. (1959): Organizing action of prenatally administered testosterone propionate on the tissues mediating mating behavior in the female guinea pig. *Endocrinology*, 65:369–382.

76. Plapinger, L., and McEwen, B. S. (1973): Ontogeny of estradiol-binding sites in rat brain. I. Appearance of presumptive adult receptors in cytosol and nuclei. *Endocrinology*, 93:1119–1128.

77. Plapinger, L., McEwen, B. S., and Clemens, L. E. (1973): Ontogeny of estradiol-binding sites in rat brain. II. Characteristics of a neonatal binding macromolecule. *Endocrinology*, 93:1129–1139.

78. Quadagno, D. M., Briscoe, R., and Quadagno, J. S. (1977): Effect of perinatal gonadal hormones on selected nonsexual behavior patterns: A critical assessment of the nonhuman and human literature. *Psychol. Bull.*, 84:62–80.

79. Quadagno, D. M., and Ho, G. K. W. (1975): The reversible inhibition of steroid-induced sexual behavior by intracranial cycloheximide. *Horm. Behav.*, 6:19–26.

80. Quadagno, D. M., Shryne, J., Anderson, C., and Gorski, R. A. (1972): Influence of gonadal hormones on social, sexual, emergence, and open field behaviour in the rat, *Rattus norvegicus. Anim. Behav.*, 20:732–740.

81. Quadagno, D. M., Shryne, J., and Gorski, R. A. (1971): The inhibition of steroid induced sexual behavior by intrahypothalamic actinomycin-D. *Horm. Behav.*, 2:1–10.

82. Raisman, G., and Field, P. M. (1973): Sexual dimorphism in the neuropil of the preoptic area of the rat and its dependence on neonatal androgens. *Brain Res.*, 54:1–29.

83. Raynaud, J. P., and Moguilewsky, M. (1977): Steroid competition for estrogen receptors in the central nervous system. *Prog. Reprod. Biol.*, 2:78–87.

84. Reinisch, J. M. (1976): Effects of prenatal hormone exposure on physical and psychological development in humans and animals: With a note on the state of the field. In: *Hormones, Behavior, and Psychopathology,* edited by E. J. Sachar, pp. 69–94. Raven Press, New York.

85. Resko, J. A. (1977): Fetal hormones and development of the central nervous system in primates. In: *Regulatory Mechanisms Affecting Gonadal Hormone Action,* Vol. 3, edited by J. A. Thomas and R. L. Singhal, pp. 139–168. Univ. Park Press, Baltimore.

86. Salaman, D. F. (1974): The role of DNA, RNA and protein synthesis in sexual differentiation of the brain. *Prog. Brain Res.*, 41:349–362.

87. Salaman, D. F. (1977): Effect of colchicine on androgen-induced sexual differentiation of the brain. *J. Endocrinol.*, 72:54P–55P.

88. Sar, M., and Stumpf, W. E. (1975): Distribution of androgen-concentrating

neurons in rat brain. In: *Anatomical Neuroendocrinology,* edited by W. E. Stumpf and L. D. Grant, pp. 120–133. Karger, Basel.

89. Sheridan, P. J., Sar, M., and Stumpf, W. E. (1974): Autoradiographic localization of ³H-estradiol or its metabolites in the central nervous system of the developing rat. *Endocrinology,* 94:1386–1390.

90. Shirama, K., Takeo, Y., Shimizu, K., and Maekawa, K. (1975): Inhibitory effect of 5-hydroxytryptophane on the induction of persistent estrus by androgen in the rat. *Endocrinol. Jpn.,* 22:575–579.

91. Staudt, J., and Dörner, G. (1976): Structural changes in the medial and central amygdala of the male rat, following neonatal castration and androgen treatment. *Endokrinologie,* 67:296–300.

92. Stumpf, W. E., Sar, M., and Keefer, D. A. (1975): Atlas of estrogen target cells in rat brain. In: *Anatomical Neuroendocrinology,* edited by W. E. Stumpf and L. D. Grant, pp. 104–119. Karger, Basel.

93. Sutherland, S. D., and Gorski, R. A. (1972): An evaluation of the inhibition of androgenization of the neonatal female rat brain by barbiturate. *Neuroendocrinology,* 10:94–108.

94. Swanson, H. H. (1967): Alteration of sex-typical behaviour of hamsters in open field and emergence tests by neonatal administration of androgen or oestrogen. *Anim. Behav.,* 15:209–216.

95. Swanson, H., and van der Werff ten Bosch, J. J. (1964): The "early-androgen" syndrome, its development and the response to hemi-spaying. *Acta Endocrinol.,* 45:1–12.

96. Takeshita, H. (1976): Evidence for soluble estradiol receptors in the amygdala of male and female rats. *Yonago Acta Med.,* 20:125–141.

97. Tarttelin, M. F., Shryne, J. E., and Gorski, R. A. (1975): Patterns of body weight change in rats following neonatal hormone manipulation: A "critical period" for androgen-induced growth increases. *Acta Endocrinol.,* 79:177–191.

98. Terkel, A. A., Shryne, J., and Gorski, R. A. (1973): Inhibition of estrogen facilitation of sexual behavior by the intracerebral infusion of actinomycin-D. *Horm. Behav.,* 4:377–386.

99. Thiessen, D., and Rice, M. (1976): Mammalian scent gland marking and social behavior. *Psychol. Bull.,* 83:505–539.

100. Turner, J. W. (1975): Influence of neonatal androgen on the display of territorial marking in the gerbil. *Physiol. Behav.,* 15:265–270.

101. Vertes, M., and King, R. J. B. (1971): The mechanism of oestradiol binding in rat hypothalamus: Effect of androgenization. *J. Endocrinol.,* 51:271–282.

102. Vreeburg, J. T. M., van der Vaart, P. D. M., and van der Schoot, P. (1977): Prevention of central defeminization but not masculinization in male rats by inhibition neonatally of oestrogen biosynthesis. *J. Endocrinol.,* 74:375–382.

103. Wade, G. N. (1972): Gonadal hormones and behavioral regulation of body weight. *Physiol. Behav.,* 8:523–534.

104. Westley, B. R., and Salaman, D. F. (1976): Role of oestrogen receptor in androgen-induced sexual differentiation of the brain. *Nature,* 262:407–408.

105. Westley, B. R., and Salaman, D. F. (1977): Nuclear binding of the oestrogen receptor of neonatal rat brain after injection of oestrogens and androgens: Localization and sex differences. *Brain Res.,* 119:375–388.

106. Whalen, R. E., Gorzalka, B. B., DeBold, J. F., Quadagno, D. M., Ho, K. W. G., and Hough, J. C., Jr. (1974): Studies on the effects of intracerebral actinomycin-D implants on estrogen-induced receptivity in rats. *Horm. Behav.,* 5:337–343.

107. Witelson, S. F. (1976): Sex and the single hemisphere: Specialization of the right hemisphere for spatial processing. *Science,* 193:425–427.

108. Wollman, A. L., and Hamilton, J. B. (1967): Prevention by cyproterone acetate of androgenic, but not of gonadotrophic, elicitation of persistent estrus in rats. *Endocrinology,* 81:350–354.

Ontogeny of Receptors and Reproductive Hormone Action,
edited by T. H. Hamilton, J. H. Clark, and W. A. Sadler.
Raven Press, New York 1979.

Development of Steroid Receptor Systems in the Rodent Brain

Neil J. MacLusky, Ivan Lieberburg, and Bruce S. McEwen

The Rockefeller University, New York, New York 10021

Maturation of the reproductive neuroendocrine system of mammals involves two interconnected processes—sexual maturation and sexual differentiation. Sexual maturation is measured by the appearance in development of such processes as: pituitary responsiveness to gonadotropin-releasing hormone (GnRH); negative feedback control of gonadotropin secretion by gonadal steroids; sexual behavior facilitated by gonadal hormones; and ovulation (for review, see 22). In lower mammals, sexual differentiation is indicated by the relative amounts of masculine and feminine sexual behavior that can be elicited after appropriate hormone treatment in adulthood. In addition, in rodents, sexual differentiation is indicated by the ability or inability of the hypothalamopituitary axis to produce surges of luteinizing hormone (LH) following estrogen replacement therapy (see Gorski, *this volume*). Both sexual differentiation and many aspects of sexual maturation appear to depend on the existence in neural tissue of steroid hormone receptor systems. The ontogeny of receptor systems in the brain is, therefore, a critical developmental event. This chapter examines steroid receptor systems in the maturing brain of the albino rat and their possible relationships to sexual differentiation and sexual maturation.

FINAL DIVISIONS OF NEURONS IN THE RAT BRAIN AND THE APPEARANCE OF DIFFERENTIATED CHARACTERISTICS

The appearance of differentiated characteristics of neurons follows the final cell division of neuroblasts. In the hypothalamus and preoptic area of the rat the final cell divisions occur between 14 and 16 days of gestation (13). Shortly thereafter, around days 17 to 18, GnRH can first be detected in hypothalamic tissue (5). Then between day 17 and day 20, estrogen receptors are first detectable (see below). In the hippocampus, neurons of Ammon's horn undergo final cell divisions prenatally. Neurons of the dentate gyrus continue to divide and differentiate into postnatal life, for perhaps as much as a month (1). Glucocorticoid receptors, which are a prominent feature of the adult hippocampus (10), are detectable as early after birth as we are able to look in the rat (postnatal day 3) (29). One aspect of our so far preliminary results (J. Gerlach, B. B. Turner and B. S. McEwen, *unpublished observations*) is that in both the Ammon's horn and dentate gyrus of the 3-day rat the oldest neurons, found on the inside of the former structure and the outside of the latter, are the ones that contain the most receptors.

ANDROGEN RECEPTORS IN THE NEONATAL RAT AND MOUSE BRAIN

The point at which androgen receptors first appear in the rodent brain remains unknown. They can be detected in the mouse brain at low concentrations as early as 2 days before birth (9). In the rat brain at postnatal days 3 to 4 they are present at a level approximately one-tenth of that observed in the gonadectomized adult rat (15, and N. J. MacLusky, I. Lieberburg, and B. S. McEwen, *unpublished observations*). Subsequently, the receptor levels increase gradually over the first 4 weeks of life (14,15). This profile differs slightly from that observed in the mouse brain, in which maturation of the androgen receptor system appears to be somewhat more rapid (3). In both rat and mouse the neonatal brain androgen receptors exhibit properties typical of receptors from adult androgen target tissues. They sediment in low ionic strength glycerol gradients at approximately 8S, they bind DNA cellulose, and they exhibit high affinity for the antiandrogens cyproterone or its 17α-acetate, and low affinity for the synthetic estrogen diethylstilbestrol (3,8,14,15, and N. J. MacLusky, I. Lieberburg, and B. S. McEwen, *unpublished observations*). It is noteworthy that, in both rat and mouse, high levels of androgen receptors do not appear in the developing cerebral cortex, in marked contrast to the situation with the estrogen and progestin receptor systems (see below).

ONTOGENY OF ESTROGEN RECEPTORS IN THE RAT BRAIN

Early attempts to demonstrate estrogen receptors in the brains of neonatal rats were complicated by the presence of high levels of a serum estrogen-binding protein, α-fetoprotein, which disappears during postnatal development (33). This protein interferes with the identification of brain estrogen receptors by reducing the availability of free estrogen to the receptors and by masking the relatively small amounts of receptor binding present. These two problems can be overcome by the use of procedures that distinguish between the receptors and α-fetoprotein on the basis of differences between them in their molecular properties and their affinities for certain synthetic estrogens.

It is now clear that the neonatal rat brain contains estrogen receptors with properties closely similar to those of adult estrogen receptors (4,20,23,43).

The distribution of these receptors in the neonatal brain, however, differs from that found in adulthood. Besides those regions of the brain in which estrogen receptors are concentrated in the adult (hypothalamus, preoptic area, and amygdala), the neonatal brain contains, in addition, high levels of receptors in the cerebral cortex (4,20,23,43).

These observations have prompted considerable interest in the maturation of estrogen receptor systems within the brain. Studies from this and another laboratory (20,26,27,34) have established that this maturation involves two well-defined phases. The initial phase, extending up to postnatal day 5, is characterized by a rapid increase in receptor concentrations throughout the brain, from undetectable levels 2 to 4 days before birth. The second, somewhat slower, phase, which occurs between postnatal days 5 and 25, transforms the neonatal receptor distribution into the adult pattern through a series of dynamic changes in estrogen receptor levels in different brain regions. In the cerebral cortex, receptor concentrations first increase to a peak at day 10 and then decline rapidly between days 10 and 15. Hypothalamic and pituitary receptor levels also increase up to day 10, but fall only slightly between days 10 and 25. In contrast, in the preoptic area, receptor levels increase steadily throughout the first 25 days of life, whereas in the amygdala and midbrain they do not change significantly from postnatal day 3 onward (26,27).

An important feature of this developmental pattern is the timing of the initial rapid perinatal appearance of receptors and the differentiation of the cerebral cortical receptors between postnatal days 10 and 15. Both of these events occur against a background of greatly elevated plasma estrogen concentrations (7,28,41), suggesting that there is a relationship between plasma estrogen and observed tissue estrogen receptor levels. Such a relationship might result from either occupation of the receptors by endogenous estrogen, rendering them unavailable for assay with radiolabeled ligands, or estrogenic regulation of receptor synthesis or degradation.

The question of the degree to which endogenous estrogen occupies the receptors is complicated by the presence of large quantities of α-fetoprotein in blood and cerebrospinal fluid over the first few weeks of life (31–33). Although plasma estrogen concentrations in the immediate perinatal period and during the second week of life are high relative to those found in adulthood (7,28,41), this does not necessarily imply that in young animals substantial amounts of free estrogen are available to the tissue receptor sites themselves. It is well established that in immature rats much larger doses of estradiol 17β are required to elicit responses in estrogen target tissues than are needed in adulthood. This relative insensitivity of immature rats to estradiol 17β has been ascribed to the influence of α-fetoprotein, which sequesters much of the administered hormone in a bound, and therefore presumably physiologically inactive, form (2,20,33). Using an equilibrium gel-filtration technique, Greenstein et al. (11,12) have shown directly that the percentage of plasma estradiol 17β that remains unbound is more than fourfold lower on the fifth day of life than in adulthood. It, therefore, seems possible that in neonatal animals free plasma estrogen concentrations do not normally reach levels high enough to significantly occupy central nervous estrogen receptor sites.

In order to test this hypothesis, we have used a recently developed estrogen-receptor-exchange assay (38) to measure the number of estrogen–receptor complexes translocated to brain cell nuclei in female rats at 3, 10, and 25 days of age. The results of these experiments may be summarized as follows. At all of the ages studied, cell nuclear-bound receptor complexes were not observed in the cerebral cortex. In contrast, in "limbic" brain tissues (hypothalamus, preoptic area, septum, and amygdala) although nuclear-bound receptors were not detectable at day 3, they were present at low levels (4 to 6% of the nuclear saturation capacity) at days 10 and 25 (26). A number of conclusions may be drawn from these data. First, it seems clear that in neonatal female rats receptor sites within the brain are not normally exposed to high free-estrogen concentrations. This, in turn, suggests that the large changes in numbers of available estrogen receptors occurring over the first few weeks of life are not the result of occupation of the receptors by endogenous estrogen. Second, the results indicate that at postnatal day 10 there is a divergence between the cerebral cortical and limbic receptor systems. Although at this age both the cortex and the limbic brain contain high levels of estrogen receptor sites, only in the limbic tissues are the receptors occupied and translocated to the cell nucleus to a measurable extent. The reason for this divergence remains uncertain. It is not due to a failure of the cell nuclear receptor translocation mechanism, since receptor complexes can be recovered from cortical cell nuclei of 10-day-old rats following estrogen treatment (26). One possibility is related to the ability of cells within the limbic brain, but not the cortex, to synthesize estrogen locally from circulating plasma androgen (16,36,42). Testosterone is present at low, but measurable, levels in the plasma of neonatal female rats (7,17). Estrogen synthesized from this testosterone within the limbic brain would not necessarily be prevented from reacting with nearby receptor sites by α-fetoprotein (20).

Finally, the observation that estrogen receptors within the brain are occupied only at a low level, if at all, during the early postnatal period in females suggests that it is unlikely that estrogen plays a major role in controlling the development of these receptor sites. Preliminary studies from this laboratory on the effects of neonatal ovariectomy have been consistent with this idea. Ovariectomy on the fifth day of life does not affect cell nuclear estrogen receptor capacity in the cerebral cortex of limbic tissues measured at either days 10 to 11 or 21 to 22, indicating that ovarian estrogen, at least, is not essential for differentiation of the receptor systems in these brain regions (26). A similar situation exists in the rat uterus, in which ovariectomy at 2 days of age does not affect the rapid synthesis of estrogen receptor sites, occurring between birth and postnatal day 10 (6). However, these data alone are not sufficient to establish unequivocally whether or not estrogen is involved, since it may be necessary only for extremely small quantities of estrogen to reach the tissues, and there remains the possibility of contributions from maternoplacental hormones prenatally and adrenal estrogens postnatally (41).

PROGESTIN RECEPTORS IN THE DEVELOPING RAT BRAIN

Of the three classes of gonadal steroid receptors present within the brain and pituitary, the progestin receptors remain the least well characterized. This stems primarily from the fact that it is not possible to selectively label progestin receptors using the natural progestins. Very recently, a means has become available by which such selective labeling can be achieved, in the form of a powerful synthetic progestin 17,21-dimethyl-17-nor-pregna-4,9-diene-3,20-dione or R5020, which can be used as a specific "tag" for progestin receptor sites (30,35). Using this compound we have identified putative progestin receptors in soluble cytoplasmic fractions from the adult rat brain and pituitary gland. The properties of these sites appear similar to those of progestin receptors from the rat uterus: they are progestin specific, exhibit a high affinity for ^{3}H-R5020 (Kd $\approx$ 0.3 nM), and sediment in low ionic strength glycerol-sucrose gradients at $\approx$7S. There appears, however, to be considerable variation between different brain regions in the sensitivity of the progestin receptor systems to *in vivo* estrogen treatment. In the hypothalamus and preoptic area, as in the pituitary and uterus, estrogen treatment results within 24 to 48 hr in an increase in progestin receptor concentrations. By contrast, in the remainder of the brain, estrogen treatment does not significantly increase progestin receptor levels. Regional studies in adrenalectomized and ovariectomized female rats suggest that the highest concentrations of these "estrogen-insensitive" progestin receptors are found in the cerebral cortex, followed by the amygdala (24).

In considering the ontogeny of the brain and pituitary progestin receptor systems, it is necessary, as in the adult, to distinguish between the presence of the receptors *per se* and the ability of the receptor systems to respond to estrogen treatment. In the pituitary, progestin receptor levels are low throughout the first 25 days of life. Even in animals as young as postnatal day 3, however, estrogen treatment can significantly increase pituitary progestin receptor concentrations, and by day 10 the maximum response of the pituitary progestin receptor system to estrogen is fully equal to that found in adult female rats. The situation in the brain is quite different. Although in 3-day-old female rats progestin receptor sites can be identified within the brain (their regional distribution being similar to that found in nonestrogen-primed adult females), the concentration of these sites is not affected by estrogen treatment in any brain region, including the hypothalamus and preoptic area.

The ability of the hypothalamic and preoptic area systems to respond to estrogen in terms of increased progestin receptor levels appears only gradually over the first 3 weeks of life, approaching adult levels by days 20 to 25 (25).

Thus three basic patterns emerge from the ontogeny of estrogen and progestin receptor systems within the rat brain. In the neonate, the cerebral cortex contains both receptor systems. Subsequently, cortical estrogen receptors virtually disappear over the second week of life, whereas the number of cortical progestin receptors remains high into adulthood. In the limbic brain, both receptor systems are present throughout postnatal life. There are differences, however, between regions of the limbic brain in their response to estrogen. In the amygdala estrogen treatment does not affect progestin receptor levels in either neonates or adults. In contrast, in the hypothalamus and preoptic area a response of the progestin receptor system to estrogen gradually becomes demonstrable as the animal matures.

RELATIONSHIP OF STEROID RECEPTOR SYSTEMS TO SEXUAL MATURATION

Behavioral and neuroendocrine studies of the developing rat make it clear that functional responses to hormones first take place well after the appearance of gonadal steroid receptors. Spontaneous ovulation and behavioral cyclicity occur in the female rat around day 41, but precocious displays of lordosis and ear wiggling can be elicited by estrogen plus progesterone somewhat before day 19, and precocious ovulation is elicited by estrogen around day 25 (see 22 for review). In male rats, mounting, intromission, and ejaculation (in that order) appear between postnatal days 45 and 55; administration of testosterone in silastic capsules on day 14 advances the appearance of these behaviors by approximately 20 days (40). It seems likely, therefore, that the appearance of these hormone-sensitive events is largely limited by neurological and physical maturation rather than by hormone sensitivity, although variations in the degree to which hormone stimulation is successful may be related to receptor levels.

One aspect of this notion of hormone sensitivity is, of course, the question of what triggers the onset of puberty (see ref. 19 for review and discussion). Although this is well beyond the scope of our present data, there is one relevant point, namely, that negative feedback effects of estradiol in the newborn female and male rat are very difficult to elicit even though androgens such as 5 α-dihydrotestosterone are effective suppressors of gonadotropin secretion before day 10 (see ref. 22 for review). The fact that negative feedback by estradiol is apparent on postnatal day 25, when α-fetoprotein levels have become very low, suggests that this estrogen-binding protein plays some role in the maturation of negative estrogen feedback in the rat, possibly by retarding estradiol's access to estrogen receptors (see ref. 22 for discussion).

This is supported by recent evidence of Andrews and Ojeda (2) that 11β methoxy, 17α ethynyl estradiol-17β (RU2858), an estrogen that does not bind to α-fetoprotein, is a very effective suppressor of gonadotropin secretion before postnatal day 13.

With respect to progesterone sensitivity, there is very little evidence so far to suggest whether the ability of estrogen to induce progesterone receptors plays a limiting role in the onset of behavioral and neuroendocrine responsiveness. In studies on the ontogeny of lordosis behavior in the rat (39), progesterone was always given after estrogen treatment. If further experiments in our laboratory support the initial findings of developmental variations in inducibility of progesterone receptor by estrogen, the investigation of this synergism will become an important topic for behavioral and neuroendocrine studies in developing animals.

ANDROGENS AND SEXUAL DIFFERENTIATION IN THE MALE RAT

Sexual differentiation in the rat occurs between days 17 or 18 of gestation and postnatal days 5 or 6. The early phase involves the differentiation of reproductive tract structures, and the later phase, the differentiation of the brain (see Gorski, *this volume*). As early as day 16 or so of gestation, the Leydig cells of the testes increase in size and secretory activity, and this continues throughout the first week of postnatal life (see ref. 22 for review). That this functional activity indicates the secretion of testosterone, at least in postnatal life, is indicated by measurements of plasma and testicular testosterone levels (37). Recently, we have measured by radioimmunoassay after chromatography the testosterone levels in plasma from male and female rats and have obtained results indicating a better than 10-fold elevation of testosterone in males compared to females (17). Between postnatal days 1 and 8, levels of testosterone in females are barely above the limits of detectability (20 to 30 pg/ml), whereas levels in males are between 350 and 500 pg/ml.

An important step in our understanding of the mechanism of testosterone action in the developing male brain was the finding that estrogens as well as testosterone are effective inducers of masculinization of the brain, whereas 5-α-dihydrotestosterone is ineffective (see 22 for review). This finding together with the reports of Naftolin and Ryan that the newborn rat brain contains the enzyme system that converts androgens to estrogens has been the departure for a number of studies on the role of testosterone-derived estradiol and estrogen receptors in the process of sexual differentiation (see ref. 22 for review). Our own contributions to this subject may be briefly summarized as follows. Following *in vivo* treatment of 3-day-old rats with an inhibitor of aromatization, androst-1,4,6-triene-3,17-dione (ATD), or an antiestrogen, nitromiphene citrate (CI-628), there is a better than 75% reduction of

[3]H-estradiol recovered from cell nuclear receptor sites in the hypothalamus, preoptic area, and amygdala after the administration of ^{3}H-testosterone without altering levels of ^{3}H-testosterone or of ^{3}H-dihydrotestosterone in the cell nuclear fraction (18). Under virtually identical experimental conditions, ATD and CI-628 attenuate the effects on sexual differentiation of endogenous testosterone in males and of exogenous testosterone or its propionate ester in females (21). It therefore seems as if aromatization of testosterone is necessary if sexual differentiation of the rat brain is to occur. Additional experiments are underway to determine whether the androgen receptors, which are present in the brain at the time of sexual differentiation, play some role in the process as well.

ACKNOWLEDGMENTS

The work in this chapter was supported by U.S. Public Health Service Grant NS 07080 and by The Rockefeller Foundation Institutional Grant RF 70095, both to Bruce S. McEwen. Neil J. MacLusky has been supported during the course of this work by postdoctoral fellowships from the Science Research Council, Great Britain, and from the Camille and Henry Dreyfus Foundation, New York. We are grateful to J-P. Raynaud of Roussel-Uclaf, Romainville, France, for kindly providing samples of unlabeled and ^{3}H-labeled R5020 and to Freddi Berg for editorial assistance in preparation of the manuscript.

REFERENCES

1. Altman, J., and Das, G. D. (1966): Autoradiographic and histological studies of postnatal neurogenesis. I. A longitudinal investigation of the kinetics, migration and transformation of cells incorporating tritiated thymidine in neonate rats, with special reference to postnatal neurogenesis in some brain regions. *J. Comp. Neurol.,* 126: 337–389.
2. Andrews, W. W., and Ojeda, S. R. (1977): On the feedback actions of estrogen on gonadotropin and prolactin release in infantile female rats. *Endocrinology,* 101: 1517–1523.
3. Attardi, B., and Ohno, S. (1976): Androgen and estrogen receptors in the developing mouse brain. *Endocrinology,* 99:1279–1290.
4. Barley, J., Ginsburg, M., Greenstein, B. D., MacLusky, N. J., and Thomas, P. J. (1974): A receptor mediating sexual differentiation? *Nature (Lond.),* 252:259–260.
5. Chiappa, S. H., and Fink, G. (1977): Releasing factor and hormonal changes in the hypothalamic-pituitary-gonadotropin and -adrenocorticotropin systems before and after birth and puberty in male, female and androgenized female rats. *J. Endocrinol.,* 72:211–224.
6. Clark, J. H., and Gorski, J. (1970): Ontogeny of the estrogen receptor during early uterine development. *Science,* 169:76–78.
7. Dohler, K. D. K., and Wuttke, W. (1975): Changes with age in levels of serum gonadotropins, prolactin, and gonadal steroids in prepubertal male and female rats. *Endocrinology,* 97:898–907.
8. Fox, T. O. (1975): Androgen- and estrogen-binding macromolecules in developing mouse brain: Biochemical and genetic evidence. *Proc. Natl. Acad. Sci. USA,* 72: 4303–4307.

9. Fox, T. O., Vito, C. C., and Weiland, S. J. (1978): Estrogen and androgen receptor proteins in embryonic and neonatal brain; hypotheses for sexual differentiation and behavior. *Am. Zool.,* 18:525–537.
10. Gerlach, J., and McEwen, B. S. (1972): Rat brain binds adrenal steroid hormone: Radioautography of hippocampus with corticosterone. *Science,* 175:1133–1136.
11. Greenstein, B. D., Puig-Duran, E., and MacKinnon, P. C. B. (1977): Measurement of the unbound estradiol-17β fraction in sera of developing female rats by a miniature method of steady-state gel filtration. *J. Endocrinol.,* 72:56P.
12. Greenstein, B. D., Puig-Duran, E., and Franklin, M. (1977): Accurate, rapid measurement of the fraction of unbound estradiol and progesterone in small volumes of undiluted serum at 37°C by miniature steady-state gel filtration. *Steroids,* 30: 331–341.
13. Ifft, J. D. (1972): An autoradiographic study of the time of final division of neurons in rat hypothalamic nuclei. *J. Comp. Neurol.,* 144:193–204.
14. Kato, J. (1976): Cytosol and nuclear receptors for 5αdihydrotestosterone and testosterone in the hypothalamus and hypophysis, and testosterone receptors from neonatal female rat hypothalamus. *J. Steroid Biochem.,* 7:1179–1187.
15. Kato, J. (1976): Ontogeny of 5αdihydrotestosterone receptors in the hypothalamus of the rat. *Ann. Biol. Anim. Biochim. Biophys.,* 16:467–469.
16. Lieberburg, I., and McEwen, B. S. (1975): Estradiol 17β: A metabolite of testosterone recovered in cell nuclei from limbic areas of neonatal rat brains. *Brain Res.,* 85:165–170.
17. Lieberburg, I., Krey, L. C., and McEwen, B. S. (1979): Sex differences in serum testosterone and in exchangeable brain cell nuclear estradiol during the neonatal period in rats. *Brain Res. (in press).*
18. Lieberburg, I., Wallach, G., and McEwen, B. S. (1977): The effects of an inhibitor of aromatization (1,4,6-androstatriene-3,17-dione) and an antiestrogen (CI628) on *in vivo* formed testosterone metabolites recovered from neonatal rat brain tissues and purified cell nuclei. Implications for sexual differentiation of the rat brain. *Brain Res.,* 128:176–181.
19. McCann, S. M. (1976): Development and maturation of the hypothalamo-hypophyseal control of the reproductive system. *Ann. Biol. Anim. Biochim. Biophys.,* 16: 279–289.
20. McEwen, B. S., Plapinger, L., Chaptal, C., Gerlach, J., and Wallach, G. (1975): Role of fetoneonatal estrogen binding proteins in the associations of estrogen with neonatal brain cell nuclear receptors. *Brain Res.,* 96:400–406.
21. McEwen, B. S., Lieberburg, I., Chaptal, C., and Krey, L. C. (1977): Aromatization: important for sexual differentiation of the neonatal rat brain. *Horm. Behav.,* 9:249–263.
22. McEwen, B. S. (1978): Sexual maturation and differentiation: The role of the gonadal steroids. *Prog. Brain Res.,* 48:291–307.
23. MacLusky, N. J., Chaptal, C., Lieberburg, I., and McEwen, B. S. (1976): Properties and subcellular inter-relationships of presumptive estrogen receptor macromolecules in the brains of neonatal and prepubertal female rats. *Brain Res.,* 114:158–165.
24. MacLusky, N. J., and McEwen, B. S. (1978): Oestrogen modulates progestin receptor concentrations in some rat brain regions but not in others. *Nature (Lond.),* 274:276–278.
25. MacLusky, N. J., and McEwen, B. S. (1979): *Manuscript in preparation.*
26. MacLusky, N. J., Chaptal, C., and McEwen, B. S. (1979): The development of estrogen receptor systems in the rat brain and pituitary: postnatal development. *Brain Res. (in press).*
27. MacLusky, N. J., Lieberburg, I., and McEwen, B. S. (1979): The development of oestrogen receptor systems in the rat brain: gestational appearance and perinatal development. *Brain Res. (in press).*
28. Meijs-Roelofs, H. M. A., Uilenbroek, J. T-J., De Jong, F. H., and Welschen, R. (1973): Plasma estradiol and its relationship to serum follicle-stimulating hormone in immature female rats. *J. Endocrinol.,* 59:295–304.

29. Olpe, H-R., and McEwen, B. S. (1976): Glucocorticoid binding to receptor-like proteins in rat brain and pituitary: ontogenetic and experimentally induced changes. *Brain Res.,* 105:121–128.
30. Philibert, D., and Raynaud, J-P. (1974): Progesterone binding in the immature rabbit and guinea pig uterus. *Endocrinology,* 94:627–632.
31. Plapinger, L., McEwen, B. S., and Clemens, L. E. (1973): Ontogeny of estradiol-binding sites in rat brain. II. Characteristics of a neonatal binding macromolecule. *Endocrinology,* 93:1129–1139.
32. Raynaud, J-P., Mercier-Bodard, C., and Baulieu, E. E. (1971): Rat estradiol binding plasma protein (EBP). *Steroids,* 18:767–788.
33. Raynaud, J-P. (1973): Influence of rat estradiol binding plasma protein (EBP) on uterotrophic activity. *Steroids,* 21:249–258.
34. Raynaud, J-P., and Moguilewsky, M. (1976): Ontogenese des recepteurs des oestrogenes chez le rat. In: *Système Nerveux, Activité Sexuelle et Reproduction,* edited by A. Soulairac, J. P. Gautray, J. P. Rousseau, and J. Cohen, pp. 85–92. Masson et Cie, Paris.
35. Raynaud, J-P. (1977): R5020, a tag for the progestin receptor. In: *Progesterone Receptors in Normal and Neoplastic Tissues,* edited by W. L. McGuire, pp. 9–21. Raven Press, New York.
36. Reddy, V. V. R., Naftolin, F., and Ryan, K. J. (1974): Conversion of androstenedione to estrone by neural tissues from fetal and neonatal rats. *Endocrinology,* 94:117–121.
37. Resko, J. A., Feder, H. H., and Goy, R. W. (1968): Androgen concentrations in plasma and testes of developing rats. *J. Endocrinol.,* 40:485–491.
38. Roy, E. J., and McEwen, B. S. (1977): An exchange assay for estrogen receptors in cell nuclei of the adult rat brain. *Steroids,* 30:657–669.
39. Sodersten, P. (1975): Receptive behavior in developing female rats. *Horm. Behav.,* 6:307–317.
40. Sodersten, P., Damassa, D. A., and Smith, E. R. (1977): Sexual behavior in developing male rats. *Horm. Behav.,* 8:320–341.
41. Weisz, J., and Gunsalus, P. (1973): Estrogen levels in immature female rats: True or spurious—ovarian or adrenal? *Endocrinology,* 93:1057–1065.
42. Weisz, J., and Gibbs, C. (1974): Conversion of testosterone and androstenedione to estrogens *in vitro* by the brains of female rats. *Endocrinology,* 94:616–620.
43. Westley, B. R., and Salaman, D. F. (1977): Nuclear binding of the estrogen receptor of neonatal rat brain after injection of estrogens and androgens: localization and sex differences. *Brain Res.,* 119:375–388.

Ontogeny of Receptors and Reproductive Hormone Action,
edited by T. H. Hamilton, J. H. Clark, and W. A. Sadler.
Raven Press, New York 1979.

Estrogen Receptors and the Activation of RNA Polymerases by Estrogens in the Central Nervous System

Ernest J. Peck, Jr., Ann L. Miller, and Katrina L. Kelner

Department of Cell Biology, Baylor College of Medicine, Houston, Texas 77030

This chapter will briefly review the current knowledge of estrogen receptors in the central nervous system. Their translocation from cytoplasmic to nuclear compartments under conditions of elevated serum estrogen will be discussed. In addition, the relationship between translocation and the activation of RNA polymerase activities in the hypothalamus will be examined. Finally, we will suggest that, while receptors for estrogens are quite similar in uterine and central nervous tissues, there exist differences in the mechanism of action of these receptors at the level of the nucleus.

CYTOPLASMIC ESTROGEN RECEPTORS IN THE CENTRAL NERVOUS SYSTEM

The first reports of cytoplasmic estrogen receptors in the brain were by Eisenfeld (6,7) and Kahwanago et al. (16,17) and demonstrated that ³H-estradiol associated with cytosol fractions of hypothalamus to form macromolecular complexes as assayed by gel filtration or gradient centrifugation. These complexes were destroyed by the proteolytic enzymes, chymotrypsin and papain, but ribonuclease and deoxyribonuclease were without effect. In keeping with the sensitivity of uterine estrogen receptors to sulfhydryl reagents, complex formation was inhibited by *p*-chloromercuriphenylsulfonic acid (17). Specificity of complex formation was demonstrated via competition studies, which showed that unlabeled estradiol and clomiphene displaced labeled estradiol from cytosol complexes, whereas progesterone and testosterone had no effect on complex formation.

Ginsburg and co-workers (11) subsequently measured the variation in number of cytoplasmic receptors of hypothalamus and pituitary during the estrous cycle. Cytoplasmic receptors decreased dramatically between metestrus and proestrus to reach a nadir on the day of proestrus, suggesting that the binding of endogenous estrogens and subsequent translocation of receptor estrogen complexes to the nuclear compartment is a physiological

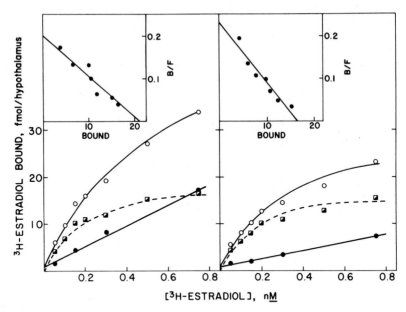

FIG. 1. Saturation and Scatchard analysis of hypothalamic estrogen receptors. Cytosol fractions of hypothalami from castrate adult female rats were incubated with various concentrations of ^{3}H-estradiol in the presence or absence of excess diethylstilbestrol. Following equilibrium, R_c-^{3}H-E complexes were precipitated with protamine sulfate and washed three times with Tris buffer, pH 7.4 (*left panel*) or Tris buffer containing 1.0% Tween 80 (*right panel*). Washed precipitates were extracted with absolute ethanol and the extracts analyzed by scintillation spectrometry. ○: total ^{3}H-estradiol bound; ●: nonspecific binding of ^{3}H-estradiol; □: specific binding of ^{3}H-estradiol.

process. Values of K_d, 0.35 nM and 0.25 nM for hypothalamic and pituitary receptor estrogen complexes, respectively, were similar to reports of others.

Cytoplasmic estrogen receptors from the uterus, pituitary, and brain have been compared and found to be quite similar. The similarity of sedimentation coefficients (8S in low ionic strength media), specificities of complex formation, and affinities for ^{3}H-estradiol have suggested that receptors from these various target organs are identical. Figure 1 is a saturation and Scatchard analysis of the binding of ^{3}H-estradiol to cytosolic receptors prepared from the preoptic-basal medial hypothalamus (K. L. Kelner and E. J. Peck, *unpublished observations*). Using protamine sulfate to precipitate cytoplasmic receptor estrogen complexes (R_cE) and mild detergent washes to reduce nonspecific binding, such receptors can be assayed in reproducible fashion. The apparent K_d is $\sim 10^{-10}$ M, and the high-affinity sites number about 20 fmoles per hypothalamus. These values indicate that although the number of receptors is much lower, their affinity for estrogens is quite similar to that for uterine receptors. It should be noted that no comprehensive survey exists to establish that estrogen receptors in various target regions of the brain are the same. The possibility exists that differences in the physiologic action of estrogens on discrete brain regions may result from differences in the character of their respective estrogen receptors. This is an interesting question that remains to be answered.

TRANSLOCATION OF ESTROGEN RECEPTORS TO THE NUCLEUS

Zigmond and McEwen (23) injected ovariectomized rats with ^{3}H-estradiol and examined various subcellular compartments for radioactivity 2 hr later. Purified nuclear fractions contained the highest levels of label, with the preoptic-basal medial hypothalamus and amygdala retaining the largest amount of label in their nuclear fractions. In examining the time course of nuclear residence, retention of ^{3}H-estradiol was maximal for the first 30 min to 1 hr and declined significantly by 4 hr.

The subcellular localization of ^{3}H-estradiol after *in vitro* incubation of the hypothalamus was examined by Chader and Villee (2,3). After 1 hr in 10^{-10} M ^{3}H-estradiol, the radioactivity extractable from various fractions was authentic estradiol and was localized primarily in the nuclear fraction, not the mitochrondrial, microsomal, or soluble fractions of the hypothalamus. Only nuclei retained radioactivity during subsequent dialysis, indicating the presence of high-affinity complexes in this subcellular compartment.

In 1972 the interaction of ^{3}H-estradiol with the hypothalamus *in vitro* was studied in our laboratory using unlabeled diethylstilbestrol to control for nonspecific interactions (5). Nuclear pellets were isolated from hypothalami after exposure to various concentrations of ^{3}H-estradiol with or without excess unlabeled diethylstilbestrol. The apparent dissociation constant for hypothalamic receptor estrogen complexes *in situ* was estimated at $\sim 10^{-9}$ M, with 10 to 50 fmoles of sites per hypothalamus.

Using the immature female rat and the ^{3}H-steroid exchange procedure, we subsequently studied the accumulation and retention of nuclear receptor estrogen complexes (R_nE) following injection of estradiol (1). Physiologic (0.5 μg) doses of estradiol increased the measurable nuclear complexes from about 5 fmoles/100 mg tissue in saline-injected controls to about 20 fmoles/100 mg tissue. After injection of a pharmacologic (25 μg) dose of estrogen a maximum of 30 to 40 fmoles of nuclear complex were demonstrable per 100 mg tissue. The dissociation constant of R_nE complexes was about 0.4×10^{-9} M. The time course for nuclear accumulation and retention of receptor estrogen complexes was similar to that of the uterus. Thus maximal R_nE levels were observed within 1 hr and remained above those of saline-injected controls for at least 6 hr. The majority of these binding sites were in the basal hypothalamus-median eminence region, where concentrations of receptor ranged up to 250 fmoles/100 mg tissue. A note of warning is necessary on this point, however, since this area includes hypophyseal cells of the pars tuberalis. Thus these high levels may reflect the receptor content of cells not of central nervous origin.

[3]H-steroid exchange of R_nE complexes from the adult animal has proven more difficult because of high levels of nonspecific binding in nuclear fractions of the adult hypothalamus. To circumvent this difficulty, we have recently used protamine sulfate precipitation of nuclear-chromatin fractions combined with detergent washes of the resultant pellet to allow the quantitative measure of R_nE complexes by exchange (K. L. Kelner and E. J. Peck, *unpublished observations*). With this procedure as many receptors are measured in nuclei of hypothalami from estrogen-treated adult animals as exist in the cytoplasmic compartment of saline-injected controls (see below).

Gradient centrifugation of a macromolecule that is extracted from hypothalamic nuclei with 0.4 M KCl and that binds estradiol with high affinity supports the idea that estradiol retained in the nuclear fraction after *in vivo* or *in vitro* exposure is bound to translocated cytoplasmic receptors (21,22). In addition, Cidlowski and Muldoon (4) have shown that injected estradiol reduces the number of cytoplasmic estradiol binding sites. However, proof of translocation requires the simultaneous study of receptor content in cytoplasmic and nuclear compartments as a function of time after estrogen injection. Figure 2 (K. L. Kelner and E. J. Peck, *unpublished observations*) shows that

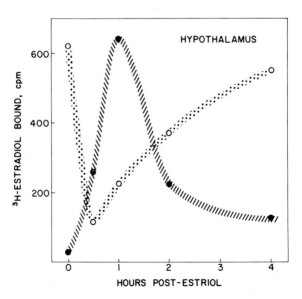

FIG. 2. Translocation of estrogen receptors in the hypothalamus. Castrate adult female rats were sacrificed at various times after injection with 5 μg estriol. Cytosol and nuclear fractions of hypothalami were prepared and total receptor in each determined by ^{3}H-estradiol exchange with or without excess diethylstilbestrol ([^{3}H-estradiol] = 5 nM) at 30°C for 60 min. Cytoplasmic and nuclear fractions were precipitated with protamine sulfate and the precipitates washed, extracted and analyzed as in Fig. 1. ○: cytoplasmic binding; ●: nuclear binding.

with the disappearance of hypothalamic cytoplasmic receptors following estrogen treatment, there is an attendant increase in R_nE as measured by ^3H-estradiol exchange. This examination of cytoplasmic and nuclear fractions of the adult ovariectomized rat establishes translocation of cytoplasmic receptor as a fact.

Despite the similarities between uterine and hypothalamic estrogen receptors, there may be differences in the way these macromolecules activate cellular processes. The majority of cytoplasmic receptors of uterine cytosol bind to DNA-cellulose and are readily and rapidly converted to the "activated" 5S state *in vitro* (10). In contrast, cytosol receptor from the hypothalamus exists in two forms, and only 30 to 50% of total cytoplasmic receptor is capable of binding to DNA-cellulose (10). There also appears to be little conversion of the hypothalamic cytosolic receptor to the 5S state *in vitro*. In addition, nuclear 5S receptors are observed within minutes after ^3H-estradiol injection in rat uteri, whereas in the hypothalamus these 5S complexes only appear after a 30-min to 1-hr lag (18,19). The absence of *in vitro* conversion of cytosol receptor from the 4S to the 5S form and the delay in appearance of 5S receptor in nuclei of the hypothalamus suggest that receptor activation, nuclear translocation, and/or the processing of nuclear receptor estrogen complexes differs between targets of the central nervous system, such as the hypothalamus, and peripheral targets, such as the uterus.

GONADAL STEROIDS AND NERVOUS SYSTEM FUNCTION

The central nervous system responds to estrogens in a multitude of ways. Gonadal steroids direct the homeostatic regulation of gonadotropin secretion, modify behavioral components important to reproduction, and in addition, direct the sexual differentiation and maturation of the central nervous systems. Although a number of biochemical effects of estrogens on central nervous tissue have been described, definitive conclusions about cause-and-effect relationships between specific biochemical changes and estrogen-induced alterations in neural function do not exist. Reviews by Gorski and McEwen (14,20), as well as recent reports of Feder (8,9) are recommended as background to this subject. In this chapter only the effects of estrogens on hypothalamic RNA polymerases will be discussed.

Estrogens increase RNA synthesis in target tissues. This stimulation of RNA synthesis appears to involve two temporally separate events: an early stimulation of DNA-dependent RNA polymerase II, as first described by Glasser et al. (12), followed by a secondary rise in both RNA polymerase I and II (13,15). Our investigation of RNA polymerase I and II activities in estrogen-stimulated uteri has shown that the transient early rise in RNA polymerase II correlates with early uterine responses but not with the uterine growth response (15). Growth in peripheral targets requires the secondary

TABLE 1. *Activation of hypothalamic RNA
polymerase by estrogen*

Time after estrogen	Activity as % Saline control	
	Polymerase I	Polymerase II
20 min	108±14	123± 8
40 min	89±11	121±11
60 min	100± 7	142±10
90 min	94±12	120± 6
2 hr	96± 4	101± 9
12 hr	106± 8	92±13
24 hr	87±10	108±15

Groups of six animals were injected with 5 μg estradiol in saline or with saline alone and sacrificed at the times indicated. Hypothalamic nuclei were prepared and endogenous RNA polymerase activities I and II were assayed essentially as described by Hardin et al. (15). Activities of polymerase I and II in saline-injected control were 1.21 ± 0.06 and 1.43 ± 0.04 pmoles uridine triphosphate incorporated/μg DNA/10 min, respectively.

rise in both polymerases, which in turn depends on the retention of receptor-estrogen complexes in the nuclear compartment for prolonged periods.

We recently initiated investigations of hypothalamic RNA polymerase I and II to ask the following: do neurons respond to estrogens in the same manner as uterine cells? Since neurons do not "grow" after maturation and since nervous tissue targets respond rapidly to transient changes in gonadal steroids, perhaps only an initial rise in RNA polymerase II, associated with the synthesis of specific mRNAs, occurs in neural tissues after exposure to steroids. The data in Table 1 are from experiments in which endogenous nuclear RNA polymerase I and II activities were assayed as a function of time after estrogen treatment. Polymerase II activity undergoes a transient activation that parallels the time course of receptor translocation. Polymerase I, on the other hand, is not responsive to estrogen treatment. Control experiments with frontal or occipital cortex, brain regions without measurable estrogen receptor, have demonstrated no change in either RNA polymerase activity at any time after estrogen treatment. Thus RNA polymerase II activation is specific for steroid target regions of the central nervous system. The early rise in RNA polymerase II suggests that new species of mRNA are synthesized in response to estrogen and/or that multiple copies of mRNAs are being produced. No dramatic or sustained alteration in RNA polymerase I is observed, consistent with our hypothesis that estrogens act as specific inducers, but not as growth promoters, in brain targets of adult animals.

CONCLUSIONS

The existing data demonstrate that estrogen receptors occur in circumscribed regions of the central nervous system and that, with respect to affinity and specificity, these appear identical to those of peripheral targets such as the uterus. In addition, these receptors reside in the cytoplasmic compartment of hypothalami from immature or adult castrate females not exposed to estrogen. Finally, upon exposure to estrogens these cytoplasmic receptors translocate to nuclear compartments with a time course similar to that for peripheral estrogen receptors. On the basis of these observations, one must conclude that estrogen receptors in the brain are the same or very similar to those of the uterus.

Certain observations of others have suggested differences, however, between estrogen receptors of the central nervous system and those of the uterus. Specifically, Fox and Johnston (10) and Linkie (18,19) have observed differences in the capacity of estrogen receptors of brain to bind to DNA and to be "activated" to the 5S state. Such observations suggest that processes involved with translocation and/or nuclear processing of estrogen receptors may differ between central nervous targets and peripheral targets such as the uterus. Thus receptors of central nervous targets may differ in the manner in which they activate nuclear processes. Our observation that RNA polymerase II shows only a transient activation (or burst) after estrogen treatment while RNA polymerase I is not altered by estrogen at all supports this suggestion. Thus retention of estrogen receptors and/or their processing within the nucleus may be different in targets of the central nervous system, and this difference may reflect the incapacity of neuronal systems to proliferate in the face of estrogen stimulation.

ACKNOWLEDGMENTS

This work was supported by grants from the NIH (HD 08389 and NS 11753), as well as by a Career Development Award to E.J.P. (HD 00022).

REFERENCES

1. Anderson, J. N., Peck, E. J., Jr., and Clark, J. H. (1973): Nuclear estrogen receptor complex: Accumulation, retention and localization in the hypothalamus and pituitary. *Endocrinology*, 93:711–717.
2. Chader, G. J., and Villee, C. A. (1970): Uptake of oestradiol by the rabbit hypothalamus: specificity of binding by nuclei *in vitro. Biochem. J.*, 118:93–97.
3. Chader, G. J., and Villee, C. A. (1971): Estrogen receptors in the hypothalamus. In: *Influence of Hormones on the Nervous System, Proceedings of the International Society of Psychoneuroendocrinology*, edited by D. H. Ford, pp. 17–24. Karger, Basel.
4. Cidlowski, J. A., and Muldoon, T. G. (1974): Estrogenic regulation of cytoplasmic receptor populations in estrogen-responsive tissues of the rat. *Endocrinology*, 95: 1621–1629.
5. Clark, J. H., Campbell, P. S., and Peck, E. J., Jr. (1972): Receptor estrogen complex in the nuclear fraction of the pituitary and hypothalamus of male and female immature rats. *Neuroendocrinology*, 77:218–228.
6. Eisenfeld, A. J. (1969): Hypothalamic estradiol binding molecules. *Nature*, 224: 1202–1203.

7. Eisenfeld, A. J. (1970): ³H-estradiol: *in vivo* binding to macromolecules from the rat hypothalamus, anterior pituitary and uterus. *Endocrinology*, 86:1313–1318.
8. Feder, H. H., Landau, I. T., and Walker, W. A. (1979): Anatomical and biochemical substrates of the action of estrogens and anti-estrogens on brain tissues that regulate female sex behavior of rodents. In: *Endocrine Control of Sexual Behavior*, edited by C. Beyer, pp. 317–340. Raven Press, New York.
9. Feder, H. H., and Marrone, B. L. (1977): Progesterone: its role in the central nervous system as a facilitator and inhibitor of sexual behavior and gonadotropin release. *Ann. N.Y. Acad. Sci.*, 286:331–354.
10. Fox, T. O., and Johnston, C. (1974): Estradiol receptors from mouse brain and uterus: binding to DNA. *Brain Res.*, 77:330–336.
11. Ginsburg, M., MacLusky, N. J., Morris, I. D., and Thomas, P. J. (1972): Cyclical fluctuation of oestradiol receptors in hypothalamus and pituitary. *J. Physiol. (Lond.)*, 224:72P.
12. Glasser, S. R., Chytil, F., and Spelsberg, T. C. (1972): Early effects of oestradiol-17β on the chromatin and activity of the deoxyribonucleic acid-dependent ribonucleic acid polymerases (I and II) of the rat uterus. *Biochem. J.*, 130:947–957.
13. Glasser, S. R., and Spelsberg, T. C. (1973): Differential modulation of estrogen induced activtiy of uterine nuclear RNA polymerases (I and II) by cycloheximide and actinomycin D. *Endocrinology*, 92 (suppl.), A88.
14. Gorski, R. A., and Clemens, L. G. (1970): Action of gonadal steroids. In: *Handbook of Neurochemistry, Vol. 4, Control of Mechanisms in the Nervous System*, edited by A. Lajtha, pp. 429–449. Plenum Press, New York.
15. Hardin, James, W., Clark, James H., Glasser, Stanley R., and Peck, Ernest J., Jr. (1976): RNA polymerase activity and uterine growth: Differential stimulation by estradiol, estriol, and nafoxidine. *Biochemistry*, 15:1370–1374.
16. Kahwanago, I., Heinrichs, W. L., and Hermann, W. L. (1969): Isolation of oestradiol receptors from bovine hypothalamus and anterior pituitary gland. *Nature*, 223:313–314.
17. Kahwanago, I., Heinrichs, W. L., and Hermann, W. L. (1970): Oestradiol "receptors" in the hypothalamus and anterior pituitary gland. Inhibition of estradiol binding by SH group blocking agents and clomiphene citrate. *Endocrinology*, 86:1319–1326.
18. Linkie, D. M. (1975): *In vivo* nuclear transformation of estrogen receptor complex in hypothalamus, pituitary, and uterus. *Endocrinology*, 96:67 (suppl.).
19. Linkie, D. M. (1977): Estrogen receptors in different target tissues: similarities of form—dissimilarities of transformation. *Endocrinology*, 101:1862–1870.
20. McEwen, B. S., Denef, C. J., Gerlach, J. L., and Plapinger, L. (1974): Chemical studies of the brain as a steroid hormone target tissue. In: *The Neurosciences: Third Study Program*, edited by F. O. Schmitt and F. G. Worden, pp. 599–620. MIT Press, Cambridge, Massachusetts.
21. Mowles, T. F., Ashkanazy, B., Mix, E., Jr., and Sheppard, H. (1971): Hypothalamic and hypophyseal estradiol binding complexes. *Endocrinology*, 89:484–491.
22. Vertes, M., and King, R. J. B. (1971): The mechanism of oestradiol binding in rat hypothalamus—effect of androgenization. *J. Endocrinol.*, 51:271–282.
23. Zigmond, R. E., and McEwen, B. S. (1970): Selective retention of estradiol by brain cell nuclei in specific regions of the ovariectomized rat. *J. Neurochem.*, 17:889–899.

Ontogeny of Receptors and Reproductive Hormone Action,
edited by T. H. Hamilton, J. H. Clark, and W. A. Sadler.
Raven Press, New York 1979.

Regulation of the Onset of Steroid Hormone Synthesis in Fetal Gonads

Fredrick W. George,* Kevin J. Catt,† and Jean D. Wilson*

*Department of Internal Medicine and the Eugene McDermott Center for Growth and Development, The University of Texas Southwestern Medical School, Dallas, Texas 75235; and †Endocrinology and Reproductive Research Branch, National Institute of Child Health and Human Development, Bethesda, Maryland 20014

Since the early studies of Jost (8–10) it has been clear that the secretion of androgen by the fetal testis plays a vital role in the differentiation of the male internal and external genitalia. In fetal rabbit (11,15), sheep (1), and man (14), the onset of testosterone formation by the testis occurs just prior to the onset of male differentiation of the urogenital tract. Further support for the primary role of androgens in male phenotypic differentiation has been established by the characterization of several single-gene mutations that cause abnormal hormone synthesis or action and prevent normal male sexual development (7). In contrast to the well-established role of androgen in male phenotypic development, little is known about the endocrine function of the fetal ovary and the role, if any, of ovarian estrogen in female phenotypic development.

The factors that regulate the onset of endocrine function of the fetal gonads have never been elucidated. The biochemical pathways by which steroid hormones are synthesized from cholesterol in the ovary and testis of the mature animal are summarized schematically in Fig. 1. First, the side chain of cholesterol is cleaved to form pregnenolone. Most of the available evidence indicates that this reaction is regulated by gonadotropins. Pregnenolone in turn undergoes at least five subsequent enzymatic reactions to form testosterone; in the rabbit the rate-limiting reaction between the formation of pregnenolone and testosterone is the 3β-hydroxysteroid dehydrogenase step that oxidizes the A ring of the steroid. In both ovary and testis, furthermore, a portion of testosterone is converted to 17β-estradiol. In the present report we will summarize evidence to indicate that the enzymatic pathways are different in the day 19 rabbit embryo (when male phenotypic differentiation has become irreversible). At this time the testis contains the enzymatic capacity to convert pregnenolone to testosterone, whereas the ovary can convert testosterone to 17β-estradiol but apparently cannot synthesize testosterone. Whether the pregnenolone substrate for testosterone

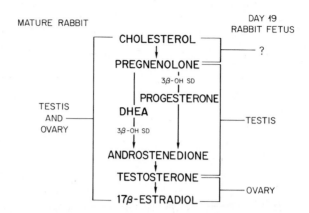

FIG. 1. Schematic representation of the pathway for steroid hormone synthesis in the ovary and testis of the mature rabbit and the day 19 rabbit fetus.

synthesis is formed within the gonad by direct synthesis or is derived from the plasma after synthesis elsewhere (placenta or fetal adrenals) is unclear.

METHODS

New Zealand white rabbits were mated during a 2-hr period in the late afternoon of a day that was arbitrarily designated as day 0 for the purpose of timing the pregnancy. At various times during the pregnancy the does were killed, and the embryos were removed and placed on ice. The gonads were dissected from the fetuses and used immediately for acute incubation studies.

Incubation of fetal tissues with [7-^{3}H]pregnenolone, and the separation of the metabolites of the reaction by gradient elution column chromatography have been described (15). The activity of 3β-hydroxysteroid dehydrogenase-$\Delta^{4,5}$-isomerase was estimated by measuring the rate of conversion of [7-^{3}H] dehydroepiandrosterone to androstenedione and testosterone (13), and estrogen synthesis was estimated by measuring the conversion of [1,2,6,7-^{3}H] testosterone to 17β-estradiol (13). For organ culture studies, day 16 or 17 undifferentiated fetal gonads were cultured in a chemically defined medium without serum throughout the period of enzymatic differentiation as described (5). The responsiveness of fetal gonads to gonadotropins was assessed in paired incubations of gonads in the presence or absence of 10 IU human chorionic gonadotropin (hCG) (5). The measurement of gonadotropin binding in rabbit fetal gonads and of testosterone and adenosine-3′,5′-cyclic monophosphate (cyclic AMP) by radioimmunoassay have also been described (2,4).

RESULTS

Due to technical reasons, it is difficult to study the side-chain cleavage reaction of cholesterol to pregnenolone in the small amounts of tissue available in early fetal gonads. Therefore, by necessity, we have focused on the terminal pathway, namely, the formation of testosterone and 17β-estradiol from pregnenolone. To be certain that the end-products of pregnenolone metabolism are the same in the fetal as in mature gonads, in the initial experiments rabbit fetal gonads were incubated with [7-³H]pregnenolone or [1,2,6,7-³H]testosterone, and all the radioactive metabolites were isolated by a combination of celite column and thin-layer chromatography (13,15). In these studies (results not shown), testosterone was the major 19-carbon steroid formed from pregnenolone by the fetal testis, whereas no significant formation of 19-carbon steroids could be demonstrated in the fetal ovary at any age (15); testosterone was converted to 17β-estradiol in significant amounts by the ovary under conditions in which no 17β-estradiol formation could be demonstrated in the fetal testis (13). Furthermore, in the fetal testis the 3β-hydroxysteroid dehydrogenase reaction was shown to be rate limiting in the conversion of pregnenolone to testosterone (15).

The developmental sequence of the onset of endocrine function in the fetal ovary and testis is illustrated in Fig. 2. The capacity of the fetal ovary to

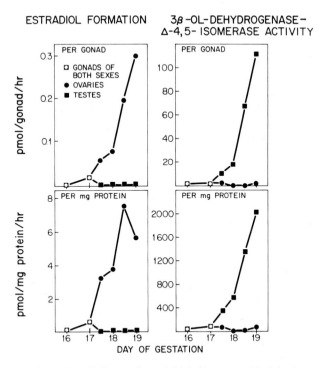

FIG. 2. Estrogen formation **(left panel)** and 3β-hydroxysteroid dehydrogenase activity **(right panel)** in gonads from fetal rabbits varying in age from 16 to 19 days of gestation. Half of the gonads were incubated with [1,2,6,7-³H]testosterone and the estradiol formed was purified as described. The remaining gonads were incubated with [7-³H]dehydroepiandrosterone to determine 3β-hydroxysteroid dehydrogenase activity. (Reprinted from Ref. 13.)

convert testosterone to 17β-estradiol develops at the same time as the ability of the fetal testis to form testosterone from pregnenolone. We have not studied the factors that regulate the appearance of estrogen formation in the fetal ovary, and all subsequent studies have been focused on the onset of testosterone synthesis in the fetal testis.

To determine if differentiation of the testicular enzymes that convert pregnenolone to testosterone is controlled by extragonadal hormones, the developmental pattern of 3β-hydroxysteroid dehydrogenase activity was characterized in gonads explanted in organ culture prior to enzymatic differentiation. Although the gonads showed little growth, they appeared to be well maintained under these conditions by histological criteria (5). In addition, histological differentiation of interstitial cells from organ-cultured testes was similar to that which occurred in *in vivo* control testes of the appropriate age (5). As shown in Fig. 3, regardless of whether gonads were cultured from day 16 or 17 of gestation, 3β-hydroxysteroid dehydrogenase activity appeared at the expected time, corresponding to day 18 *in vivo*. After 4 days in organ culture, the enzyme activity was approximately one-half that ob-

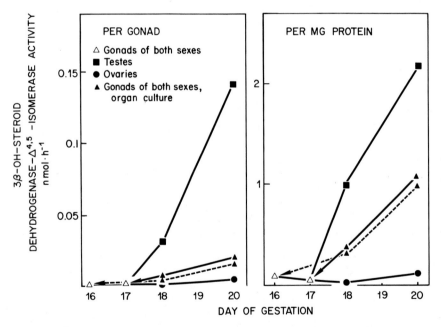

FIG. 3. Developmental pattern of 3β-hydroxysteroid dehydrogenase-$\Delta^{4,5}$-isomerase activity *in vivo* and in organ culture. Gonadal explants from day 16 (—–△—–) and day 17 (—△—) fetuses were cultured according to the method previously described and harvested at times corresponding to days 18 and 20 *in vivo*. Cultured and *in vivo* control gonads were incubated with 5 μM [7-³H]dehydroepiandrosterone and 3β-hydroxysteroid dehydrogenase activity was determined as described (5). (Reprinted from ref. 5.)

served in the corresponding control testes when expressed per mg protein. This is the value that would be predicted, since only about one-half of the indifferent gonads cultured were destined to be testes. Thus the enzymatic capacity to synthesize testosterone from pregnenolone develops independently of extragonadal hormones. However, in net terms, testosterone must ultimately be derived from cholesterol, and these studies did not provide information as to what regulates overall testosterone synthesis.

The timing of the appearance of 3β-hydroxysteroid dehydrogenase activity in the fetal testis is almost identical to the increase in testosterone content and the development of high-affinity gonadotropin receptors in the tissue (Fig. 4). The close correlation between the appearance of these gonadotropin receptors and the appearance of testosterone in the fetal testis suggested that testosterone synthesis might be mediated from its initiation by gonadotropins of placental or fetal pituitary origin. In contrast, gonadotropin binding is negligible in fetal ovaries at all ages (Fig. 4, ref. 2).

To determine whether testosterone formation is regulated by gonadotropin at this time, fetal gonads of various stages of development were incubated in the presence or absence of hCG, and the samples were then analyzed for testosterone and cyclic AMP content by radioimmunoassay (Fig. 5). An increase in cyclic AMP formation in response to hCG stimula-

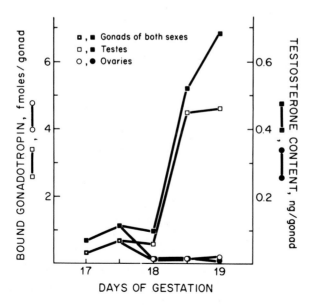

FIG. 4. hCG binding and testosterone content of fetal rabbit gonads between days 17 and 19 of gestation. (Reprinted from ref. 2.)

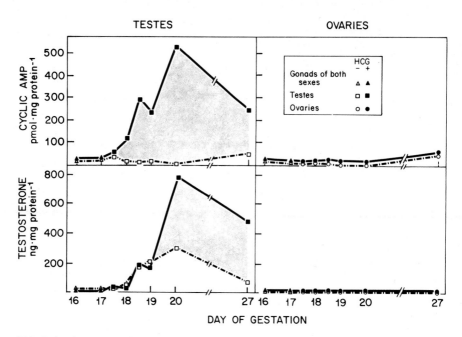

FIG. 5. *In vitro* cyclic AMP and testosterone responses of fetal rabbit gonads to hCG. Fetal rabbit gonads at various stages of development were incubated in the presence or absence of 10 IU hCG. Cyclic AMP and testosterone were measured by radio-immunoassay as described (2,4). (Reprinted from ref. 5.)

tion was first demonstrable in testes from the day 17.5 fetus, and the maximal response was seen on day 20. Cyclic AMP formation in the presence or absence of hCG was undetectable in ovaries at any age. This development of the cyclic AMP response to gonadotropin closely approximates the appearance of specific gonadotropin receptors in the fetal rabbit testis and also corresponds to the appearance of 3β-hydroxysteroid dehydrogenase activity. The finding in regard to testosterone content of testis in the presence and absence of hCG was different. The increase in testosterone content corresponded closely to the previously described increase in 3β-hydroxysteroid dehydrogenase, appearance of the gonadotropin receptors, and development of cyclic AMP responsiveness to hCG stimulation (5); however, enhancement of testosterone content by hCG was not demonstrable until after day 19 of gestation, many hours after the development of the hCG receptors.

To document this apparent dichotomy more precisely, the testosterone content, 3β-hydroxysteroid dehydrogenase activity, cyclic AMP content, and gonadotropin binding were all measured during a detailed time study of fetal testes between days 16 and 20 of gestation (Fig. 6). The patterns of development of gonadotropin binding and 3β-hydroxysteroid dehydrogenase activity were almost identical, first appearing at day 17.5 and increasing strikingly between days 19 and 20. Once again, the development of the cyclic

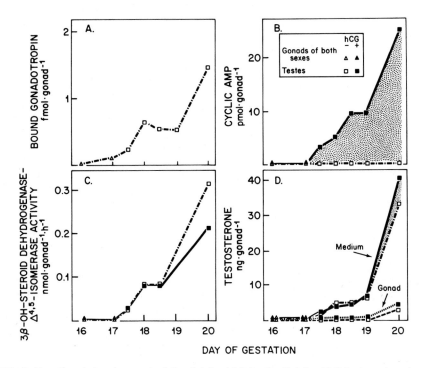

FIG. 6. Functional development of the fetal rabbit testis. Fetal rabbit testes at various stages of development were incubated in the presence or absence of 10 IU hCG as described (5). At the end of the incubation period the testes were separated from the media, and the media were boiled and then stored at −70°C prior to analysis. Gonads were either used directly or stored frozen. **A:** Gonadotropin binding by fetal testes. Binding studies were carried out on fetal testes taken from control incubations in the absence of hCG. **B:** Cyclic AMP response to hCG. Cyclic AMP was measured by direct radioimmunoassay of the incubation medium (4). **C:** 3β-Hydroxysteroid dehydrogenase activity. After incubation in the presence or absence of hCG, the testes were transferred to medium containing 5 μM [7-³H]dehydroepiandrosterone, incubated for 2 hr, and analyzed as described (13). **D:** Testosterone content of the medium and gonads. After incubation in the presence or absence of hCG, the gonads were separated from the media, and both were analyzed for testosterone as described (11). (Reprinted from ref. 5.)

AMP response to added gonadotropin (Fig. 6B) was synchronous with the increase in gonadotropin binding (Fig. 6A). Although a rise in both 3β-hydroxysteroid dehydrogenase activity and testosterone content were apparent in day 17.5 fetal testes, both parameters were unresponsive to hCG stimulation until day 20 (Fig. 6C,D).

DISCUSSION

Although the precise mechanism by which gonadal hormone secretion is initiated is still unclear, the present studies have provided insight into gonadal events occurring at the time of phenotypic sexual differentiation.

The finding that the fetal rabbit ovary has the enzymatic capacity to convert 19-carbon steroids to estrogens and that this capacity is acquired by the fetal ovary at the same time that the fetal testis of the rabbit acquires the capacity to form testosterone from pregnenolone is of fundamental interest. On the basis of histological studies, it has been presumed that the ovary differentiates later than the testis (6), but it is now clear that in terms of the acquisition of the enzymatic activities that allow these tissues to serve as distinct endocrine organs, differentiation of the indifferent gonad into an ovary or testis must take place almost simultaneously.

Whether the fetal ovary actually synthesizes estrogens from endogenous precursors has yet to be determined, but the fact that acquisition of unique enzyme profiles occurs simultaneously in testis and ovary suggests to us the possibility that the same factor(s) may be regulating the process in the two tissues. This could be the result of programming that is dictated by the genes themselves or some other aspect of embryogenesis; alternatively, it is possible that endocrine factors regulate the process. It seems reasonable to hypothesize that the enzymatic programming and the timing of the appearance of the enzymes are determined at a genetic level and that hormones are not involved in the process. The fact that the embryonic rabbit ovary acquires the enzymatic capacity to synthesize estrogen from 19-carbon steroids does not necessarily imply that estrogen synthesis by the ovary is important for female phenotypic differentiation. Indeed, additional work would have to be done to determine whether adequate endogenous substrate for estrogen synthesis is present in the embryonic ovary to result in significant rates of formation in the tissue. The concept that enzymatic differentiation of the fetal testis is due to inherent genetic programming rather than to hormonal or substrate induction at the time of gonadal differentiation is supported by the finding that enzymatic differentiation and histological development of fetal rabbit testes occur in organ culture in the absence of hormones or serum.

The fact that the enzymatic and structural machinery for testosterone synthesis from pregnenolone develops in the cultured fetal rabbit testis does not resolve the question of whether net testosterone synthesis at this critical period in male development is, itself, independent of hormonal regulation. Indeed, given the appropriate machinery for the conversion of pregnenolone to testosterone, the question must then be resolved as to what regulates the availability of pregnenolone to serve as substrate for androgen synthesis. On the one hand, these steroids are present in the placenta and the fetal adrenal and might reach the testis in sufficient quantity to serve as substrate for testosterone synthesis. On the other hand, gonadotropins of pituitary (or placental) origin might, as in the mature Leydig cell, regulate the enzymes responsible for the side-chain cleavage of cholesterol to pregnenolone in the developing testis.

The finding that gonadotropin receptors appear in the fetal testis concomitant with the appearance of the 3β-hydroxysteroid dehydrogenase-$\Delta^{4,5}$-isomerase reaction on day 17.5 of gestation is compatible with the possibility that gonadotropins might be involved in the onset of testosterone synthesis. We have attempted to resolve this question by investigating the functional integrity of the gonadotropin receptors in the gonads at various stages of development.

The simultaneous development of gonadotropin receptors and *in vitro* cyclic AMP responses to hCG indicates that as soon as receptors are detectable in the tissue, functional interaction with adenylate cyclase is established and the receptor-cyclase complex is responsive to hormonal activation. Thus, the genetic expression of receptor sites and adenylate cyclase appear to be linked during maturation of the fetal Leydig cell.

Although from their earliest appearance the gonadotropin receptors are capable of mediating the enhancement of cyclic AMP formation in the presence of hCG, these receptors do not appear to be coupled functionally to the testosterone biosynthetic pathway until later in gestation. This finding suggests to us that at its initiation testosterone synthesis in the male embryo is independent of gonadotropin control. If this were the case, then the formation of the male phenotype would be independent of control by gonadotropins of whatever origin (placental or pituitary), and the imposition of gonadotropin control later in gestation would be essential only for the maturation of the basic male anatomy.

However, this interpretation cannot be viewed as conclusive. In the mature rat testis, occupancy of only 1% of the gonadotropin receptors can result in maximal stimulation of testosterone synthesis (3,12). It remains possible that between days 18 and 19 of gestation, a limited number of gonadotropin receptors are functionally coupled to testosterone synthesis but are saturated with endogenous gonadotropins and hence cannot be stimulated further by exogenous hCG. To resolve this issue, it will be necessary to analyze the nature and quantity of circulating and bound gonadotropins in the fetal rabbit testis during this critical phase of development.

SUMMARY

Several aspects of the functional development of the testis and ovary have been characterized in the fetal rabbit. First, the enzymatic capacity of the fetal ovary to form 17β-estradiol from testosterone develops at the same time in embryogenesis as the enzymatic capacity to form testosterone develops in the fetal testis. Second, the enzymatic differentiation of the fetal testis occurs at the expected time in organ culture, suggesting that this aspect of development is not hormone induced. Third, gonadotropin receptors appear in testis concomitant with the initial rise in testosterone content, but are virtually absent in ovaries throughout gestation. However, the onset of testosterone synthesis and the initial rise in testosterone content in the fetal testis do not appear to be coupled to gonadotropin stimulation, since hCG did not stimulate testosterone formation at a time when there was a 200-fold stimulation of cyclic AMP formation.

ACKNOWLEDGMENTS

This work has been supported by Grant AMO3892 from the National Institutes of Health. F. W. George was the recipient of a fellowship under NIH Training Grant TI CA 5200.

REFERENCES

1. Attal, J. (1969): Levels of testosterone, androstenedione, estrone and estradiol-17β in the testes of fetal sheep. *Endocrinology,* 85:280–289.
2. Catt, K. J., Dufau, M. L., Neaves, W. B., Walsh, P. C., and Wilson, J. D. (1975): LH-hCG receptors and testosterone content during differentiation of the testis in the rabbit embryo. *Endocrinology,* 97:1157–1165.
3. Catt, K. J., and Dufau, M. L. (1973): Spare gonadotropin receptors in rat testis. *Nature* [*New Biol.*], 244:219–221.
4. Dufau, M. L., Watanabe, K., and Catt, K. J. (1973): Stimulation of cyclic AMP production by the rat testis during incubation with hCG in vitro. *Endocrinology,* 92:6–11.
5. George, F. W., Catt, K. J., Neaves, W. B., and Wilson, J. D. (1978): Studies on the regulation of testosterone synthesis in the rabbit fetal testis. *Endocrinology,* 102: 665–673.
6. Gilman, J. (1948): The development of the gonads in man, with a consideration of the role of fetal endocrines and the histogenesis of ovarian tumors. *Contrib. Embryol.,* 32:83–131 (Carnegie Institute Washington, Publ. 575).
7. Goldstein, J. L., and Wilson, J. D. (1974): Hereditary disorders of sexual development in man. In: *Birth Defects,* edited by A. G. Motulsky and W. Lentz, pp. 165–173. Proceedings of the 4th International Conference (Vienna), Excerpta Medica Foundation, Amsterdam.
8. Jost, A. (1953): Problems of fetal endocrinology: The gonadal and hypophyseal hormones. *Recent Prog. Horm. Res.,* 8:379–418.
9. Jost, A. (1960): The role of fetal hormones in prenatal development. *Harvey Lect.,* 55:201–226.
10. Jost, A. (1970): Hormonal factors in the sex differentiation of the mammalian foetus. *Phil. Trans. R. Soc. Lond. B,* 259:119–130.
11. Lipsett, M. B., and Tullner, W. W. (1965): Testosterone synthesis by the fetal rabbit gonad. *Endocrinology,* 77:273–277.
12. Mendelson, C., Dufau, M., and Catt, K. (1975): Gonadotropin binding and stimulation of cyclic adenosine 3′:5′-monophosphate and testosterone production in isolated Leydig cells. *J. Biol. Chem.,* 250:8818–8823.
13. Milewich, L., George, F. W., and Wilson, J. D. (1977): Estrogen formation by the ovary of the rabbit embryo. *Endocrinology,* 100:187–196.
14. Siiteri, P. K., and Wilson, J. D. (1974): Testosterone formation and metabolism during male sexual differentiation in the human embryo. *J. Clin. Endocrinol. Metab.,* 38:113–125.
15. Wilson, J. D., and Siiteri, P. K. (1973): Developmental pattern of testosterone synthesis in the fetal gonad of the rabbit. *Endocrinology,* 92:1182–1191.

Ontogeny of Receptors and Reproductive Hormone Action,
edited by T. H. Hamilton, J. H. Clark, and W. A. Sadler.
Raven Press, New York 1979.

Prenatal Effect of the Estrogenic Hormone on Embryonic Genital Organ Differentiation

Ching Sung Teng and Christina T. Teng

Department of Cell Biology, Baylor College of Medicine, Houston, Texas 77030

During the past decades extensive research has been done on the study of the normal growth pattern and development of genital organs at the postnatal period. However, there is little knowledge available about the age-related onset of biological events in these organs in the embryonic stage and even less information about prenatal hormone influence on differentiation and future development. We have selected the chick urogenital organ as a model system for these studies.

Between the fifth and eighth days of incubation, the urogenital system of the developing chick passes through a neutral stage during which both the genetic male and female possess a pair of müllerian ducts. In the male embryo, the right and left ducts start involution on the eighth day of incubation and disappear by the 13th day. In the female embryo, the right duct starts to regress on the ninth day of incubation and, by the time of hatching, appears as a tiny rudiment. The left duct continues to develop and becomes a functional hen oviduct (37,62).

In the male embryo, the right and left gonads develop into the testis, whereas in the female, the right gonad regresses and the left gonad develops into an ovary (9,11). The synthesis and secretion of sex steroids in the gonads of the chick embryo have been reported to start during the early stages of development (17,21,56). Both the left and right gonads of the female embryo are capable of estrogenic and androgenic hormone secretion and, consequently, are subject to the gonadotropin effect (51). The gonadal secretion of sex steroids (57,64), or other factors (32), are believed to be responsible for the growth or regression of the müllerian ducts. How gonadal secretion affects the differentiation of the embryonic sex tracts is still unknown.

Our recent observations have indicated the existence of estrogenic hormone receptors and acceptors capable of responding to the estrogenic hormone in the embryonic sex tract (46–49). Consequently, it is important to know the estrogenic effect on embryonic organ differentiation.

Previous reports indicated that the administration of estrogenic hormones to the fertilized chick eggs before sex organ differentiation affects the embryonic female and male reproductive tract development (8,12,24,42,59). However, these experiments did not demonstrate a sufficient hormonal effect on retention and responses of the müllerian ducts. It is, thus, important to reinvestigate this developmental problem with a more refined technique in order to gain new insights.

This chapter concentrates on the study of the morphological, histological, and biochemical events in the genital organ and their response to estrogenic hormones. These biological events are essential for the development and function of the genital organs.

RESULTS AND DISCUSSION

Morphological and Histological Change after Diethylstilbestrol Treatment

Diethylstilbestrol Administration and the Prevention of Müllerian Duct Regression

The conventional method for administering hormones to chick embryo, by applying hormone solution to the chorioallantoic membrane or into the allantois, risks a high rate of embryonic mortality (approximately 50 to 60% after 2 days of administration). The technique of administration described in this study is based on a temperature-dependent process of diffusing diethylstilbestrol (DES) into the embryo. The optimal temperature for enhancing hormonal efficiency is 50 to 75°C. Figure 1 and Table 1 show the normal patterns of development at the 15th day of incubation; the female right müllerian duct has regressed, whereas the left duct remains. After a chick embryo receives one treatment of DES at the fifth day of incubation, the right duct of the 15- or 18-day-old female embryo is retained. The retention persists after birth (Fig. 1d). Both ducts show an increase in size and weight. The optimal dose of DES used to prevent the right duct from regressing is 40 mg DES/ml. The treatment under these described conditions has no effect on the overall rate of mortality of the embryos.

Previous experiments demonstrated a partial retention of the right duct after hormone administration (24,42). This observation led to the question

$\longrightarrow$

FIG. 1. The retention of the chick müllerian duct after treatment of DES at fifth day of incubation. **a:** Fifteen-day female embryo (without DES). **b:** Fifteen-day female embryo (with DES, 40 mg/ml). **c:** Eighteen-day female embryo (with DES, 40 mg/ml). **d:** Six-month-old hen (with DES, 10 mg/ml). **e:** Fifteen-day genetic male embryo (with DES, 40 mg/ml). **f:** An abnormal twelve-day genetic female embryo (without DES). **a,b,c** and **e:** ×2.3. **f:** ×2.85. **d:** ×0.515, r, right; I, left; o, ovary; T, testis; Md. müllerian duct; Od, oviduct.

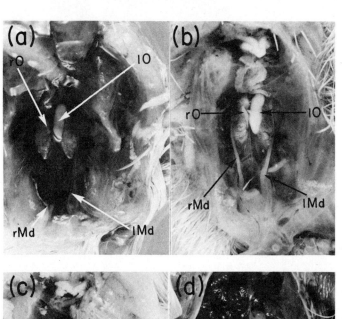

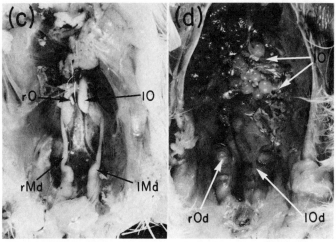

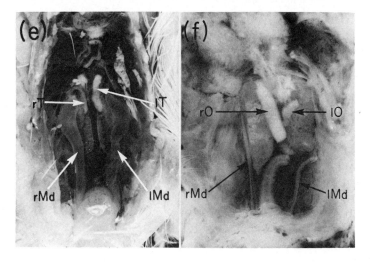

TABLE 1. *The effect of temperature and the dose of DES during the treatment of embryo on the development of female left and right müllerian ducts*

	Temperature (°C)	DES (mg/ml)	Length (mm/duct)		Wet weight (mg/duct)		Survival rate (%)
			Left	Right	Left	Right	
(a)	Control	—	10.5 ± 1.2	5.1 ± 0.3	2.5 ± 0.2	0.59 ± 0.1	75
	37	40	11.3 ± 1.2	9.5 ± 0.4	4.2 ± 0.3	4.3 ± 0.3	74
	41	40	12.4 ± 1.3	9.8 ± 0.5	3.8 ± 0.3	3.8 ± 0.3	73.5
	50	40	13.2 ± 1.4	12.8 ± 0.9	5.0 ± 0.3	4.5 ± 0.3	74
	75	40	13.4 ± 1.4	13.3 ± 1.2	5.4 ± 0.4	4.6 ± 0.4	73
(b)	50	0.5	11.6 ± 1.2	7.1 ± 0.3	3.6 ± 0.2	1.8 ± 0.2	74
	50	1.0	12.2 ± 1.3	7.9 ± 0.4	3.1 ± 0.2	2.3 ± 0.2	73
	50	5.0	12.4 ± 1.2	9.4 ± 0.5	4.0 ± 0.3	2.9 ± 0.2	72
	50	10	12.6 ± 1.3	9.6 ± 0.5	4.5 ± 0.3	3.5 ± 0.3	73
	50	20	12.8 ± 1.4	10.4 ± 1.2	4.9 ± 0.4	3.9 ± 0.3	73
	50	40	13.2 ± 1.4	13.0 ± 1.3	5.4 ± 0.4	4.6 ± 0.3	73

The technique for DES treatment was initially reported by Seltzer (41) and Van Tienhoven (55) and modified as follows: fertilized white Leghorn chick eggs after incubation for 5 days were placed in the solution (with various concentrations of DES, as specifically indicated, in 70% of ethanol at various temperatures) so that the bottom 1 to 1½ inch was immersed in the DES solution for 10 sec. The control eggs received the same treatment with the solution without DES. After the treatment was completed, the eggs were returned to the incubator. When the embryos reached the 15th day of incubation, they were dissected midventrally and the müllerian ducts were removed for measurement. Each value represents the mean ± SD of 80 measurements.

whether exogenous hormones differed in their properties from endogenous hormones or were administered by an inadequate method. The improved administration technique provided an efficient way for complete retention of the right duct, the frequency of right-duct retention being 99% out of a total sample of 1,500 treated embryos.

In male embryos, a similar type of müllerian duct retention has been observed. In addition its gonadal development resembles the female pattern; the left testis develops into an ovotestis and the right testis regresses (Fig. 1e). This finding corroborates previous observations by Gaarenstroom (16) and Willier et al. (58).

The regression of the müllerian duct is a common phenomenon found in the embryos of reptiles, birds, and mammals (7,38,43,62). At present, whether the regression of the müllerian ducts is caused by the testosterone action (30,57,64) or is the result of a nonsteroidal müllerian-inhibiting factor (20,32) is still under debate. The mechanism by which estrogenic hormone prevents the müllerian ducts (both male ducts and the female right duct) from regressing, as presented in this report, is also still unknown. If the observed early gonadal secretion is confined in the cell clusters of the medullary cords (39), it is likely that DES exerts its inhibiting effect on the differentiation of this group of cells, thereby inactivating the cellular secretion. On the other hand, it is possible that a high DES content in the müllerian duct counteracts the effect of testosterone, which initiates the hydrolytic enzymes for cellular lysis (40). It is unlikely that DES exerts its effect through the pituitary level since the hypothalamic-pituitary axis is not connected until the 13th day of incubation (66). We lack sufficient information to determine which possibility is valid.

Factors Influence the Retention of Müllerian Duct

Table 2 indicates that the age of the embryo is a critical factor for the retention of the right duct. If DES is administered later than the fifth day of incubation, the retention of the müllerian duct is incomplete. To obtain a complete retention the optimal age for treatment is from the third to fifth day of incubation for both male and female embryos. This observation suggests that certain releasing factors from the gonads influence organ death in this stage. The following observations support this theory. First, Woods et al. (65) have reported the presence of plasma testosterone in the blood of both sexes on day 5.5 of incubation. The testosterone level peaks at the 13th day of incubation when the male duct regression is completed. It also was demonstrated that the plasma testosterone level in females increased to the highest level on day 15.5 of incubation, the same date for a complete regression of the female right duct. The high level of testosterone in the plasma at this stage is probably the result of a high rate of testosterone secretion from both ovaries of the female chick (8,22). Second, Groen-endijk-Huijbers and Burggraaff (20) observed the effect on müllerian duct

TABLE 2. *The effect of age of the embryos on response to DES treatment*

Day of treatment	Length (mm/duct)		Wet weight (mg/duct)	
	Left	Right	Left	Right
Control (without DES)	10.5 ± 1.2	5.1 ± 0.3	2.5 ± 0.2	0.59 ± 0.1
2	13.5 ± 1.3	13.3 ± 1.3	5.5 ± 0.4	5.0 ± 0.4
5	13.2 ± 1.4	13.0 ± 1.3	5.4 ± 0.4	4.6 ± 0.3
6	11.4 ± 1.2	10.9 ± 1.0	4.4 ± 0.4	3.8 ± 0.3
7	11.6 ± 1.2	9.9 ± 1.0	4.2 ± 0.3	3.5 ± 0.3
8	11.3 ± 1.2	8.7 ± 0.9	4.3 ± 0.3	3.5 ± 0.3
9	14.1 ± 1.3	8.2 ± 0.9	4.9 ± 0.4	3.8 ± 0.3
11	15.5 ± 0.14	8.2 ± 0.9	6.7 ± 0.5	3.3 ± 0.3
12	15.2 ± 0.15	7.5 ± 0.5	6.3 ± 0.5	2.6 ± 0.2
13	15.3 ± 0.14	5.9 ± 0.5	5.2 ± 0.4	1.8 ± 0.1

The embryos were treated with DES (40 mg/ml) at 50°C at various stages of development as indicated. After the treatment the eggs were returned for incubation until they reached the age of the 15th day, then the müllerian ducts were dissected from the embryos and used for measurement. The data represents the means ± SD of 25 measurements.

regression of implanting testicular fragments from increasingly aged male chicks into the abdominal wall of a 4-day-old chick embryo. They found a 100% regression with testes ranging in age from 5 days of incubation to 7 days posthatching. The percentage of regression dropped to 10 to 0% with 10-day- to 3-week-old testes. The nature of the regression factor in the testis is unknown.

Although the existing findings do not constitute evidence for the hypothesis that androgenic hormones are responsible for müllerian duct regression in the chick embryo, they do suggest that the hormones initiate the production of certain factors for organ regression. It also might be possible that testosterone binds to a nonsteroidal factor and subsequently exerts the inhibitory effect. Another possible explanation for the regression of the right duct may be its own genomic programming for the organ death (see review, 25). Based on our current observation, this possibility seems to be unlikely. We are in favor of the hypothesis that female right duct regression is controlled by the right ovotestis; in the early stage of development, an unknown factor secreted from the ovotestis could cause right duct regression through a local effect. Recently, we have discovered a 12th-day female embryo (one out of 1.5×10^5 embryos dissected) with a normal right ovary and a left ovotestis. Consequently, this embryo has an intact right müllerian duct and a regressing left müllerian duct (Fig. 1f). This observation emphasizes the importance of ovarian effect on the sex tract differentiation.

We have considered the possibility that the genetic makeup of a given

TABLE 3. *The context of DES in body fluid of the embryo after the treatment*

Age of embryo (day)	DES content (ng/ml)					
	Amniotic fluid			Serum		
	Control	DES-treated (10 mg/ml)	DES-treated (40 mg/ml)	Control	DES-treated (10 mg/ml)	DES-treated (40 mg/ml)
5	ND	12.6 ± 0.66	12.8 ± 0.77	ND	13.2 ± 0.70	13.4 ± 0.66
6	ND	12.4 ± 0.66	12.8 ± 0.74	ND	12.2 ± 0.64	13.2 ± 0.68
7	ND	12.2 ± 0.65	12.6 ± 0.73	ND	11.4 ± 0.53	12.4 ± 0.66
8	ND	2.6 ± 0.24	3.4 ± 0.25	ND	2.5 ± 0.23	3.5 ± 0.30
9	ND	1.2 ± 0.15	2.2 ± 0.16	ND	1.5 ± 0.15	2.8 ± 0.17
12	ND	0.9 ± 0.06	1.7 ± 0.15	ND	0.5 ± 0.07	2.0 ± 0.08
15	ND	ND	ND	ND	ND	ND
18	ND	ND	ND	ND	ND	ND

Fertilized eggs after incubation for 5 days were treated with DES (10 or 40 mg/ml). The amniotic fluid and blood serum were collected daily from the embryo after the treatment. The method for radioimmunoassay of DES was according to the procedure of Teng and Teng (51). Radioactive DES solution was prepared by diluting the stock solution with phosphate-buffered saline containing 0.1% gelatin. DES was measured by radioimmunoassay with an antiserum against DES-succinate-HSA (Radioassay System Lab., Carson, Cal.). Each value represents the mean ± SD of three determinations.

ND, nondetectable.

species affects the retention of the right duct, as suggested by Morgan and Kohlmeyer (33). Using this technique of administration, a complete retention was also demonstrated in other species of domestic fowl, e.g., Japanese silky, brahma, New Hampshire red, polish buff, and quail (*unpublished data*).

After DES treatment it is important to investigate the content of the hormone and the duration of its presence in the embryo. These factors could influence the retention as well as the growth response of the ducts. The information presented in Table 3 indicates that right after treatment the DES content in the amniotic fluid and blood serum ranged from 12 to 13 ng/ml; the DES content decreased to 1.0 to 2.0 ng/ml after 7 days of treatment and became undetectable after 10 days of treatment.

Precocious Differentiation of Genital Tract after DES Treatment

The histological observation presented in Fig. 2 shows the invagination of the epithelium of the left and right duct and the development of ovalbumin-secreting gland from the columnar epithelium after DES treatment. The precocious differentiation of the tubular gland cells was observed in the epithelial layer in both ducts. The thickness of the epithelial and stromal layers was increased. This observation is similar to a previous observation of

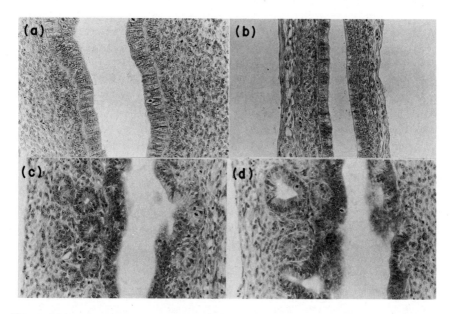

FIG. 2. Light micrograph of 15-day-old female chick müllerian ducts with or without DES treatment. **a:** Left duct (without DES). **b:** Right duct (without DES). **c:** Left duct (with DES, 40 mg/ml treatment). **d:** Right duct (with DES, 40 mg/ml treatment). **a–d:** ×1050.

the immature chick oviduct after DES administration for 6 days (36). However, the stromal thickness in the müllerian duct after DES treatment is far less than that in the immature chick, indicating that a genomic preprogrammed full responsiveness toward the estrogenic effect in the genital tract is not reached in the early stage of development. On the other hand, DES might cause a certain cell type to proliferate and might inhibit (or have no effect on) other types. This possibility is discussed in the following section.

Changes in Biochemical Parameters due to DES Treatment in the Sex Tract

DNA, RNA, and Proteins

The prenatal DES treatment initiates a precocious stimulation of müllerian duct growth. Total RNA content in the left and right ducts was increased approximately 83 and 551%; the total protein content increased about 78 and 357% in the left and right duct, respectively. However, the DNA content in the left duct was not significantly increased (approximately 5 to 10% above control). DNA increased in the right duct about 218% (Table 4).

The smaller response of the left duct to estrogen stimulation for DNA synthesis could be explained by the following possibilities: (a) The early prenatal estrogen treatment prolongs the estrogenic effect on the embryonic target tissue, thus possibly causing the suppression of cell division as has been demonstrated in the epithelium of mice uterine tissue (27,31). (b) Estrogen selectively causes certain types of cells in the epithelium to differentiate and inhibits the mitotic rate of other types of cells. This possibility has been observed in the immature mice müllerian vagina (13,14,45). (c) The developmental timing for cell division in the sex tract is not reached until after the late stage of organogenesis. Therefore, an early estrogenic effect could only cause the existing cells to grow and proliferate rather than to multiply.

The characteristic nature of the RNAs induced by prenatal DES treatment has been studied in our laboratory. We have analyzed the ovalbumin messenger RNA (mRNAov) by hybridization of cDNA, transcribed from pure mRNAov, to a vast excess of its template, and we were able to detect the induction of mRNAov in the 15-day female chick müllerian duct after having received prenatal DES treatment. With an optimal dose of DES treatment, 285 molecules of mRNAov/cell could be induced in the müllerian duct, compared to an undetectable amount of message in the duct without DES stimulation (2).

The induction of mRNAov in the embryonic target organ parallels the active translation of ovalbumin in the precociously stimulated epithelium. A drastic increase in the target organ-specific transcriptional product (ovalbumin) has been detected by immunodiffusion technique as presented in Fig. 3. The increase in ovalbumin production in the müllerian duct after

TABLE 4. The effect of DES on the development of the female left and right müllerian ducts

Müllerian duct	DNA[a] (µg/duct)		RNA[b] (µg/duct)		Protein[b] (µg/duct)	
	(a)	(b)	(a)	(b)	(a)	(b)
Left duct	10.5 ± 0.45	11.2 ± 0.52	13.73 ± 0.54	25.11 ± 0.92	202 ± 4.4	360 ± 5.2
Right duct	3.4 ± 0.22	10.8 ± 0.35	3.6 ± 0.32	23.45 ± 0.81	69.8 ± 2.2	319 ± 4.6

The müllerian ducts were excised from the 15-day-old embryos treated with DES at the fifth day of incubation. DNA was determined by the diphenylamine reaction (18) with embryonic chick DNA as a standard. RNA was determined by the procedure of Munro and Fleck (35). Protein was determined by the procedure of Lowry et al. (28), with bovine serum albumin as a standard.
(a), Control without the hormone; (b), treatment with the hormone.
[a] Each value represents the mean ± SD of 10 determinations.
[b] Each value represents the mean ± SD of three determinations.

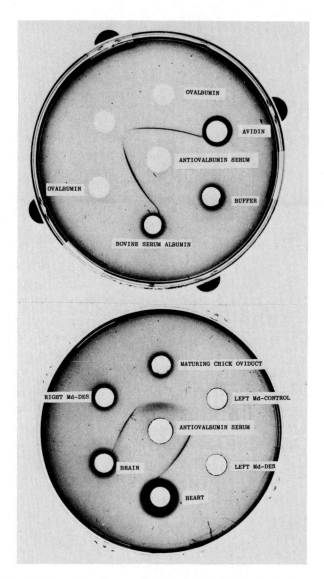

FIG. 3. Immunodiffusion of antiserum raised against ovalbumin. Center well contains antiserum. The cytoplasmic protein (105,000 × g supernatant) from control or DES-treated left or right müllerian duct (or other embryonic organs) was individually applied on each well as indicated. The immunodiffusion was carried out in 4°C for 18 hr. The precipitin lines were observed opposite wells containing cytoplasmic protein of DES-treated left and right müllerian ducts.

DES treatment could account for the increase of protein synthesis reported in this section.

Cytoplasmic and Nuclear Estrogen Receptor Concentration

After administration of DES to the female embryos on the fifth day of incubation, the concentration of estrogen receptor in the cytoplasm was measured by dextran/charcoal technique (Fig. 4). The specific binding of cytosol to [^{3}H]estrogen in the range of 10 to 100 nM is illustrated in Fig. 4a. A Scatchard plot based on the specific-binding data for estradiol indicated the presence of the saturatable binding sites (*steep parts of the curves*) and the nonsaturatable binding sites (*flat parts of the curves*) (Fig. 4b).

The cytoplasmic estrogen receptor concentration in the saturatable part is 0.63, 0.40, and 0.28 nM/mg of cytosol protein for increasing doses of DES treatment. If, according to our previous observation, the concentration of the cytoplasmic receptor in the 15-day müllerian duct cytosol is 1.57 nM/mg of cytosol protein, then the receptor concentration in the DES-treated duct represents approximately 18 to 39% of the nontreated duct (46). The value of the dissociation constant (K_D) calculated from the slopes of the Scatchard plots is approximately 3.0 nM, which is equivalent to that of the control value (47). An attempt was made to remove the possible existence of endogenous estrogen by incubating the cytosol with dextran/charcoal at 4, 20, and 45°C; this treatment did not increase the additional saturatable binding sites (*unpublished data*).

After DES treatment, the nonsaturatable binding sites in cytosol were greatly increased compared to the nontreated in the müllerian duct (Fig. 4b). A large proportion of this receptor was also observed in the oviduct of the laying hen (5). At present, the role of this class of receptor in the hormonal action is still unknown and needs to be investigated.

Nuclei isolated from the DES-treated müllerian duct were measured for the endogenous estrogen binding sites which are 0.584, 0.620, 0.570, and 0.612 pmoles/mg of DNA for the respective doses of DES treatment. These values are approximately equivalent to the value (0.536 pmoles/mg of DNA) of the nontreated müllerian duct nuclei (Table 5).

The reduction of the concentration of cytoplasmic high-affinity estrogen binding sites in the embryonic target organ could be due to the following possibilities: (a) the prenatal DES treatment had a permanent effect on the initial synthesis and replenishment of the receptor; (b) DES treatment altered the receptor binding sites, from a high-affinity to a low-affinity nonsaturatable type; and (c) DES occupied the high-affinity binding sites rendering the current technique insensitive for detection. However, this last possibility is unlikely since any cytoplasmic receptor binding to estrogen would result in its translocation into the nucleus. Our result does not show

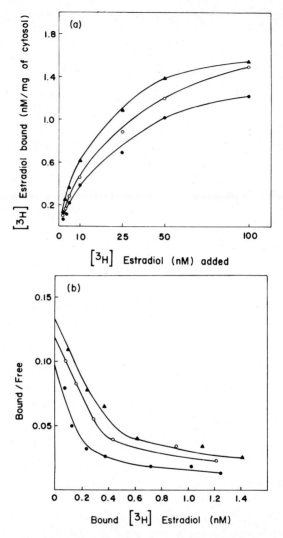

FIG. 4. Binding curve and Scatchard plot of cytosol from müllerian duct-received DES treatment. The procedure for the preparation of cytosol and the estrogen-binding sites in this fraction were according to the procedures described in Teng and Teng (48). In each experiment, two batches of cytosol in separate tubes were incubated simultaneously, one with [³H]estradiol, and the other with [³H]estradiol together with a 100-fold excess of the nonradioactive estradiol. The specific binding is the difference between the total and the nonspecific binding. **a:** Specific cytoplasmic [³H]estradiol binding as a function of added estrogen. **b:** Scatchard plot derived from the specific binding. The following symbol represents the cytosol from the müllerian duct of 15-day-old female chick receiving various doses of DES treatment at the fifth day of incubation. ●, DES 10 mg/ml; ○, DES 5 mg/ml; ▲, DES 0.5 mg/ml.

TABLE 5. *Endogenous nuclear estradiol-binding sites in the müllerian duct after DES treatment*

Dose of DES treated (mg/ml)	Specific [³H]estradiol bound (pmoles/mg of DNA)
Control	0.536 ± 0.043
0.5	0.584 ± 0.037
1.0	0.620 ± 0.038
5.0	0.570 ± 0.042
10.0	0.613 ± 0.044

Nuclei were prepared from the 15-day-old embryos that were treated with various doses of DES at the fifth day of incubation. The nuclei were incubated with 13 nM [³H]estradiol with or without a 100-fold excess of nonlabeled estradiol for 1 hr at 37°C. The nuclear preparation and the specific binding of the estrogen in the nuclei were according to Teng and Teng (48) and Anderson et al. (1). The results presented were the specific binding of mean ± SD of three preparations.

any significant increase in nuclear binding sites due to prenatal DES treatment.

It is still unknown whether the reduction of cytoplasmic receptors could cause the malfunction of the sex tract or whether the limited number of nuclear estrogen binding sites is insufficient to maintain normal physiological function of the embryonic organ. A similar observation indicated that neonatal estrogenization of female mice reduces uterine responsiveness in the adult (34,67), and the reduction is related to the reduction of estrogen binding in uterine tissue so treated (54).

Estrogenic Effect on Embryonic Gonadal Function

Inhibition of Gonad Development

Table 6 indicates that prenatal DES treatment inhibits the development of both the ovary and the testis. In the ovary, the DNA, RNA, and protein were inhibited about 36 to 41, 52 to 53, and 45 to 47%, respectively. Similar inhibitions observed in the testis were 57, 40, and 45%, respectively.

A similar inhibition of ovarian (or testicular) growth and function has been demonstrated in the prepubital rodents after neonatal estrogen treatment (15,44,67). This sex-steroid treatment causes an irreversible change in the hypothalamic regulatory mechanisms of pituitary gonadotropic function, resulting in a permanent suppression of gonadal growth and function (4,19). On the other hand, the early embryonic gonad's secretion has been shown to play a key role in the development of the gonads (6,26). The effect of DES on ovarian and testicular secretion in the embryo could be another contributing factor.

TABLE 6. *The effect of DES on the development of embryonic gonads*

Gonads (age of embryo)	DNA (μg/gonad)			RNA (μg/gonad)			Protein (μg/gonad)		
	(a)	(b)	Inhibition %	(a)	(b)	Inhibition %	(a)	(b)	Inhibition %
Ovary (15)	21.0 ± 2.15	12.4 ± 1.25	41	44 ± 4.3	21 ± 2.6	53	410 ± 1.0	210 ± 2.2	47
Ovary (18)	23.0 ± 1.44	14.7 ± 1.33	36	48 ± 4.0	23 ± 2.3	52	558 ± 6.5	361 ± 5.4	45
Testis (18)	25.2 ± 1.56	10.8 ± 1.08	57	25 ± 2.6	15 ± 1.8	40	352 ± 4.5	193 ± 2.0	45

The gonads were excised from the 15- (or 18-)day-old embryos treated with DES (40 mg/ml) at the fifth day of incubation. The procedures for DNA, RNA, and protein determination were described in Table 4. Each value represents the mean ± SD of three determinations. (a), Control without the hormone; (b), treatment with the hormone.

The inhibition of DNA and RNA synthesis in the gonads is reflected in a current observation that after receiving estrogen for 3 hr, the ornithine decarboxylase (ODC) activity in the 15-day-old chick gonads was inhibited by approximately 40%. If DES treatment was started on the fifth day of incubation and the ODC activity measured at the 15th day, the activity was observed to be 50% lower (50). The close correlation of ODC activity with the enhancement of nucleic acids synthesis has been studied in many developmental systems (3,10).

Gonadal Secretion

The sexual reversal in the avian embryonic testis by estrogenic hormone treatment has long been observed by Wolff (60,61) and Lutz-Ostertag (29). Their histological observation indicated a proliferation of the cortical tissue in the testis, and further observation by Haffen demonstrated that 3 months after hatching the oocytes were developed in the cortex of the intersexual male (23). However, very little is known about the endocrine function of this sexually altered male. Applying our DES treatment technique, the testis in the male chick embryo was feminized and mimicked the female pattern of development with the regression of the right testis and the proliferation of the left one (Fig. 1e). The estrogen secretion in the left testis was induced by DES and increased from a nondetectable level to 320 pg/10 gonads, whereas the testosterone secretion was inhibited from 35 to 12 pg/10 gonads, a decrease of 66% (Table 7). The decrease in testosterone secretion probably corresponds to the regression of testicular medulla as reported by Wolff and Ginglinger (63).

The ovarian secretion for estrogen was reduced from 1,950 to 1,050 pg/10 gonads, a decrease of 46%; the testosterone secretion was reduced from 120 to 50 pg/10 gonads, a decrease of 58% (Table 7). In the section on

TABLE 7. *Steroid secretion in gonads after DES treatment*

Gonads	Estradiol (pg/10 gonads)			Testosterone (pg/10 gonads)		
	(a)	(b)	Inhibition %	(a)	(b)	Inhibition %
Ovary	1,950 ± 150	1,050 ± 92	46	120 ± 12	50 ± 2.2	58
Testis	ND	320 ± 60	—	35 ± 4.3	12 ± 1.6	66

The gonads were excised from the 15-day-old embryos treated with DES (40 mg/ml) at the fifth day of incubation. Ten testes and ovaries were separately cultured in a Falcon model 300/tissue culture dish containing 1 ml Hanks' balanced salt solution in an incubator circulated with $O_2 + CO_2$ (95:5) at 41°C for 4 hr. The steroids released to the medium were determined by radioimmunoassay as described in Teng and Teng (51). Each value represents the mean ± SD of three determinations.
(a), Control without the hormone; (b), treatment with the hormone; ND, nondetectable.

inhibition of gonad development (see above), we demonstrated that after DES treatment the DNA, RNA, and protein content in the ovary is almost half that of the control, which is probably why the steroids production is reduced by approximately half. Furthermore, we have investigated the possible effect of prenatal DES treatment on the gonadal responsiveness toward gonadotropic hormone. Control or DES-treated gonads were cultured separately in the presence of 20 IU of human choriogonadotropin (hCG) per 10 gonads in culture for 2 hr and their steroid production was measured. We found that early DES treatment changed the testicular responsiveness to hCG. The estrogen production increased from none to 1,000 pg/10 gonads, and the testosterone production was reduced from 620 to 340 pg/10 gonads. In ovary, the estrogen production was reduced from 5,100 to 4,000 and the testosterone production decreased from 1,990 to 1,300 pg/10 gonads (52,53). However, if the steroid production was calculated in terms of per cell basis, there was no significant difference between the control and the DES-treated gonads, indicating that the biochemical events responsible for steroidogenesis are probably not damaged by DES treatment.

CONCLUDING REMARKS

A technique for estrogenic hormone DES administration has been established that enhances the uptake of the hormone in the embryos without risking embryonic mortality. This technique provides efficiency and constancy in hormone administration, which is beneficial for studying embryonic organ development at the biochemical level. The prenatal effect of DES on sex tract differentiation can be described as twofold: first, the regression of the right müllerian duct was prevented, and second, both right and left ducts were stimulated for growth and precocious differentiation of epithelial gland cell. Decreases in the concentration of cytoplasmic estrogen receptor in the müllerian duct corresponded to increases in the concentration of DES administered. After treatment with DES, the cytoplasmic estrogen receptors were characterized by high-affinity binding proteins plus low-affinity nonsaturatable binding proteins. The nuclear estrogen binding sites remained at a constant level after DES treatment.

The prenatal DES administration caused a sexual reversal in the male embryo, with the gonadal development resembling the female pattern. The growth of the testis and ovary was inhibited. The testicular secretion of the estrogenic hormone was greatly increased and accompanied by a reduction of testosterone secretion. In the ovaries, both estrogen and testosterone secretions were inhibited. The cause of these alterations in the pattern of sexual differentiation due to prenatal DES treatment is yet unknown and needs to be further investigated.

ACKNOWLEDGMENT

We thank R. Metivier, D. Kilgo, and R. Downing for assistance. This work was supported by National Institutes of Health Grants HD-08218, HD-09467, and AG-00523.

REFERENCES

1. Anderson, J., Clark, J. H., and Peck, E. J., Jr. (1972): Oestrogen and nuclear binding sites: Determination of specific sites by [³H]oestradiol exchange. *Biochem. J.*, 126: 561–567.
2. Andrews, G., and Teng, C. S. (1978): Studies on sex-organ development: prenatal effect of oestrogenic-hormone on ovalbumin messenger RNA induction in chick Müllerian duct. *Biochem. J. (in preparation.)*
3. Bachrach, U. (1973): *Function of Naturally Occurring Polyamines.* Academic Press, New York.
4. Barraclough, C. A. (1966): Modifications in the CNS regulation of reproduction after exposure of prepubertal rats to steroid hormones. *Recent Prog. Horm. Res.,* 22:503–539.
5. Best-Belpomme, M., Mešter, J., Weintraub, H., and Baulieu, E. E. (1975): Oestrogen receptor in chick oviduct: Characterization and subcellular distribution. *Eur. J. Biochem.,* 57:537–547.
6. Bouin, P., and Ancel, P. (1903): Sur la signification de la glande interstitielle du testicule embryonnaire. *C.R. Soc. Biol. (Paris),* 55:1682.
7. Brode, M. J. (1928): The significance of the asymmetry of the ovaries of the fowl. *J. Morphol. Physiol.,* 46:1–57.
8. Burns, R. K. (1955): *Urogenital System Analysis of Development,* pp. 462–491. Saunders, Philadelphia.
9. Burns, R. K. (1961): Role of hormones in the differentiation of sex. In: *Sex and Internal Secretion,* edited by W. C. Young, pp. 76–158. Williams & Wilkins, Baltimore.
10. Cohen, S. S. (1971): Introduction to the polyamines. Prentice-Hall, Englewood Cliffs, New Jersey.
11. Domm, L. V. (1927): New experiments on ovariectomy and the problem of sex inversion in the fowl. *J. Exp. Zool.,* 48:31–173.
12. Domm, L. V. (1955): Recent advances in knowledge concerning the role of hormones in the sex differentiation of birds. In: *Recent Studies in Avian Biology,* pp. 309–325. Univ. of Illinois Press, Urbana, Illinois.
13. Forsberg, J. G. (1969): The development of atypical epithelium in the mouse uterine cervix and vaginal fornix after neonatal oestradiol treatment. *Br. J. Exp. Pathol.,* 50:187–195.
14. Forsberg, J. G. (1970): An estradiol mitotic rate inhibiting effect in the Müllerian epithelium in neonatal mice. *J. Exp. Zool.,* 175:369–374.
15. Frick, J., Chang, C. C., and Kincl, F. A. (1969): Testosterone plasma levels in adult male rats injected neonatally with estradiol benzoate or testosterone propionate. *Steroids,* 13:21–27.
16. Gaarenstroom, J. H. (1939): Action of sex hormones on the development of the Müllerian duct of the chick embryo. *J. Exp. Zool.,* 82:31–42.
17. Galli, F., and Wasserman, G. F. (1973): Steroid biosynthesis by gonads of 7- and 10-day-old chick embryos. *Gen. Comp. Endocrinol.,* 21:77–83.
18. Giles, K. W., and Myers, A. (1965): An improved diphenylamine method for the estimation of deoxyribonucleic acid. *Nature,* 206:93.
19. Gorski, R. A. (1966): Localization and sex differentiation of the nervous structures which regulate ovulation. *J. Reprod. Fertil.,* 1(Suppl.):67–88.
20. Groenendijk-Huijbers, M. M., and Burggraaff, J. M. (1974): Experimental studies

on the capability of embryonic and young chick testes to regress embryonic chick oviducts. *Anat. Anz. Bd.,* 135:43–46.

21. Guichard, A., Cedard, L., and Haffen, K. (1973): Aspect comparatif de la synthese de stéroides sexuels par les gonades embryonnaires de poulet à différents stades du développement. *Gen. Comp. Endocrinol.,* 20:16–18.

22. Guichard, A., Cedard, L., Mignot, Th. M., Scheib, D., and Haffen, K. (1977): Radioimmunoassay of steroids produced by cultured chick embryonic gonads: Differences according to age, sex, and side. *Gen. Comp. Endocrinol.,* 32:255–265.

23. Haffen, K. (1969): Quelques aspects de l'intersexualité experimentale Chez le poulet (*Gallus gallus*) et la caille (*Coturnix coturnix japonica*). *Bull. Biol. Fr. Belg.,* 103:401–417.

24. Hamilton, T. H. (1961): Studies on the physiology of urogenital differentiation in the chick embryo. 1. Hormonal control of sexual differentiation of Müllerian ducts. *J. Exp. Zool.,* 146:265–274.

25. Hamilton, T. H. (1963): Hormonal control of Müllerian duct differentiation in the chick embryo. *Proc. 13th Int. Ornithol. Congr.,* Ithaca, N.Y., p. 1004–1040.

26. Jost, A. (1947): Recherches sur la differenciation sexuelle de l'embryon de lapin. III. Rôle des gonades foetales dans la differenciation sexuelle somatique. *Arch. Anat. Microsc. Morphol. Exp.,* 36:271–315.

27. Lee, A. E. (1972): Cell division and DNA synthesis in the mouse uterus during continuous oestrogen treatment. *J. Endocrinol.,* 55:507–513.

28. Lowry, O. H., Rosebrough, N. J., Farr, A. L., and Randall, R. J. (1951): Protein measurement with the folin phenol reagent. *J. Biol. Chem.,* 193:265–275.

29. Lutz-Ostertag, Y. (1964): Action du stilboestrol sur les canaux de Müller de l'embryon de caille (*Coturnix coturnix japonica*). *C.R. Acad. Sci. (Paris),* 259: 879–881.

30. Lutz-Ostertag, Y. (1974): Nouvelles preuves de l'action de la testosterone sur le développement des canaux de Müller de l'embryon d'oiseau en culture *in vitro. C.R. Acad. Sci. (Paris),* 278:2351–2353.

31. Martin, L., Finn, C. A., and Trinder, G. (1973): Hypertrophy and hyperplasia in the mouse uterus after oestrogen treatment: An autoradiographic study. *J. Endocrinol.,* 56:133–144.

32. Maraud, R., Stoll, R., and Couland, H. (1970): Données Nouvelles sur le role du testicule et d'hypophyse dans la differenciation sexuelle du poulet. *Bull. Assoc. Anat.,* 148:442–449.

33. Morgan, W., and Kohlmeyer, W. (1957): Hens with bilateral oviducts. *Nature,* 180:98.

34. Mori, T. (1975): Effects of postpuberal oestrogen injections on mitotic activity of vaginal and uterine epithelial cell in mice treated neonatally with oestrogen. *J. Endocrinol.,* 64:133–140.

35. Munro, H. N., and Fleck, A. (1966): The determination of nucleic acids. *Methods Biochem. Anal.,* 14:113–176.

36. O'Malley, B. W., McGuire, W. L., Kohler, P. O., and Korenman, S. G. (1969): Studies on the mechanism of steroid hormone regulation of synthesis of specific proteins. In: *Recent Progress in Hormone Research,* edited by E. B. Astwood, pp. 105–160. Academic Press, New York.

37. Price, D., Zaaijer, J. J. P., Ortiz, E., and Brinkmann, A. O. (1975): Current views on embryonic sex differentiation in reptiles, birds, and mammals. In: *Trends in Comparative Endocrinology,* edited by J. W. Barrington. pp. 173–195. *Am. Zool.* 15, Suppl. 1.

38. Raynaud, A., Pieau, C., and Raynaud, J. (1970): Étude histologique comparative de l'allongement des canaux de Müller, de l'arret de leur progression en direction caudale et de leur destruction, Chez les embryons males de diverses espèces de reptiles. *Ann. Embryol. Morphog.,* 3:21–47.

39. Scheib, D., and Haffen, K. (1968): Sur la localization histoenzymologique de la 3β-hydoxysteroide deshydrogénase dans les gonades de l'embryon de poulet; apparition et spécificite de l'activité enzymatique. *Ann. Embryol. Morphog.,* 1:61–72.

40. Scheib, D. (1976): Mécanismes cellulaires et determinisme hormonal de la regrés-

sion de canaux de Müller chez l'embryon de poulet: Etude cytologique et biochimique. In: *Mécanismes de la Rudimentation des Organes Chez les Embryons de Vertébrés,* Colloques Internationaux Centre National de la Recherche Scientifique, Paris.

41. Seltzer, W. (1956): The Method of Controlling the Sex of Avian Embryo, Improving Embryo Hatchability and Improving Viability of the Hatched Chick. U.S. Patent 2,734,482.
42. Snedecor, J. G. (1949): A study of some effects of sex hormones on the embryonic reproductive system and comb of the white Leghorn chick. *J. Exp. Zool.,* 110:205–246.
43. Swift, C. H. (1915): Origin of the definitive sex-cells in the female chick and their relation to the primordial germ cells. *Am. J. Anat.,* 19:441–470.
44. Takasugi, N. (1970): Testicular damages in neonatally estrogenized adult mice. *Endocrinol. Jpn.,* 17:277–281.
45. Takasugi, N., and Bern, H. A. (1964): Tissue changes in mice with persistent vaginal cornification induced by early postnatal treatment with estrogen. *J. Natl. Cancer Inst.,* 33:855–865.
46. Teng, C. S., and Teng, C. T. (1975): Studies on sex-organ development: Isolation and characterization of an oestrogen receptor from chick Müllerian duct. *Biochem. J.,* 150:183–190.
47. Teng, C. S., and Teng, C. T. (1975): Studies on sex-organ development: ontogeny of cytoplasmic oestrogen receptor in chick Müllerian duct. *Biochem. J.,* 150:191–194.
48. Teng, C. S., and Teng, C. T. (1976): Studies on sex-organ development: oestrogen-receptor translocation in the developing chick Müllerian duct. *Biochem. J.,* 154:1–9.
49. Teng, C. S., and Teng, C. T. (1977): Studies on sex-organ development: changes in chemical composition and oestradiol-binding capacity in chromatin during the differentiation of chick Müllerian ducts. *Biochem. J.,* 172:361–370.
50. Teng, C. S., and Teng, C. T. (1978): Studies on sex-organ development: hormonal effect on ornithine decarboxylase activity in the developing gonads of chick embryos. *Biochem. J. (in preparation).*
51. Teng, C. T., and Teng, C. S. (1977): Studies on sex-organ development: the hormonal regulation of steroidogenesis and adenosine 3:5-cyclic monophosphate in embryonic-chick ovary. *Biochem. J.,* 162:123–134.
52. Teng, C. T., and Teng, C. S. (1977): The effect of diethylstilbestrol on the embryonic gonadal steroid production and its responsiveness to gonadotropic hormone. *J. Cell. Biol.,* 75:190a.
53. Teng, C. T., and Teng, C. S. (1978): Studies on sex-organ development: the effect of diethylstilbestrol on the embryonic gonadal steroids production and its responsiveness to gonadotropic hormone. *Biochem. J. (in preparation).*
54. Terenius, L., Meyerson, B. J., and Palis, A. (1969): The effect of neonatal treatment with 17β-oestradiol or testosterone on the binding of 17β-oestradiol by mouse uterus-vagina. *Acta Endocrinol. (Kbh.),* 62:671–678.
55. Van Tienhoven, A. (1957): A method of "controlling sex" by dipping of eggs in hormone solutions. *Poult. Sci.,* 36:628–632.
56. Weniger, J. P., and Zeis, A. (1971): Biosynthèse d'oestrogènes par les ébauches gonadiques de poulet. *Gen. Comp. Endocrinol.,* 16:391–397.
57. Willier, B. H. (1939): The embryonic development of sex. In: *Sex and Internal Secretions,* edited by E. Allen, C. H. Danforth, and E. A. Doisy, pp. 64–144. Williams & Wilkins, Baltimore.
58. Willier, B. H., Gallagher, T. F., and Koch, F. C. (1937): The modification of sex development in the chick embryo by male and female sex hormones. *Physiol. Zool.,* 10:101–113.
59. Witschi, E. (1939): Modification of the development of sex in lower vertebrates and in mammals. In: *Sex and Internal Secretions,* edited by E. Allen, C. H. Danforth, and E. A. Doisy, pp. 145–226. Williams & Wilkins, Baltimore.
60. Wolff, Et. (1936): L'évolution après l'éclosion des poulets mâles transformés en intersexués par l'hormone femelle injectée aux jeunes embryons. *Arch. Anat. Histol. Embryol.,* 23:1–28.

61. Wolff, Et. (1939): Action du diethylstilboestrol sur les organes génitaux de l'embryon de poulet. *C.R. Acad. Sci. (Paris)*, 208:1532–1533.

62. Wolff, Et. (1949): L'Evolution des canaux de Müller de l'embryon d'Oiseau aprés castration precoce. *C.R. Soc. Biol.*, 143:1239–1241.

63. Wolff, Et., and Ginglinger, A. (1935): Sur la transformation des poulets mâle en intersexués par injection d'hormone femelle-folliculine-aux embryons. *Arch. Anat. Histol. Embryol.*, 20:219–278.

64. Wolff, Et., Lutz-Ostertag, Y., and Haffen, K. (1952): Sur la régression et la nécrose *in vitro* des canaux de Müller de l'embryon de poulet sous l'effet de substances hormonales. *C.R. Soc. Biol.*, 146:1793–1795.

65. Woods, J. E., Simpson, R. M., and Moore, P. L. (1975): Plasma testosterone levels in the chick embryo. *Gen. Comp. Endocrinol.*, 27:543–547.

66. Woods, J. E., and Weeks, R. L. (1969): Ontogenesis of the pituitary-gonadal axis in the chick embryo. *Gen. Comp. Endocrinol.*, 13:242–254.

67. Wrenn, T. R., Wood, J. R., and Bitman, J. (1969): Oestrogen responses of rats neonatally sterilized with steroids. *J. Endocrinol.*, 45:415–420.

A

Actinomycin D, 375
  androgenization attenuation
    by, 379
  effects on
    estrogen effects on VLDL,
      339
    uterine estrogen response,
      87-88
Adenovirus 2, introns of, 7
Adenylyl cyclase
  assay of, 177-178
  in corpus luteum, hormone
    effects on, 179-186
  estradiol, 189-193
  follicular coupling of, 173,
    174,177
  in granulosa cell luteinization,
    160
  physiologic regulation of, 177
  in Sertoli cell, FSH effects
    on, 207-224
α-Amanitin, 121
  androgenization attenuation
    by, 379
Androgen(s)
  effects on
    follicles, 166,167,173-174
    uterine estrogen response,
      83,88
  fetal secretion of, 411
  ovarian cell receptors for,
    167-169
  ovarian secretion of, 169-170
  in sexual differentiation in
    rat, 399-400

Androgen-binding protein (ABP),
  FSH regulation of, 207
Androgen receptors, in brain,
  394
Androgenization, agents
  attenuating, 379
Androst-1,4-triene-3,17-dione
  (ATD), effect on brain dif-
    ferentiation, 399-400
Antiestrogens
  activity of, in duck liver,
    362-363
  androgenization attenuation by,
    379
  effects on breast cancer, 281
  uterine interaction with, 88-99
  estradiol effects, 124-127
AP_3 proteins, role in progesterone receptor activity, 40
Apolipoproteins, estrogen
  ingestion and, 331
ApoVLDL-II, purification and
  characterization of, 332-333
A-protein, of S. aureus,
  estradiol-receptor complex
  binding to, 15
Arachidonic acid, 253
Aromatization inhibitors, androgenization attenuation by, 379
ATP, role in progesterone
  receptor activity, 29
ATP-pyrophosphate exchange
  activity
  in progesterone-receptor
    complex, 23-30
  assay, 24-25